Vascular Wilt Diseases of Plants

Basic Studies and Control

NATO ASI Series

Advanced Science Institutes Series

A series presenting the results of activities sponsored by the NATO Science Committee, which aims at the dissemination of advanced scientific and technological knowledge, with a view to strengthening links between scientific communities.

The Series is published by an international board of publishers in conjunction with the NATO Scientific Affairs Division

A **Life Sciences** B **Physics**	Plenum Publishing Corporation London and New York
C **Mathematical and Physical Sciences** D **Behavioural and Social Sciences** E **Applied Sciences**	Kluwer Academic Publishers Dordrecht, Boston and London
F **Computer and Systems Sciences** G **Ecological Sciences** H **Cell Biology**	Springer-Verlag Berlin Heidelberg New York London Paris Tokyo

Series H: Cell Biology Vol. 28

Vascular Wilt Diseases of Plants

Basic Studies and Control

Edited by

E. C. Tjamos

Agricultural University of Athens
Laboratory of Phytopathology
118 55 Votanikos, Athens, Greece

C. H. Beckman

University of Rhode Island
Department of Plant Sciences
Kingston, RI 02881-0804, USA

Springer-Verlag
Berlin Heidelberg New York London Paris Tokyo
Published in cooperation with NATO Scientific Affairs Division

Proceedings of the NATO Advanced Research Workshop on The Interaction of Genetic and Environmental Factors in the Development of Vascular Wilt Diseases of Plants, held at Cape Sounion, Athens, Greece, May 22–27, 1988

ISBN-13: 987-3-642-73168-6 e-ISBN-13: 987-3-642-73166-2
DOI: 10.1007/987-3-642-73166-2

Library of Congress Cataloging-in-Publication Data. NATO Advanced Research Workshop on the Interaction of Genetic and Environmental Factors in the Development of Vascular Wilt Diseases of Plants (1988 : Athens, Greece) Vascular wilt diseases of plants : basic studies and control / edited by E. C. Tjamos, C. H. Beckman. p. cm.—(NATO ASI series. Series H, Cell biology ; vol. 28) "Published in cooperation with NATO Scientific Affairs Division." "Proceedings of the NATO Advanced Research Workshop on the Interaction of Genetic and Environmental Factors in the Development of Vascular Wilt Diseases of Plants, held at Cape Sounion, Athens, Greece, May 22–27, 1988"–Verso t.p. Includes index.
ISBN-13: 987-3-642-73168-6 (U.S.)
1. Wilt diseases—Congresses. 2. Fungi, Phytopathogenic—Control—Congresses. 3. Bacteria, Phytopathogenic—Control—Congresses. I. Tjamos, E. C., 1943- . II. Beckman, Carl H. III. North Atlantic Treaty Organization. Scientific Affairs Division. IV. Title. V. Series. SB741.W44N37 1988 632'.32—dc 20
89-6357

Softcover reprint of the hardcover 1st edition 1989

2131/3140-543210 – Printed on acid-free paper

PREFACE

This book is a collection of papers presented at a NATO Advanced Research Workshop held at Cape Sounion, Greece, in May of 1988. The primary objectives of the workshop were to bring together researchers who are involved in studies of vascular diseases of plants caused by bacterial and fungal pathogens, and whose primary involvement and approaches range from the ecological to the organismal, cellular and molecular. It was anticipated that the kind of interaction that would occur in an intimate meeting of several days duration, such as a NATO Workshop offers, would produce a broader appreciation and understanding of various aspects of wilt diseases, new levels of collaboration and, in the longer term, a more sound and broad base of measures for practical disease control.

In all, eighteen lectures and twenty-seven research papers were presented. The first day was devoted to the life cycles of the pathogens, their colonization of root surfaces, entry into roots, and colonization of the vascular systems of plants. The second day was devoted to recognition phenomena, structural changes and physiological and biochemical interactions by which the plant defends itself and those by which the pathogen counters these processes and causes disease. On day three, these studies were carried to the genetic and molecular levels of understanding, principally for bacteria, but with exciting progress and prospects with fungi as well. Day four began the process of integrating these understandings with physical and biological components of the environment in ecological terms. Finally, day five was devoted to discussing comprehensive systems of control in light of the understandings gained during the week.

The workshop was successful in its intended purpose of enhancing confluent thought at the ecological, organismal and molecular levels and, indeed, several collaborative ventures that cross these disciplinary lines were agreed upon or are being explored. For example, the use of immunohistochemistry has allowed far more precise studies of the colonization of epidermal and cortical tissues by species of Fusarium and Verticillium. The question arises, can these techniques not also be used to examine the colonization of roots by competing organisms found in suppressive soils with or without chemical additives, including fertilizers, that may enhance or reduce apparent suppressiveness? Again, colonization of epidermal and cortical tissues of roots by these species is commonplace and general along certain portions of roots, yet successful vascular colonization leading to apparent disease is a relatively rare event in comparison. Either suitable vascular infection courts or the conditions suitable for vascular infection must be of rare occurence. Can immunohistochemical procedures be used to help identify those places and conditions under which infection can take place, and if so, can we take steps to further reduce the frequency of successful vascular infection and colonization by pathogenic forms? Can genetic factors within the host that contribute to the rarity of such an event also be enhanced? Can the use of agronomic procedures to reduce the likelihood of such infection events be used to protect major gene resistance from evolving races of the pathogen and thus prolong the effective time span of their usefulness?

It is with these kinds of questions in mind that we ask the reader to approach this book.

The organization and the successful running of an ARW would be impossible without the help of many colleagues and assistants but we can name a few. We express our thanks to Prof. L. Sequeira and Dr. P. Talboys for their involvement in the initial stages of the organization of the meeting. We are also indebted to Prof. J. DeVay, Prof. G. Pegg, Prof. A. Matta, Prof. J. Louvet, Dr. C. Alabouvette and Dr. P. Psallidas for their valuable advice during the organization of the Workshop. Special thanks are due to the chairpeople who so successfuly chaired the sessions. We thank Prof. C. Panagopoulos, the President of the Hellenic Phytopathological Society for his welcome address with which he opened the Workshop. We also thank Miss Asteria Karadimas for preparing the excellent typescripts from which this volume was copied.

On behalf of those who participated, profited from and enjoyed this exceptional kind of scientific meeting, we wish to thank the Scientific Affairs Division of NATO for sponsoring the Workshop and for the financial grant that allowed the participants to gather from great distances and spend several fruitful days together in a setting so conducive to the process of exchange and understanding, and particularly to Dr. Graig Sinclair of the Division for his encouragement and advice about the Workshop during all stages from its conception to its fruition.

Spring 1989 E.C. Tjamos C.H. Beckman

CONTENTS

LIFE CYCLES OF PATHOGENS, COLONIZATION OF HOSTS AND PATHOGENESIS

RECOGNITION AND RESPONSE BETWEEN HOST AND PARASITE AS DETERMINANTS IN RESISTANCE AND DISEASE DEVELOPMENT

GENETIC AND MOLECULAR BASES FOR RESISTANCE OR SUSCEPTIBILITY OF THE HOST AND VIRULENCE OR AVIRULENCE OF THE PARASITE

A COMPREHENSIVE SYSTEM OF CONTROL MEASURES

DYNAMICS OF COLONIZATION OF PLANT ROOTS BY VERTICILLIUM DAHLIAE AND OTHER FUNGI

O.C. Huisman[1] and J.S. Gerik[2]
Department of Plant Pathology
University of California
Berkeley, CA 94720
USA

INTRODUCTION

In pathogenesis, a critical first step is the initial contact and colonization of the host by the pathogen. Little is known about this initial host-parasite interaction for the soil-borne pathogens involved in vascular wilt diseases: particularly under natural conditions. The lack of information is in large part due to the location where these initial interactions occur. The soil is not a media that facilitates direct observation. Additionally, since many fungi colonize the roots of plants, investigations focused on a single organism in the presence of all the others have been difficult. A number of features have been established. The fungal pathogens persist in soil for extended periods of time as dormant propagules. For Verticillium dahliae Kleb., longevity of microsclerotia extends for years to decades. These propagules will germinate in response to root exudates (Rovira 1965; Schreiber and Green 1963). Since exudates are greatest in the zone of elongation, it is suggestive that at least some of the initial contact occurs near this site. Little is known about what happens between germination in response to root exudates and the detection of the pathogen in the vascular system.

Several lines of evidence suggest that V. dahliae has an epi-

2. USDA-ARS
1636 E. Alisal
Salinas, CA 93905
USA

NATO ASI Series, Vol. H28
Vascular Wilt Diseases of Plants

phytic phase on roots. Lacey and Horner (1966), in greenhouse studies with artificially infested soils, showed that V. dahliae could be recovered consistently from immune and susceptible host roots. Evans and Gleeson (1973), examining the roots of a wide range of young plants grown in the greenhouse, found colony densities of V. dahliae on roots were directly related to inoculum densities in soil. Colonies were small, randomly scattered along the root surface and probably confined to the rhizoplane since they were very sensitive to brief mercury treatment.

Work in our laboratory has been directed at examining the initial colonization of plant roots by V. dahliae. Colonization of roots by other fungi was also examined in order to put the findings with V. dahliae into better perspective. Following is a presentation of our findings (Gerik and Huisman 1988; Huisman 1988).

BASIC METHODS

Susceptible and tolerant cotton cultivars were grown for several years in a field at the West Side Field Station in central California, USA. The field had initially been set up with subplots with differing inoculum levels established by addition of tomato debris naturally infested with V. dahliae (Ashworth et al 1979). Soil cores (2 cm diam) were collected to a depth of 30 cm at intervals during the season. The soil cores were used to obtain inoculum density data, root density data, and root segments used to evaluate their colonization by fungi. Additionally, data were collected on the incidence of systemic infection (bases of foliar symptoms).

Colonization of roots by fungi was evaluated by two procedures: a bio-assay and an immuno-histochemical assay. In the bio-assay, thouroughly washed fresh root segments were plated on a semi-selective medium that severely restricted colony size and permitted easy identification of site of origin of the colony on the root (Huisman 1988). When desired, roots were segregated with respect to cortical appearance from no evident damage to severe cortical browning and collapse. Data were collected on root length plated, colony frequency of identifiable fungi (V. dahliae, V. tricorpus Isaac, Fusarium oxysporum Schlecht. (amed. Sny. and Hans.) Penicillium miczynski Zaleski, a Gliocladium sp. Corda, Aspergillus mellus Yukawa, Stemphyllium botryosum Wallr, Cladosporium cladosporioides de Vries, Doratomyces stemonitis, Trichoderma harzianum Rifai, Gliocladium roseum Bain., Rhizoctonia spp., Pythium ultimum Trow, P. aphanidermatum (Edson) Fitzp., and an unidentified Pythium sp.) which grew from the roots and the site of origin (colony center) relative to the root apex (if applicable) or to the end of the root segment. Additionally, for Verticillium, the colony frequency on individual root segments was also noted.

The immuno-histochemical assay (Gerik et al 1987), a modified ELISA assay, utilized a specific antibody prepared in rabbits against soluble proteins extracted from either V. dahliae or F. oxysporum. Intact root segments, either fresh or preserved with glutaraldehyde, were exposed first to the antibody, then to anti-rabbit antibodies conjugated to alkaline phosphatase, and finally to a histochemical substrate for alkaline phosphatase. The chromogenic, insoluble product of the enzymatic hydrolysis of the product deposited at the site of antibody binding and outlined the hyphae of the target fungus in blue. Data were kept on colony appearance, size, and location relative to the root apex.

Unless specifically stated otherwise, all colonization data were obtained from roots collected from the field plots. Since we have found colonization of the cortex to be distinct from colonization of the vascular system, the term colonization is used to refer to the former and the term infection to refer to the latter, although technically, both can be viewed as an infection.

FACTORS AFFECTING COLONIZATION OF VERTICILLIUM

Verticillium dahliae and most of the other fungi grew as distinct localized colonies from cotton roots plated on the growth restrictive media. The colonization pattern was similar to that reported by Evans and Gleeson (1973) for V. dahliae on young greenhouse grown plants. The colonies of Verticillium, as well as those of other fungi, were randomly scattered along the root and were small, being only a few millimeters in length. Colonies were apparently superficial for they were quite sensitive to brief hypochlorite treatment; a 5 to 15 second exposure reduced colony frequencies by half.

The small size and localized nature of the colonies permitted a quantitative evaluation of the effect of a number of variables on colonization of roots by Verticillium and several of the other fungi. A direct linear relationship was found between soil inoculum densities of Verticillium and colony frequency per cm root for both cotton and tomato. Colony densities (colonies per cm root per unit inoculum density) on cotton were 0.0017 and 0.0058, respectively, for V. dahliae and V. tricorpus.

Neither host susceptibility nor current infection status had any influence on colony densities. No significant differences in colony densities were found between highly susceptible cultivars (SJ-2 and 70-110) and very tolerant ones (SJ-4 and SJ-5). Systemic invasion of cotton plants did not lead to systemic invasion of the fibrous root system. The same localized colonization pattern and colony densities were observed on roots collected late in the season from plants exhibiting no vascular necrosis as on roots from wilt defoliated plants. This is consistent with the known acropetal movement of V. dahliae in plants (Beckman 1964; Beckman et al 1976). Once in the vascular system, the pathogen moves primarily with the transpiration streas as microconidia, transversing end plates by hyphal growth. With such movement, only roots lying between the point of vascular penetration and the tap root would be expected to be systemically invaded. With only a few systemic invasions per plant, such roots would represent only a tiny fraction of the total root system.

Soil temperature had a marked effect on the colonization of roots by V. tricorpus but not on that of V. dahliae. In the greenhouse, V. tricorpus did not colonize roots of cotton at temperatures from 27 to 31 C but readily colonized roots at 20 to 23 C. In contrast, V. dahliae colonized roots at all these temperatures. These results are in agreement with the reported effect of temperature on growth for these two fungi (Isaac 1949). Colonization patterns and colony densities on roots from plants grown in the greenhouse in field soil were similar to those observed for field roots.

Roots of cotton were colonized by V. dahliae and V. tricorpus throughout the growing season. With V. tricorpus colony densities on roots were significantly lower during the summer months than during the spring or fall. Much of this variation can be explained on the basis of soil temperature. Changes in temperature with depth and time spanned the range (20 to 30 C) critical to colonization of roots by V. tricorpus, and colony densities were inversely related to overall soil temperature with a short lag (about one week). Colony frequencies of V. dahliae on roots increased in late spring and early summer and then were relatively constant. The changes in frequency of V. dahliae on roots paralled changes in inoculum density in the soil with a 3 to 4 week lag and colony density per unit inoculum was remarkably constant through the growing season. For the first part of the season, colonies were randomly scattered alone the root length since the frequency of multiple colonies on individual root segments closely fit a Poisson distribution. This distribution is consistent with a random distribution of inoculum in the soil as a result of soil mixing by the various cultural operations. In early summer, colonies exhibited a high degree of clustering on roots and no longer fit a Poisson distribution.

The increases in inoculum density in June along with the subsequent increases in root colonization can be accounted for by decomposion and dispersal of infested crop residues from the preceding season. Such increases in inoculum density are typical in California in late spring for cotton (Ashworth et al 1976). Although the results indicate the occurence of dispersal, its mechanism is not clear. Without dispersal, no increase in colony density on roots would be expected, since, upon decomposition, inoculum should remain in pockets of exceptionally high densities. Such an inoculum cluster should be seen as a single colony, perhaps somewhat larger in size. Colony centers would have to be at least 3 to 5 mm apart to be seen as separate colonies in the bio-assay. No increase in colony size was evident at any time during the season. Limited dispersal should lead to a high degree of colony clustering on roots. This appeared to be the case, for the observed appearance of colony clustering coincided with the increase in colony frequency on roots in early summer. Apparently, during July, sufficient dispersion of propagules took place to effect a separation of 3 to 5 mm, the minimum distance needed to resolve adjacent colonies on roots, but not sufficient to achieve random distribution in the soil. The precise manner of dispersal is not clear. Cultural disturbance of the soil was negligible at that time. The large size of microsclerotia make it unlikely that dispersion was effected by capilary movement of water, while production and movement of conidia would have lead to a breakdown of the relation between colonies on roots and inoculum in soil - unsupported by the data. Soil fauna could act as agents of dispersal and their activity would be preferentially associated with decaying tissue (Price 1976).

Both Verticillium species readily colonized roots of a variety of plants in the field. These findings are consistent with reports on colonization of diverse plant roots by V. dahliae under greenhouse conditions (Evans and Gleeson 1973). Distinct differences in the colonization frequency of roots occurred for a given Verticillium species among the plants (Table 1). However, the differences were not large since the lowest to the highest values represented only a ten fold spread, similar to the spread observed by Evans and Gleeson (1973). The colonization frequencies of V. dahliae and V. tricorpus did not change in the same proportion among the plant species (Table 1). Thus, the colonization frequency of roots would appear to be a unique property of each plant species-fungal species interaction.

Large discrepancies were observed between root colonization and systemic infection of cotton by V. dahliae. It was possible to estimate the mean total number of cortical colonies of V. dahliae on the root system of a single plant in the upper 30 cm of soil in the same plots for which systemic infection data (foliar symptoms and vascular necrosis) were available. Total colonies

Table 1. Colonization of plant roots by fungi: Colony densities and apparent sphere of influence

	Colony density on roots (10^{-3})[a]	Sphere of influence (radius in mm)[b]	Soil under influence (%)[c]
V. dahliae on:			
Cotton	1.7	0.23	0.7
Cotton (Evans)[d]	1.8	0.24	
Tomato	6.4	0.43	2.9
Tomato (Evans)[d]	2.7	0.29	
Sowthistle	2.2	0.25	
Wild turnip	0.5	0.13	
Lambsquarter	1.1	0.16	
Alfalfa	1.3	0.20	
Pistachio	0.8	0.14	
Barley (Benson)[e]	1.0	0.15	5.0
V. tricorpus on:			
Cotton	5.8	0.43	5.0
Tomato	4.2	0.35	2.2
Sowthistle	14	0.67	
Wild turnip	23	0.86	
Lambsquarter	5.3	0.40	
Cotton with:			
V. dahliae	1.7	0.23	
V. tricorpus	5.8	0.43	
P. ultimum	0.3	0.12	
Pythium sp.	1.7	0.23	
Rhizoctonia sp.	620	5.1	
Gliocladium sp.	0.7	0.14	
Trichoderma sp.	0.6	0.13	

[a] Colonies cm $root^{-1}$, $propagules^{-1}$, g soil. All values should be multiplied by 10^{-3}.
[b] Values given represent the radius of a cylinder of soil (measured from the center of the root - root diam = 0.10 mm) around the root which contains the number of propagules required to account for the observed colony density of Verticillium on roots.
[c] Calculated from measured root densities (3.3 cm root per cm^3 soil for cotton, 4.6 for tomato and 20 for barley) and the indicated size of the sphere of influence around roots.
[d] Calculated from data presented by Evans and Gleeson.
[c] Calculated from data presented by Benson and Ashworth.

were estimated by combining data on colony density on roots with root densities in soil and with plant densities. Systemic infection data were linearly related to soil inoculum densities when corrected for random multiple infections thus permitting estimates of mean number of infections per plant. When these values were compared, a ratio of around 5,000 cortical colonies to each systemic infection was consistently observed. Apparently, cortical colonization only rarely leads to a systemic invasion of the vascular system. But since total colonies per plant numbered in the thousands to tens of thousands, actual incidence of infection was quite high.

COLONIZATION BY OTHER FUNGAL SPECIES

Colonization of cotton roots by saprophytic fungi (Gliocladium spp. and Trichoderma spp.) was identical to that observed for Verticillium. The colony frequencies per unit propagule density for these fungi were similar to that observed for V. dahliae; being around 0.001 colonies per cm root per unit inoculum density.

Colonization of roots by Verticillium and the saprophytic fungi did not lead to visible damage to root tissue. The colony densities on all categories of roots, ranging from those with a clean white cortex to those with severe cortical browning and collapse were equivalent throughout the season.

Colonization of roots by Rhizoctonia was in most respects similar to that of the other fungi. However, two major differences were noted. Colonies of Rhizoctonia were more frequently (four fold) associated with roots showing extensive browning and cortical collapse than with other roots. This observation is consistent with the ability of Rhizoctonia, especially R. solani, to induce cortical lesions. Also, colony densities were two orders of magnitude higher (0.6 colonies per cm root per unit inoculum density) than those for the other fungi.

Colonization of cotton roots by Pythium species differed from that of the other fungi. Colony lengths on root were much longer (30 mm), being about an order of magnitude longer than those of other fungi (3 mm). Colony densities were quite cyclical and were closely correlated with soil moisture, being highest immediately after an irrigation and then declining drastically during subsequent weeks. Immediately after an irrigation, there was no marked association of colonies of *Pythium* with roots exhibiting tissue damage. Subsequently, colony densities on roots dropped dramatically as a result of dilution by new root growth and by a shift of remaining colonies into the roots with brown steles and finally into dead roots. This pattern, along with the unusual length of the colonies and the known ability of *Pythium* to invade and kill roots, was indicative the *Pythium* species were involved in root death. No major differences were detected among the colonization patterns of *P. ultimum*, *P. aphanidermatum* and the unidentified *Pythium* species.

The colonization data permitted an estimate of the effective sphere of influence of roots in the soil. With a known colony density of a fungus on roots in soil for which the inoculum density is known, the radius of a cylinder of soil around the root which contains the number of propagules required to account for the amount of root colonization observed can be calculated. Such calculations yielded very small interaction values for most fungi, 0.1 to 0.6 mm (Table 1). This distance should be viewed as an average value since there will be individual variations among propagules in their capacity to respond to exudates. Additionally, the distance only refers to an influence that culminates in the establishment of a cortical colony and thus is an operational one. It is plausable that propagules further out respond to exudates but are unable to initiate colonization. In spite of these limitations, the figures show that the influence of small cotton roots is limited and at best is less than a millimeter. One exception to this was *Rhizoctonia* which yielded interaction distances in excess of 5 mm. When the root densities in soil are taken into consideration, it is possible to calculate the fraction of soil that comes under the influence of the rhizosphere during the growing season with respect to the propagules of a given fungus. Such calculations yielded typical values of 1 to 5 percent (Table 1). These low values suggest that even if roots induced germination without propagule replacement, as might be the case for the 'causal' colonizers, it is unlikely to have a major impact on inoculum densities in soil.

COLONIZATION DYNAMICS

The fifteen species of fungi cited above accounted for over 95% of the fungi detectable with the bio-assay. The most abundant were F. oxysporum (33%), followed by P. miczynski (32%) and V. dahliae (13%) while each of the others were 5% or less. Two distinct morphological types were recognized for F. oxysporum, one formed sclerotial-type structures on PDA and was designated as a sclerotial type while the other sporulated readily on PDA and was designated as a sporodochial type.

An examination of colony frequency as a function of distance from the root apex (determined for all the fungi except Pythium and Rhizoctonia) revealed that essentially all colonization occurred within one centimeter of the root apex. Most of the fungi observed colonized the root within the first millimeter behind the apex and attained their maximum colony frequency within 2 to 4 mm of the apex. V. dahliae, V. tricorpus and the Gliocladium sp. differed in that little colonization was detectable in the first few mullimeters, however, maximum densities were achieved within 5 to 10 mm of the apex. Since the bio-assay did not permit resolution of the very early phases of root colonization, the immuno-histochemical assay was used to directly examine the location of colonies for V. dahliae. The results confirmed the earlier findings. No colonization was observed in the first millimeter behind the root tip. Colony densities increased gradually over the next few millimeters, reaching maximal values at 6 to 8 mm from the apex and remained at those levels at more distal locations.

In the field roots studied, the zone of elongation extended about 0.25 mm behind the apex. Thus, the initial point of contact by V. dahliae is well behind the region of undifferentiated tissues. This is a significant point, for Garber and Houston (1966) observed that under laboratory conditions, V. dahliae was able to enter the vascular system of cotton roots through undifferentiated tissued but not through differentiated cortical tissue unless massive quantities of inoculum were applied. This is consistent with the calculations showing that the vast majority of the cortical colonies of this pathogen do not lead to systemic invasion of the vascular system.

The fungi could be divided into two basic groups depending on their longevity on the root. For one group (F. oxysporum sporodochial type, P. miczynski, V. dahliae, V. tricorpus, A. mellus, and the Gliocladium sp.), colony densities on the more distal root tissue remained at or near the high levels attained in the first few millimeters of root behind the apex. Although colony length increased and covered larger portions of the root, colony densities on older root tissue were the same as those attained in the first centimeter. For the other group (F. oxysporum sclerotial type, S. botryosum, C. cladosporioides, G. roseum, and D. stemonitis), colony frequencies in root tissue 10 to 20 mm behind the tip declined to a third or less from the maximal values attained in the first few millimeters. In many cases, frequencies were below detectable levels in older root tissue. These results support the findings by others. Parkinson et at (1963) reported that while many fungi can initiate root colonization, only a portion become stable residents on the root surface. Taylor and Parkinson (1961) concluded from their work that successive colonization of a growing root by fungi was a result of new colonization by fungi from the soil rather than from longitudinal growth of the fungi down the root from previous colonies. The data suggest that fungal colonization of cotton roots in a field situation occurs immediately behind the root apex and that little if any subsequent new colonization occurs.

COLONIZATION SITES

Direct examination, in the root, of colonies of V. dahliae stained with the immuno-histochemical assay revealed a number of interesting features. Hyphae at the root surface were usually seen only near the center of the colony. Colony appearance was consistent with growth of hyphae from the root surface, through the cortex, and toward the stele. Approximately 1 mm from the center of the colony, the hyphae were exclusively located deep within the cortex near the stele and usually completely surrounded it. Preferential growth in the inner cortex could be an adaptive feature. The faster colonization rates observed for other fungi should provide strong competition to V. dahliae for colonization of the outer cortex. Beckman and Talboys (1981) have pointed out the selective advantage of this strategy for avoiding competition from other organisms at the root surface. V. dahliae was conspicuously absent from the area of lateral root emergence. Garber and Houston (1966) also reported sucn an absence. Mace et al (1974) have reported high concentrations of gossypol-like compounds at these sites and such compounds may inhibit fungal activity there.

Measured colony lengths averaged 7 mm, considerably longer than the 2.3 mm determined by means of the bio-assay (Huisman 1988) and the 2 mm reported by Evans and Gleeson (1973). The discrepancy is probably related to preferential growth in the inner cortex by the fungus. In the bio-assay, which relies on growth of fungi from the root unto the agar surface, competition from other fungi located in the outer cortex may have precluded detection of V. dahliae when root breaks occured more than 1 mm from the center of the colony. Following initial colonization, colonies of V. dahliae continued to grow. Colony lengths increased with distance from the apex, averaging 2 and 17 mm in length at distance of, respectively, 3 and 30 mm behind the apex.

There was some suggestion that V. dahliae may inflict some damage to the cortex. In approximately 20% of the colonies, a scattering of individual cortical cells were found to be completely ramified with hyphae. When viewed at low magnification, these cells superficially resembled mycorrhizal arbuscules. In about 5% of the colonies, some weakening of vascular tissue was evident based on the ease with which vascular bundle disorganization could be induced with mild pressure of a cover slip. Such weakening was only observed in association with hyphae of V. dahliae and could provide a point of entry into the vascular system. The frequency, however, was far in excess of the number of vascular infections expected on the basis of the observed systemic infection rate (Huisman 1982). Such entries, however, are not precluded if the fungus could be delimited in the vascular system after entry.

Colonies of F. oxysporum stained by the immuno-histochemical assay differed from those of V. dahliae. Hyphae were primarily restricted to the root surface and the outer cortex and were absent from the inner cortex next to the stele. The isolated, heavily stained cortical cells seen with V. dahliae were not present in root tissue stained with antibodies to F. oxysporum.

ROOT COLONIZATION MODEL

The root colonization pattern observed for V. dahliae, as well as that of other fungi, indicates that colonization is a very early event in the life of the root. Root tips play a central role in the colonization process. Root tips, specifically the zone of elongation, are regions or high exudate release (Rovira 1965; Schroth and Snyder 1961). Given the growth rate of roots (3 to 10 mm per day, Huisman 1982) and the apparent short interaction distance (less than a millimeter for most fungi), a dormant propagule in the soil will need to respond quickly to the exudates diffusing from the tip. Response times much longer than a few hours would make it difficult for the fungus to contact the moving tip. Longer germination times should lead to contact with areas behind the tip. Dormant propagules in the soil probably receive their strongest exudate signal from the root tip. This statement is deduced from the following considerations. Exudation rates from root tissue distal to the zone of elongation are much lower than near the tip. Additionally there is an increasingly active removal of exudates by the growing microbial community in the rhizosphere. This should lead to a severe shrinking of the radial extent of the rhizosphere which probably approaches the rhizoplane several centimeters behind the apex (Newman and Watson 1977). Thus, it is unlikely that dormant propagules which do not respond to a passing root tip will do so at a later time. Therefore, the sphere of influence calculated for V. dahliae and the other fungi (Table 1) probably represents that of the root tip rather than of the root proper. Root colonization can then be viewed as consisting of two distinct phases. The first involves the response of dormant propagules to host exudates, germination and initial contact with the root, all at or near the tip. The second involves the establishment and growth of the fungus on or in the root cortex and would extend for many centimeters behind the root apex. While many fungi are capable of the first phase, far fewer are successful in the second. Reductions in the population of those less well adapted to growth on the root are already evident at 10 mm behind the tip.

The combined data presented here and that cited lead to the following conclusions. V. dahliae is primarily a colonizer of the root cortex and is adapted to a wide host range. Almost all plants thus far examined (57 out of 63 species), including both dicotyledonous and monocotyledonous plants, were colonized. The colonization rate is surprizingly uniform, with only a twenty fold difference separating the most and least colonized plants and is comparable to that of many other saprophytic root colonizing fungi. Whether a plant is susceptible or immune to systemic infection appears unrelated to the ability of the fungus to colonize the root cortex. The principle difference between wilt susceptible and immune plants would appear to be the ability by the fungus to systemically invade the vascular system of the former. Verticillium clearly belongs to the 'stable' group of root colonizers and not to the 'causal' group as defined by Parkinson et al (1963). Although it is similar to other root colonizers, V. dahliae does differ from them in two important respects. First, it is comparatively slow in initiating contact with passing roots. Whereas most fungi reached maximal colony densities a few millimeters behind the apex, V. dahliae did not do so until about a centimeter behind the tip. Secondly, V. dahliae primarily colonized the inner cortex adjacent to the stele, an area not yet recognized as being exploited by the saprophytic colonizers. Perhaps this special internal niche allows it the luxury of a comparatively slow response to the passing root tip.

The interaction between V. dahliae and roots of susceptible hosts can be divided into two distinct phases. The first is the colonization of the cortex and the second entry into the vascular system. The first phase can be viewed as an epiphytic phase, although, technically we have invasion of host tissue. However, this process is a common feature for essentially all plants this fungus encounters and, although it may be a precursor to, does not appear related to pathogenic activity. As presented above, we now have a reasonable outline of the events involved in and the factors affecting this first phase.

With respect to the second phase, we are still pretty much at the speculative stage. Direct observation of the process of vascular penetration will be difficult. With thousands of cortical colonies for each successful systemic invasion, it is analogous to looking for a needle in a haystack. Most workers studying the physiology of disease have sidestepped this phase by introducing the pathogen directly into the vascular system. A few meager clues are available. Entry through the undifferentiated tissue in the zone of elongation, although an attractive hypothesis, is not consistent with the data currently available. Nor is entry through wounds generated by lateral root emergence. A scenario in which all colonies adjacent to the stele have a ready capability of killing cells in or otherwise transversing the endodermis also seems unlikely. A slow entry prossess can not be ruled out, for this would allow time for the host to marshall defenses and limit the pathogen to the site of penetration for 99.98% of the cases, as is required for such a scheme. The observation of limited but significant (5% incidence) apparent vascular weakening would be consistent with such a process. Alternatively, with viable colonies already in place at the edge of the stele, some breach of this boundary at the site of such a colony by another entity (pathogen, nematode) or process (soil movement during drying/wetting cycles, cultivation) could provide a ready entry point into the vascular system.

REFERENCES

Ashworth LJ Jr, Huisman OC, Grogan RC and Harper DM (1976) Copper-induced fungistasis of microsclerotia of Verticillium albo-atrum and its influence upon infection of cotton in the

field. Phytopathology 66:970-977
Ashworth LJ Jr, Huisman OC, Harper DM, Stromberg LK and Basset D (1979) Verticillium wilt disease of cotton: influence of inoculum density in the field. Phytopathology 69:483-489
Beckman CH (1964) Host responses to vascular infection. Ann Rev Phytopathol 2:231-252
Beckman CH and Talboys PW (1981) Anatomy of resistance, In: Mace ME, Bell AA and Beckman CH (eds) Fungal wilt diseases of plants. Academic Press, New York, p 640
Beckman CH, Vandermolen GE and Mueller WC (1976) Vascular structure and distribution of vascular pathogens in cotton. Physiol Plant Pathol 9:37-94
Benson DW and Ashworth LJ Jr (1976) Survival of Verticillium albo-atrum in nonsuscept roots and residue in field soil. Phytopathology 66:883-887
Evans G and Gleeson AC (1973) Observations on the origin and nature of Verticillium dahliae colonizing plant roots. Aust J Biol Sci 26:151-161
Garber RH and Houston BR (1966) Penetration and development of Verticillium albo-atrum in the cotton plant. Phytopathology 56:1121-1126
Gerik JS, Lommel SA and Huisman OC (1987) A specific serological staining procedure for Verticillium dahliae in cotton roots. Phytopathology 77:261-265
Gerik JS and Huisman OC (1988) Study of field-grown cotton roots infected with Verticillium dahliae using an immunoenzymatic staining technique. Phytopathology (in Press)
Huisman OC (1982) Interrelations of root growth dynamics to epidemiology of root-invading fungi. Ann Rev Phytopathol 20:303-327
Huisman OC (1988) Seasonal colonization of roots of field-grown cotton by Verticillium dahliae and Verticillium tricorpus. Phytopathology (in Press)
Huisman OC (1988) Colonization of field-grown cotton roots by pathogenic and saprophytic fungi. Phytopathology (in Press)
Isaac I (1949) A comparative study of pathogenic isolates of Verticillium. Trans Br Mycol Soc 32:137-157
Isaac I (1953) A further comparative study of pathogenic isolates of Verticillium: V. nubilum Pethylr. and V. tricorpus sp. nov. Trans Br Mycol Soc 36:180-195
Lacey ML and Horner CE (1966) Behavior of Verticillium dahliae in the rhizosphere and on roots of plants susceptible, resistant, and immune to wilt. Phytopathology 56:427-430
Mace ME, Bell AA and Stipanovic RD (1974) Histochemistry and isolation of gossypol and related terpenoids in roots of cotton seedlings. Phytopathology 64:1297-1302
Newman EI and Watson A (1977) Microbial abundance in the rhizosphere: a computer model. Plant and Soil 48:17-56
Parkinson D, Taylor GS and Pearson R (1963) Studies on the root region. I. The development of fungi on young roots. Plant and Soil 19:332-349
Price DE (1976) Passage of Verticillium albo-atrum propagules through the alimentary canal of the bulb mite. Phytopathology 66:46-50
Rovira AD (1965) Plant root exudates. Bot Rev 35:35-57
Schreiber LR and Green RJ Jr (1963) Effect of root exudates on

germination of conidia and microsclerotia of Verticillium albo-atrum inhibited by the soil fungistasis principle. Phytopathology 53:260-264
Schroth MN and Snyder WC (1961) Effects of host exudates on chlamydospore germination of the bean root rot fungus, Fusarium solani f. phaseoli. Phytopathology 51:389-393
Taylor GS and Parkinson D (1961) The growth of saprophytic fungi on root surfaces. Plant and Soil 15:261-267

COLONIZATION OF THE VASCULAR SYSTEM OF PLANTS BY FUNGAL WILT PATHOGENS: A BASIS FOR MODELING THE INTERACTIONS BETWEEN HOST AND PARASITE IN TIME AND SPACE

C.H. Beckman
Department of Plant Sciences
University of Rhode Island
Kingston, RI 02881
USA

INTRODUCTION

When we undertake the study of phenomena as dynamically complex as wilt diseases, we are faced with the problem not only of obtaining the information we seek, but also of organizing many bits of information to show significant relationships. This is true because to understand the dynamics of a phenomenon we must use reductive methods to dissect the processes into finer, more detailed, measurable bits, whereas to understand the complexity we must somehow integrate the resulting information to show the relationships within the whole sweep of events. A system of realistic inter-connected models is required. It is my purpose, first, to propose that we all join together in an ongoing process of modeling wilt diseases of plants, and, second, to propose a basis for modeling that has evolved and that places known events within a framework of interconnected time and space frames. The overall framework is derived from our understanding of the disease cycle involving a soil-borne, fungal, vascular parasite and its host.

In 1957, Talboys reported studies on host-parasite interactions in Verticillium wilt of hop, in which he distinguished two phases in the progress of the disease: a determinative phase, during which the interacting processes of the host and the pathogen establish the specific conditions of vascular invasion that are prerequisite for the development of a specific syndrome; and

NATO ASI Series, Vol. H28
Vascular Wilt Diseases of Plants
Edited by E. C. Tjamos and C. Beckman

an expressive phase, in which continued activity of the fungus, more or less limited by the conditions established in the determinative phase, results in the development of visible symptoms in the aerial parts of the plant.

Talboys then expanded the concept and proposed that the host-parasite interactions in vascular diseases could be divided into three distinct phases: 1) the Primary Determinative Phase, in which the success or failure of the pathogen in penetrating the outer tissues and entering the vascular elements of the host is determined, 2) the Secondary Determinative Phase, in which the success of failure of the pathogen in colonizing the vascular system is determined, and 3) the Expressive Phase, in which symptom development in the host plant occurs. We will use this concept as the basis for our model (Fig. 1). We can add to it several additional segments of the life cycle of the pathogen: 4) the ramification of senescing or dead host tissues by the pathogen, 5) the production of resting structures, 6) the release of resting structures into the soil, 7) the renewal of resting structures by saprophytic growth on organic debris and root exudates, and 8) the renewal of resting structures by transient infections of the roots of host and non-host plants. This represents the broad outline of the model, which we may choose to modify during the course of workshop. More specific sub-models can be added and updated to show new understanding of the interactions that take place between the host and parasite at various times during the disease cycle.

My specific assignment is to review our understanding of the interactions between host and parasite that determine whether systemic colonization is achieved by the pathogen or is prevented by the host, that is, Determinative Phase II in Fig. 1. The sub-model that I will present is an attempt to relate specific happenings and interactions that occur during a specified time span and specified space within the vascular tissues of a host that is infected with a fungal, vascular parasite. The model is a hypothetical one because it is based on results obtained from both Verticillium and Fusarium wilts of different cultivars of tomato - with a bit of cotton or banana thrown in for good measure.

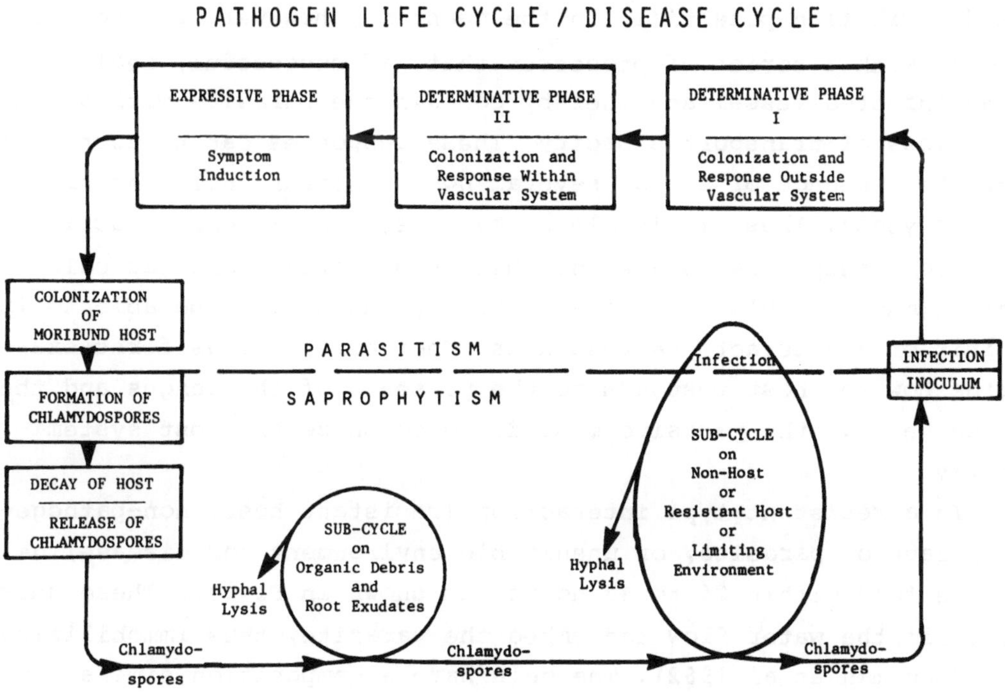

Fig. 1. The life cycle of Fusarium oxysporum, including its saprophytic and parasite growth and successive phases of colonization and pathogenesis within the host plant. Reprinted, by permission, from Beckman (1987).

Let us start with banana because the vessels in banana roots are large and long and straight. It is therefore possible, by means of longitudinal sections and light microscopy, to observe the course of happenings in terms of both time and space with relative ease and accuracy, as shown in Fig. 2. Within a few minutes after inoculating banana (or tomato or cotton) roots with a suspension of fungal spores and spore-size tracer particles (Beckman et al 1962) the inoculum can be found adhering to various trapping sites including, finally, the vessel endings. Within 24 hours the fungal spores will have germinated, penetrated the membranes in the grids of vessel endings, and within 48 to 72 hours will have produced secondary microconidia and reestablished a distribution potential. In the meantime, however, the host responds with a series of processes that, if successful, seal off the infected vessel and thereby contain the parasite with a minimum loss of transport capacity. These responses can be turned on by the presence of any of several non-pathogens including brewer's yeast. Thus, it is clear that the plant is easily activated to a responsive state and that an effective vascular colonizer must be able to avoid activating the host or be able to inhibit or counteract the responses. Therefore, let us first examine how the host responds to the presence of the fungus and then observe what the parasite must do to colonize the host systemically.

In a resistant-type interaction (resistant host, non-pathogenic race of parasite, or unsuitable environment for disease) gel plugs form within 24 to 48 hours, as shown in Fig. 2. These gels cut off the water flow and embed the parasite, thus immobilizing it (Beckman et al 1962). The gels have a composition similar to primary cell walls, but lack cellulose (Vander Molen et al 1983). They become infused with stress metabolites that make them highly resistant to degradation (Beckman 1969). By 2 to 3 days after inoculation parenchyma cells that ensheath the vessels (contact cells) produce outgrowths (tyloses) through the pits invessel walls that completely wall off the infected vessel above the trapping site. The walls of tyloses that interface with the infection also become infused with stress metabolites so that the gel and tylose occlusion is "lignified" and highly resistant to physical and chemical degradation. This sequence of events occurs and infections are sealed off in many host plants, in response to many introduced organisms, under most environmental conditions. It fails when a susceptible host is infected with a suitable pathogenic race of a vascular parasite under environmental conditions that favour disease.

When resistance fails (Fig. 2 left) the gels apparently become weakened and shear under transpirational tension (Beckman et al 1962). Tylose development, although initiated normally, becomes inhibited after the second day (Fig. 3). Thus, when the gels shear, the secondary spores of the parasite are drawn upward in the transpiration stream to the next trapping site. A series of such surges results in systemic fungal distribution (Beckman et al 1962,and confirmed as in Fig. 4). A 1966 model (Fig. 2) therefore included as host factors for resistance the induction of gels and tyloses, presumably by hormones produced by the host (Sequeira 1965) in response to infection and their infusion, at the infection interface, with stress metabolites that serve to lignify the interface (phenolics) and to inhibit the growth of the parasite (phytoalexins). Mace will discuss the occurrence of stress metabolites in greater detail in the next section.

Likely pathogen factors that could overcome the resistance included a gel degrader (s), presumably endo-forms of carbohydrases (Cooper and Wood 1975, 1980), an unknown growth inhibitor, and the capacity of the parasites to grow and reproduce rapidly within the vascular environment.

In the meantime, Mace (1963) showed that a major phenolic 3-hydroxytyramine was compartmented in specialized vascular parenchyma cells of banana, released following infection, became oxidized and polymerized with host gels and walls, and in the process, served to mediate the synthesis of IAA (Mace and Solit 1966). Matta showed a similar happening in tomato infected with Fusarium (1970). Sequeira (1965) and Kleinhe-Borchers (1982) showed IAA accummulation, apparently of host origin, in tomatoes infected with Pseudomonas and Fusarium. Furthermore, Mussel et al (1982) showed that cell wall preparations from tomato contain an enzyme that can generate ethylene from 1-aminocyclopropane-1-carboxylic acid. The latter substance apparently occurs as an intermediate in the pathway from methionine to ethylene in all plants. The enzyme was released from the cell walls by purified PG from V. dahliae. Thus, the attack of a pathogen would automatically trigger the release of ethylene and IAA which can, in turn, trigger responses in host tissues somewhat removed from the immediate site of infection. These responses include isodiametric growth, i.e., the kind of growth associated with tylose formation, which is favoured by high concentrations of IAA and ethylene relative to cytokinins (Apelbaum and Burg 1972) and the formation of vascular gels, shown to be induced by ethylene (Vander Molen et al 1983).

Tjamos and Smith (1974), studying Verticillium wilt of tomato, showed the time course of appearance of terpenoid substances,

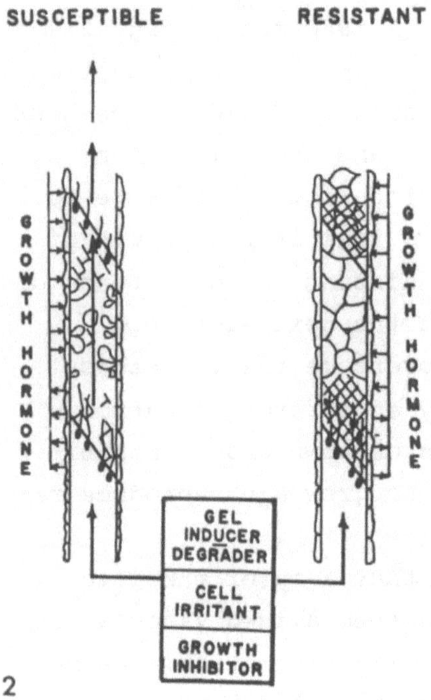

2

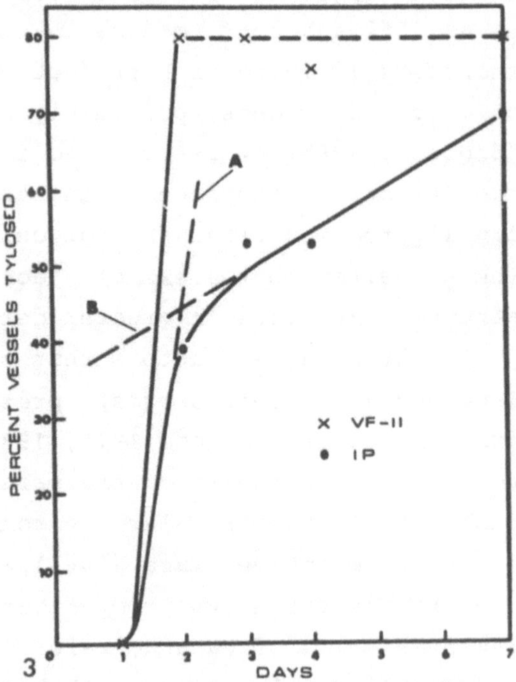

3

Fig. 2. Diagrammatic representation of hypothetical metabolites of host and parasite that act to produce a resistant or a susceptible reaction. These include a gel inducer and/or degrader, a cell irritant, a growth inhibitor, and growth-promoting hormone (s). Reprinted, by permission, from Beckman (1966).

Fig. 3. Percentage of xylem vessels, above visible trapping sites, in which tyloses (occluding or not) were initiated in relation to time in days after inoculation with Fusarium oxysporum f.sp. lycopersici race 1. In the resistant isoline Improved Pearson VF-11, the rate of tylose initiation was rapid, and all infected vessels, up to the 80% observable limit, contained tyloses 2 days after inoculation. The situation in the susceptible isoline Improved Pearson was somewhat more complex. A projection of the curve between day 1 and day 2 (line A) suggests that initiation of tyloses in the susceptible IP isoline was at first comparable to that in the resistant VF-11 isoline. A projection of the curve for tylose initiation in IP between day 2 and day 7 (line B) suggests that an inhibitory factor came into play in the Fusarium-infected susceptible isoline on or soon after day 2. Enlargement of tylose initials in the susceptible IP isoline was also visibly retarded during this period. Reprinted, by permission, from Beckman et al (1972).

identified the sesquiterpenoid rishitin (Fig. 5), and showed it to be inhibitory to the pathogen. Cooper and Wood (1975) examined the carbohydrate-cleaving enzymes that were induced in Fusarium and Verticillium when grown on tomato cell wall preparations, including the manner in which their synthesis was regulated. They also (1980) determined the activity of two of these enzymes, Endo-pectin lyase and Endo-pectin galacturanase, over a period of 20 days within infected tomato tissue that had been carefully sampled to exclude as much extraneous tissue as possible (Fig. 6). This was an important advance because it recognized that if meaningful quantitative data were to be obtained, the selection of samples and sampling times would have to be appropriate to the questions being asked.

By 1980 then, specific components of the interaction were being identified and quantified. Harrison (1987) sought to identify a growth inhibitor in Fusarium-infected vascular tissue of tomato that could account for the inhibition of tyloses demonstrated earlier (Fig. 3). Using the method of sampling devised by Cooper and Wood (1980) and vascular discolouration as determinants of appropriate samples, he could, at different time periods within 28 days after inoculation, find no evidence of a growth inhibitor produced by the parasites. Rather, of two strong growth inhibitors, one was tentatively identified as rishitin, the other probably a phenolic and both probably of host origin. Rishitin was found to be present in the infected tissues at 40 µg per gram of fresh weight, well above levels that strongly inhibited and killed tomato cells grown in culture. The other inhibitor appears to have an ortho-dihydroxy phenolic structure (unpublished). Thus, a 1984 model needed to be revised to show that the growth inhibitors (and probably a proposed respiratory inhibitor) are of host origin (Fig. 7). They are apparently part of the family of stress metabolites produced by the host in response to infections. Presumably, they should serve either as "lignifiers" or fungal inhibitors (phytoalexins), or both.

These results suggest that the stress metabolites that normally play a role in "locking up" the host-parasite interface at the trapping site, have, somehow, been carried forward into the region above the trapping site where tyloses should be developing. Rather than locking up the interface, they appear to lock up the developing tyloses. Obviously some earlier event in

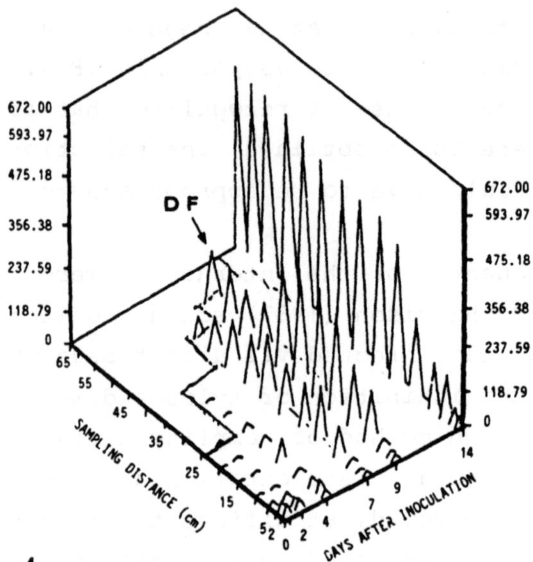

4

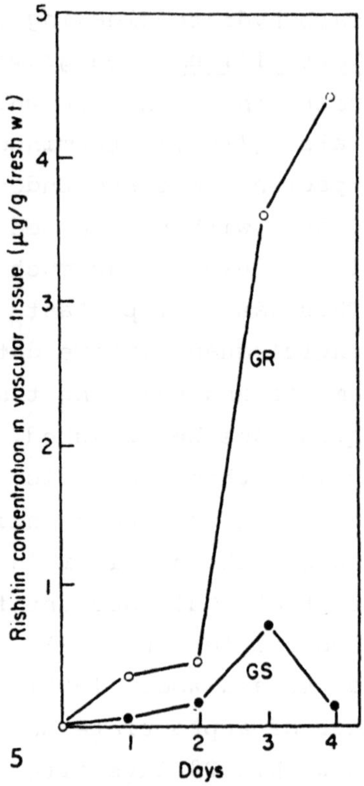

5

Fig. 4. Measurements and pattern of upward distribution of Fusarium oxysporum f.sp. vasinfectum within the vascular systems of stems of the susceptible cotton cultivar Rowden at various times after inoculation by means of a single hypodermic injection of a spore suspension into the lower hypocotyl region. The data show both the numbers and extent of vascular infections and the time course of the progressive advance (line D) of the pathogen in the susceptible cultivar. Reprinted, by permission, from Harrison (1981).

Fig. 5. Accumulation of rishitin in resistant (GR) and susceptible (GS) tomato stem segments inoculated with spore suspension of Verticillium albo-atrum. Reprinted, by permission, from Tjamos and Smith (1974).

the sequence of defense reactions has been disrupted. The number of presumed pathogen factors has, with the removal of growth and respiratory inhibitors, been reduced to two. The first is the production of hydrolytic enzymes. The second is the capacity to produce secondary spores above the trapping site where the parasite has a potential for renewed distribution.

If the titer of hydrolytic enzymes in infected tomato is much

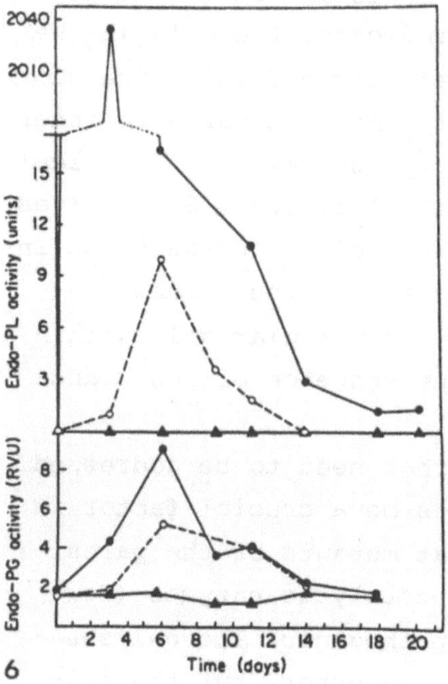

6

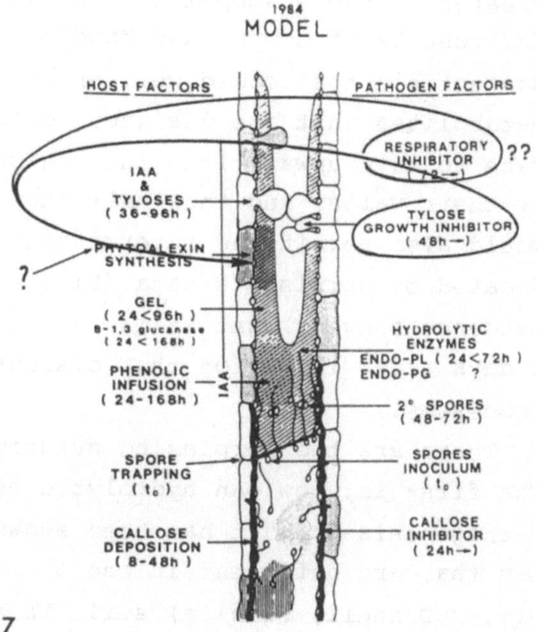

7

Fig. 6. Activity of endo-pectin lyase (endo-PL) and endopolygalacturonase (endo-PG) in stems of healthy (▲) and Verticillium-infected cuttings of the susceptible (●) and resistant (O) isolines of the tomato cultivar Gem. Cuttings from 6-week old plants were inoculated with approximately 5 x 10^6 spores per milliliter and kept in nutrient solution at 21 C. Each determination was from an extract of four plants. Reprinted, by permission, from Cooper and Wood (1980).

Fig. 7. Time-space model of host parasite interactions in an infected vascular element and the surrounding contact parenchyma cells. The time frame is the period from t_o through t_7 (the 7-day period after initial infection), and the space frame corresponds to S_o and S_1 as described in Fig. 8. Shown are a spore-trapping site, which delays the advance of the pathogen, and, to the left, components of the host defense system that subsequently serve to localize and neutralize the infection. To the right are the pathogen factors that may circumvent or overcome the host responses and permit systemic distribution of the parasite within the plant. Times noted are hours after t_o, the time of initial infection. The symbol < indicates an increasing response during the period indicated; the symbol ⟶ indicates a continuing response. PL= pectin lyase; PG= polgalacturonase. The large arrows indicate that the growth and proposed respiratory inhibitors, initially presented as pathogen factors, are in reality host factors that presumeably (question mark) are intended to function as phytoalexins or lignifiers.

greater in the susceptible than in the resistant cultivar, as the results of Cooper and Wood (1980) indicate, the gels in the susceptible host could well shear prematurely and allow the stress metabolites that are designed to lignify the host-parasite interface to move upward into the region of tylose development, lignify their walls, and thus halt the growth of tyloses. Such surges would also permit the intermittent advance of the parasite as indicated by Harrison's data (Fig. 4). Thus, numerous lines of evidence suggest that it is the failure of vascular gels within 2 days afrer infection that disrupts the sequence of resistance processes.

There are two perplexing questions that need to be addressed. The first is: how can hydrolytic enzymes be a crucial factor in pathogenesis when it has been shown that mutants of the parasites that are deficient in one or more pectolytic enzymes (Howell, McDonnell, Puhalla) still remain pathogenic? The gel shearing hypothesis may well turn out to be in error, but the evidence is, in my judgement, still too strong to be abandoned. Vascular gels are composed of large complex molecules in a distended state, as opposed to the dense structure of cell walls. In such a distended state, the gels might induce numerous different enzymes simultaneously - some of which, particularly hemicellulases, have not been sought. The cleavage of relatively few internal bonds within any of these diverse macromolecules could substantially reduce the resistance of the gels to shearing under transpirational pull. The elimination of one or two of these enzymes from the battery of those that a vascular parasite is able to produce (Cooper and Wood 1975) therefore need not prevent systemic colonization.

The second question is this: how can the same organism produce such different titers of enzyme (Fig. 6) in resistant and susceptible hosts? Cooper and Wood (1980) found similar activities of endo-PL in Gem resistant and susceptible stem segments infiltrated with Verticillium propagules. They concluded that the difference in enzyme titer in infected plants reflected a difference in fungal growth within the plants, rather than effects of different cell walls on enzyme synthesis or activity.

There is another finding that could account for the differences in enzyme titer found by Cooper and Wood. Vascular parenchyma cells of tomato that are immediately adjacent to vessels (and therefore share common pits) were examined following inoculation of the vessels with Fusarium. These contact cells in a resistant cultivar showed a heavy deposit of callose within 12 to 24 hours whereas those of a susceptible cultivar showed a considerably weaker response. When the resistant and susceptible cultivars were inoculated with root flora only, both deposited callose strongly. It appears, therefore, that a virulent race of Fusarium somehow influences callose deposition adversely, and does so within a few hours after inoculation.

Callose is known to be highly impermeable to aqueous solutes (Heslop-Harrison 1966). So too are wall coating responses that have been reported by Street et al (1986) and Robb et al (1987) to be initiated within 8 to 12 hours after inoculation of vascular tissues of tomato with Verticillium albo-atrum. Furthermore, extensive callose deposition permitted only occasional invasion of vascular parenchyma cells in the resistant host, whereas a weaker callose deposition permitted extensive tissue invasion in the susceptible host. These differences will be dealt with in detail in the next section, but it is clear that, because of these differences in the level of response, the nutrient base for the parasite is much greater in the susceptible than in the resistant host. Here then are early effects on the system of host responses that could materially alter the nutritional environment of the parasite within resistant vs. susceptible hosts, alter the level of hydrolytic enzyme synthesis within these hosts, and, in a cascade of events, alter the rate at which gels are cleaved, alter their resistance to shear forces, alter the distribution of stress metabolites, alter the rate of tylose development, and alter the extent to which secondary spores of the parasite can, by successive spore passage, colonize the host. An updated model of host-parasite interactions, based on these considerations is presented in Fig. 8. In it are included and rearranged the host and pathogen factors now judged to be interactive in space frames 0 and 1 and within the first few hours to seven days after introduction of a parasite into vascular elements of the host plant. This model may remain intact, however,

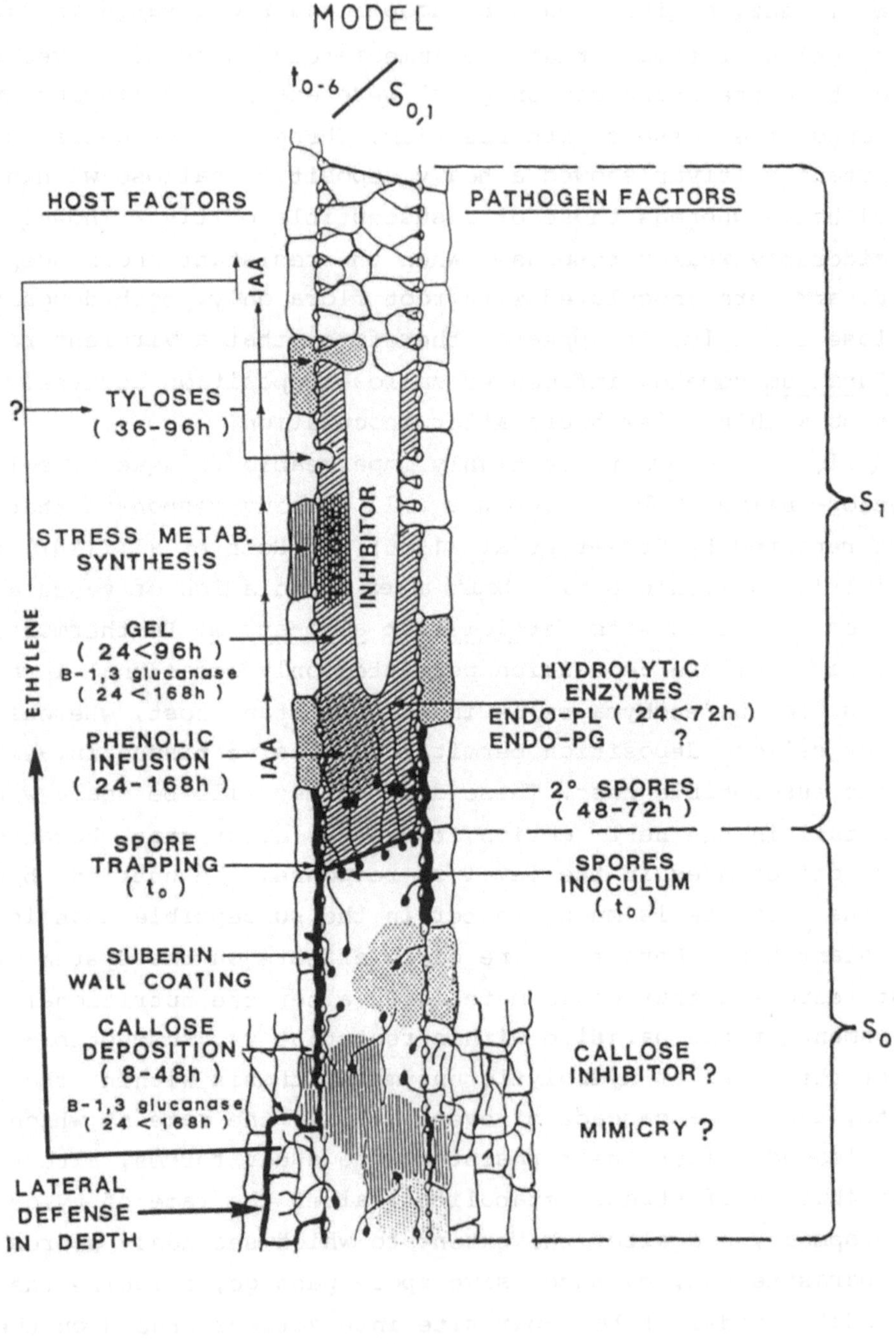

Fig. 8. Diagrammatic representation of the interactions between a host and a potential pathogen within an initially infected vessel that delimits Space$_0$ (S_0) and the next vessel above (S_1) during a time 0 to 6 days (t_{0-6}) after inoculation and showing the approximate time periods within which these events have been established to occur.

only until the discussion begins!

REFERENCES

Apelbaum A and Burg SP (1972) Effect of ethylene on cell division and deoxyribonucleic acid synthesis in Pisum sativum. Plant Physiol 50:117-124

Beckman CH (1966) Cell irritability and localization of vascular infections of plants. Phytopathology 56:821-824

Beckman CH (1969) The mechanics of gel formation by swelling of simulated plant cell wall membranes and perforation plates of banana root vessels. Phytopathology 59:837-843

Beckman CH (1987) The Nature of Wilt Diseases of Plants. APS Press, St. Paul

Beckman CH, Elgersma DM and MacHardy WE (1972) The localization of Fusarial infections in the vascular tissue of single dominant-gene resistant tomato. Phytopathology 62:1256-1260

Beckman CH, Halmos S and Mace ME (1962) The interaction of host, pathogen, and soil temperature in relation to susceptibility to Fusarium wilt of banana. Phytopathology 52:134-140

Cooper RM and Wood RKS (1980) Cell wall degrading enzymes of vascular wilt fungi. III. Possible involvement of endo-pectin lyase in Verticillium wilt of tomato. Physiol Plant Pathol 16:285-300

Harrison NA (1981) Propagule build-up and distribution of Verticillium dahliae and Fusarium oxysporum f.sp. vasinfectum in wilt-resistant and-susceptible cotton (Gossypium)cultivars. M.S. Thesis, University of Rhode Island, Kingston, RI U.S.A.

Harrison NA (1987) Growth inhibitors associated with Fusarium wilt of tomato. Physiol Mole Plant Pathol 30:401-420

Harrison NA and Beckman CH (1981) Time/space relationships of colonization and host response in wilt-resistant and wilt-susceptible cotton (Gossypium) cultivars inoculated with Verticillium dahliae and Fusarium oxysporum f.sp. vasinfectum. Physiol Plant Pathol 21:15-35

Heslop-Harrison J (1966) Cytoplasmic continuities during spore formation in flowering plants. Endeavour 25:65-72

Howell CR (1976) Use of enzyme-deficient mutants of Verticillium dahliae to assess the importance of pectolytic enzymes in symptom expression of Verticillium wilt of cotton. Physiol Plant Pathol 9:279-283

Kleinke-Borchers A (1982) Auswirkung der Interaktion zwischen Meloidogyne incognita und Fusarium oxysporum f.sp. lycopersici auf den Phytohormonhaushalt in Tomatenpflanzen. Z Pflanzenkr Pflanzenschutz 89:132-143

Mace ME (1963) Histochemical localization of phenols in healthy and diseased banana roots. Physiologia Plantarum 16:915-925

Mace ME and Solit E (1966) Interactions of 3-indoleacetic acid and 3-hydroxytyramine in Fusarium wilt of banana. Phytopathology 56:245-247

Matta A (1970) Mechanism of indoleacetic acid accumulation in Fusarium wilt of tomato. Ann Acad Sci Fenn, Ser. A4 168:58-60

McDonnell K (1958) Absence of pectolytic enzymes in a pathogenic strain of Fusarium oxysporum f.sp. lycopersici. Nature (London) 182:1025-1026
Mussell H, Stiwell P and Peck S (1982) The possible origin of ethylene in Verticillium wilt of tomato. Phytopathology 72: 968 Abstract
Puhalla JE (1981) Genetic considerations in the genus Fusarium. Pages 291-305 in: Fusarium: Diseases, Biology and Taxonomy. Nelson PE, Tousoun TA and Cook RJ (eds). Pennsylvania State University Press, University Park
Robb J, Powell DA and Street PFS (1987) Time course of wall-coating secretion in Verticillium-infected tomatoes. Physiol Mole Plant Pathol 31:217-226
Sequeira L (1965) Origin of indoleacetic acid in tobacco plants infected by Pseudomonas solanacearum. Phytopathology 55:1232-1236
Street PFS, Robb J and Ellis BE (1986) Secretion of vascular coating components by xylem parenchyma cells of tomatoes infected with Verticillium albo-atrum. Protoplasma 132:1-11
Talboys PW (1957) The possible significance of toxic metabolites of Verticillium albo-atrum in the development of hop wilt symptoms. Trans Brit Mycol Soc 40:415-427
Tjamos EC and Smith IM (1974) The role of phytoalexins in the resistance of tomato to Verticillium wilt. Physiol Plant Pathol 4:249-259
Vander Molen GE, Labavitch JM, Strand LL and DeVay JE (1983) Pathogen-induced vascular gels: Ethylene as a host intermediate. Physiol Plant 59:573-580

HOST COLONIZATION AND PATHOGENESIS IN PLANT DISEASES CAUSED BY FASTIDIOUS XYLEM-INHABITING BACTERIA

M.J. Davis
Fort Lauderdale Research and Education Center
Institute of Food and Agricultural Sciences
University of Florida
3205 College Avenue
Fort Lauderdale, FL 33314
USA

INTRODUCTION

The diversity of plant pathogenic bacteria capable of colonizing plant xylem is astounding. They are represented in almost every genus containing bacterial plant pathogens and incite diseases characterized by symptoms such as wilts, cankers, soft rots, leaf spots, galls, blights, declines, stunts, and leaf marginal necroses (Hayward 1974; Van Alfen 1982). Included among these xylem colonizers are bacteria with exacting nutritional and physiological requirements for growth as evidenced by their inability to grow <u>in vitro</u> on media commonly used for cultivation of plant pathogenic bacteria (Davis et al 1981). These bacteria are collectively referred to as fastidious xylem-inhabiting bacteria (FXB).

FXB were discovered originally in the early 1970's in association with Pierce's disease (PD) of grapevines (Goheen et al 1973; Hopkins and Mollenhauer 1973), phony disease of peach (PDP) (Hopkins et al 1973; Nyland et al 1973), and ratoon stunting disease (RSD) of sugarcane (Gillaspie et al 1973; Maramorosch et al 1973; Teakle et al 1973). Previously, the etiological agents of these diseases were presumed to be viruses. In addition to PD, PPD, and RSD, other diseases that are known to be caused by or associated with FXB include: Bermudagrass stunting disease, alfalfa dwarf, plum leaf scald, leaf scorches of elm, oak, mulberry, sycamore

NATO ASI Series, Vol. H28
Vascular Wilt Diseases of Plants
Edited by E. C. Tjamos and C. Beckman

and almond, periwinkle wilt, ragweed stunt, and citrus blight (Davis et al 1984; Wells et al 1987). Other xylem-inhabiting bacteria which are somewhat less fastidious have been found to cause Sumatra disease of cloves (Hunt et al 1987) and a wilt of bent grass (Roberts et al 1981). Some of the aforementioned diseases are caused by the same pathogen. For example, one pathogen causes PD, alfalfa dwarf, almond leaf scorch and possibly citrus blight, while another causes both PDP and plum leaf scald. Furthermore, the host range of the different bacteria, when examined in nature or by experimental inoculation of plants, is usually very wide (Frietag 1951; Hopkins and Adlerz to be published; Raju et al 1983; Steindl 1957). However, most of the hosts remain symptomless when colonized. Known hosts of the PD bacterium alone include members of at least 28 families of monocotyledonous and dicotyledonous plants.

Following their discovery, initial attempts to isolate FXB in axenic culture failed until the PD bacterium was first isolated (Davis et al 1978). Subsequently, different media formulations were developed that support growth of all known FXB (Chen et al 1982). These accomplishments have greatly facilitated further studies on the organisms and the diseases they cause. The majority of FXB have been taxonomically classified within two genera. The Gram-positive FXB causing RSD and Bermudagrass stunting disease were found to be related to less fastidious plant pathogenic coryneform bacteria, and a formal proposal to classify these two organisms as Clavibacter xyli subsp. xyli and C. xyli subsp. cynodontis, respectively, and to reclassify the related coryneform bacteria in the same genus was made (Davis et al 1984). The remainder of the FXB, including the causal agents of PD and PDP, are Gram-negative, and most have been shown to be related to one another and to be more distantly related to Xanthomonas sp. (Wells et al 1987). The classification of these organisms into a new genus as Xylella fastidiosa was formally proposed.

Advances in research on FXB have made it possible for more attention to be focused directly at understanding their interactions with plants. Recent studies, primarily concerning RSD, PD and PDP, have provided new information on how FXB colonize and incite disease in their respective hosts and have identified factors that are involved in limiting this process. In the present paper, these studies are reviewed.

TRANSMISSION AND ESTABLISHMENT WITHIN THE HOST

Many plant pathogenic bacteria can survive, and sometimes multiply, outside their host plants for extended periods of time and are transmitted to another host plant by wind, rain, insects, or mechanical means. Some may be transmitted on or within seed. Bacterial pathogens invade their hosts in a variety of ways; ingress may begin either through natural openings or wounds (Goodman 1982; Huang 1986). Many bacteria have flagella which confer motility and thus, enable them to reach infection courts. Xylem-inhabitors may reach the xylem directly through wounds or, after breaching initial barriers to infection, move intercellularly through plant tissues until they reach the xylem (Billings 1982; Nelson and Dickey 1970). Once the xylem is invaded, bacterial growth within tracheary elements may eventually result in systemic colonization of the plant. Bacterial egress from the xylem and invasion of surrounding tissues often precedes the terminal, and possibly the most damaging, stages in some diseases (Nelson and Dickey 1970; Preece 1982). Characteristic symptoms of some diseases, such as cankers and soft rots, may only be expressed at this stage. These symptoms may not be as indicative of xylem-inhabiting bacteria as are other symptoms, such as wilting.

Unlike many bacterial plant pathogens, the FXB rarely survive in nature outside their hosts for extended periods of time (Davis et al 1981). Furthermore, they require direct access or deposition to the xylem through wounds in order to become established within their hosts. Since FXB do not have flagella and are nonmotile, they cannot actively seek infection courts. Once established within the xylem, FXB are limited to this tissue since they apparently lack the capacity to actively invade surrounding tissues. Gram-negative FXB, such as those causing PD and PDP, are transmitted by xylem-feeding insects which deposit the bacteria directly into tracheary elements of the xylem through wounds created during feeding (Purcell 1982a, 1982b). Potential vectors include all insects with piercing, sucking mouthparts that are chiefly xylem-feeders. Leafhoppers (Cicadellidae) of the subfamily Cicadellinae and spittlebugs (Cercopidae) are known vectors. Although the FXB apparently propagate within the foregut of these insects, they are eliminated with the foregut lining when the insects molt and are not transovarially passed from one insect generation to another; thus, the insects are short-term intermediate hosts. In contrast, Gram-positive FXB appear to be transmitted by mechanical means principally through the agency of man. Chewing animals, such as rats, and insects are also suspected of being occasional vectors of C. xyli subsp. xyli in sugarcane, although the pathogen in usually transmitted on the blades of implements used to harvest and cultivate sugarcane crops (Steindl 1961). Grafting or vegetative propagation of infected host material can also contribute to the spread of FXB; however, transmission in true seed is not known to occur.

The mechanism of transmission of FXB becomes a central issue when considering the possible effects of environmental factors on initial establishment of infection within the host plant. When insect vectors are the immediate inoculum source, they not only deliver the bacteria to the infection court but also modulate environmental and other factors, such as inoculum quality and concentration, which in turn may affect transmission efficiency. Transmission of some FXB is less complex in as much as it does not involve an intermediate insect host. Since C. xyli subsp. xyli is transmitted directly between plants by mechanical means, it is usually subjected to the environment outside of the host for only a short period of time. The amount of inoculum transmitted, however, is likely to vary substantially at any given time but may be largely dependent on the sugarcane cultivar that is serving as the inoculum source because the size of C. xyli subsp. xyli populations supported by different sugarcane cultivars can differ considerably (Davis et al 1988; Gillaspie et al 1976). Furthermore, infection frequency appears to be largely dependent upon both inoculum dose and cultivar (Harrison and Davis 1986). Generally, more resistant cultivars support smaller populations of C. xyli subsp. xyli and are also less likely to become infected at lower inoculum concentrations.

Many factors can affect the successful establishment of bacterial plant pathogens within their hosts (Billing 1982; Goodman 1982). Environmental factors, such as moisture, nutrition, temperature, and light, often play an important role in the predisposition of the host to infection. However, little is known about the importance of these factors in the establishment of infection by FXB.

SYSTEMIC COLONIZATION OF THE HOST PLANT

Typically after establishment within tracheids or vessels, FXB invade other tracheary elements and, thereby, systemically colonize their respective hosts. Zimmermann and McDonough (1978) contend that sufficient injury is caused when a vessel is invaded by a pathogen to disrupt negative tension of the water column within the vessel and create an embolism composed of water vapor and air. In turn, the embolism renders the individual vessel dysfunctional. A major implication of this theory is that embolisms are the primary cause of xylem dysfunction in vascular diseases, unless vessels somehow retain or regain their function upon invasion. Furthermore, it would follow that vascular pathogens multiply solely in dysfunctional tracheary elements and are, therefore, not carried through the plant by the transpiration stream. Conceivably, embolisms might help to create an environment conducive to pathogen development within individual tracheary elements. Whether pathogens are carried by the transpiration stream or not, most need to breach pit membranes in order to move from one tracheary element to another. Consequently, pit membranes might provide the most serious impediment to the intraxylar movement of these pathogens, unless their progress is also impeded by obstacles formed in response to infection such as gums and tyloses.

Colonization of plants by FXB has been examined in a number of studies using either transmission or scanning electron microscopy (Davis et al 1983b; Huang et al 1986; Kao and Damann 1980;

Lowe et al 1976; Mollenhauer and Hopkins 1974; Weaver et al 1977; Worley and Gillaspie 1975). Frequently, the bacteria are spread throughout the lumen of tracheary elements. Sometimes the bacteria are appressed to the walls of the lumen which might indicate the presence of an embolism but might also be due to bacterial aggregation on the walls in the absence of embolisms. The bacterial distribution suggests that they are in nonfunctional tracheary elements and apparently have not been swept in the transpiration stream toward pit membranes or the upper end of the cell. Frequently, vessels that are completely plugged with bacteria are found adjacent to vessels that are devoid of bacteria; evidence that pit membranes provide a significant obstacle to bacterial movement. FXB apparently must breach pit membranes while moving from one tracheary element to another; however, the mechanism by which this is accomplished remains unclear for most FXB. Swelling of pit membranes and dissolution of the middle lamella has been observed in thin sections of vessels infested by some FXB (see Fig. 3 in Davis et al 1983). However, the actual breaching of pit membranes by FXB has rarely been observed. At least some FXB may dissolve small holes in pit membranes through which they pass, as evidenced in micrographs of C. xyli subsp. xyli cells in sugarcane and sorghum-sudangrass (Weaver et al 1977; Worley and Gillaspie 1975).

Little is known about the nutritional and physical requirements of FXB in their hosts (Davis et al 1981). If tracheary elements are dysfunctional during colonization, bacterial nutrients would have to be derived from materials present in the tracheary elements at the time they became dysfunctional or from surrounding cells. Failure of the PD bacterium to grow in xylem sap exudates (Davis et al 1980) suggests that the nutrient composition of sap alone is insufficient for growth of the PD bacterium and other FXB. The lack of growth of the PD bacterium in xylem sap exudates might have been due to insufficient quantities of some nutrients. It has been suggested that continuous bathing of xylem-inhabiting bacteria with nutrients carried in the transpiration stream might provide them with adequate quantities for growth (Davis et al 1981; Van Alfen 1982). This of course would be contingent upon tracheary elements being functional for a substantial period of time after invasion, which may not be the situation.

FXB can usually colonize tracheary elements throughout a host plant; however, the extent of colonization in different parts of the plant may vary. Environmental conditions, host genotype, and tissue maturity are important factors governing the distribution of FXB in their hosts. Furthermore, their distribution apparently influences the type of disease symptoms that develop.

Leaf marginal necrosis, a characteristic symptom of some diseases caused by these bacteria, usually develops first in the older, more mature leaves. Hopkins and Thompson (1984) found that the concentrations of the PD bacterium in grapevine tissues correlated very well with PD symptom development. The proportion of xylem vessels containing bacteria approached 100% in 5-mm lengths of both stem and leaf tissues of both muscadine and bunch grape; whereas, lower percentages were found in petiole tissues. Leaf marginal necrosis symptoms developed when the percentage of infested leaf vessels approached the maximum. The development of populations of the PD bacterium was related to susceptibility of the particular grape cultivar to injury due to PD. The accumulation of PD bacterial occlusions were found to occur 2 months earlier in late spring in a susceptible bunch grape cultivar than in resistant muscadine cultivars, although nearly all vessels within stems and leaves of each cultivar were eventually infested.

The most obvious symptom of PDP in peach trees is a reduction in annual growth. Leaf marginal necrosis is not a symptom associated with PDP but is characteristic of plum leaf scald, another disease caused by the same bacterium. Populations of the bacterium in both peach and plum trees are usually much larger in roots than in shoots; however, greater proportions of the populations were found in plum shoots than in peach shoots (Wells et al 1980). The populations of PDP bacterium in different <u>Prunus</u> sp. were generally greatest in the late spring months.

In sugarcane, the predominant external symptom associated with RSD is stunting. Populations of <u>C. xyli</u> subsp. <u>xyli</u> are generally larger in shoots than in roots (Steindl 1961) and larger in mature stalk tissues than in immature stalk tissues and leaves (Teakle et al 1975; Bailey 1977). The development of C. xyli subsp. xyli populations in sugarcane stalks is directly correlated with tissue maturity and occurs at approximately the same rate throughout most of the growing season in cultivars differing in resistance to injury due to RSD (Davis, Dean and Harrison, to be published). In more susceptible cultivars, however, larger populations of the RSD bacterium develop earlier in the growing season and also to a greater extent by the end of the growing season. Harrison and Davis (to be published) used a novel tissue-blot enzyme immunoassay to detect vascular tissues infested by C. xyli subsp. xyli in cross sections of sugarcane stalk internodes and found that the number of infested vascular bundles diminished with increasing distance from the base of stalks in all cultivars examined, and that the proportion of infested vascular bundles was greater in the central region of stalks than in more peripheral regions. Thus, colonization appears to be directly correlated to tissue maturation both longitudinally and laterally within stalks.

Although different FXB may individually colonize plants of the same host, their virulence in a particular host may differ. The periwinkle wilt bacterium and the PD bacterium, both classified as X. fastidiosa, systemically colonized both grapevine and periwinkle following inoculation (Davis et al 1983b). However, the periwinkle wilt bacterium did not incite symptoms in grapevine, and the PD bacterium incited slight chlorosis in periwinkle but not typical periwinkle wilt symptoms. Similarly, C. xyli subsp. xyli and C. xyli subsp. cynodontis systemically colonized both sugarcane and Bermudagrass plants, but C. xyli subsp. xyli did not incite stunting in Bermudagrass and C. xyli subsp. cynodontis did not incite internal discolouration symptoms of RSD in sugarcane (Davis et al 1980, 1983a). Furthermore, the frequency of successful inoculation of sugarcane and Bermudagrass plants was consistently greater following inoculation with the subspecies of the pathogen normally associated with each plant species.

Differences in virulence and loss of virulence have been detected among some strains of the PD bacterium after they have been subjected to axenic culture for various periods of time (Hopkins 1984). An avirulent strain, a weakly virulent strain, and a strongly virulent strain of the PD bacterium were all able to multiply in grape petioles following inoculation; however, the growth rate was directly correlated with virulence (Hopkins 1985). Weakly and strongly virulent strains eventually reached the same maximum population size; whereas, the population size reached by the avirulent strain was tenfold less than the virulent strains. When grape internodes were inoculated with the strains, only the virulent strains were able to move systemically from the inoculated internode. Interestingly, strains of C. xyli subsp. xyli and C. xyli subsp. cynodontis have been maintained continuously by serial passage in axenic culture for longer than a year without losing their ability to systemically colonize their respective hosts (Davis unpublished), suggesting that the Gram-positive FXB might be more stable in respect to virulence than the Gram-negative FXB.

RESISTANCE TO COLONIZATION

Localization of FXB by vascular occlusions produced in response to infection may limit the extent of colonization and thus, serve as a resistance mechanism to disease. Mollenhauer and Hopkins (1976) found that gums and tyloses formed 7 to 8 times more frequently in resistant muscadine and wild grapevines than in susceptible bunch grapevines. The gums were often observed to encapsulate the PD bacteria. Huang et al (1986) observed that a morphological change of the PD bacterium from rippled cell walls to smooth cell walls occurred in invaded xylem vessels, and this was more rapid in muscadine grapevines than in bunch grapevines. Also, encapsulation of the smooth cell-walled bacteria by an electron dense substance was more frequently observed in muscadine than in bunch grapevines. Histochemical tests indicated that vascular plugs associated with PD are primarily composed of pectic substances with some gums and tannins. A partially pectinaceous gel-like material often occludes tracheary elements in sugarcane with RSD and occurs in both infested and noninfested cells (Kao and Damann 1978, 1980). However, extensive vascular gelation in cultivars susceptible to RSD injury may indicate that occlusion by gels is not an effective resistance mechanism and may actually contribute more to pathogenesis in these cultivars. The formation of tyloses has not been associated with RSD of sugarcane. Presumably, the "electron dense substance", "matrix material", "gums" and "gels" referred to in different studies on vascular colonization by FXB are often composed of the same or very similar materials of plant origin. Vascular gelation appears to be a general response phenomenon following infection by fungal wilt pathogens (VanderMolen et al 1977) and may also be generally associated with vascular infections by bacteria.

Xylem anatomy alone may play an important role in localization of FXB. Resistance to RSD in sugarcane cultivars is directly correlated to the extent of branching of metaxylem vessels in the nodal regions of stalks (Teakle et al 1978). Conversely, susceptibility to RSD is directly correlated to the number of vascular bundles with vessels that pass through nodal regions without terminating (Teakle et al 1975). Harrison and Davis (to be published) found a direct correlation between the number of pathogen-infested vascular bundles in cross-sections of sugarcane stalk internodes and susceptibility to injury due to RSD. In the most susceptible cultivars, the proportion of infested vascular bundles in internode cross sections from the basal part of stalks averaged between 50% and 70%; whereas, in the most resistant cultivar, less than 5% of the bundles were colonized. Furthermore, a strong linear relationship existed between the number of infested vascular bundles and the size of C. xyli subsp. xyli populations in adjacent internode tissues, suggesting that susceptibility to RSD is governed by accessibility of vascular tissues to C. xyli subsp. xyli rather than the ability of the pathogen to multiply within infested vascular tissue.

PATHOGENESIS

Tissue degradation, alterations in cell wall permeability,

obstruction of water movement, and cytopathological effects due to toxins, have all been suggested as possible mechanisms of pathogenesis in diseases caused by xylem-inhabiting bacteria. Obstruction of water movement within the plant is the principal mechanism associated with many of the diseases caused by FXB and other xylem-inhabiting pathogens. However, the diseases incited by FXB are often more insidious or subtle than those caused by other xylem-inhabiting bacteria.

Early studies on insect and graft transmission of the PD and PDP agents demonstrated that the agents were xylem-limited (Houston et al 1947; Hutchins 1933). Esau (1948) examined the pathological anatomy of grapevines and alfalfa (alfalfa dwarf disease) infected with the PD agent and peach infected with PDP agent and found that deposition of gums in tracheary elements of the xylem were among the first changes associated with development of PD, alfalfa dwarf and PDP. The formation of gums was also accompanied by the excessive formation of tyloses in vessel of grapevine and peach but not alfalfa. As histochemical strains which would have resolved bacteria were not used in the study, the etiological agents were overlooked. It is possible that bacterial aggregates may have been mistakenly identified as gums in some instances. Subsequently, Mollenhauer and Hopkins (1976) observed that bacterial aggregates, gums, and tyloses block vessels in grapevines with PD. Hopkins (1981) later concluded that there was sufficient occlusion by bacterial plugs throughout petioles or leaf veins to cause leaf marginal necrosis due to water stress. Plugging and discolouration of the xylem are features associated with RSD (Steindl 1961) and were recognized as diagnostic symptoms prior to the discovery of the causal agent. However, Artschweager (1960) also noted that lignification of the phloem was another feature associated with RSD.

Several studies have assessed the effect of FXB on water relationships within infected plants. Goodwin et al (1988a,b) found that at dawn and midday, leaves of grapevines exhibiting marginal necrosis symptoms of PD developed a lower water potential and lower cell turgor than healthy leaves. An increase in midday stomatal resistance was also associated with leaf marginal necrosis. Hydraulic conductivity through nodes of sugarcane stalks is reduced in some sugarcane cultivars infected with C. xyli subsp. xyli, and the extent of the reduction is directly correlated with the susceptibility of cultivars to injury due to RSD (Valarini and Tokeshi 1981; Davis ana Harrison 1987). Evert (1987) found that both the hydraulic conductivity of stems and the water potential of leaves were lower in peach trees with PDP than in healthy trees, and suggested that symptoms of PDP are caused by xylem blockage. However, Andersen and French (1987) did not find significant reductions in the hydraulic conductivity of both peach roots and stems and concluded that the relatively minor reductions in midday water potential observed in their study may be insufficient to entirely explain PDP symptoms.

Andersen and French (1987) suggested that altered root metabolism and the involvement of growth regulators may be a factor in PDP expression, based on differences in the nutrient status of healthy and diseased trees, especially elevated Na and Al in diseased trees, and the previous finding that the application of gibberellic acid to peach trees with PDP resulted in the suppression of the disease (French and Stassi 1978). Application of growth regulators to grapevines also influences the expression of PD symptoms (Hopkins 1985). Indoleacetic acid, kinetin, and gibberellic acid suppressed the disease and retarded foliar senescence;whereas, ethylene and abscisic acid, two growth regulators known to enhance foliar senescence, slightly enhanced symptom development. Furthermore, reduced accumulation of PD bacteria in leaves was associated with symptom suppression by indoleacetic acid and kinetin, and cultivar susceptibility to PD mediated the influence of growth regulators on disease development.

A number of physiological changes due to PD have been detected in diseased leaves as compared to healthy leaves including reduced photosynthesis which was at least partly due to inhibition of nonstomatal-related aspects of CO_2 fixation, higher concentrations of abscisic acid, glucose, fructose, Ca^{2+} and Mg^{2+}, and lower concentration of K^+ (Goodwin et al 1988b). Additionally, chlorophyll reduction in diseased leaves was associated with lower concentrations of sucrose and starch, and increases in electrolyte leakage, lipid peroxidation and superoxide anion.

These changes are evidence that leaf marginal necrosis in PD is due to accelerated leaf senescence brought on by prolonged water stress.

The production of phytotoxins or extracellular polysaccharides (EPS) by FXB and their involvement in pathogenesis of diseases caused by the bacteria has not been conclusively demonstrated. Although phytotoxin production by the PD bacterium was reported (Lee et al 1982), more recent evidence indicates that the toxic substances are actually components of the culture medium (Goodwin et al 1988). Unlike some plant pathogenic bacteria, FXB apparently do not produce substantial amounts of EPS. Attempts to visualize EPS associated with C. xyli subsp. xyli in sugarcane tissue by electron microscopy after staining with ruthenium red (Weaver et al 1977) and to isolate EPS from supernatants after washing cells of the bacterium from culture (Davis unpublished) have both been unsuccessful. Furthermore, FXB have not been found to produce noticeable amounts of EPS in axenic culture.

CONCLUSION

The FXB are an intriguing group of endophytic bacteria. In the evolutionary process, they have become adapted to an extremely narrow ecological niche, namely the xylem of plants, and have a very limited capacity to survive outside the shelter of this environment. Where insect vectors are involved in the plant-to-plant transmission of FXB, their ecological niche extends to the foregut of specific kinds of xylem-feeding insects which serve as intermediate hosts. Although the ecological niche of the FXB is narrow in respect to the tissues they inhabit, their host range is apparently very wide. However, they incite appreciable disease in a few of their plant hosts.

For the most part, both symptom expression and disease severity appear to be directly related to the extent or rate of xylem colonization by FXB in host plants. Conversely, disease resistance appears to be directly related to a limited bacterial distribution among tracheary elements of the xylem which results in a smaller overall bacterial populations in more resistant hosts. Relative differences in resistance to injury have been observed among some host genotypes that support similar populations of a pathogen; thus, host tolerance affects the degree of resistance expressed. Resistance to colonization may be due to preformed or induced structural barriers within the xylem. Pit membranes limit the spread of FXB from one tracheary element to another. In some hosts, vascular occlusions may also limit colonization and are formed in response to infection by FXB. The relatively importance of different structural barriers in limiting colonization may vary depending upon the particular host-pathogen interaction involved.

Prolonged water stress due to obstruction of water movement in the xylem, appears to be the primary cause of plant injury in at least some of the diseases caused by FXB. The events leading to xylem dysfunction are not well understood. Although bacterial masses often fill the entire lumen of tracheary elements and thus, appear to be responsible for blocking water movement, such elements may have been dysfunctional before this occurred. The involvement of embolisms or vascular gelation in rendering tracheary elements dysfunctional is suspected but not proven. Micronutrient and growth regulator imbalances might also contribute to pathogenesis, but whether they are incited directly by FXB or indirectly due to other disease conditions, such as water stress, is uncertain at the present time.

REFERENCES

Anderson PC and French WJ (1987) Biophysical characteristics of peach trees infected with phony peach disease. Physiol Mole Plant Pathol 31:25-40

Artschweager E(1960) Sieve tube lignification in sugarcane and its significance in relation to ratoon stunting disease. Proc Int Soc Sugar Cane Technol 10:1077-1082

Bailey RA (1977) The systemic distribution and relative occurrence of bacteria in sugarcane varieties affected by ratoon stunting disease. Proc S Afr Sugar Technol Assoc 51:55-56

Billing E (1982) Entry and establishment of pathogenic bacteria in plants. In: Rhodes-Roberts ME and Skinner FA (eds) Bacteria and Plants. Academic Press, New York London, p 51

Chen TA, Wells JM and Liao CH (1982) Cultivation in vitro: Spiroplasmas, plant mycoplasmas, and other fastidious, walled prokaryotes. In: Mount MS and Lach GH (eds) Phytopathogenic Prokaryotes, vol II. Academic Press, New York London, p 417

Davis MJ, Dean JL and Harrison NA (1988) Quantitative variability of Clavibacter xyli subsp. xyli populations in sugarcane cultivars differing in resistance to ratoon stunting disease. Phytopathology 78:462-468

Davis MJ, Dean JL and Harrison NA (to be published) Distribution of Clavibacter xyli subsp. xyli in stalks of sugarcane cultivars differing in resistance to ratoon stunting disease. Plant Disease

Davis MJ, Gillaspie AG Jr, Harris RW and Lawson RH (1980) Ratoon stunting disease of sugarcane: Isolation of the causal bacterium. Science 210:1365-1367

Davis MJ, Gillaspie AG Jr, Vidaver AK and Harris RW (1984) Clavibacter: A new genus containing some phytopathogenic coryneform bacteria, including Clavibacter xyli subsp. xyli sp. nov., subsp. nov. and Clavibacter xyli subsp. cynodontis subsp. nov., pathogens that cause ratoon stunting disease of sugarcane and bermudagrass stunting disease. Int J Syst Bact 34: 107-117

Davis MJ and Harrison NA (1987) Hydraulic conductivity in sugarcane clones as related to resistance to ratoon stunting disease. In: Civerolo EL, Collmer A, Davis RE and Gillaspie AG (eds) Plant pathogenic bacteria. Martinus Nijhoff, Boston, p 613

Davis MJ, Lawson RH, Gillaspie AG Jr and Harris RW (1983a) Properties and relationships of two xylem-limited bacteria and a mycoplasma-like organism infecting Bermuda grass. Phytopathology 73:341-346

Davis MJ, Purcell AH and Thomson SV (1978) Pierce's disease of grapevine: Isolation of the causal bacterium. Science 199: 75-77

Davis MJ, Purcell AH and Thomson SV (1980) Isolation media for the Pierce's disease bacterium. Phytopathology 70:425-429

Davis MJ, Raju BC, Brlansky RH, Lee RF, Timmer LW, Norris RC and McCoy RE (1983b) Periwinkle wilt bacterium: Axenic culture, pathogenicity and relationships to other Gram-negative, xylem-inhabiting bacteria. Phytopathology 73:1510-1515

Davis MJ, Whitcomb RF and Gillaspie AG Jr (1981) Fastidious bacteria of plant vascular tissue and invertebrates (including so-called rickettsia-like bacteria). In: Starr MP, Stolp H, Truper HG, Balows A and Schlegel HG (eds) The Prokaryotes - a handbook on habitats, isolation, and identification of bacteria. Springer-Verlag, Berlin Heidelberg, p 2172

Esau K (1984) Anatomic effects of the viruses of Pierce's disease and phony peach. Hilgardia 18(12):423-464

Evert DR (1987) Influence of phony disease of peach on stem hydraulic conductivity and leaf xylem pressure potential. J Amer Soc Hort Sci 112(6):1032-1036

Freitag JH (1951) Host range of the Pierce's disease virus of grapes as determined by insect transmission. Phytopathology 41:920-934

French WJ and Stassi DL (1978) Response of phony-infected peach trees with gibberellic acid. Hort Sci 13:158-159

Gillaspie AG Jr, Flax G and Koike H (1976) Relationship between numbers of diagnostic bacteria and injury by ratoon stunting disease in sugarcane. Plant Disease Reporter 60:573-575

Gillaspie AG Jr, Davis RE and Worley JF (1973) Diagnosis of ratoon stunting disease based on the presence of a specific microorganism. Plant Disease Reporter 57:987-990

Goheen AC, Nyland G and Lowe SK (1973) Association of a rickettsia-like organism with Pierce's disease of grapevines and alfalfa dwarf and heat therapy of the disease in grapevines. Phytopathology 63:341-345

Goodman RN (1982) The infection process. In: Mount MS and Lacy GH (eds) Phytopathogenic Prokaryotes, vol. I. Academic Press, New York London, p 31

Goodwin PH, DeVay JE and Meredith CP (1988) Roles of water stress and phytotoxins in the development of Pierce's disease of the grapevine. Physiol Mole Plant Pathol 32:1-15

Goodwin PH, DeVay JE and Meredith CP (1988) Physiological response of *Vitis vinifera* cv. "Chardonnay" to infection by the Pierce's disease bacterium. Physiol Mole Plant Pathol 32:17-32

Harrison NA and Davis MJ (1986) Infectivity titrations of *Clavibacter xyli* subsp. *xyli* and sugarcane cultivars differing in susceptibility to ratoon stunting disease. Plant Disease 70: 556-558

Harrison NA and Davis MJ (to be published) Colonization of vascular tissues by *Clavibacter xyli* subsp. *xyli* in stalks of sugarcane cultivars differing in susceptibility to ratoon stunting disease. Phytopathology

Hayward AC (1974) Latent ifections by bacteria. Ann Rev Phytopathol 12:87-110

Hopkins DL (1981) Seasonal concentration of the Pierce's disease bacterium in grapevine stems, petioles, and leaf veins. Phytopathology 71:415-418

Hopkins DL (1984) Variability of virulence in grapevine among isolates of the Pierce's disease bacterium. Phytopathology 74: 1395-1398

Hopkins DL (1985) Physiological and pathological characteristics of virulent and avirulent strains of the bacterium that causes Pierce's disease of grapevine. Phytopathology 75:713-717

Hopkins DL (1985) Effects of plant growth regulators on development of Pierce's disease symptoms in grapevine. Plant Disease 69:944-946

Hopkins DL and Adlerz WC (to be published) Natural hosts of *Xylella fastidiosa* in Florida. Plant Disease

Hopkins DL and Mollenhauer HH (1973) Rickettsia-like bacterium associated with Pierce's disease of grapes. Science 179:298-300

Hopkins DL, Mollenhauer HH and French WJ (1973) Occurrence of a rickettsia-like bacterium in the xylem of peach trees with phony disease. Phytopathology 63:1422-1423

Hopkins DL and Thompson CM (1983) Seasonal concentration of the Pierce's disease bacterium in 'Carlos' and 'Welder' muscadine grapes compared with 'Schuyler' bunch grapes. Hort Sci 19(3): 419-420

Houston BR, Esau K and Hewitt WB (1947) Mode of vector feeding

and the tissues involved in the transmission of Pierce's disease in grape and alfalfa. Phytopathology 37:247-253
Huang J-S (1986) Ultrastructure of bacterial penetration in plants. Ann Rev Phytopathol 24:141-157
Huang P-Y, Milholland RD and Daykin ME (1986) Structural and morphological changes associated with the Pierce's disease bacterium in bunch and muscadine grape tissues. Phytopathology 76:1232-1238
Hunt P, Bennett CPA, Syamsu H and Nurwenda E (1987) Sumatra disease in cloves induced by a xylem-limited bacterium following mechanical inoculation. Plant Pathology 36:154-163
Hutchins LM (1933) Apparent localization of phony disease virus in the woody cylinder. Phytopathology 29:12
Kao J and Damann KE Jr (1978) Microcolonies of the bacterium associated with ratoon stunting disease found in sugarcane xylem matrix. Phytopathology 68:545-551
Kao J and Damann KE Jr (1980) In situ localization and morphology of the bacterium associated with ratoon stunting disease sugarcane. Can J Bot 58:310-315
Lee RF, Raju BC, Nyland G and Goheen AC (1982) Phytotoxin(s) produced in culture by the Pierce's disease bacterium. Phytopathology 72:886-888
Lowe SK, Nyland G and Mircetich SM (1976) The ultrastructure of the almond leaf scorch bacterium with special reference to topography of the cell wall. Phytopathology 66:147-151
Maramorosch K, Plavsic-Banjac B, Bird J and Liu LJ (1973) Electron microscopy of ratoon stunted sugarcane: Microorganisms in xylem. Phytopathol Z 77:270-273
Mollenhauer HH and Hopkins DL (1974) Ultrastructural study of Pierce's disease bacterium in grape xylem tissue. J Bacteriol 119:612-618
Mollenhauer HH and Hopkins DL (1976) Xylem morphology of Pierce's disease-infected grapevines with different levels of tolerance. Physiol Plant Pathol 9:95-100
Nelson PE and Dickey RS (1970) Histopathology of plants infected with vascular bacterial pathogens. Ann Rev Phytopathol 8:259-280
Nyland G, Goheen AC, Lowe SK and Kirkpatrick HC (1973) The ultrastructure of a rickettsia-like organism from a peach tree affected with phony disease. Phytopathology 63:1275-1278
Preece TF (1982) The progression of bacterial disease within plants. In: Rhodes-Roberts ME and Skinner FA (eds) Bacteria and Plants. Academic Press, New York London, p 71
Purcell AH (1982a) Evolution of insect vector relationships. In: Mount MS and Lacy GH (eds) Phytopathogenic Prokaryotes, vol I. Academic Press, New York London, p 121
Purcell AH (1982b) Insect vector relationships with procaryotic plant pathogens. Ann Rev Phytopathol 20:397-417
Raju BC, Goheen AC and Frazier NW (1983) Occurrence of Pierce's disease bacteria in plants and vectors in California. Phytopathology 73:1309-1313
Roberts DL, Vargas JM Jr, Detweiler R, Baker KK and Hooper GR (1981) Association of a bacterium with a disease of Toronto creeping bentgrass. Plant Disease 65:1014-1016
Steindl DRL (1957) Host range of the sugarcane ratoon stunting disease virus. J Aust Inst Agric Sci 23(3):238

Steindl DRL (1961) Ratoon stunting disease. In: Martin JP, Abbott EV and Hughes CG (eds) Sugarcane diseases of the world, vol I. Elseview Publ Co, Amsterdam, p 433

Teakle DS, Appleton JM and Steindl DRL (1978) An anatomical basis for resistance of sugarcane to ratoon stunting disease. Physiol Plant Pathol 12:83-91

Teakle DS, Smith PM, Hayward AC and Steindl DRL (1975) Diagnosis of RSD by electron microscopy of sugar-cane tissue diffusates. Proc Queensl Soc Sugar Cane Technol 115-116

Teakle DS, Smith PM and Steindl DRL (1973) Association of a small coryneform bacterium with the ratoon stunting disease of sugarcane. Australian J Agric Res 24:869-874

Teakle DS, Smith PM and Steindl DRL (1975) Ratoon stunting disease of sugarcane: Possible correlation of resistance with vascular anatomy. Phytopathology 65:138-141

Valarini PJ and Tokeshi H (1981) Evaluation of ratoon stunting disease resistance by water flow of sugarcane stalks. Summa Phytopathol 7(3/4):45-46

Van Alfen NK (1982) Wilts: Concepts and mechanisms. In: Mount MS and Lacy GH (eds) Phytopathogenic Prokaryotes, vol I. Academic Press, New York London, p 459

VanderMolen GE, Beckman CH and Rodehorst E (1977) Vascular gelation: a general response phenomenon following infection. Physiol Plant Pathol 11:95-100

Weaver L, Teakle DS and Hayward AC (1977) Ultrastructural studies on the bacterium associated with the ratoon stunting disease of sugarcane. Australian J Agric Res 28:843-852

Wells JM, Raju BC, Hung HY, Weisburg WG, Mandelco-Paul L and Brenner DJ (1987) Xylella fastidiosa gen. nov., sp. nov: Gram-negative, xylem-limited,fastidious plant bacteria related to Xanthomonas spp. Int J Syst Bact 37:136-143

Wells JM, Weaver DJ and Raju BC (1980) Distribution of recketTsia-like bacteria in peach, and their occurrence in plum, cherry, and some perennial weeds. Phytopathology 70:817-820

Worley JF and Gillaspie AG Jr (1975) Electron microscopy in situ of the bacterium associated with the ratoon stunting disease in sudangrass. Phytopathology 65:287-295

Zimmerman MH and McDonough J (1978) Dysfunction of the water system. In: Horsfall JG and Cowling EB (eds) Plant disease, an advanced treatise, vol 3. Academic Press, New York San Francisco London, p 117

PATHOGENESIS IN VASCULAR DISEASES OF PLANTS

G.F. Pegg
Department of Horticulture
University of Reading
Earley Gate
Reading, RG6 2AU
U.K.

VASCULAR PATHOGENS

Prokaryote and eukaryote vascular pathogens represent a wide but unique range of micro-organisms which exist in the host xylem during the pathogenic phase of the life cycle and from there induce symptoms of wilt and /or necrosis. The major fungal and bacterial species involved are shown in Table 1. The term 'vascular pathogen' has tended to be used synonymously with wilt pathogen. Where wilt symptoms represent a transitory phase of the disease however, and foliar necrosis is dominant, the list of organisms can be extended greatly to include species such as *Xanthomonas campestris* on *Brassica olearacea*, *Erwinia amylovora* on *Pyrus* and other hosts, and *Stereum purpureum* on *Prunus*. Several pathogens; *Fusarium* formae, *Ophiostoma ceratocystis*, *Clavibacter michiganensis* subsp. *insidiosus* and *Erwinia stewartii* exhibit specificity in their natural host range. *Verticillium dahliae*, *V. albo-atrum* and *Pseudomonas solanacearum* by contrast infect over 300 host plants including many of the main field and plantation crops in temperate and tropical countries.

ENTRY INTO THE XYLEM

All of the pathogens listed in Table 1 can be regarded as soil-borne but some, e.g. *Ceratocystis fagacearum* and *Ophiostoma ulmi* tend not to be free living in soil and like *Clavibacter*

NATO ASI Series, Vol. H28
Vascular Wilt Diseases of Plants
Edited by E.C. Tjamos and C. Beckman

michiganensis subsp. sepedonicus compete poorly with soil organisms in the absence of host tissue. Unlike Verticillium and Fusarium which possess either resistant mycelia, microsclerotia or chlamydospores, the gram negative, (Erwinia and Pseudomonas) and gram positive (Clavibacter) rods do not form spores and survive mostly in plant debris as bacterial cells.

Bacteria, unlike fungi, have no infective energy and enter the plant passively through broken roots or as in the case of E. stewartii and race 2 of P. solanacearum, by insect vectors into wounds (Zea mays) or floral nectaries (Musa paradisica). Kelman and Sequeira (1965) showed that P. solanacearum could enter unwounded roots of tomato at the point where lateral roots emerge from the primary root. Bacteria hydrolyse a parenchymatous sheath which surrounds the emerging root. Wallis and Truter (1978) showed that a massive build up of bacterial numbers in the infection court was a prelude to entry into the xylem. Bacterial cells were finally released into the vessel via ruptured tyloses derived from the adjacent colonised parenchyma cells which had extended through vessel pits.

Fungal pathogens behave in an identical manner to the vascular prokaryotes. It is likely that most infections occurring in the soil do so via exposed vessels in broken roots for which adequate water would be essential. Such damage could be part of the normal condition of an ostensibly healthy plant growing in soil where mechanical attrition from rock particles, and the activity of the soil micro and macro fauna provide infection courts at microsites representing some 20% of the fine root system.

THE PATHOGEN ENVIRONMENT

The xylem of angiosperms consists of tracheids and vessels stacked end to end forming continuous tubes varying in length from 2-5 cm in herbaceous plant stems, (Scholander et al 1957) to several meters in Ulmus americana (Banfield 1941). In terminal shoots, vessel elements are greatly reduced, from 5 cm in Ulmus to a few mm in herbaceous plants. The xylem element proper

Table 1. The Major Vascular Pathogens of Higher Plants

Pathogen	Host
Fusarium oxysporum	
f.sp. *lycopersici*	*Lycopersicon esculentum*
f.sp. *cubense*	*Musa paradisica*
f.sp. *tracheiphilum*	*Glycine max*
f.sp. *melonis*	*Cucumis melo*
f.sp. *niveum*	*Citrullus lanatus*
f.sp. *conglutinans*	*Brassica oleracea*
f.sp. *vasinfectum*	*Gossypium hirsutum*
f.sp. *elaedis*	*Elaeis guineensis*
f.sp. *albedinis*	*Phoenix dactylifera*
f.sp. *lini*	*Linum usitatissimum*
f.sp. *udum*	*Cajanus cajan*
f.sp. *pisi*	*Pisum sativum*
f.sp. *cucumini*	*Cuminum cyminum*
f.sp. *batatas*	*Ipomea batatas*
f.sp. *ciceris*	*Cicer arietinum*
f.sp. *redolens*	Numerous
Clavibacter michiganensis	
subsp. *michiganensis*	*Lycopersicon esculentum*
subsp. *insidiosus*	*Medicago sativa*
subsp. *nebraskensis*	*Zea mays*
subsp. *sepedonicus*	*Solanum tuberosum*
Curtobacterium flaccumfaciens	
pv. *flaccumfaciens*	*Phaseolus* spp. and *Vigna*
Erwinia stewartii	*Zea mays*
E. tracheiphila	*Citrullus*, *Cucumis* and *Cucurbia* spp.
E. chrysanthemi pv. *chrysanthemi*	*Chrysanthemum* spp. and *Euphorbia pulcherrima*
E. chrysanthemi pv. *dianthicola*	Several including *Lycopersicon esculentum*, *Begonia intermedia*, *Solanum tuberosum* and *Dianthus* spp.
Pseudomonas solanacearum	Numerous including Solanaceae and Musaceae

can be envisaged as a tube of pecto-hemicellulosic membrane - the primary cell wall, lined peripherally by annular, helical, or reticulate bands or a perforated sheet of lignified cellulose. In the case of internal mature vessels the membrane is sandwiched between lignin strengthening and the membrane is exposed in both lumina in pit pairs. It is the fate of these pit pairs which determines the course of the vascular disease. Gymnosperms with short tracheids and bordered pits have no vascular pathogens. Water can move freely by mass flow longitudinally through the vascular system or laterally into the stele, depending on the osmotic potential of the adjoining xylem parenchyma. Solutes, host metabolites and pathogen macromolecules, on the other hand, are limited in lateral movement by the molecular pore size of the intact membrane (Stamm and Wagner 1961). Van Alfen (1982) has shown in lucerne (*Medicago sativa*) that pore sizes in petiolar pit membranes are smaller than those in the main stem.

It is in the xylem fluid that the pathogen must obtain all nutrients essential for normal metabolism, and especially carbon for wall synthesis. Xylem fluid in herbaceous plants and trees contains low concentrations of sugars, inorganic salts and amino acids (Wood 1961; Singh and Smalley 1969; Dixon and Pegg 1972). Several possibilities exist for enhanced nutrient availability. Until vascular elements become completely occluded, the pathogen is bathed in a replacement solution which will vary in composition due to lateral enrichment at different parts of the shoot. The flow rates of xylem sap *c*. 0.3 mm/sec (Zimmerman and Brown 1971) found in healthy plants are more than adequate to satisfy the nutrient requirement of most vascular bacteria and fungi. This would not be so if xylem flow was greatly reduced during pathogenesis when alternative sources of nutrients would be required.

Hemicellulose and pectolytic enzyme of the pathogen (Cooper and Wood 1975) may play a primary role in nutrition by degrading the pit membrane. This can have several effects, i) to make available directly carbon residues for growth and metabolism, ii) to alter the permeability of the pit membrane both to host and pathogen molecules and iii) to provide direct access to xylem parenchyma cells.

Visual evidence for the degradation of the pit membrane has been provided by Beckman and co-workers for *Fusarium* (Pegg 1985) and *Pseudomonas solanacearum* (Wallis and Truter 1978) and *Corynebacterium michiganense* (Wallis et al 1973). Numerous accounts of changes in the chemical composition of xylem fluid following infection confirm the altered permeability of the pit membrane; in fungi, Dixon and Pegg (1969); Elgersma (1969); Booth (1969) Singh et al (1971); and bacteria, Braithwaite and Dickey (1971); Coplin et al (1974). Bacteria may obtain a major source of nutrients from parenchyma cells via breached pit membranes, both in intact cells and as tyloses. (Wallis and Truter 1978; Wallis et al 1973).

SYMPTOM PATTERNS

The classic wilt symptoms resulting from loss of leaf turgor are seen in shoots of most herbaceous plants affected by bacteria and fungi, and in leaflets of trees affected by Ceratocystis and Ophiostoma. Depending on the virulence of the isolate, the age and nutritional status of the plant and environmental conditions, all of which determine the speed and severity of symptoms; most wilt symptoms develop as an acropetal sequence. Grieve (1939) and Wellman (1941) described epinasty as the first symptom of P. solanacearum and Fusarium wilts of tomato. This is followed in fairly rapid succession by flaccidity, chlorosis and necrosis of the terminal leaflet of the lowest leaf, and repeated on successive leaflets and leaves. Acute chlorosis is followed by the death and often abscission of the leaf. Adventitious root production in the lower internodes is characteristic of tomato and other non-woody herbaceous hosts. Plants generally are stunted and show greatly reduced leaf area. Selman and Pegg (1957) described a chronic stunting in the absence of flaccidity when tomato seedlings where inoculated at the cotyledon stage with V. albo-atrum.

In some diseases, e.g.'mal seco' disease of citrus, Verticillium infection of lucerne, Dutch Elm disease and progressive wilt of hop; leaves and young shoots may desiccate in a manner reminiscent of lyophilisation. The very localised nature of fungal tomato wilt is illustrated by unilateral stunting or wilting, which in extreme cases may be frequently restricted to a half leaf lamina. Specialised symptoms as are seen in "tiger striping" in hop leaves from plants infected by V. albo-atrum with areas of interveinal chlorosis and necrosis, and specific patterns of stomatal behaviour; silver leaf of fruit trees, foliar curling in cotton wilt and adventitious leaf production in Dutch Elm disease, none of which can be reproduced by simulated water deficits.

Various definitions and interpretations of virulence and pathogenicity have been made in the literature. In this chapter, pathogenicity is referred to conventionally as a state of damage to the plant resulting from infection by (by definition) a virulent pathogen. Virulence is thus a quantitative expression of pathogenicity - the degree of symptom intensity. Where discriminating host genotypes are available to pathogenic isolates (or races) such as the Ve and I genes in tomato to Verticillium and Fusarium, virulence can be seen not as an innate character of a pathogen but the result of an interaction with the host. Thus Fusarium oxysporum f. sp. lycopersici race 1 is virulent on I_1- genotypes and avirulent (non-pathogenic) on I_1+ or I_2+ cultivars. Aggressiveness - an obsolete term - is synonymous with virulence. Difficulty has arisen in the literature from a failure of research workers precisely to define the use of the above terms and susceptibility. Susceptibility is used for the most part to describe a plant's propensity to develop symptoms - to become diseased. Some authors have used it to describe infection independently of symptom intensity. Since there is no immunity to vascular wilt infection, even in the most resistant cultivars and since symptomless weed - 'non-hosts' of vascular pathogens frequently show limited colonisation, the use of the term in this context without qualification is misleading.

THE PRODUCTION OF MACROMOLECULES BY THE PATHOGEN

The large molecular weight metabolites produced by wilt pathogens discussed in this section are considered separately from enzyme macromolecules. Vascular bacteria, unlike fungi, secrete a complex heteropolysaccharide, the EPS, which surrounds the cell, buffering it from rapid hydration and dehydration, concentrating nutrients and extracellular enzymes. It has long been associated with pathogenicity and virulence. Husain and Kelman (1958) attributed wilting in tobacco caused by race 1 of P. solanacearum to vascular occlusion by EPS production in vivo. These and other authors have claimed that virulence is a function of EPS production since rough mutants lacking EPS were avirulent and were also susceptible to binding and encapsulation on host cell surfaces, unlike the highly virulent smooth, wild type. (Ayers et al 1979; Bradshaw-Rouse et al 1981; Duvick and Sequeira 1984). One of the most investigated systems is the Clavibacter (Corynebacterium) michiganensis subsp. insidiosus - lucerne system.

Spencer and Gorin (1961) described the EPS as a large, neutral sugar polymer with repeating units of L-fucose, D-galactose, D-glucose, and pyruvate. Lucerne shoots treated with EPS showed wilt symptoms comparable to field symptoms. Ries and Strobel (1972 a and b) claimed that the active toxin was a non-specific 5×10^6 kd glycoprotein, which was produced in infected plants based on the recovery of a ^{14}C galactose-labelled moiety. The most comprehensive investigation of the structure and possible function of the C. michiganensis subsp. insidiosus EPS has been by Van Alfen and co-workers. In a series of studies (Van Alfen and Allard-Turner 1979; Van Alfen and Turner 1975a; Van Alfen and McMillan 1982; Van Alfen et al 1987a and b) they isolated three components of the culture filtrate EPS, fractions I, II and III. Two of these (I and II) had a basic repeating unit of gal:gluc:fuc, in the ratio of 1:1:2 and 2 mg ketoacid-pyruvate (Gorin et al 1980). In addition, fraction I contained 13% nitrogen in the form of 9 polypeptides ranging in size from 12.75 to 75.75 Kd. This glycoprotein, an aggregate of fraction II, had a large but indeterminate molecular mass. Fraction II was found to have a mass of 5×10^6 d, identical to the size of the Ries and Strobel molecule but contained no peptide. The third fraction, (III) had neutral sugar ratios of approximately 1 man:1 gluc:5 fuc:5 rham:13 gal, and a molecular mass of 22 Kd. An interesting feature of this work was that the sugar ratios, with the exception of mannose in fractions I and II were remarkably stable in cultures of different strains and when cultured on different media. Fraction III varied considerably in sugar composition and ratio in EPS both from different strains and on different media.

Working with model dextran and labelled markers, Van Alfen and co-workers found that pores in the pit membrane were larger in stems than in petiole junctions. Moreover the frequency of vessel endings was greater in leaflets. The quantity of dextran required to stop vascular flow was 8.0 picomoles in petiole junction and 0.4 picomoles in leaflet veins. Using purified labelled EPS fractions, fraction III was found to pass through all membranes and accumulated in leaves. Fraction II could pass through the stem vessel pit membrane but occluded membranes in the terminal shoots and petiole junctions, thereby causing wilt. The large aggregate glycoprotein fraction I was unable to pass through stem vessel pits and coated the membrane. There seems little doubt that the EPS of C. michiganensis subsp. insidiosus plays a major role in reducing vessel conduction, creating a water deficit in the shoots. The evidence for its production in vivo is still circumstantial. Van Alfen's work could not confirm the existence of a glycoprotein of the same molecular weight as reported by Ries and Strobel (1972a). The possibility of fraction III functioning as a toxin sensu stricto on laminar membranes has not been discounted but its presence in diseased tissue with fractions I and II awaits confirmation.

Although culture filtrate products of C. michiganensis subsp. insidiosus are non-specific in inducing wilt, Straley et al (1974) claimed that the glycopeptide of Ries and Strobel (1972b) induced wilting selectively in susceptible, compared with resistant, lucerne shoots. A similar result was reported by Goodman et al (1974) for the EPS amylovorin produced by Erwinia amylovora. This largely acid galactan with glucose, mannose, galacturonic acid, pyruvate and succinate, has a molecular mass in the order of 165 kd. Traces of protein are also present. Susceptible cultivars or species wilt rapidly when treated with purified solutions of amylovorin, whereas resistant ones only wilt after long exposure to the polysaccharide and in proportion to the level of resistance. These results are paradoxical, since if wilting is due mainly to the blocking of pit membranes by large molecular weight fractions of the EPS, it is unlikely that pore size in pit walls particularly within cultivars of different species will vary in a simple relation to resistance. The alternative explanation is that small molecular weight fractions passing into leaf laminar tissue affect membranes and leaf turgor. A further possibility is that host enzymes in the different cultivars act differentially on the polymer. A crude glucanohydrolase preparation from tomato and purified glucanase fractions was able to effect partial hydrolysis of the neutral EPS from Pseudomonas solanacearum (Pegg, Young and Austin unpublished).

Rai and Strobel (1969) working on C. michiganensis sub. sp. michiganensis have reported the production of multiple glycopeptide toxins in culture. Based on labelling experiments and comparisons with authentic dextrans these authors claimed that one toxin fraction, in quantities too small to induce vascular occlusion, was freely translocated through stem and leaf tissue and induced wilt by its action on leaf plasma membranes. The molecular masses of three culture filtrate fractions; I, II and III were 200, 130 and 35 kd respectively. Of these, I and II would pass through stem pit membranes, based in Van Alfen's work, but would be expected to block vein pits in leaflets. Fraction III would be able to pass through all pits in stems and leaves and pass into leaf parenchyma. The relationship between these glycopeptides and the EPS of C. michiganensis sub. sp. michiganensis is not clear. The toxin hypothesis proposed by Rai and Strobel would justify further investigation in the light of the work by Van Alfen et al (1987 a and b).

Recently, an important study on mutants of Pseudomonas solanacearum (Xu et al 1987) has called for a complete reappraisal of the role of bacterial EPS in virulence and pathogenicity. In nature spontaneous mutagenesis of a fluidal (smooth) virulent wild-type (K60) resulting in loss of EPS is always associated with loss of virulence. Xu et al (1987), using Tn5 mutagenesis obtained a class of mutants of K60 which had lost the EPS but retained wild type virulence to tobacco and eggplant. To confirm that loss of EPS was due to Tn5 insertion, one of the EPS-mutants, KD700 was used in marker exchange experiments to transform K60. The majority of the transformants obtained lost EPS but retained their virulence to eggplant and tobacco. A cosmid clone from a genomic library of K60 fully restored EPS production when transferred into KD700. It is of particular interest that the growth rates in planta of KD700, the EPS complemented equivalent of KD700 and the wild-type K60 were similar. N acetyl galactosamine (GalNac) a constituent of the P. solanacearum EPS could not be detected in KD700 cultures or plants infected with KD700. A claim by Denny (1988) (unpublished) that pathogenicity of EPS-mutants is due to an EPS or some similar polysaccharide induced in the infected host plant, awaits substantiation. These findings, unique in phytopathogenic bacteria, leave unanswered the whole question of the mechanism of bacterial wilt pathogenicity. The only possible explanations at this stage of our knowledge are that bacteria themselves, their lytic products are capable of blocking leaf pit membranes, or that virulence (and pathogenicity) are related to other macromolecules induced during infection.

The potential of large molecular weight toxic metabolites of Ophiostoma (Ceratocystis) ulmi to induce many of the symptoms of Dutch elm disease, has attracted much interest over the last 20 years. Salemink et al (1965) isolated a 30 kd glycopeptide moiety with wilt-inducing properties. Van Alfen and Turner (1975b) described a large polydisperse 200-500 kd glycoprotein with 5% peptide which was thought to effect wilt by xylem occlusion. A novel protein with a mass of 13 kd probably too small to act by xylem or pit field blocking called cerato-ulmin was described by Takai (1974) and characterized by Stevenson et al (1979). Cerato-ulmin can aggregate into higher polymers by chilling or bubbling gas through a solution. A powerful argument for its involvement in pathogenesis was its detection by immunoassay in extracts from diseased elm trees (Takai et al 1983). Scheffer and Elgersma (1981) earlier detected a phytotoxic glycopeptide from O. ulmi in planta using ELISA but this was not subsequently confirmed as cerato-ulmin. Early studies on Verticillium led to the idea of the involvement of a pathogen macromolecule in the induction of symptoms, based on shoot bioassays of culture filtrate extracts. (Porter and Green 1952; Green 1954; Caroselli 1955).

Kiessig and Haller-Kiessig (1957) claimed that two components were involved in Verticillium wilt of lucerne, a thermolabile wilt toxin and a thermostable necrosis toxin. Using relatively simple purification procedures, Stoddart and Carr (1966) reported that a 100 kd protein with mild cellulase activity and a 5 - 10 kd fructosan were both involved in V. albo-atrum-induced wilt symptoms. In the U.S.S.R. Zelt'ser and Malysheva (1966), and Malysheva and Zelt'ser (1968) purified a protein lipopolysaccharide (PLP) from cultures of the cotton pathotype of V. dahliae and reproduced in detail the scorching and desiccation foliar symptom seen in Verticillium wilt of cotton. This work was confirmed by Zelt'ser and Avazkhodzhaev (1973); Keen and Long (1972) and Keen et al (1972). The molecule had a molecular mass of 3 x 10^6 d. The polysaccharide, consisting of gluc., man., gal. with smaller quantities of uronic acid, glucosamine and galactosamine, accounted for 70% of the PLP and 15% each of protein and lipid. ^{14}C-labelled PLP was found in interveinal and marginal areas of leaves and was equally active on a range of different genera. Keen et al (1972) showed that the PLP did not block the uptake of acid fuchin into leaf veins. Toxin-induced symptoms on cotton shoots were inversely related to the degree of resistance. Asamov et al (1975) found increased permeability (leakage) of artificial membranes to Na^+ and K^+ ions when treated with a polypeptide and an oligosaccharide from V. dahliae cultures consistent with the findings on cotton shoots. Mussell (1972) claimed that a series of small mwt. proteins with no enzyme activity, induced symptoms in cotton in addition to endo PG.

In a series of studies on V. dahliae on potato and tomato, Buchner et al (1982); Nachmias et al (1982), (1985) and (1987) claimed that the active toxin of V. dahliae produced on a defined culture medium was a peptide of variable composition occurring as a component of the PLP, which induced symptoms in susceptible but not in resistant (Ve^+) cultivars (Nachmias et al 1987). Keen et al (1972) found no correlation between virulence and PLP production. Nachmias et al (1987), unlike other workers, found no clear relationship between colonization and symptom expression with race 1 and race 2 in susceptible and resistant combinations. A weakness in the latter study is that one bioassay requires the toxin to be injected somewhat unnaturally into the leaf tissue. The question as to whether the peptide is produced *in planta* remains to be answered.

Symptoms of progressive wilt in hop plants of low resistance may develop as a striking pattern of foliar chlorosis and necrosis referred to as "tiger striping". The lamina does not wilt in the classical manner but remains stiff, up-curled and slowly desiccates. Interveinal stripes of chlorosis, later becoming necrotic are separated from the veins by narrow bands of green tissue. During the onset of chlorosis, stomata remain closed but as chlorosis becomes intense, the stomata open wide and remain so regardless of environmental conditions, even when the tissue becomes necrotic. The stomata in the green veinal lamina remain closed. This pattern of open stomata is reminiscent of the effect of fusicoccin. Talboys (1978) proposed the idea of two-component toxin. Hop shoots placed in a solution of 0.001M PEG 1500 and 10 ppm cepa (2-chloroethyl phosphonic acid) to generate ethylene developed precise symptoms of tiger striping in which the pattern of stomatal behaviour was reproduced exactly. Cepa or PEG used alone did not reproduce typical symptoms. PEG 1000 had no effect, while a larger molecule, PEG 6000 induced general flaccidity atypical of normal symptoms. Cronshaw and Pegg (1976) demonstrated the effect of ethylene in predisposing tomato tissue to the effects of a carbohydrate polymer and pectolytic enzymes. Austin (1979) isolated a 70 - 100 kd polydisperse galacto-glucomannan from *V. albo-atrum* cultures. This molecule interacted with ethylene to simulate normal wilt symptoms (Pegg and Austin unpublished). The reason for acropetal symptom development in hop was explained by Talboys as increased sensitivity of older leaves to the toxin and/or, to ethylene generation.

THE INVOLVEMENT OF GROWTH REGULATORS

Growth substances have been the subject of general reviews by Sequeira (1963, 1973) and Pegg (1976 a and b) (1981 a and b).

Many symptoms of vascular pathogenesis, epinasty, adventitious root production, xylem hyperplasia, tylosis, stunting, foliar chlorosis, petiole abscission, adventitious leaf production are indicative of enhanced or disturbed plant growth substances (PGS) production. All of the five classes of growth regulator, viz. auxins, cytokinins, ethylene, gibberellins and abscisic acid, have been implicated in pathogenesis in addition to toxins such as fusicoccin which show growth substance-like properties in their mode of action. This account is confined to the role of PGS in symptom development excluding their possible involvement in resistance.

A feature common to the fungal and bacterial wilt diseases is hyperauxiny, which has been described for Oak wilt (Ceratocystis fagacearum), Beckman et al (1953); Fergus and Wharton (1957); Verticillium wilt of tomato (V. albo-atrum), Pegg and Selman (1959); bacterial wilt of tobacco (Pseudomonas solanacearum), Sequeira (1964); Nishimura (1960); and Fusarium wilt of banana (F.o. cubense) Mace and Solit (1966).

Research on auxins in pathogenesis has centerd around the questions of host or pathogen origin of IAA, the role of auxin in symptom development and the interaction of auxins with other metabolic systems. The dependency of ethylene on the supply of IAA (Jones and Kende 1979) and the interaction of IAA in phenol metabolism and with pectolytic enzymes make an interpretation of its role in vascular wilts difficult. The most detailed studies on PGS in host pathogen relations have been on Pseudomonas spp. (Sequeira 1965; Smidt and Kosuge 1978). Sequeira (1965) attempted to assess the relative contribution of host and bacterium to the free pool of IAA in Nicotiana tabacum infected by Pseudomonas solanacearum. Sequeira and Williams (1964) showed that a virulent (fluidal) strain, (K60) synthesised IAA via the oxidative deamination pathway and indolyl pyruvic acid. The inability of the pathogen to utilise tryptamine was exploited by feeding 2-^{14}C tryptamine to healthy and infected plants and extracting IAA at different times after infection. The recovery of ^{14}C IAA demonstrated that up to 6 days after inoculation, the increase in IAA concentration could be accounted for solely by increased host synthesis. As a cause or a feed-back result of this increased synthesis, Pegg and Sequeira (1968) found large de novo increases in phenylalanine DOPA tyrosine and tryptophan 48 hours after infection, indicative of a stimulation of the shikimic acid pathway. Only one report in the literature has positively associated IAA production with virulence. Using mutants of P. savastanoi deficient and hyperproductive for IAA, Smidt and Kosuge (1978) showed that the IAA minus mutants were avirulent on Oleander and high IAA mutants induced severe gall formation compared with wild types. Paradoxically an avirulent mutant of P. solanacearum (B.) produced five-fold increases in IAA in vivo compared with the virulent K60 (Sequeira and Williams 1964; Phelps and Sequeira 1968).

A key branch point metabolite in the shikimate pathway is chorismic acid which can shunt aromatic synthesis to phenolic substances, melanin and lignin, under the control of chorismate mutase; or to anthranilic acid, tryptophan and IAA, mediated by anthranilate synthetase. Recent work on V. albo-atrum-infected tomato using ^{3}H and ^{14}C intermediates, showed that both enzyme systems were stimulated, together with higher levels of tryptophan and an increased turnover of IAA into bound auxin (Pegg unpublished). The significance of IAA conjugates in vascular wilt disease has not been infestigated in detail but binding to amides, glycosidic esters and glycoproteins may protect against the action of oxidases. IAA synthesis in Verticillium, Fusarium and the tomato appears to be via the Trypt→ I Py A→ IAAld→ IAA route. Matta and Gentile (1965) considered that pathogen-synthesised IAA in Fusarium wilt of tomato, was only a minor contribution to the increased pool.

There is no evidence in any vascular wilt disease, that hyperauxiny is reflected in an increase in plant height or leaf area. IAA may be seen much more in an interactive physiological role, principally in the stimulation of ethylene biosynthesis (Jones and Kende 1979) and binding pectin methyl esterase (PME) to rhamnogalacturans in the pit membrane thus regulating the degree of methylation of the carboxyl group (Glaziou 1957; Glaziou and Inglis 1958; Jansen et al 1960). At low concentrations the degree of de-esterification of pectinic acid is reduced and wall plasticity maintained. At higher concentration PE and PME are less bound, permitting free carboxyl groups to bridge with Ca^{2+} to form more durable calcium pectates. This differential effect may explain in part the confusion of a primary role for IAA in resistance or susceptibility. IAA is known to mediate proton exflux from the cell in much the same manner as fusicoccin and at the transcription level to involve the synthesis of new mRNAs. An interpretive role for IAA at this sub-cellular level in terms of its involvement in complex pathological systems based on gross levels in healthy and infected tissue cannot be made. IAA and ethylene will both induce tylose formation (Pegg 1959, 1976b) most probably by inducing the synthesis of an ATPase directing the energetic flow of protons through the plasmalemma together with the removal of calcium (Rayle 1973; Cleland and Rayle 1977) thereby plasticising the paravascular cell wall. Whether IAA functions independently, or merely by its involvement in ethylene synthesis remains to be shown.

The findings of Wiese and DeVay (1970) are of special interest in interpreting the contribution of IAA to cotton wilt symptoms. Two strains of V. dahliae inducing mild non-defoliating (SS4) and severe defoliating wilt of cotton (T9) both induced hyperauxiny. Defoliation and wilt severity, however, were only correlated with ethylene production which showed a five-fold increase in T9 infected plants compared with two-fold in SS4 infected plants relative to controls. Similarly ABA levels were doubled in T9 infected plants. No differences in IAA production, however, were found between the two strains.

The interaction between phenols, oxidative enzymes and IAA metabolism have been described in detail (Pegg 1981a; Bell and Mace 1981; Beckman 1987). Phenols have been implicated in the metabolism of IAA by activation or inhibition of IAA oxidase. For the most part monohydric phenols such as coumaric acid or tyrosine activate, while orthodihydric phenols, for example caffeic and chlorogenic acid, inhibit the enzyme (Pegg 1976a). A further interesting link between phenols in paravascular cells and IAA biosynthesis is the involvement of a catecholase in the condensation of a quinone with tryptophan to form a quinonimine which then converts to IAA (Gordon and Paleg 1961). This was confirmed by Matta and Gentile (1965) and Mace and Solit (1966). Most references to phenolics are related to resistance responses and will not be considered here.

Ethylene

Ethylene is implicated in all vascular diseases and is the simplest but most powerful of the growth hormones. Its involvement in the physiology and biochemistry of host and pathogen is also one of great complexity. Early studies by Girardin (1864) described chlorosis, necrosis and defoliation of *Tilia europea* caused by ehtylene toxicity from a leaking gas main. These symptoms closely paralleled those of *Ophiostoma ulmi* in *Ulmus* spp. Ethylene inhibits a wide range of cellular and tissue functions including: mitosis, cellular and tissue DNA synthesis, cell expansion, and induced leaf, flower and bud abscission and chlorophyll breakdown. It also stimulates root and root hair initiation; enzyme, protein and RNA synthesis; respiratory changes, foliar epinasty, induction of phytoalexins, resistance and tyloses in addition to many other whole-plant effects (Abeles 1973). Some effects of ethylene are duplicated by IAA and cytokinin and may be reversed by IAA and ABA (Pegg 1981a).

From this list of physiologically conflicting effects it will be immediately apparent that no simple explanation of its role in vascular wilts can be forthcoming.

Ethylene as a phytotoxin

Dimond and Waggoner (1953) first concluded that epinasty in Fusarium wilt of tomato was ethylene mediated. A *prima facie* case for ethylene in necrosis and foliar abscission has been presented for *V. dahliae*-infected cotton (Wiese and DeVay 1970). Ethylene has also been associated with bacterial wilt of banana (Freebairn and Buddenhagen 1964; Bonn et al 1975). Wilt of radish seedlings caused by *Fusarium* has been attributed wholly to ethylene production by the pathogen based on the suppression of ethylene and symptoms in inoculated seedlings treated with an ethylene inhibitor (Cohen et al 1986).

Bacterial and fungal pathogens and their hosts all produce ethylene. Roots damaged during inoculation also release ethylene which will modify the infection response where low levels of inoculum are used. Gentile and Matta (1975) and Pegg and Cronshaw (1976a,b) showed ethylene production as a pulse approximately 9 - 12 days after inoculation of tomato with Fusarium oxysporum f. sp. lycopersici, Verticillium albo-atrum and Pseudomonas solanacearum respectively. In bacterial wilt an avirulent B1 strain produced higher levels of ethylene in culture (1.8×10^{-22} mol min^{-1} $cell^{-1}$) compared with the virulent wild type (9.83×10^{-24} mol min^{-1} $cell^{-1}$). A comparison of bacterial cell numbers in the leaf with ethylene output compared to bacterial cells and ethylene produced in culture showed that elevated levels of ethylene in the diseased plant could only result from enhanced host production. Gentile and Matta (1975) using detatched tomato leaves, claimed that ethylene production was associated with loss of leaf turgor. Pegg and Cronshaw (1976a) using drought-stressed whole plants found no changes in ethylene output 24 hours after the onset of leaf wilt. Ethylene is only associated with early stages of leaf symptoms. When leaves become chlorotic and necrotic, ethylene evolution falls rapidly. Leaf chlorosis is associated with a stimulation of chlorophyllase which may result from ethylene-induced mRNA for the enzyme.

Tomato plants subjected to continuous exposure of 0.6 vpm ethylene in air for prolonged periods up to 136 hours developed acute epinasty, chlorosis and subsequently loss of leaf turgor and necrosis. Of these only epinasty and leaf abscission corresponded precisely to symptoms of Verticillium or Fusarium wilts. After prolonged exposure, dark purple spots developed on leaves reminiscent of managanese toxicity, but not seen as a features of normal disease. Current evidence suggests that epinasty, adventitious root production and foliar abscission result from host ethylene production. An interactive role for ethylene as a toxin-bicomponent however, is not excluded (Talboys 1978).

Ethylene-enhanced tissue sensitivity

Leaves removed from whole tomato plants previously exposed

to 1:5 ppm ethylene:air concentrations for 12 hours and bioassayed with pectolytic enzyme or polysaccharide fractions from cultures showed greatly enhanced symptoms. A link between pectolytic enzymes, ethylene and di or polyvalent ions was established by Austin and Pegg (unpublished). Leaves of cotton and tomato from ethylene-treated plants were allowed to take up 5 mM Ca^{2-} Mg^{2+} or PO_4^{3-}. Wilt and necrosis symptoms developed identical to natural infection by Verticillium albo-atrum. Neither ions nor ethylene alone at the concentration and times used had any effect on leaves. Mussell (1973) claimed that endo PG produced in vitro by V. dahliae would induce symptom development in cotton. Purified PG, however, was only active in the presence of Ca^{2+} or Mg^{2+} ions at 5 mM and above. Similarly Albersheim et al (1960) have shown a Ca^{2+} dependence for endo pectate lyase and endo PG. A further connection with divalent cations is seen in the movement of calcium during V. albo-atrum infection of tomato. Calcium levels in stem tissue on a dry matter basis are lower in infected plants than healthy at least 3 days prior to the appearance of symptoms. In infected leaves, however, calcium accumulated significantly immediately before symptom appearance a finding consistent with the Ca^{2+} bioassay result (Pegg 1985). Monovalent ion mobility was unaffected by ethylene or by infection.

From the foregoing a possible mechanism exists for symptom development, involving Ca^{2+}, pectolytic enzymes and ethylene. Lund (1973); Mussell and Strand (1977); Mussell and Strouse (1972) have shown that endo PG, endo PL can liberate wall-bound peroxidase and generate ethylene during hydrolysis of the pit membrane. The gross loss of calcium in the stem may also render the membrane in this region more susceptible to attack. Vander-Molen et al (1983) found that endo PG, endo PL and 1,4-β-xylanase produced by Fusarium oxysporum f. sp. cubense, induced gels when introduced into petioles of Ricinus communis. Gels had previously been described by Ludwig (1952); Beckman and Halmos (1962). Leaves treated with 3 ppm ethylene in the absence of the enzyme also induced gels. This process was inhibited in the presence of an ethylene inhibitor. The possible importance of gels in resistance has been reviewed by Beckman and Talboys (1981). Their role in symptom expression, however, is only likely to be minor and ephemeral, since they can be sheared by a strong transpiration pull.

Abscisic Acid

The first report of an endogenous growth inhibitor (inhibitor-β) associated with stunting in a vascular wilt (Pegg and Selman 1959) was subsequently confirmed as cis-abscisic acid (ABA). Although claims have been made for increases in ABA in P. solanacearum-infection of tobacco as the cause of stunting (Steadman and Sequeira 1969), the actual evidence for its role in growth inhibition is weak. Pegg (1981b) demonstrated that 1,000 μg ABA applied to a healthy tomato leaf caused only slight inhibition on the target and immediate subtending leaves. The endogenous content of ABA in these leaves however, was 2,000 μg kg^{-1} fresh wt., an eight-fold increase over more severely stunted infected leaves. The primary role for ABA would appear to be as a stomatal regulator of transpiration. When leaf turgor falls, ABA is released from a bound state and mediates an influx of K^+ ions to the guard cell thereby closing the stoma (Wright and Hiron 1969). In any wilt disease, therefore, loss of turgor would lead to an increase in ABA.

ABA is also well recognised as an inducer of abscission. Although Wiese and DeVay (1970) reported an increased level of ABA associated with V. dahliae strain T9 infection, this result could be explained on the basis of wilt-induced ABA. Ethylene, however, is a more powerful abscission factor in cotton than ABA. The relative roles of ABA and ethylene in abscission must still be conjectural in the absence of critical experiments.

THE ROLE OF SMALL MOLECULAR WEIGHT METABOLITES

Fusicoccin

Fusicoccum amygdali the Mediterranean wilt and canker pathogen of peach and almond, produces a family of carbotricyclic diterpene glucosides, a major representative of which is fusicoccin mwt. 670 (Graniti 1964; Ballio et al 1964, 1976; Marré et al 1974). The evidence for its role as an *in vivo* wilt toxin *sensu stricto* is overwhelming (Marré 1979); its mode of action is to open stomata irreversibly leading to terminal desiccation - an opposite effect to that of ABA. Although many wilt pathologists regard the mode of action of fusicoccin as an exception, or a special case it is salutary to note that the stomata in hop leaves showing acute interveinal necrosis behave in an identical way to leaves of peach or almond infected by *F. amygdali* (Talboys 1978). Fusicoccin exhibits many properties of the growth substances cytokinins, gibberellins and in particular IAA. Like IAA, fusicoccin causes a reduction of pH in the bathing solution, an acid growth effect. This is due to the binding (like IAA) to specific proteins on the plasma membrane stimulating a K^+, Mg^{2+} ATPase which provides the energy for the extrusion of protons from the guard cell. The electro potential of the cell is maintained by a corresponding influx of K^+ ions co-transported with sugars and amino acids. The effect of this is to cause the stomata to open. A recent claim from the USSR based on NMR analysis that fusicoccin has been recovered from healthy plants as a naturally-occurring substance, awaits further confirmation (Muromtsev, personal communication).

Fusaric acid

Fusaric acid (FA), (5-*n*-butyl picolinic acid) mwt. 179 has been the subject of a longstanding debate on its role in Fusarium wilt pathogenesis and the virulence of *Fusarium* formae. Gaümann (1957), presented evidence for FA as an *in vivo* toxin based on the stimulation of some of the shoot symptoms of acute *F. o*. f. sp. *lycopersici*-wilt of tomato at 10^{-7} M and 20 µg/g fresh weight concentrations. Support for its involvement in some aspects of pathogenesis is based on its strong chelating properties and its positive identification in *Fusarium*-infected plants; tomato (Kern and Klupfel 1956); banana (Page 1959), and flax, tomato and watermelon (Davis 1969). The case against FA is that it is not produced by *F.o*. f. sp. *conglutinans* and when introduced into cabbage plants is degraded rapidly without inducing symptoms (Heitfuss et al 1960). Similarly Page (1961) found the FA deficient mutants of F. oxysporum were pathogenic and remained FA minus when reisolated from the host.

Some at least of the inconsistencies in the literature may reflect the response of the pathotypes on defined media. Good correlation has been found between virulence and FA production in vivo when host cell walls have been used as the carbon source, e.g. F. o. f. sp. lini (Trione 1960); F. o. f. sp. niveum on watermelon seedlings (Davis 1969) and F. o. f. sp. lycopersici on tomato cell walls (Kern 1972). Gaümann and co-workers have described at length the possible effects of FA copper complexes in pathogenesis. Notwithstanding its questionable role in symptom expression, FA is likely to play an essential role in the pathogen's iron nutrition and be synthesised as a siderophore in proportion to the iron availability in the host. Thus in a low Fe^{2+} environment such as the xylem fluid, FA may be synthesised to sequester iron for the pathogen's metabolism with corresponding secondary effects on the host plant (Neilands 1973). Similarly FA could play a defensive role for the pathogen where iron was at damaging concentrations, depending ultimately on the compartmenting of an FA-Fe complex in the hypha. The Fe nutrition of Fusarium-infected plants would make a rewarding study in relation to FA production in vivo and the tolerance of isolates or species to Fe^{2+}.

Phytonivein

F.o. f. sp. niveum produces in addition to FA a sterol, empirical formula $C_{29}H_{46}O_2$, mwt. 426, which has been called phytonivein (Hiroe and Nishimura 1956). The substance is produced in culture and in diseased watermelon shoots. At concentrations as low as 1×10^{-5}M (4.26 ppm) phytonivein induces irreversible wilting in watermelon shoots. Although no experiments were carried out on the possibility of toxin-induced vascular occlusion - a weakness in many toxin claims and investigations - the toxin is a good candidate for the mechanism of wilting in watermelon and fulfils the criteria for a vivo-toxin. Like FA it is thermostable, dialysable and non-volatile. It also acts as a ligand for metallic ions. In this capacity it could also serve in the other role as a pathogen siderophore.

THE ROLE OF PATHOGEN ENZYMES

Fungal and bacterial vascular pathogens produce a range of o-glycosidases in culture (Table 2) and *in vivo* which have been implicated in pathogenesis in various ways.

Table 2. O-glycosyl glycosidases produced by vascular wilt pathogens

Enzyme	V.aa	F.o.l.	V.d.	C.fag.	O.ulmi	P.sol.	E.chrys.*
Pectin esterase	+	+	+	-	-	-	-
Polygalacturonase	+	+	+	+	+	-	-
Pectin lyase	+	+	+	-	-	-	+
Glucosidase	+	+	+	-	-	+	-
Galactosidase	+	+	-	-	-	-	-
Cellulase	+	+	+	+	+	+	-
Galactanase	+	-	-	-	-	-	-
Xylanase	+	+	-	-	-	-	-
Arabinanase	+	+	-	-	-	-	-

* *Verticillium albo-atrum*, *Fusarium oxysporum* f.sp. *lycopersici*, *Verticillium dahliae*, *Ceratocystis fagacearum*, *Ophiostoma ulmi*, *Pseudomonas solanacearum*, *Erwinia chrysanthemi*

The xylem fluid as a substrate is relatively low in nutrients (Pegg 1981a) and the production of extracellular enzymes may represent a means of providing the pathogen with essential or supplementary nutrients by degrading and removing carbon fragments from the pit membrane. The pit field can represent nearly 50%

of the surface area of the vessel wall to which the pathogen is exposed (VanderMolen et al 1987).

The structure of the pit membrane representing the low cellulosic pectinaceous middle lamella fraction of the primary cell wall is complex. For detailed reviews see Talmadge et al (1973); Keegstra et al (1973); Albersheim (1975); McNeil et al (1984) and Keon et al (1987). The principle polymer is a rhamnogalacturonan consisting of chains of 1,4-α-covalently linked galacturonic acid interspersed with 1,2-α-linked rhamnose. From this spine, highly branched side chains of polymers of the neutral sugars 1,3-α, and 1,5-α-L arabinofuranose (araban) and 1,4-β-D-galactopyrenose (galactan) or both (arabinogalactan) emerge all constituting the pectin matrix. Where cellulose occurs it is hydrogen bonded to the pectin via hemicellulose polymers of 1,4-β-linked glucose with side chains of xylose (xyloglucan). In monocots, xylose is a principle component of the primary wall together with non-cellulosic 1,4-β and 1,3-β glycans. A consideration of this structure suggests that many enzymes may be involved in its degradation. While a principle role has been established for endo PG and endo PL in splitting the main rhamnogalacturonan spine, the importance in pathogenesis of hemicellulases and enzymes attacking the neutral sugar polymers is more conjectural. Similarly the significance of esterases such as PME depend upon the degree of methylation of uronic acid and hence its availability to forms of PG and PL. The calcium status of the plant will exert a profound effect on pectic enzyme activity. The resistance of galacturonic acid to cleavage by PG and PL increases when Ca is ionically bound to free carboxyl radicals forming a rigid Ca-pectate. Edgington and Walker (1958) first, and later Corden (1965) showed the effect of Ca deficiency of pathogenesis in Fusarium wilt of tomato. Physiological conditions affecting Ca mobility or availability, such as high humidity, low transpiration and cation antagonism in calcium non-limiting nutrient conditions, however, may determine the course or severity of enzyme attack.

Dimond (1970) proposed five criteria for the involvement of pathogen pectinases in wilt diseases; their presence in diseased but not healthy plant extracts and vascular fluids; the ability of pathogens to produce them on host tissue; in vivo chemical evidence for the desappearance of substrate; and microscopic evidence for the degradation of middle lamella and the positive staining of infected xylem tissue with ruthenium red. Cooper (1984) further proposed a molecular variation of Koch's postulates whereby wall changes and disease symptoms should be reproduced by treatment with pure enzyme preparations. This latter requirement, however, excludes the complex, possibly essential reactions determining pathogenesis in planta. Interpretation is also made difficult by the fact that PG is a naturally-occurring enzyme of many healthy plants, although usually localised in function (Besford and Hobson 1972; Huberman and Goren 1979).

Scheffer and Walker (1953) were among the first to study pectolytic enzymes in a vascular disease (F. o. f. sp. lycopersici in tomato). Many workers subsequently described these enzymes in culture filtrates of pathogens (Pegg 1981a) and in extracts from diseased plants. Two features of the early work led to confusing results or misinterpreted findings based on a lack of appreciation of the importance of the culture medium-the enzyme substrate. Thus, McDonnell (1958); Mann (1962); Talboys and Busch (1970); Weise and DeVay (1970) failed to correlate virulence and pathogenicity with pectolytic enzyme activity using mineral salts media with pectin or polypectate as the enzyme inducer. In a different approach, Husain and Dimond (1960); Patil and Dimond (1967 and 1968) demonstrated catabolite repression of PG by glucose, which explained earlier negative results with enzymes, where high levels of glucose and sucrose had been used in the culture medium. Jones et al (1972) in a signal contribution, showed that F. o. f. sp. lycopersici grown on a salts medium with tomato cell walls as the sole carbon source produced pectinases, hemicellulase and cellulase in a specific sequence. Each enzyme released the monomeric inducer of the next enzyme in the sequence. This was: PG → α-galactosidase α-arabinofuranosidase → β-xylosidase and β-galactosidase and cellulase. They showed moreover that walls of resistant and susceptible plants were hydrolysed equally providing no basis for resistance. Cooper and Wood (1973, 1975) in cultural studies on Fusarium and Verticillium demonstrated the importance of low levels of these specific inducers. Using a slow and controlled rate of inducer release into the medium, higher levels of enzyme were found than had previously been recorded. When the appropriate monomer for enzyme induction (e.g. galacturonic acid for PG and PL) exceeded the level needed for optimal growth, strong inhibition occurred. The sequence of Fusarium enzymes produced between 2 and 9 days was: endo PG → endo PL arabinanase → α-D galactosidase → β-xylanase → cellulase. The use of cell walls while simulating more closely the activity of the pathogen in the host may still give misleading results, since non-lignified cellulose and other polysaccharides not present in the pit membrane are available as enzyme substrates. Only PG and PL of the total complement of enzymes induced on cell walls were active in cell wall degradation. These have been successfully detected in xylem fluid and tissue extracts from diseased plants (Deese and Stahmann 1962 a and b; Patil and Dimond 1968; Mussell and Green 1970; Heale and Gupta 1972; Cooper and Wood 1980). A major problem in in vivo studies, however, has been the low recovery of enzymes from diseased plants. Experiments on cotton (Keen and Erwin 1971) and tomato (Blackhurst and Wood 1963) showed that only a small percentage of PG added to tissue macerates could be detected. This suggests that the enzyme is irreversibly bound in xylem vessels or is inactivated by host proteins from tomato cell walls (Albersheim and Anderson 1971).

By comparison with vascular fungi all wall-degrading enzymes (CWDE) of comparable bacterial pathogens have received little attention. Garibaldi and Bateman (1971) purified PL from carnation and Philodendron isolates of Erwinia chrysanthemi. The carnation isolate had 2 and the Philodendron isolate 4 isozymes each with a Ca^+ and high pH requirement, while all the isozymes degraded pectin and polypectate, one found in each isolate failed to macerate potato, carrot or cucumber tissue or cause cell leakage. This work was confirmed in genetic transformation studies by Keen et al (1984) and Collmer (1986) in which DNA coding for PL was transferred to E. coli. Some, but not all, transformants were capable of macerating and killing tissue. Collmer et al (1986) also confirmed that only 3 of 5 isozymes from E. chrysanthemi were necessary for pathogenicity in potato. The role of PL in wilt pathogenicity induced by E. chrysanthemi where no vascular occlusion occurs other than by bacteria of EPS is entirely conjectural. The detection of PL in vivo and evidence for its diffusion through the EPS would need to be resolved before implicating it in pathogenesis.

The conflicting evidence for the determinative role of cell wall degrading enzymes (CWDE) in wilt symptom induction, has been reviewed by Talboys (1968); Dimond (1970); Mussell and Strand (1977) and Pegg (1981a). Endo PL and PG have been found more in Verticillium susceptible than resistant plants. As symptoms in infected plants became severe, the activity of both enzymes declined (Cooper and Wood 1980). These authors could find no increase in CWDE in root-inoculated plants although vessel lumina contained gels and had degraded pit membranes. There is general agreement, however, as with bacterial wilts, that effective vascular occlusion provides a plausible explanation for wilting. The confusion centers around the degree of vascular occlusion found in diseased plants and the preceis or exclusive role that enzymes play in this. Beckman (1987) has presented equivocal evidence for a permanent role for gelation. Phenolic infusion of gels (Beckman et al 1974) would increase their stability. Pierson et al (1955); Chi and Hansen (1964) and Cooper and Wood (1980) working with Fusarium induced gelation by CWDE. McIntyre (1965) and Beckman (1969), however, failed to do so. Viruses and root wounding, in the absence of enhanced host constitutive pectic enzymes, both induce vascular gelation presumably via ethylene as reported by VanderMolen et al (1983). Similar controversy extends to the ability of CWDE in solution to induce wilt symptoms in excised shoots. Blackhurst (1963) found that wilting induced in tomato plants by culture filtrates of V. albo-atrum, high in PG was lost after autoclaving the solution. Cooper and Wood (1980) induced wilting in tomato with purified PG and PL and Mussell (1973), produced flaccidity and necrosis in cotton leaves in PG infused shoots. However, Keen and Erwin (1971), working on the identical cotton system found no association with PG and wilting.

Attemps to resolve the problem using mutants of Verticillium and Fusarium deficient in one or more CWDE have met with variable and conflicting results. McDonnell (1958) and Puhalla and Howell (1975) developed mutant deficient strains which retained the pathogenicity of the wild type and the enzyme deficiency on reisolation. Howell (1976) showed that V. dahliae mutants deficient for PME, PG and PL or PG and PL induced normal symptoms in cotton and remained stable. Durrands and Cooper (1988a,b) found that mutants of V. albo-atrum deficient in one or more pectinases or their isozymes showed reduced virulence in pathogenicity tests compared with the wild type. Colonisation, however, was unaffected with the exception of one mutant which lacked the ability to secrete enzymes.

Similarly, in contrast to the earlier studies of Talboys and Busch (1970); Carder et al (1987) working on a large population of hop V. albo-atrum isolates, found an overall good correlation between PG and PL production and the virulence of two pathotypes inducing mild, fluctuating and severe (progressive) wilt symptoms. The more virulent pathotype secreted more PG and PL in the medium. The relationship between the large variation in PG and PL produced in each population and virulence was not examined.

In summary, the evidence for the involvement of one or more CWDE in virulence and pathogenicity is strong but not conclusive. If CWDE are responsible for gels and gums it is unlikely that they contribute to the melanoid electron dense coating found on vessel walls and pit membranes, described by Pegg et al (1976) and others. Such a deposit would be an effective barrier to water movement even where no obvious occlusion in the vessel lumen was apparent. Since nearly every facet of the wilt syndrome may be reproduced by agencies other than fungi and bacteria, it is likely that alternative and not mutually exclusive pathways or mechanisms may exist. Thus, even if an apparent prime facie case can be presented for CWDE and wilting, the effects of other host and pathogen molecules described in this chapter need to be taken into account. Not the least of these are ethylene and wall bound oxidases, both of which may be generated (but not exclusively) by pectolytic enzymes.

HOST GLUCANOHYDROLASES AND VASCULAR PATHOGENESIS

The existence of glucosidases in higher plants with activities similar to those normally associated with microorganisms, has been known for many years. (Grassman et al 1934; Clarke and Stone 1962; Abeles et al 1970). Fleming (1922), characterized lysozyme (muramidase) from pawpaw (Carica papaya L.). This enzyme for which no apparent substrate exists is part of a complex of powerful hydrolases found also in Ficus carica L. and Ananas comosus. The role of these enzymes in host parasite relations has received little attention. Pegg and Vessey (1973) studying in vitro autolysis and in vivo lysis of V. albo-atrum in tomato, found chitinase (endo 1,4-β-N acetyl glucosamine hydrolase) activity in healthy shoots which increased following infection with hop and tomato isolates of Verticillium. Pegg and Young (1981); Pegg and Young (1982) subsequently characterised different chitinases from the pathogen and the tomato, both of which were constitutive enzymes. Chitin per se is not found in tomato or other higher plants but the monomer N-acetylglucosamine (NAG) exists in a bound form. The presumption has been made that chitinase is a non-specific enzyme (like many other glycosidases) probably acting in the tomato on NAG rich glycoproteins on cell membranes.

Young and Pegg (1981) characterised three isozymes of a constitutive 1,3-β-glucanase from tomato. The enzymes were all endo enzymes active only on long chain 1,3-β glucan polymers. Laminaripentaose and small oligomers were unaffected as were carbohydrates with β-1,4 and β-1,6 linkages. Exo 1,3-β-glucanase and 1,4-β-glucosidase from Verticillium were all active on the smaller β-1-3 oligomers and on lichenan with β, 1-3 and β, 1-4 linkages. The finding that Pseudomonas solanacearum and virus infection of tomato also led to increased chitinase and glucanase (Pegg 1977) suggests that a stimulus other than chitin or 1-3-β-glucan or their monomers can be involved in enzyme induction. The major emphasis with host glycosidases has been as a defence reaction (Pegg 1977; Young and Pegg 1982) in relation to the phenomenon of in vivo lysis (Blackhurst and Wood 1963; Wilhelm and Taylor 1965; Taylor and Flentje 1968; Dixon and Pegg 1969).

More recently Metraux and Boller (1986) have reported antifungal activity of chitinase from Phaseolus.

Ethylene is a powerful inducer of many plant enzymes (Abeles et al 1970; Abeles 1973; Pegg 1976a). Ethylene treatment of excised Phaseolus petioles led to a 50-fold increase in chitinase and glucanase activities but not peroxidase (Abeles et al 1970). In tomato, ethylene treatment of whole plants, however, led to increased synthesis of 1,3-β glucanase in the leaf but a decrease in stems. Chitinase was inhibited by ethylene in both leaf and stem (Pegg 1976b). Chitinase experiments were carried out on detached plant parts.

Young and Pegg (1982) found that hydrolysis of V. albo-atrum wall required a combination of host 1,3-β-endo glucanase and chitinase together with either a fungal exo 1,3-β-glucanase or 1,4-β-glucosidase. Although the concept of a host lytic defence mechanism is attractive and is in agreement with ethylene-induced resistance to V. albo-atrum (Pegg 1976b) this effect is not related to single gene Ve^+ resistance. Chitinase and glucanase are both induced to a greater extent in susceptible cultivars than resistant. Moreover, the Ve^- isoline recognised autoclaved cells of the pathogen which from a simple root dip inoculum leads to a linear production of enzyme over 21 days (Pegg and Young 1981). The magnitude of such a response suggests that host enzymes could be involved in the recognition response of the pathogen or its products. In line with this argument, pathogen extra-cellular polysaccharides similar in composition to hyphal wall polymers may be recognised by the host either before or after modification by host enzymes. It is these oligomers which may trigger off other host responses which led either to susceptibility or resistance (Pegg 1985).

The full range of host enzymes induced by bacterial and fungal vascular pathogens needs to be determined. N acetyl glucosaminidase and 1,4-β-glucosidase are also present in tomato as constitutive enzymes and increase following infection. The probability that pathogen metabolites from cultures are further modified in planta during bioassays and natural infection is very high. This could explain many of the apparent inconsistencies in the literature where "toxins" have appeared unstable in simulation experiments or have failed to be recovered following feeding experiments. The current state of our knowledge is too limited to attempt a model involving different host and pathogen metabolites.

PHENOL METABOLISM IN THE DISEASED HOST

Changes in phenol metabolism represent one of the early events in most vascular diseases. Browning and phenolic staining of xylem tissue and paravascular elements is one of the classic pointers to vascular infection. Much debate has arisen as to whether changes in phenol metabolism represent a cause or an effect of pathogenesis, and whether this is part of a resistance response. Although it is known what reactions are affected by phenolic substances and their condensation products, the case for a role in resistance is mostly, if not wholly, speculative.

Browning in tomato is thought to be due to caffeic acid derivatives (Mace et al 1972); in elm, (+)-catechin (Gagnon 1967), cotton, catechin and gallocatechin (Bell and Howell 1976) and banana, dopamine (Mace 1964; Mace and Wilson 1964). Monohydric phenols commonly encountered in tracheal exudate from infected plants are coumaric acid and p-hydroxybenzoic acid and dihydric caffeic acid, chlorogenic acid catechol DOPA and scopoletin. Mace (1963) first described specialised phenol storing cells in banana root xylem parenchyma where dopamine derived from phenylalanine was stored in a free and unoxidised state. The ortho-dihydric catechin and gallocatechin were found in similar cells in cotton roots (Mace and Howell 1974) and caffeic acid derivatives in tomato (Beckman et al 1974; Mueller and Beckman 1976). Sequeira (1969) described the synthesis of scopolin and scopoletin in tobacco infected with Pseudomonas solanacearum. Phenolic substances, like auxins are derived mostly from the shikimic acid pathway.

Two pre-requisites are required for the observed increase in phenol metabolism: a) stimulated phenol synthesis and/or b) the liberation of free phenols from glycoside or ester conjugates by the release or synthesis of the appropriate enzymes. Godin (1955) has shown that increased phenol synthesis resulted from a stimulation of the pentose pathway. Both increased pentose phosphate activity and phosphoenol pyruvate turnover occur in wilt diseases increasing phosphoshikimate. Little information is available from vascular wilts specifically, on the turnover and stimulation of the pathway. Phenylalanine ammonia lyase PAL responsible for the deamination of phenylpropane derivatives is a key enzyme controlling phenolic acid synthesis. The turnover of this and the branch point enzymes chorismate mutase and anthranilate synthetase, would be major controlling stages. <u>Trans</u> cinnamic acid and <u>p</u>-coumaric acid act as feed-back inhibitors of PAL (Kosuge 1969). Similarly cinnamic, ferulic, vanillic, <u>p</u>-hydroxybenzoic and protocatechinic acids all inhibited the first stage in the aromatic pathway dehydroshikimic acid → shikimic acid. Here then is a clear link between phenol and auxin metabolism. Although ethylene has been linked to phenol metabolism through its induction of PAL (Riov et al 1969), Pegg (1976b) found a reduction in PAL and peroxidase following ethylene gassing of whole tomato plants.

The second source of free phenols is from glycosidic or ester conjugates. Davis et al (1953) found a constitutive 1,4-β-glucosidase in <u>F. o.</u> f. sp. <u>lycopersici</u> and a similar enzyme in infected but not healthy plants. <u>V. albo-atrum</u> has a constitutive 1,4-β-glycosidase and <u>exo</u> 1,3-β-glucanase both of which are non-specific on 1 → 3, 1 → 4, and 1 → linkages. (Young and Pegg 1981; Reese et al 1967). Davis et al (1953) demonstrated the increase in free phenols by the increase in quinone products in diseased plants compared to healthy, following the addition of polyphenol oxidase (PPO) from <u>Agaricus</u>. Briel (1967) and Pollock and Drysdale (1976) found increases in total soluble phenolics in <u>Fusarium</u> and <u>V. albo-atrum</u> infected tomatoes concluding that it represented an expression of pathogenesis in susceptible plants and was unrelated to resistance (Matta et al 1969).

Numerous workers have reported increases in PPO in diseased tissue since Waggoner and Dimond's (1956) first account (See

Pegg 1981a). Ethylene gassing of intact tomato plants at low concentration led to a big increase in PPO but not peroxidase (Pegg 1976b), illustrating a further metabolic link between phenols and growth substances. Although Beckman (1987) has suggested that the end point of these phenolic reactions could be in the lignification of tissue as a putative defence reaction, there is no supportive experimental evidence. In contrast, Davis and Dimond (1954) found fewer lignified elements in the xylem of diseased compared with healthy, and proposed that the sink of oxidised (brown) phenolic products deflected precursors away from lignin synthesis. Soluble phenolic compounds as acids or oxidation products are at best only weakly fungitoxic. Their role instead is most probably as non-specific inhibitors or inactivators of enzymes. Cell wall degrading enzymes (CWDE) have been shown to be inactivated <u>in vitro</u> by oxidised phenols (Deverall and Wood 1961; Bateman 1967; Patil and Dimond 1967). Free radical intermediates of quinones react readily with NH_2 and OH groups. This property presents a paradox with regard to the effect of product repression on enzymes such as PPO. How important free or oxidised phenols are in inhibiting CWDE during pathogenesis compared with wall-bound or cytoplasmic protein inhibitors remains to be demonstrated.

SUMMARY

As can be seen from this brief review of the voluminous literature on vascular wilt pathogenesis the complexity of the different metabolic systems is formidable and no single formula for pathogenesis is possible, to embrace the range of pathogens and hosts listed in Table 1. Indeed a major difficulty in attempting to construct a working model of pathogenesis (excluding resistance) is the profusion of contending factors. Most, if not all, of the biochemical studies that have been conducted to date could be profitably re-examined in the context of the host xylem environment. Thus the conviction that has existed for some 30 years that bacterial wilt symptoms were based on a simple water deficit resulting from the occlusion of xylem vessels by EPS will now have to be reappraised in the light of the P. solanacearum mutant studies of Xu et al (1987).

The work of Van Alfen and Dunniway working with bacteria and fungi has shown convincingly in their systems that leaf water deficits in vascular wilts are real and result primarily from an occlusion of the water supply in the minor, terminal veins of leaves. These findings, however correct, do not account for the range of "wilt" symptoms that cannot be simulated by a simulated water deficit or drought hardening. There is an abundance of evidence to show that small molecules affect leaf and stem paravascular parenchyma before the main conducting system is blocked.

There is a growing awareness that the nutritional status of the xylem fluid will have a controlling influence on the quality and quantity of metabolites produced by the pathogen. A major question still awaiting resolution is whether virulence represents an innate property of a bacterial or fungal pathotype or whether virulence or avirulence may be the result of recognition by the host genotype. Since virulence is not always related to pathogen growth and biomass the probability of detecting novel metabolites in planta is high. The role of the host in differentially modifying pathogen products through a complex of enzymes, superficially resembling those of the pathogen has scarcely been examined and would be particularly rewarding for bacterial diseases.

The very nature of vascular pathogenesis remains a paradox, whether the xylem represents a preferred habitat or one to which the pathogen is confined. In this context the symptomless condition of the non-host, where no real barrier to colonisation exists, or the recognition or non-recognition of the vascular parasite, represents the next biochemical challenge.

REFERENCES

Abeles FB (1973) Ethylene in Plant Biology. Academic Press, New York

Abeles FB and Forrence LE (1970) Temporal and hormonal control of β-1,3-glucanase in Phaseolus vulgaris. Plant Physiology 45:395-400

Albersheim P (1975) The walls of growing plant cells. Scient Amer 232:81-95

Albersheim P and Anderson AJ (1971) Host-pathogen Interactions III. Proteins from plant cell walls inhibit enzymes secreted by plant pathogens. Proc Nat Acad Sci USA 68:1815-1819

Albersheim P, Neukom H and Deuel H (1960) Über die bildung con ungesättigten abbauprodukten durch ein pektinabbauendes Enzym. Helvt Chim Acta 43:1422-1426

Asamov DK, Tursunkulova RK, Isaev PI, Ostroshchenko OS and Stepanichenko NN (1975) The action of phytotoxic substances of the fungus Verticillium dahliae on the permeability of a synthetic phospholipid membrane. Khim Pir Soedin 5:670 (Russian) Chem Nat Compds 11:713-714 Transl (1976)

Austin S (1979) Interaction of ethylene with Verticillium alboatrum culture metabolites and with ions in symptom expression in tomato. PhD thesis Univ London

Ayres AR, Ayres SB and Goodman RN (1979) Extracellular polysaccharide of Erwinia amylovora: a correlation with virulence. Appl Environ Microbiol 38:659-666

Ballio A, Chain EB, DeLeo P, Erlanger BF, Kauri M and Tonolo A (1964) Fusicoccin; a new wilting toxin produced by Fusicoccum amygdali. Nature 203:297

Ballio A, D'Alessio V, Randazzo G, Bottalico A, Graniti A, Sparapano L, Bosnar B, Casinovi CG and Gribanovski-Sassu O (1976) Occurrence of fusiccocin in plant tissues infected by Fusicoccum amygdali Del. Physiol Plant Path 8:163-169

Banfield WM (1941) Distribution by the sap stream of spores of three fungi that induce vascular wilt diseases of elm. J Agric Res 62:637-681

Bateman DF (1967) Alteration of cell wall components during pathogenesis by Rhizoctonia solani. In: Mirocha CJ and Uritani I (eds) The Dynamic Role of Molecular Constituents in Plant-Parasite Interaction. APS, St Paul Minnesota

Beckman CH (1969) The mechanics of gel formation by swelling of simulated plant cell wall membranes and perforration plates of banana root vessels. Phytopathology 59:837-843

Beckman CH (1987) The Nature of Wilt Diseases of Plants. Am Phyt Soc Press, St Paul Minnesota

Beckman CH and Halmos S (1962) Relation of vascular occluding reactions in banana roots to pathogenicity of root-invading fungi. Phytopathology 52:893-897

Beckman CH, Kuntz JE, Riker AJ and Berbee JG (1953) Host responses associated with the development of oak wilt. Phytopathology 43:448-454

Beckman CH, Mueller WC and Mace ME (1974) The stabilisation of artificial and natural cell wall membranes by phenolic infusion and its relation to wilt disease resistance. Phytopathology 64:1214-1220

Beckman CH and Talboys PW (1981) Anatomy of resistance. In: Mace ME, Bell AA and Beckman CH (eds) Fungal Wilt Diseases of Plants. Acad Press, New York London Toronto Sydney, 487-518

Bell AA and Howell CR (1976) Report ot the disease and pathogen physiology. In: Bell AA (ed) Proc Beltwide Cotton Prod Res Conf

Bell AA and Mace ME (1981) Biochemistry and Physiology of resistance. In: Mace ME, Bell AA and Beckman CH (eds) Fungal Wilt Diseases of Plants. Acad Press, New York London Toronto Sydney, 431-477

Besford RT and Hobson GE (1972) Pectic enzymes associated with the softening of tomato fruit. Phytochem 11:2201-2205

Blackhurst FH (1963) Induction of Verticillium wilt disease symptoms in detached shoots of resistant and susceptible tomato plants. Ann Appl Biol 52:79-88

Blackhurst FH and Wood RKS (1963) Resistance of tomato plants to Verticillium albo-atrum fungal lysis. Trans Brit Mycol Soc 46:385-392

Bonn WG, Sequeira L and Upper CD (1975) Technique for the determination of the rate fo ethylene production by Pseudomonas solanacearum. Plant Physiology 56:688-691

Booth JA (1969) Gossypium hirsutum tolerance to Verticillium albo-atrum infection. X, Amino acid exudation from aseptic roots of tolerant and susceptible cotton. Phytopathology 59:43-46

Bradshaw-Rouse JJ, Whatley MH, Coplin DL, Woods A, Sequeira L and Kelman A (1981) Agglutination of Erwinia stewartii strains with a corn agglutin: correlation with extracellular polysaccharide production and pathogenicity. Appl Environ Microbiol 42(2):344-350

Braithwaite CWD and Dickey RS (1971) Permeability alterations in detached carnation leaf tissue inoculated with Pseudomonas caryophylli and Corynebacterium sp. Phytopathology 61:317-321

Briel W van den (1967) The occurrence of acid hydrolysable phenolics in relation to Fusarium wilt disease of tomato plants. Neth J Plant Path 73:126-128

Buchner V, Nachmias A and Burnstein Y (1982) Isolation and partial characterization of a phytotoxic glycopeptide from a protein-lipopolysaccharide complex produced by a potato isolate of Verticillium dahliae. Fed Euro Biochem Socs Letters 138:261-264

Carder JH, Hignett RC and Swinburne TR (1987) Relationship between the virulence of hop isolates Verticillium albo-atrum and their in vitro secretion of cell wall degrading enzymes. Physiol Mol Plant Path 31:441-453

Caroselli NE (1955) Investigations of toxins produced in vitro by the maple wilt fungus Verticillium sp. Phytopathology 45: 183 (Abstr)

Chi CC and Hanson EW (1964) Mechanism of wilting by Fusarium in red clover. Phytopathology 54:646-653

Clarke AE and Stone BA (1962) β-1,3-Glucanhydrolases from the grape vine (Vitis vinifera) and other plants. Phytochem 1: 175-188

Cleland RE and Rayle DL (1977) Revaluation of the effect of calcium ions or auxin-induced elongation. Plant Physiology 60: 709-712

Cohen R, Riov J, Lisker N and Katan J (1986) Involvement of ethylene in herbicide-induced resistance to Fusarium oxysporum f. sp. melonis. Phytopathology 76:1281-1285

Collmer A (1986) The molecular biology of pectic enzyme production and bacterial soft rot pathogenesis. In: Bailey JA (ed) Biology of Plant Pathogen Interactions, Plenum Press New York

Collmer A, Schoedel C, Roeder DL, Ried JL and Rissler JF (1986)

Molecular cloning in Escherichia coli of Erwinia chrysanthemi genes encoding multiple forms of pectate lyase. J Bact 161: 913-920

Coplin D, Sequeira L and Hanson RS (1974) Pseudomonas solanacearum: virulence of biochemical mutants. Canad J Microbiol 20: 519-529

Corden ME (1965) Influence of calcium nutrition on Fusarium wilt of tomato and polygalacturonase activity. Phytopathology 55:222-224

Cooper RM (1984) The role of cell wall degrading enzymes in infection and damage. In: Plant Diseases Infection Damage and Loss. Wood RKS and Jellis JG (eds) Blackwell Scientific, Oxford, 13-27

Cooper RM and Wood RKS (1973) Induction of synthesis of extracellular cell-wall degrading enzymes in vascular wilt fungi. Nature 246:309-311

Cooper RM and Wood RKS (1975) Regulation of synthesis of cell wall degrading enzymes by Verticillium albo-atrum and Fusarium oxysporum f. sp. lycopersici. Physiol Plant Path 5:135-156

Cooper RM and Wood RKS (1980) Cell wall degrading enzymes of vascular wilt fungi. III Possible involvement of endopectin lyase in Verticillium wilt of tomato. Physiol Plant Path 16: 285-300

Cronshaw DK and Pegg GF (1976) Ethylene as a toxin synergist in Verticillium wilt of tomato. Physiol Plant Path 9:33-44

Davis D (1969) Fusaric acid in selective pathogenicity of Fusarium oxysporum. Phytopathology 59:1391-1395

Davis D and Dimond AE (1954) The source and role of phenols in Fusarium wilt symptoms. Phytopathology 44:485-486

Davis D, Waggoner PE and Dimond AE (1953) Conjugated phenols in the Fusarium wilt syndrome. Nature 172:959

Deese DC and Stahmann MA (1962a) Formation of pectic enzymes by Verticillium albo-atrum on susceptible and resistant tomato stem tissues and on wheat bran. Phytopath Z 46:53-70

Deese DC and Stahmann MA (1962b) Pectic enzymes in Fusarium-infected susceptible and resistant tomato plants. Phytopathology 52:255-261

Deverall BJ and Wood RKS (1961) Chocolate spot of beans (Vicia faba L.)-interactions between phenolase of host and pectic enzymes of the pathogen. Ann Appl Biol 49:473-487

Dimond AE (1970) Biophysics and biochemistry of the vascular wilt syndrome. Ann Rev Phytopath 8:301-322

Dimond AE and Waggoner PE (1953) The cause of epinastic symptoms in Fusarium wilt of tomatoes. Phytopathology 43:663-669

Dixon GR and Pegg GF (1969) Hyphal lysis and tylose formation in tomato cultivars infected by Verticillium albo-atrum. Trans Brit Mycol Soc 53:109-118

Dixon GR and Pegg GF (1972) Changes in amino-acid content of tomato xylem sap following infection with strains of Verticillium albo-atrum. Ann Bot 36:147-154

Durrands PK and Cooper RM (1988a) Selection and characterisation of pectinase-deficient mutants of the vascular wilt pathogen Verticillium albo-atrum. Physiol Mol Plant Path 32:343-362

Durrands PK and Cooper RM (1988b) The role of pectinases in vascular wilt diseases as determined by deficient mutants of Verticillium albo-atrum. Physiol Mol Plant Path 32:363-371

Duvick JP and Sequeira L (1984) Interaction of Pseudomonas solanacearum with suspension-cultured tobacco cells and tobacco leaf cell walls in vitro. Appl Environ Microbiol (48) 1:199-205

Edgington LV and Walker JC (1958) Influence of calcium and boron nutrition on development of Fusarium wilt of tomato. Phytopathology 48:324-326

Elgersma DA (1969) Resistance mechanisms of elms to Ceratocystis ulmi. Med van den Phytopath dienst, Lab Willie Commelin Scholten 77:1-84

Fergus CL and Wharton DC (1957) Production of pectinase and growth-promoting substance by Ceratocystis fagacearum. Phytopathology 11:635-636

Fleming A (1922) On a remarkable bacteriolytic element found in tissues and secretions. Proc Roy Soc (B) 93:306-317

Freebairn HT and Buddenhagen IW (1964) Ethylene production by Pseudomonas solanacearum. Nature 202:313-314

Gagnon C (1967) Histochemical studies on the alteration of lignin and pectic substances in white elm infected by Ceratocystis ulmi. Can J Bot 45:1619-1623

Garibaldi A and Bateman DF (1971) Pectic enzymes produced by Erwinia chrysanthemi and their effects on plant tissue. Physiol Plant Path 1:25-40

Gaümann E (1957) Fusaric acid as a wilt toxin. Phytopathology 47:342-357

Gentile IA and Matta A (1975) Production of and some effects of ethylene in relation to Fusarium wilt of tomato. Physiol Plant Path 5:27-37

Girardin JPL (1864) Jahresber Agrikult-Chem Versuchssta Berlin 7:199

Glasziou KT (1957) The effect of auxins on the binding of pectin methyl-esterase to cell wall preparations. Aust J Biol Sci 10:426-434

Glasziou KT and Inglis SD (1958) The effect of auxins on the binding of pectin methylesterase to cell walls. Aust J Biol Sci 11:127-141

Godin P (1955) Production of phenolic substances from various ternary compounds by P. brevi-compactum. J Microbiol 21:94-102

Goodman RN, Huang JS and Huang PY (1974) Host-specific phytotoxic polysaccharide from apple tissue infected by Erwinia amylovora. Science 183:1081-1082

Gordon SA and Paleg LG (1961) Formation of auxin from tryptophan through action of polyphenols. Plant Physiology 36:838

Gorin PAJ, Spencer JFT, Lindberg B and Lindh F (1980) Structure of the extra-cellular polysaccharide from Corynebacterium insidiosum. Carbohydr Res 79:313-315

Graniti A (1964) In: Kiraly Z and Ubriszy G (eds) Host-Parasite Relations in Plant Pathology. Budapest, 211-217

Grassmann W, Zechmeister L, Bender R and Toth G (1934) Über die Chitinspaltung durch Emulsin-Präparate III Mitteil über enzymatiche spaltung von Polysacchariden. Ber Deutch Chem Ges 67: 1-5

Green RJ Jr (1954) A preliminary investigation of toxins produced in vitro by Verticillium albo-atrum. Phytopathology 44: 433-437

Grieve BJ (1939) Epinastic response induced in plants by Bacte-

rium solanacearum EFS. Ann Bot (Lond)3:587-600
Heale JB and Gupta DP (1972) The mechanism of vascular wilting induced by Verticillium albo-atrum. Trans Brit Mycol Soc 58: 19-28
Heitfuss R, Stahmann MA and Walker JC (1960) Oxidative enzymes in cabbage infected by Fusarium oxysporum f. conglutinans. Phytopathology 50:370-375
Hiroe I and Nishimura S (1956) Pathochemical studies on watermelon wilt. I On the wilt toxin Phytonivein produced by the causal fungus. Am Phytopathol Soc Japan 29:161-164
Howell CR (1976) Use of enzyme-deficient mutants of Verticillium dahliae to assess the importance of enzymes in symptom expression of Verticillium wilt of cotton. Physiol Plant Path 9:279-283
Huberman M and Goren R (1979) Exo-and-endo cellular cellulase and polygalacturonase in abscission zones of developing orange fruits. Physiol Plantarum 9:279-283
Husain A and Kelman A (1958) The role of pectic and celluloty-tic enzymes in pathogenesis by Pseudomonas solanacearum. Phytopathology 48:377-386
Husain A and Dimond AE (1960) Role of cellulolytic enzymes in pathogenesis by Fusarium oxysporum f. sp. lycopersici. Phytopathology 50:329-331
Jansen EF, Jang R and Bonner J (1960) Binding of enzymes to Avena coleoptile cell walls. Plant Physiology 35:567-574
Jones JF and Kende H (1979) Auxin-induced ethylene biosynthesis in subapical stem sections of etiolated seedlings of Pisum sativum L. Planta 146:649-656
Jones TM, Anderson AJ and Albersheim P (1972) Host-pathogen interactions IV. Studies on the polysaccharide degrading enzymes secreted by Fusarium oxysporum f. sp. lycopersici. Physiol Plant Path 2:153-166
Keegstra K, Talmadge KW, Bauer WD and Albersheim P (1973) The structure of plant cell walls. III A model of the walls of suspension-cultured sycamore cells based on the interconnections of the macromolecular components. Plant Physiology 51: 188-197
Keen NT, Dahlbeck D, Staskawicz B and Belser W (1984) Molecular cloning of pectate lyase genes from Erwinia chrysanthemi and their expression in Escherichia coli. J Bact 159:825-831
Keen NT and Erwin DC (1971) Endopolygalacturonase: evidence against involvement in Verticillium wilt of cotton. Phytopathology 61:198-203
Keen NT, Long M and Erwin DC (1972) Possible involvement of a pathogen produced protein-lipopolysaccharide complex in Verticillium wilt of cotton. Physiol Plant Path 2:317-331
Keen NT and Long M (1972) Isolation of a protein-lipopolysaccharide complex. Physiol Plant Path 2:307-315
Keissig R and Haller-Keissig R (1957) Beitrage zur Kenntinis einer infektiosen Welkekranheit der Luzerne (Verticillium albo-atrum Reinke et Bert.). Phytopath Z 31:185-222
Kelman A and Sequeira L (1965) Root to root spread of Pseudomonas solanacearum. Phytopathology 55:304-309
Keon JPR, Byrde RJW and Cooper RM (1987) Some aspects of fungal enzymes that degrade plant cell walls. In: Pegg GF and Ayers PG (eds) Fungal Infection of plants. Cambridge, 134-159

Kern H (1972) Phytotoxins produced by Fusaria. In: Wood RKS, Ballio A and Graniti A (eds) Phytotoxins in Plant Disease. Academic Press, New York, 35-48

Kern H and Klupfel D (1956) Die bildung von fusarinsäure durch Fusarium lycopersici in vivo. Experientia 12:181-182

Kosuge T (1969) The role of phenolics in host response to infection. Ann Rev Phytopathol 7:195-111

Ludwig RA (1952) Studies in the physiology of hydromycotic wilting in the tomato plant. Tech Bull MacDonald Agric Coll 20

Lund BM (1973) Effect of bacteria on ethylene production. In: Byrde RJW and Cutting CV (eds) Fungal Pathogenicity and the Plant's Response. Academic Press, London New York, 69-84

Mace ME (1963) Histochemical localisation of phenols in healthy and diseased banana roots. Physiol Plant Path 16:915-925

Mace ME (1964) Peroxidases and their relation to vascular browning in banana roots. Phytopathology 54:1033-1034

Mace ME and Howell CR (1974) Histochemistry and identification of condensed tannin precursors in the roots of cotton seedlings. Canad J Bot 52:2423-2426

Mace ME and Solit E (1966) Interactions of 3-indoleacetic acid and 3-hydrox-tyramine in Fusarium wilt of banana. Phytopathology 56:245-247

Mace ME, Veech JA and Beckman CH (1972) Fusarium wilt of susceptible and resistant tomato isolines: spore transport. Phytopathology 61:627-630

Mace ME and Wilson EM (1964) Phenol oxidases and their relation to vascular browning in Fusarium invaded banana roots. Phytopathology 54:840-842

Malysheva KM and Zel'tser SS (1968) Protein-lipid-polysaccharide complex of culture liquid and mycelium of Verticillium dahliae Kleb, the causative agent of Verticillium wilt in cotton. Doklady Akad nauk SSSR 179:231-234

Mann B (1962)Role of pectic enzymes in the Fusarium wilt syndrome of tomato. Trans Brit Mycol Soc 45: 169-178

Marrè E (1979) Fusicoccin: a tool in plant physiology. Ann Rev Plant Physiol 30:273-288

Marrè E, Columbo R, Lado R and Rasi-Caldogno F (1974) Correlation between proton extrusion and simulation of cell enlargement. Effects of fusicoccin and of cytokinins on leaf fragments and isolated cotyledons. Plant Sci Letters 2:139-150

Matta A and Gentile IA (1965) Sul meccanismo di accumulo dell' acido-indolil-acetico in piante di Pomodor infette da Fusarium oxysporum f. sp. lycopersici. Phytopath Mediterr 4:129-137

Matta A , Gentile IA and Gia I (1969) Accumulation of phenols in tomato plants infected by different forms of Fusarium oxysporum. Phytopathology 59:512-513

McDonnell K (1958) Absence of pectolytic enzymes in a pathogenic strain of Fusarium oxysporum f. sp. lycopersici. Nature London 182:1025-1026

McIntyre GA (1965) Absence of pectic plugs in tomato cuttings treated with polygalacturonases of Verticillium albo-atrum. Phytopathology 55:1067 (Abstr)

McNeil M, Darvill AG, Fry SS and Albersheim P (1984) Structure and function of primary cell walls of plants. Ann Rev Biochem 53:625-663

Metraux JP and Boller T (1986) Local and systemic induction of chitinase in cucumber plants in response to viral, bacterial and fungal infections. Physiol Mol Plant Path 28:161-169
Mueller WC and Beckman CH (1976) Ultrastructure and development of phenolic-storing cells in cotton roots. Can J Bot 54:2074-2082
Mussell HW (1972) Toxic proteins secreted by cotton isolates of Verticillium albo-atrum. In: Wood RKS, Ballio A and Graniti A (eds) Phytotoxins in Plant Disease. Academic Press, New York, pp 443-445
Mussell HW (1973) Endopolygalacturonase: Evidence for involvement in Verticillium wilt of cotton. Phytopathology 63:62-70
Mussell HW and Green RJ (1970) Host colonisation and polygalacturonase production by two tracheomycotic fungi. Phytopathology 60:192-195
Mussell HW and Strand LL (1977) Pectic enzymes: involvement in pathogenesis and possible relevance to tolerance and specificity. In: Solheim B and Raa J (eds) Cell Wall Biochemistry Related to Specificity in Host-Plant Pathogen Interactions. Universitetsforlaget, Oslo, 31-70
Mussell HW and Strouse B (1972) Characterisation of two polygalacturonases produced by Verticillium albo-atrum. Canad J Biochem 50:625-632
Nachmias A, Buchner V and Burnstein Y (1985) Biological and immunochemical characterisation of a low molecular weight phytotoxin isolated from a protein-lipopolysaccharide complex produced by a potato isolate of Verticillium dahliae Kleb. Physiol Plant Path 26:43-55
Nachmias A, Buchner V and Krikun J (1982) Comparison of protein-lipopolysaccharide complexes produced by pathogenic and non-pathogenic strains of Verticillium dahliae Kleb. from potato. Physiol Plant Path 20:213-221
Nachmias A, Buchner V, Tsror L, Burnstein Y and Keen N (1987) Differential phytotoxicity of peptides from culture fluids of Verticillium dahliae Races 1 and 2 and their relationship to pathogenicity of the fungi on tomato. Phytopathology 77: 506-510
Neilands JB (1973) Microbial iron transport compounds (siderochromes) In: Eickhorn G (ed) Bioinorganic Chemistry. Elsevier, Amsterdam, pp 167-202
Nishimura S (1960) Pathochemical studies on watermelon wilt. 7, The physiology of Fusarium wilt of watermelon plant. Trans Tottori Soc Ag Sci 12:13-17
Page OT (1959) Fusaric acid in banana plants infected with Fusarium oxysporum f. sp. cubense. Phytopathology 49:230
Page OT (1961) Induced variation in Fusarium oxysporum. Canad J Bot 39:1509-1519
Patil SS and Dimond AE (1967) Induction and repression of polygalacturonase synthesis in Verticillium albo-atrum. Phytopathology 57:825
Patil SS and Dimond AE (1968) Repression of polygalacturonase synthesis in Fusarium oxysporum f. sp. lycopersici by sugars and its effect upon symptom reduction in infected tomato plant. Phytopathology 58:676-682
Pegg GF (1959) Role of indole-3-acetic acid in the development of disease symptoms in Verticillium wilt of tomato. Proc IX

Int Bot Cong Montreal Canada, p 297
Pegg GF (1976a) Auxins ethylene gibberellins and inhibitors in healthy and diseased plants. In: Heitfuss R and Williams PH (eds) Encyclopaedia of Plant Physiology, Physiological Plant Pathology. Springer, Berlin Heidelberg New York, 560-616
Pegg GF (1976b) The response of ethylene-treated tomato plants to infection by Verticillium albo-atrum. Physiol Plant Path 9:215-226
Pegg GF (1977) Glucanhydrolases of higher plants: a possible defence mechanism against parasitic fungi. In: Solheim B and Raa J (eds) Cell Wall Biochemistry Related to Specificity in Host-Plant Pathogen Interactions. Universitetsforlaget, Oslo 305-342
Pegg GF (1981a) Biochemistry and physiology of pathogenesis. In: Mace ME, Bell AA and Beckman CH (eds) Fungal Wilt Diseases of Plants. Academic Press, New York, 193-253
Pegg GF (1981b) The involvement of growth substances in the diseased plant. In: Ayres PG (ed) Effects of Disease on the Physiology of the Growing Plant. Cambridge, 149-177
Pegg GF (1985) Life in a black hole. The microenvironment of the vascular pathogen. Trans Brit Mycol Soc 85:1-20
Pegg GF and Cronshaw DK (1976a) Ethylene production in tomato plants infected with Verticillium albo-atrum. Physiol Plant Path 8:279-295
Pegg GF and Cronshaw DK (1976b) The relationship of in vitro to in vivo ethylene production in Pseudomonas solanacearum infection in tomato.Physiol Plant Path 9:145-154
Pegg GF, Gull K and Newsam RJ (1976) Transmission electron microscopy of Verticillium albo-atrum hyphae in xylem vessels of tomato plants. Physiol Plant Path 8:221-224
Pegg GF and Selman IW (1959) An analysis of the growth response of young tomato plants to infection by Verticillium albo-atrum. II The production of growth substances. Ann Appl Biol 47:222-231
Pegg GF and Sequeira L (1968) Stimulation of aromatic biosynthesis in tobacco plants infected with Pseudomonas solanacearum. Phytopathology 58:476-483
Pegg GF and Vessey JC (1973) Chitinase activity in Lycopersicon esculentum and its relationship to the in vivo lysis of Verticillium albo-atrum mycelium. Physiol Plant Path 3:207-222
Pegg GF and Young DH (1981) Changes in glycosidase activity and their relationship to fungal colonisation during infection of tomato by Verticillium albo-atrum. Physiol Plant Path 19:371-382
Pegg GF and Young DH (1982) Purification and characterisation of chitinase enzymes from healthy and Verticillium albo-atrum-infected tomato plants and from Verticillium albo-atrum. Physiol Plant Path 21:389-409
Phelps RH and Sequeira L (1968) In: Wightman F and Settefield G (eds) Biochemistry and Physiology of Plant Growth Substances. Runge, Ottawa
Pierson CF, Gothoskar SS, Walker JC and Stahmann MA (1955) Histological studies on the role of pectic enzymes in the development of Fusarium wilt symptoms in the tomato. Phytopathology 45:524-527
Pollock CJ and Drysdale RB (1976) The role of phenolic compounds

in the resistance of tomato cultivars to Verticillium albo-atrum. Phytopath Z 86:56-66
Porter CL and Green RJ (1952) Production of exotoxin in genus Verticillium. Phytopathology 42:472 (Abstr)
Puhalla JE and Howell CR (1975) Significance of endopolygalacturonase activity to symptom expression of Verticillium wilt in cotton, assessed by the use of mutants of Verticillium dahliae. Physiol Plant Path 7:147-152
Rai RV and Strobel GA (1969) Phytotoxic glycopeptides produced by Corynebacterium michiganense. II Biological properties. Phytopathology 59:53-57
Rayle DL (1973) Auxin-induced hydrogen ion secretion in Avena coleoptiles and its implications. Planta 114:63-73
Reese ET, Maguire AH and Parrish FW (1967) Glucosidases and exoglucanases. Can J Biochem 46:25-34
Ries SM and Strobel GA (1972a) A phytotoxic glycopeptide from cultures of Corynebacterium insidiosum. Plant Physiol Lancaster 49:676-684
Ries SM and Strobel GA (1972b) Biological properties and pathological role of a phytotoxic glycopeptide from Corynebacterium insidiosum. Physiol Plant Path 2:133-142
Riov J, Monselise SP and Kahan RS (1969) Ethylene induction of phenylalanine ammonia lyase in citrus fruit peel. Plant Physiology 44:631-635
Salemink CA, Rebel H, Kerling LCP and Tchernoff V (1965) Phytotoxin isolated from liquid cultures of Ceratocystis ulmi. Science 149:202-203
Scheffer RJ and Elgersma DM (1981) Detection of a phytotoxic glycopeptide produced by Ophiostoma ulmi in elm by enzyme-linked immuno-specific assay (ELISA). Physiol Plant Path 18:27-32
Scheffer RP and Walker JC (1953) The physiology of Fusarium wilt of tomato. Phytopathology 43:116-125
Scholander PF, Rudd B and Leivestad H (1957) The rise of sap in a tropical liana. Plant Physiology 32:1-6
Selman IW and Pegg GF (1957) An analysis of the growth response of young tomato plants to infection by Verticillium albo-atrum. Ann Appl Biol 45:675-681
Sequeira L (1963) Growth regulators in plant disease. Ann Rev Phytopathol 1:5-30
Sequeira L (1965) Origin of indoleacetic acid in tobacco plants infected by Pseudomonas solanacearum. Phytopathology 55:1232-1236
Sequeira L (1969) Synthesis of scopolin and scopoletin in tobacco plants infected by Pseudomonas solanacearum. Phytopathology 59:473-478
Sequeira L (1973) Hormone metabolism in diseased plants. Ann Rev Plant Phys 24:353-380
Sequeira L and Williams PH (1964) Synthesis of indoleacetic acid by Pseudomonas solanacearum. Phytopathology 54:1240-1246
Singh D, Brinkerhoff LA and Guinn G (1971) Effect of alanine on development of Verticillium wilt in cotton cultivars with different levels of resistance. Phytopathology 61:881-882
Singh D and Smalley EB (1969) Changes in amino acid and sugar constituents of the xylem sap of American Elm following inoculation with Ceratocystis ulmi. Phytopathology 59:891-896
Smidt M and Kosuge T (1978) The role of indole-3-acetic acid

accumulation by alpha-methyl tryprophan-resistant mutants of Pseudomonas savastanoi in gall formation on oleanders. Physiol Plant Path 13:203-214

Spencer JFT and Gorin PAJ (1961) The occurrence in the host plant of physiologically-active gums produced by Corynebacterium insidiosum. Canad J Microbiol 7:185-188

Straley CS, Straley ML and Strobel GA (1974) Rapid screening for bacterial wilt resistance in alfalfa with a phytotoxic glycopeptide from Corynebacterium insidiosum. Phytopathology 64:194-196

Stamm AJ and Wagner E (1961) Determining the distribution of interstructural openings in wood. For Prod J 11:141-144

Steadman JR and Sequeira L (1969) A growth inhibitor from tobacco and its possible involvement in pathogenesis. Phytopathology 59:499-503

Stevenson KJ, Slater JA and Takai S (1979) Cerato-ulmin a wilting toxin of Dutch Elm disease fungus. Phytochem 18:235-238

Stoddart JL and Carr AHJ (1966) Properties of wilt toxins produced by Verticillium albo-atrum Reinke and Berth. Ann Appl Biol 58:81-92

Takai S (1974) Pathogenicity and cerato-ulmin production in Ceratocystis ulmi. Nature 252:124-126

Takai S, Richards WC and Stevenson KJ (1983) Evidence for the involvement of cerato-ulmin, the Ceratocystis ulmi toxin, in the development of Dutch elm disease. Physiol Plant Path 23:275-280

Talboys PW (1968) Water deficits in vascular disease. In: Kozlowski T (ed) Water Deficits and Plant Growth. Vol II. Acad Press, New York, 255-311

Talboys PW and Busch LV (1970) Pectic enzymes produced by Verticillium species. Trans Brit Mycol Soc 55:367-381

Talboys PW (1978) Simulation of the symptoms of hop Verticillium wilt by model toxins. In: Pathological Wilting of Plants. Univ Madras, India, 231-246

Talmadge KW, Keegstra K, Bauer WD and Albersheim P (1973) The structure of plant cell walls. I.The macromolecular components of the walls of suspension cultured sycamore cells with a detailed analysis of the pectic polysaccharides. Plant Physiology 51:158-173

Taylor JB and Flentje NT (1968) Infection, recovery from infection and resistance of apricot trees to Verticillium alboatrum. New Zeal J Bot 6:417-426

Trione EJ (1960) Extracellular enzyme and toxin production by Fusarium oxysporum f. sp. lini. Phytopathology 50:480-482

Van Alfen NK (1982) Wilt concepts and mechanisms. In: Mount MS and Lacey GH (eds) Phytopathogenic Prokaryotes Vol 1. Acad Press, New York

Van Alfen NK and Allard-Turner V (1979) Susceptibility of plants to vascular disruption by macromolecules. Plant Physiology 63:1072-1075

Van Alfen NK and McMillan BD (1982) Macromolecular plant-wilting toxins: Artifacts of the bioassay method. Phytopathology 72: 132-135

Van Alfen NK, McMillan BD, and Dryden P (1987a) The multicomponent extracellular polysaccharide of Clavibacter michiganensis subsp. insidiosum. Phytopathology 77:496-501

Van Alfen NK, McMillan BD and Wang Y (1987b) Properties of the extracellular polysaccharide of Clavibacter michiganensis subsp. insidiosum that may affect pathogenesis. Phytopathology 77:501-505

Van Alfen NK and Turner NC (1975a) Changes in alfalfa stem conductance induced by Corynebacterium insidiosum. Plant Physiology 45:304-309

Van Alfen NK and Turner NC (1975b) Influence of a Ceratocystis ulmi toxin on water retention of elm (Ulmus americana). Plant Physiology 55:312-316

VanderMolen GE, Labavitch JM, Strand LL and DeVay J (1983) Pathogen-induced vascular gels: ethylene as a host intermediate. Physiol Plant 59:573-580

VanderMolen GE, Beckman CH and Rodehorst E (1987) The ultrastructure of tylose formation in resistant banana following inoculation with Fusarium oxysporum f. sp. cubense. Physiol Mole Plant Path 31:185-200

Waggoner PE and Dimond AE (1956) Polyphenol oxidases and substrates in potato and tomato stems. Phytopathology 46:495-497

Wallis FM and Truter SJ (1978) Histolopathology of tomato plants infected with Pseudomonas solanacearum with emphasis on ultrastructure. Physiol Plant Path 13:307-317

Wallis FM, Rijkenberg FHJ, Joubert JJ and Martin MM (1973) Ultrastructural histopathology of cabbage leaves infected with Xanthomonas campestris. Physiol Plant Path 3:371-379

Wellman FL (1941) Epinasty of tomato, one of the earliest symptoms of Fusarium wilt. Phytopathology 31:281-283

Weise MV and DeVay JE (1970) Growth regulator changes in cotton associated with defoliation caused by Verticillium albo-atrum. Plant Physiology 45:304-309

Wilhelm S and Taylor JB (1965) Control of Verticillium wilt of olive through natural recovery and resistance. Phytopathology 55:311-361

Wood RKS (1961) Verticillium wilt of tomatoes - the role of pectic and cellulolytic enzymes. Ann Appl Biol 49:120-139

Wright STC and Hiron RWP (1969) (+)-Abscisic acid, the growth inhibitor induced in detached leaves by a period of wilting. Nature 224:719-720

Xu P, Leong SA and Sequeira L (1987) Molecular cloning of genes encoding virulence in Pseudomonas solanacearum. Phytopathology 77:1772 (Abstr)

Young DH and Pegg GF (1981) Purification and characterisation of 1,3- glucan hydrolases from healthy and Verticillium albo-atrum-infected tomato plants. Physiol Plant Path 19:391-417

Young DH and Pegg GF (1982) The action of tomato and Verticillium albo-atrum glycosidases on the hyphal wall of V. albo-atrum. Physiol Plant Path 21:411-423

Zel'tser SS and Avazkhodzhaev MK (1973) The characteristics of the polysaccharide fractions from the cultural liquid of the cotton wilt causal agent. Sel'skokhoz Biol 8:66-69

Zel'tser SS and Malysheva KM (1966) Carbohydrate metabolites extracted from Verticillium in a culture medium Vilt Khlopkovodstva 96-100. Chem Absts (1968) 8:10542

Zimmerman MH and Brown CL (1971) Trees, Structure and Function. Springer, Berlin New York

BIOLOGY AND HOST-PARASITE RELATIONS OF FUSARIUM OXYSPORUM F. SP. RADICIS-LYCOPERSICI

Nicole Benhamou[1], P.-M. Charest[1] and W.R. Jarvis[2]
Department de Phytologie
Université Laval
Ste. Foy, Quebec, G1K 7P4
Canada

INTRODUCTION

Fusarium crown and root rot (or foot and root rot) was first recognized in tomatoes (Lycopersicon esculentum L.) in plastic-covered greenhouses in the Ono and Kamiiso districts of Japan in 1969, on 25 of 74 farms (33%) surveyed, with complete crop loss in some cases. By 1971, 15 of 30 greenhouses (50%) were affected in Ono (Sato and Araki 1974). In retrospect, Sato and Araki thought that the disease had appeared first in 1963. It then appeared in staked field tomatoes in California in 1971 (Leary and Endo 1971) and the causal agent was identified as a new race of Fusarium oxysporum Schlecht. f.sp. lycopersici Sacc.: Snyder & Hansen (FOL) which attacked cultivars of tomato resistant to races 1 and 2 (races 0 and 1 sensu Gabe 1975) of F. oxysporum f.sp. lycopersici (Sanchez et al 1975). In Japan, the causal organism was also identified as F. oxysporum f.sp. lycopersici (Ogura and Ban 1971a, 1971b).

SYMPTOMS

In the greenhouse, the disease is characterized by a sudden wilt just before the first fruits are ready for picking, espe-

2. Agriculture Canada, Research Station
Harrow, Ontario, N0R 1G0
Canada

NATO ASI Series, Vol. H28
Vascular Wilt Diseases of Plants
Edited by E.C. Tjamos and C. Beckman

cially on sunny days. Affected plants may recover to some extent at night or on overcast days, or when the fruit load is reduced by picking. Nevertheless, in the long run, affected plants die (Farley 1975; Jarvis et al 1975; Sato and Araki 1974; Yamamoto et al 1974). In field plants, Sonoda (1976, 1979) recognized a fast-wilt resulting in early death, and a slow-wilt with progressive acropetal leaf death. Plants with slow-wilt survived until the end of the season and produced a flush of new growth after most of the fruits were picked.

Other symptoms in mature greenhouse crops include a chocolate-brown cortical rot at soil level, often not apparent until the outer layers of stem are sliced off; a reddish-brown vascular discolouration extending upwards in the stem for 5-10 cm, and usually no more than 25 cm; severe rotting or loss of the primary seminal root, and numerous small grey-brown lesions at the point of emergence of lateral roots; proliferation of adventitious roots above the stem lesions; thinning of the stem apex; upper leaves wilting first, followed by the lower leaves, with golden-yellow discolouration from the tip, and eventually leaf death (Jarvis and Shoemaker 1978). Fruit from affected plants is flaccid and lacks its normal bright colour.

In sterilized and heavily-reinfested soil, young seedlings are severely affected, toppling over and resembling damped-off seedlings attacked by Pythium spp. (Jarvis et al 1975).

The fungus can be isolated only from within a few millimetres of the edge of the discolouration in the stem, in contrast with the condition in Fusarium wilt (Walker 1971). Occasionally, but not usually, dead or near-dead plants bear conspicuous external masses of pinkish-white mycelium and spores.

GEOGRAPHICAL DISTRIBUTION

Following the initial reports from California and Japan, the pathogen was next found in 1973 in Ohio (Farley 1975; Farley et al 1975), Ontario (Jarvis et al 1975) and again in Japan (Sato and Araki 1974; Yamamoto et al 1974). Since then, it has been reported in the U.S.A. from Florida (Sonoda 1976), New Hampshire (Nutter et al 1978). Texas (Barnes and Martyn 1985) and Pennsylvania (Ricker 1987); in Canada, from British Columbia (Anonymous 1975) and Quebec (Thibodeau and Simard 1978); and from Mexico (George in Rowe 1980); Crete (Malathrakis et al 1980); Italy (Tammienti and Lento 1986); Israel (Krikun et al 1983); France (Couteaudier et al 1985); Netherlands (Ester and Gerlagh 1986); Belgium (Vanachter et al 1983); and from U.K. (J.T. Fletcher, personal communication) (Fig. 1). Rowe (1980) showed that isolates from greenhouse and field tomatoes caused the same disease.

Fig. 1. World distribution of Fusarium oxysporum f.sp. radicis-lycopersici, 1988.

TAXONOMY

All investigators have agreed that the causal fungus is Fusarium oxysporum Schlecht., indistinguishable in morphology and cultural characteristics from other isolates of F. oxysporum (Booth 1971; Gordon 1965), but it is distinguishable in its usual close association with Lycopersicon spp., and in the symptoms it causes on them. Sato and Araki (1974), Yamamoto et al (1974) and Rowe (1980) found that it was pathogenic on cultivars of L. esculentum resistant to races 1 and 2 of FOL. In order to differentiate between the two pathogens, Rowe used four cultivars; Vendor (susceptible to race 1 of FOL (R_1), race 2 (R_2) and crown rot (CR); Ohio MR13 (Res-R_1, Susc-R_2, CR); Walter (Res-R_1R_2, Susc-CR); IRB-301-31 (a Japanese breeding line; Yamakawa and Nagata 1975) (Res-CR, Susc-R_1R_2), while Yamamoto et al (1974) used three differential cultivars in a similar way; Pondorosa (susceptible to races J1 and J2 of FOL), Okitsu 3 (Res-J1, Susc-J2) and Walter (Res-J1, J2) which were all susceptible to the crown rot pathogen. Thus Yamamoto et al (1974) designated it race J3, following the Japanese terminology of FOL races J1 and J2, both wilt pathogens.

The crown and root rot disease has an optimum temperature of about 15-18 C (Jarvis and Thorpe 1976a; Sato and Araki 1974; Sonoda 1976; Yamamoto et al 1974) in contrast to the 27 C optimum for FOL wilt expression (Walker 1971). Further, the symptoms caused by FOL and the new fungus are quite distinct, and the pathogen can be isolated only close to the lesion, whereas FOL can be isolated from high in the wilted plants (Jarvis and Shoemaker 1978). These considerations led Jarvis and Shoemaker (1978) to propose a new forma specialis: F. oxysporum Schlecht. f.sp. nov. radicis-lycopersici Jarvis and Shoemaker (FORL) to denote its distinction from FOL, its damage to roots and its close association with Lycopersicon. Leary and Sanchez (1976) and Sanchez et al (1975) found FORL unusual in its continued segregation of variants when the wild type was cloned by single-spore isolation (Leary 1972). Leary and Sanchez (1976) noted that culturing the pathogen on rich media such as potato-dextrose agar resulted in loss of pathogenicity, and they recommended its passage through the host, or starving it in distilled water agar at 4 C for one week, to restore pathogenicity. FORL has been assigned its own single-member vegetative compatibility group in Puhalla's (1985) classification of strains of F. oxysporum, and to a majority classes A of nit 1 and B of Nit M nitrate reductase loci in F. oxysporum ff.spp. (Correll et al 1987).

Belhadj et al (1987) found differences in protein composition between FOL and FORL in the over-100 kd range, as well as differences in composition of FOL and FORL from liquid cultures at 3, 4 and 5 days of growth, suggesting that the FOL and FORL have a specific physiological state of growth for protein synthesis, according to age. FORL is more tolerant of mycostatin that FOL, and FORL contains about three times more $C_{18:0}$ and twice as much $C_{18:1}$ fatty acids than FOL, whereas FOL contains twice as much $C_{16:1}$ than FORL (Madosingh and Starratt 1987).

Lectin agglutination provided evidence of three sugars on the surface of microconidia, N-acetyl-glucosamine, N-acetyl-galactosamine and mannose (Boyer et al 1986), which are probably lost in conventional cytochemical tests.

A. Boyer and P.-M. Charest (unpublished) found quantitative differences in these residues between microconidia of FOL and FORL, depending on lectin concentration in the agglutination test. The sugars were detected at the surface of the microconidium cell wall whereas N-acetyl-glucosamine residues occur here in the outer electron-dense layer of the microconidium wall, they occur only in the internal wall layer in hyphae (Chamberland et al 1985).

HISTOPATHOLOGY

Using axenic methods (Charest et al 1983a), Charest (1983a) and Charest et al (1984) found that at 22-24 C FORL colonized the root epidermis in the susceptible cultivars Bonny Best and Vendor within 12-24 hours and was usually restricted to the outer cortical zone for 96 hours. The fungus was both inter- and intracellular. Host cell walls were thickened and papillae formed in the cortex but not in the endodermis. Disintegration of host cytoplasm and cell walls was complete in the cortex within 144 hours. Invasion of the endodermis began 96-120 hours after inoculation, extending into the stele by 144 hours (Table 1). Similar observations were made by Brammall (1986) and Brammall and Higgins (1985).

Table 1. Events in the colonization of susceptible tomato roots by _Fusarium_ _oxysporum_ f.sp. _radicis-lycopersici_ at 20-22 C (Brammall 1986; Charest et al 1984)

Time after inoculation in hours	Event
12-24	Colonization of epidermis
24-48	Colonization of outer cortex; fungus both intra- and intercellular. Host wall thickening
96-120	Colonization of endodermis
96-144	Cytoplasm and walls of host cells disintegrate
144 +	Colonization of stele; cytoplasm and walls of stelar parenchyma disintegrate; hyphae invade vessels through pit membranes and middle lamellae

Cell invasion occurred through middle lamellae and pit membranes. Tyloses were found in the vessels, extending from vessel to vessel. The vessels also contained fibrillogranular material interspersed with bubbles and an osmiophilic coating material, rather similar to the condition in Fusarium wilt caused by _FOL_. Ultrastructurally, the pathogen appeared as large irregular or amoeboid bodies, multinucleate and with regular organelles, but apparently lacking a continuous cell wall (Charest 1983; Charest et al 1983). In intercellular spaces, these bodies were surrounded by partly-disintegrated host cytoplasm or bands of osmiophilic material similar to that in the vessels. There is, apparently, simultaneous disruption of both host and parasite. The use of gold labeling (Chamberland et al 1985) enabled chitin to be identified in structure associated with fungal wall formation or degradation (Charest 1983; Charest et al 1984; Chamberland et al 1985). In addition to locating the position of fungal cell wall material in encasing walls of internal cells, and of wall thickenings (Charest et al 1984), gold labeling also indicated the presence of chitin in fungal microbodies that resembled Woronin bodies, and which may have a role in plugging septa (Chamberland et al 1985).

Gold labeling also enabled Benhamou and Charest (1986) to identify sialic acid and fucose, two sugars hitherto unknown in a pathogenic fungus, and it also enabled Chamberland et al (1987) to identify galactose, N-acetyl-galactosamine and glucosides in paramural vesicles in infected tomato root cells. By contrast, Benhamou et al (1987) could find no cellulosic glucans by that technique in FORL cell walls; they suggested, however, that specific sugars might be involved in host defence reactions (Benhamou et al 1988).

The electron-dense layer in intercellular spaces and vessels was identified by Chamberland et al (1986) as host pectinesterase, concentrated in the middle lamella, wherever the pectic materials had been altered. It also occurred in the bordered pits.

In the resistant cultivar CR6, Brammall (1986) and Brammall and Higgins (1985), found that the colonization of the crown tissue rarely proceeded into the cortex, and that colonization was limited in the root cortex. Inter- and intracellular colonization of the epidermis was common in CR6, and macroconidia and chlamydospores occurred on affected tissues. Colonization stopped at the suberized exodermis, whereas in the susceptible cv. MR13 colonization proceeded beyond it.

By using sub-lethal concentrations of the herbicide glyphosate, which disrupts phenolic metabolism, Brammall (1986) was able to induce susceptibility in the resistanct cv. CR6, apparently because the incorporation of phenolic lignin precursors into callosic wall appositions near colonizing fungal cells was inhibited. The fungus was able to grow through the apposition. Glyphosate also induced susceptibility to Fusarium solani f.sp. pisi, normally not a pathogen of tomato. Glyphosate-induced susceptibility declined with increasing delay of application after inoculation. It therefore seems that, it part, resistance is a mechanical function of lignified papillae and cell walls, although the glyphosate effect was not reversed by L-phenylalanine as it has been in other cases. Papillae and cell-wall appositions also occur in susceptible cultivars in response to infection by

FORL, but perhaps too late to form an effective barrier (Brammall 1986); the difference between susceptibility and resistance seems to be a quantitative one in the elicitation of a defense response and its implementation.

CONCLUSIONS

Though morphologically indistinguishable from Fusarium oxysporum f.sp. lycopersici, F. oxysporum f.sp. radicis-lycopersici is physiologically and pathogenically distinct. It differs principally in its mode of pathogenesis on tomato, its principle but not sole host, in its temperature relations for pathogenesis, and in its lack of systemic invasion of the whole vascular system. FORL appears to be unusual among other formae speciales of F. oxysporum in that it is readily dispersed by air-borne microconidia, and is an efficient coloniser of sterile soils. It has been effectively dispersed in the northern hemisphere. The two formae specialies differ physiologically in protein composition; FORL is more tolerant to mycostatin than FOL; it contains a different complement of fatty acids; and has unusual sugars on its microconidia.

REFERENCES

Anon (1975) Soil-borne greenhouse diseases. Cross-Canada Report 1628:9-10

Barnes LW and Martyn RD (1985) Fusarium crown and root rot of greenhouse tomatoes in Texas. Plant Disease 69:1100

Belhadj R, Charest P-M and Theriault G (1987) Electrophoretic comparison of the protein composition of two closely related fungi: Fusarium oxysporum f.sp. radicis-lycopersici (FORL) and F. oxysporum f.sp. lycopersici (FOL). Can J Plant Pathol 9:272-273 (Abstr)

Benhamou N, Chamberland H, Ouellette GB and Pauzé FJ (1987) Ultrastructural localization of β-(1⟶4)-D-glucans in two pathogenic fungi and in their host tissues by means of exoglucanase-gold complex. Can J Microbiol 33:405-417

Benhamou N, Chamberland H, Ouellette GB and Pauzé FJ (1988) Detection of galactose in two fungi causing wilt diseases and

their plant host tissues by means of gold-complexed Ricinus communis agglutin. Physiol Mol Plant Pathol 33: (in Press)
Benhamou N and Charest P-M (1986) Mise en évidence ultrastructurale d'acide sialique et de fucose chez deux ascomycètes phytopathogènes par l'intermédiaire de lectines complexées à l'or colloidal. Phytoprotection 67:21-23
Booth C (1971) The genus Fusarium. Commonwealth Mycological Institute, Kew
Boyer A, Lafontaine JG and Charest P-M (1986) Étude cytologique pariétale des microconidies du Fusarium oxysporum f.sp. lycopersici (FOL) et du F. oxysporum f.sp. radicis-lycopersici (FORL) à l'aide de lectines. Phytoprotection 67:144
Brammall RA (1986) Host-parasite interactions in Fusarium crown and root rot disease of tomato. PhD Thesis, University of Toronto, Ontario
Brammall RA and Higgins VJ (1985) Histopathological observations of fungal colonization of tomato cultivars resistant or susceptible to Fusarium crown and root rot diseases. Can J Plant Pathol 7:442 (Abstr)
Brammall RA and Higgins VJ (1987) The control of Fusarium crown and root rot disease of tomato by root inoculation with alien pathogenic fungi. Can J Plant Pathol 9:274 (Abstr)
Chamberland H, Benhamou N, Ouellette GB and Pauzé FJ (1987) Detection of sugars in paramural vesicles induced in tomato root cells infected by Fusarium oxysporum f.sp. radicis-lycopersici (FORL) as revealed by gold-complexed enzymes and lectins. Can J Plant Pathol 9:274-275 (Abstr)
Chamberland H, Charest P-M, Ouellette GB and Pauzé FJ (1985) Chitinase-gold complex used to localize chitin ultrastructurally in tomato root cells infected by Fusarium oxysporum f.sp. radicis-lycopersici, compared with a chitin specific gold-conjugated lectin. Histochem J 17:313-321
Chamberland H, Charest P-M, Pauzé FJ and Ouellette GB (1986) Localisation ultrastructurale de la pectine-estérase de la tomate dans les racines du Lycopersicon esculentum infectées par Fusarium oxysporum f.sp. radicis-lycopersici (FORL). Phytoprotection 67:145 (Abstr)
Chamberland H, Ouellette GB and Pauzé FJ (1983) Use of gold labeling technique with chitin-specific lectins and chitinase to follow development of Fusarium oxysporum f.sp. radicis-lycopersici in tomato roots. Can J Plant Pathol 5:202 (Abstr)
Charest P-M (1983) Études anatomiques et ultrastructurales de Fusarium oxysporum f.sp. radicis-lycopersici Jarvis and Shoemaker dans les tissus de son hôte (Lycopersicon esculentum Mill.) et en culture pure. These PhD, Université Laval, Quebec
Charest P-M, Miville L and Pauzé FJ (1983) Note sur une technique simple pour l'obtention de plantules stériles et son application pour des études en histopathologie. Phytoprotection 64:31-34
Charest P-M, Ouellette GB and Pauzé FJ (1983) Fine structure of Fusarium oxysporum f.sp. radicis-lycopersici in root tissues of tomato. Can J Plant Pathol 5:202-203
Charest P-M, Ouellette GB and Pauzé FJ (1984) Cytological observations of early infection process by Fusarium oxysporum f. sp. radicis-lycopersici in tomato plants. Can J Bot 62:1232-1244

Charest P-M, Pauzé FJ and Ouellette GB (1981) Études ultrastructurales des interactions entre Fusarium oxysporum f.sp. radicis-lycopersici et la tomate Lycopersicon esculentum. Rep Can Hort Council, Ottawa, p 174

Correll JC, Klittich CJR and Leslie JF (1987) Nitrate nonutilising mutants of Fusarium oxysporum and their use in vegetative compatibility tests. Phytopathology 77:1640-1646

Couteaudier Y, Alabouvette C and Soulas M-L (1985) Nécrose du collet et pourriture des racines de tomate. Rev Hortic (Paris) 254:39-42

Endo RM, Leary JV and Sims KF (1973) The epidemiology and control of Fusarium crown rot of tomatoes. Proc 2nd Int Congr, Plant Pathology (Abstr 0428)

Ester A and Gerlagh M (1986) Resistentieveredeling tegen Fusariumvoetrot bij tomaat. Zaadbelangen 40:12-13

Farley J (1975) Identified: New tomato root and crown rot disease. Am Veg Grower 23(10):30

Farley J, Oakes G and Jaberg C (1975) A new greenhouse tomato root-rot disease caused by Fusarium oxysporum: a preliminary report. Research Summary 82, Ohio Agric Res Devel Center, Wooster, Ohio, pp 27-29

Gabe HL (1975) Standardization of nomenclature for pathogenic races of Fusarium oxysporum f.sp. lycopersici. Trans Br Mycol Soc 64:156-159

Gordon WL (1965) Pathogenic strains of Fusarium oxysporum. Can J Bot 43:1309-1318

Jarvis WR and Shoemaker RA (1978) Taxonomic status of Fusarium oxysporum causing foot and root rot of tomato. Phytopathology 68:1679-1680

Jarvis WR and Thorpe HJ (1976) Susceptibility of Lycopersicon species and hybrids to the foot and root rot pathogen Fusarium oxysporum. Plant Dis Reporter 60:1027-1031

Jarvis WR, Thorpe HJ and MacNeill BH (1975) A foot and root rot disease of tomato caused by Fusarium oxysporum. Can Plant Dis Surv 55:25-26

Krikun J, Nachmias A, Offenbach R and Ucko O (1983) Establishment of the crown and root rot pathogen of tomato in the Negev region. Phytoparasitica 11:3-4

Leary JV and Endo RM (1971) A fusarium induced root rot of staked tomatoes. Phytopathology 61:900

Leary JV and Sanchez LE (1976) Mutagenesis of Fusarium oxysporum: analysis of the fungal genome in relation to the expression and regulation of pathogenicity. In: Tomiyama K, Daly JM, Uritani I, Oku H and Ouchi S (eds) Biochemistry and cytology of plant-parasite interaction. Kodansha, Tokyo, Chap 2.1 pp 12-15

Madosingh C and Starratt AN (1987) Comparison of antibiotic tolerance, lipids and respiration in the tomato pathogens Fusarium oxysporum lycopersici and F. oxysporum radicis-lycopersici. J Environ Sci Health B22:731-749

Malathrakis NE, Xaphis CT and Avgelis AD (1980) Tomato diseases in Crete. Phytopathol Medit 19:71

Nutter FW, Warren CG, Wells OS and MacHardy WE (1978) Fusarium foot and root rot of tomato in New Hampshire. Plant Dis Reporter 62:976-978

Ogura H and Ban M (1971a) Fusarium oxysporum causes tomato wilt disease. I. Factors affecting the appearance of tomato wilt disease accompanied by root rot. Res Rep Kochi University, Agric Sci 20:61-69

Ogura H and Ban M (1971b) Fusarium oxysporum causes tomato wilt disease. II. Presence of F. oxysporum causes tomato wilt disease accompanied by root rot. Res Rep Kochi University, Agric Sci 20:71-77

Puhalla JE (1985) Classification on strains of Fusarium oxysporum on the basis of vegetative compatibility. Can J Bot 63:179-183

Ricker MD (1987) Fusarium crown and root rot of tomato in Pennsylvania. Plant Disease 71:469

Rowe RC (1980) Comparative pathogenicity and host ranges of Fusarium oxysporum isolates causing crown and root rot of greenhouse and field-grown tomatoes in North America and Japan. Phytopathology 70:1143-1148

Rowe RC and Coplin DL (1976) Dispersal of Fusarium oxysporum in tomato greenhouses. Proc Am Phytopath Soc 2:88 (Abstr)

Rowe RC, Farley JD and Coplin DL (1977) Air-borne spore dispersal and recolonization of steamed soil by Fusarium oxysporum in tomato greenhouses. Phytopathology 67:1513-1517

Sanchez LE, Endo RM and Leary JV (1975) A rapid technique for identifying the clones of Fusarium oxysporum f.sp. lycopersici causing crown- and root-rot of tomato. Phytopathology 65:726-727

Sato R and Araki T (1974) On the tomato root-rot disease occurring under vinyl-house conditions in southern Hokkaido. Annu Rep Soc Plant Protect N Japan 25:5-13

Sonoda RM (1976) The occurrence of Fusarium root rot of tomatoes in South Florida. Plant Dis Reporter 60:271-274

Sonoda RM (1979) Observations on Fusarium crown rot of tomato in winter of 1978-79. Res Rep Fort Pierce Fla Agraic Res Center RL 1979-10

Tammienti G and Lento G (1986) Il marciume basale de pomodoro da Fusarium oxysporum f.sp. radicis-lycopersici. Inf Fitopatol 11:59-61

Thibodeau PO and Simard M (1978) Evaluation de la résistance de cultivars de tomate de serre à la fusariose. Can Hort Council Res Rep 1978:172

Vanachter A, Van Wanbeke E and Van Assche E (1983) Potential danger for infection and spread of root disease of tomatoes in hydroponics. Acta Hortic (The Hague) 133:119-128

Walker JC (1971) Fusarium wilt of tomato. Monograph 6. American Phytopathological Society, Minneapolis

Yamakawa K and Nagata N (1975) Three tomato lines obtained by the use of chronic gamma radiation with combined resistance to TMV and Fusarium race J-3. Techn News, Inst Radiat Breeding 16

Yamamoto I, Kamada H, Kuniyasu K, Saito M and Ezuka A (1974) A new race of Fusarium oxysporum f.sp. lycopersici inducing root rot of tomato. Proc Kansai Plant Protect Soc 16:17-29

DIAGNOSIS OF BACTERIAL INFECTIONS BY IMMUNOLOGICAL METHODS

A. Sletten
Norwegian Plant Protection Institute
Division of Plant Pathology
P.O. Box 70 N-1432 As-NLH
Norway

INTRODUCTION

Plant diseases caused by bacteria may be difficult to diagnose. Correct diagnosis often cannot be made on the presence of symptoms alone, since similar pathological conditions may be caused by quite different bacterial pathogens. Thus, isolation and subsequent identification of the bacteria by means of different biochemical tests and inoculation of plants may be necessary to fulfil the requirements of Koch's postulates. Such methods may be rather time-consuming, and in an effort to simplify and speed up the detection and identification of phytopathogenic bacteria, the use of different immunological methods have come into widespread use. For many years traditional techniques, such as agglutination and immunodiffusion were the only ones applied, but in recent years new methods like fluorescent immunoassay, FIA, and enzyme-linked immunosorbent assay, ELISA, have been developed for use in the diagnosis and study of many bacterial diseases. Fluorescent immuno assay (FIA), and enzyme-linked immunosorbent assay (ELISA), were evaluated for routine diagnosis and research on the etiology of potato ring rot, (Clavibacter michiganensis subsp. sepedonicus), leaf spot and blight of begonia, (Xanthomonas campestris pv. begonia), pelargonium, (X. c. pv. pelargonii), and bacterial wilt of forage grasses (Xanthomonas campestris pv. graminis).

For a number of years we have been using FIA, recently also ELISA in our laboratory as diagnostic tools, and this report is a presentation of our experiences with these methods.

NATO ASI Series, Vol. H28
Vascular Wilt Diseases of Plants
Edited by E. C. Tjamos and C. Beckman

MATERIAL AND METHODS

Antisera were produced in rabbits from live bacterial cells. Bacteria were grown on nutrient glycerol agar slants, washed three times in phosphate-buffered saline (PBS), and injected subcutaneously at weekly intervals for five weeks. Immunoglobulin was fractioned partly by the method of Cherry (1970) and partly according to Harboe and Ingild (1983). In FIA studies the indirect method was used (Sletten 1985). For ELISA, antibodies were conjugated to alkaline phosphatase and the assay carried out by a slight modification of the method of Clark and Adams (1977).

Bacterial samples were prepared from agar slants, and washed three times in PBS. Plant samples were collected from naturally infected material, and from infection experiments. Plant tissue extracts were prepared from potato tubers according to Lelliott et al (1976) and from other plant parts by squeezing a small sample in a hand-press. Plant debris was allowed to settle before preparing slides for FIA or ELISA-plates.

RESULTS

Potato ring rot

The disease may be difficult to diagnose solely on symptoms and Gram-staining of tissue extracts. Additional isolation on nutrient media and further testing may be necessary, which is very time-consuming due to the slow growth of the bacterium. Tissue extracts may also be inoculated directly into eggplants (<u>Solanum melanogena</u>), which develop wilting symptoms within 10-40 days, depending on the concentration of the bacterium. However, by the use of FIA, diagnosis can be carried out within a couple of hours. <u>C. m.</u>subsp. <u>sepedonicus</u> was detected in tissue extracts from diseased tubers and vines. In tubers with symptoms the number of bacteria in the vascular bundles proved to be extremely high, but also in tubers without symptoms the concentration of bacteria could be considerable. Such latent infected tubers are difficult to detect by conventional methods, but by means of FIA large numbers of tubers could be screened easily in a short time, whereby low levels of infection were shown to persist for many generations of tubers (Sletten 1985). These ovservations led to the set-up of a ring rot eradication programme in Norway, in which seed potato lots are screened for infection by FIA and eggplants (Sletten 1987). FIA proved to be a very useful tool in studies of the ways of entry of the ring rot bacterium into the potato plant, the danger of it being transmitted by meristem-tip culture, and survival studies in soil and plant debris.

With a sample size of 10 µl the detection level of the ring rot bacterium could be set around 10^3 cells/ml. For routine use however, this level was increased by the power of ten. Cross-reacting bacteria were seldom found, except in badly decayed tubers, and samples contaminated with soil. In such cases positive samples should be checked by inoculating eggplants. In latent infected tubers cross-reacting bacteria were only present in low numbers, and could usually be discriminated by their less bright fluorescence, and a different cell shape and size compared with C. m. subsp. sepedonicus.

Leaf spot and blight of begonia and pelargonium, and bacterial wilt of forage grasses

Compared with C. m. subsp. sepedonicus, Xanthomonas campestris-pathovars grow fast on nutrient media, and may easily be detected by their yellow colony colour. However, the diagnosis of the different pathovars in the campestris-group can only be done with a host test, which may take considerable time, depending on how easy it is to grow the actual plant in the greenhouse.

In the study of the pathovars begonia, pelargonii and graminis and the etiology of the diseases they cause, FIA and ELISA made it possible quickly to detect the presence of these bacteria in tissue extracts from plants with or without disease symptoms. The detection level was 10^3 cells/ml both for FIA and ELISA. In ELISA, a positive sample would then have an optical density higher than 0.1 read at 405 nm. However, tissue extracts from strongly coloured begonia and pelargonium-varieties could give a high background in ELISA-readings, making it necessary to include sufficient control-samples when running an assay. Cross-reacting bacteria were rarely found, except in badly decayed plant tissue, and in samples contaminated with soil. Nor did any of the bacteria isolated by washing leaves from healthy and diseased plants react in FIA and ELISA-tests.

The three pathovars studied are mainly restricted to the vascular bundles, in addition pv. begonia and pelargonii make up a part of the leaf surface flora both on plants with or without symptoms. By the use of FIA, ELISA and plating of leaf-washings on nutrient agar, it could be shown that both bacteria had an uneven distribution within the plant. By examining one plant, a stem could be heavily infected, while in another no bacteria where dected in the vascular bundles. However, they could be present in large numbers on the surface of leaves on the same stem. These findings demonstrate that in the selection of mother-plants for the production of disease-free plantlets, great care should be taken in choosing a representative sample for indexing.

DISCUSSION

Diagnosing bacterial infections by FIA and ELISA may be a great improvement from more conventional techniques like isolations on nutrient media and pathogenicity testing. However, care should be taken to look into all interfering factors thoroughly when an assay system is set up, and in cases of doubt other tests than immunological should be included in the diagnosis. In particular, the specificity of the antibodies produced should be tested against bacteria that might contaminate the tissue extract to be investigated. Cross-reactions may be avoided by optimal dilution of the antiserum, adsorption with cross - reacting bacteria or purification of the antigen prior to immunization (Trigalet et al 1978).

ELISA is often considered less sensitive than FIA (Lopez et

al 1987), but in our trials with X. c. pv. begonia and pelargonii both methods could detect around 10^3 cells/ml. FIA is a simpler technique to perform, faster, and needs less expensive equipment than ELISA. However, when a large number of samples are assayed, ELISA is the method of choice because of its potential of automation. In addition, the results of the assay will be read as optical density, where positive-negative thresholds are set by the ELISA-reader, (Sutula et al 1986). Reading FIA however, will need personel trained in the use of a microscope, and the setting of a positive-negative threshold would be subjective.

REFERENCES

Cherry WB (1970) Fluorescent-antibody techniques. In: Blair JE et al (eds) Manual of Clinical Microbiology. Am Soc for Microbiol, pp 693-704

Clark MG and Adams AN (1977) Characteristics of the microplate method of enzyme-linked immunosorbent assay for the detection of plant viruses. J Gen Virol 34:475-483

Harboe NM and Ingild A (1983) Immunization, isolation of immunoglobulins and antibody titre determination. Scand J Immunol 17:345-351

Lelliott RA and Sellar PW (1976) The detection of latent ring rot (Corynebacterium sepedonicum (Spiek. et Kotth.) Skapt. et Burkh.) in potato stocks. EPPO Bulletin 6:101-106

Lopez MM, Cambra M, Aramburu JM and Bolinches J (1987) Problems of detecting phytopathogenic bacteria by ELISA. EPPO Bulletin 17:113-117

Sletten A (1985) The effect of Corynebacterium sepedonicum on symptoms and yield of four potato cultivars. Potato Res 28: 27-33

Sletten A (1987) Prospects for detecting and controlling potato ring rot in Norway. In: Civerdo EL et al (eds) Plant pathogenic bacteria. Proc VI Int Conf Plant Path Bacteria.M Nijhoff Publ 1987, p 884

Sutula CL, Gillett JM, Morrissey SM and Ramsdell DC (1986) Interpreting ELISA data and establishing the positive-negative threshold. Plant Disease 70:722-726

Trigalet A, Samson R and Coleno A (1978) Problems related to the use of serology in phytobacteriology. In: Proc IV Int Conf Plant Path Bacteria, pp 271-288. INRA Angers al 1987), but in our trials with *X. c.* pv. *begoniae* and *pelargonii* both methods could detect around 10^3 cells/ml. FIA is a simpler technique to perform, faster, and needs less expensive equipment than ELISA. However, when a large number of samples are assayed, ELISA is the method of choice because of its potential of automation. In addition, the results of this assay will be read as optical density, where positive-negative thresholds are set by the ELISA-reader (Sutula et al 1986); reading FIA however, will need personnel trained in the use of a microscope, and the setting of a positive-negative threshold would be subjective.

REFERENCES

Cherry WB (1970) Fluorescent antibody techniques. In: Lennette EH et al (eds) Manual of Clinical Microbiology. Am Soc Microbiol, pp 693-70[illegible]

Clark MF and Adams AN (1977) Characteristics of the microplate method of enzyme-linked immunosorbent assay for the detection of plant viruses. J Gen Virol 34:475-483

Harboe N and Ingild A (1973) Immunization, isolation of immunoglobulins and antibody titre determination. Scand J Immunol 1:345-351

Lelliott RA and Sellar PW (1976) The detection of latent ring rot (*Corynebacterium sepedonicum* (Spiek. et Kotth.) Skapt. et Burkh.) in potato stocks. EPPO Bulletin 6:101-106

López MM, Gambra M, Arambarri JM and [illegible] (1987) Problems of detecting phytopathogenic bacteria by ELISA. EPPO Bulletin 17(4):611-617

Sletten A (1985) The effect of *Corynebacterium sepedonicum* on symptoms and yield of four potato cultivars. Potato Res 28:27-33

Sletten A (1987) Prospects for detecting and controlling potato ring rot in Norway. In: Civerolo EL et al (eds) Plant Pathogenic Bacteria. Proc VI Int Conf Plant Path Bacteria. M Nijhoff Publ 1987, p 884

Sutula CL, Gillett JM, Morrissey SM and Ramsdell DC (1986) Interpreting ELISA data and establishing the positive-negative threshold. Plant Disease 70:722-726

Trigalet A, Samson R and Coleno A (1978) Problems related to the use of serology in phytobacteriology. In: Proc IV Int Conf Plant Path Bacteria, pp 271-288, INRA Angers

COLONIZATION AND PATHOGENESIS IN CHICKPEAS INFECTED BY RACES OF <u>FUSARIUM</u> <u>OXYSPORUM</u> F. SP. <u>CICERI</u>

R.M. Jiménez-Díaz, Maria J. Basallote-Ureba and Hava Rapoport
Departamento de Agronomia
Universidad de Córdoba
Apartado 3048
14080 Córdoba
Spain

INTRODUCTION

Three races of *Fusarium oxysporum* Schlecht. emend. Snyd. & Hans. f. sp. *ciceri* (Padwick) Snyd. & Hans (*Foc*), namely races 0, 5 and 6, infect chickpeas (*Cicer arietinum* L.) in Andalucia, southern Spain (Jiménez-Díaz et al unpublished). These races are different from races 1,2,3 and 4 described in India (Haware and Nene 1982). Race 0 induces a progressive foliar yellowing with vascular discolouration and late death of the plant (yellowing syndrome) mainly in local 'kabuli' cultivars, and it is not pathogenic to cultivar JG 62 which is the universal suscept to all other *Foc* races (Cabrera de la Colina et al 1985; Haware and Nene 1982; Trapero Casas and Jiménez-Díaz 1985). In contrast, races 5 and 6 induce a severe chlorosis and flaccidity, with vascular discolouration and early plant death (wilt syndrome), and they are highly virulent to most 'kabuli' and 'desi' cultivars (Cabrera de la Colina et al 1985; Jiménez-Díaz et al unpublished).

Fusarium wilt is controlled mainly by the use of resistant cultivars (Haware and Nene 1980; Nene and Reddy 1987). However, little is known about the nature of resistance as well as the processes of infection and colonization of susceptible and resistant plants by the pathogen. In our experience, resistant plants in all incompatible interactions remain symptomless although the pathogen can be recovered from root and collar tissues (Cabrera

NATO ASI Series, Vol. H28
Vascular Wilt Diseases of Plants
Edited by E. C. Tjamos and C. Beckman

de la Colina 1986; Trapero Casas and Jiménez-Díaz 1985). Ayyar and Iyer (1936) reported that roots of resistant plants have a thick layer of suberin in the periphery of the cortex, and that development of the fungus in the root after penetration was much slower in resistant plants as compared to susceptible ones, resulting in only a few infected vessels. Preliminary results with race 1 of *Foc* indicate that spore germination and mycelial growth of the fungus is stimulated by root exudates from susceptible cultivar JG 62 and inhibited by those from resistant CPS 1 (Haware and Nene 1984; Satyaprasad and Ramarao 1983). This paper presents results of research to determine and contrast the processes and histological modifications in the colonization of susceptible and resistant chickpea cultivars by races 0 and 5 of *Foc*.

MATERIALS AND METHODS

Chickpea cultivars P 2245, JG 62 and WR 315 were inoculated with single-spore isolates of races 0 and 5. P 2245 is susceptible and WR 315 is resistant to both races, while JG 62 is resistant to race 0 and susceptible to race 5. Inocula were increased in a cornmeal-sand (CMS) mixture (Nene and Haware 1980) at 23 ± 3 C for 12-14 days. Infested CMS was thoroughly mixed with an autoclaved soil mixture (1:12 w/w) and germinated seeds were sown in pots filled with that mixture. Control plants were grown in a comparable mixture of uninfested CMS and autoclaved soil. Experiments were carried out in a growth chamber adjusted to 26 ± 2 C/21 ± 2 C (day/night), 60-90% relative humidity, and a 14 hours photoperiod of fluorescent light at about 14 klux.

Colonization of the plant in compatible and incompatible interactions

Colonization by the pathogen was investigated by means of sequential isolations along the plant axis and histological observations with light microscope. For isolations, seven seedlings for each race-cultivar interaction were sampled at 4, 10, 15, 20 and 25 days after inoculation with race 5, and at 4, 10, 15, 20, 25, 30, 35 and 40 days after inoculation with race 0. The seedlings axes, free of lateral rootlets, cotyledons and leaves, were surface-disinfested, cut into 1-cm-lengths and plated onto a modified V8 juice-oxgall-PCNB agar (González-Torres 1985).

Histology

For histology, five seedlings of each race-cultivar interaction were sampled at 10, 15 and 20 days after inoculation with race 5, and at 10, 15, 20 and 40 days after inoculation with race 0. Pieces (5-mm-long) from the main root, and the first, 2nd and 5th stem internodes were fixed in FAA, dehydrated in a series of tertiary butyl alcohol (Johansen 1940) embedded in paraffin, sectioned longitudinally or transversely at 12-16 µm, and stained with tanic acid-ferric chloride-safranine-fast green (Basallote-Ureba 1987; Jensen 1962). Fifteen transverse sections were examined with phase contrast or bright field optics for the incidence (percentage of xylem bundles infected) and intensity (proportion of all vessel lumina in the xylem occupy by fungal structures) of colonization, as well as for the morphology of fungal structures and histological modifications.

RESULTS

Colonization of the plant in compatible and incompatible interactions

For the compatible interactions, wilt symptoms (cultivars P 2245, JG 62/race 5) developed by 15 days after inoculation and plants died 5-10 days later, while yellowing symptoms (cultivar P 2245/race 0) developed by 20-25 days after inoculation and plants died 10-15 days later. No symptoms developed for the incompatible interactions (cultivar WR 315/race 5, cultivars JG 62,

WR 315/race 0). The fungus grew out of plated tissue for all compatible and incompatible interactions from the first attempted isolations 4 days after inoculation, with colonies formed from tissues located at or near sites of cotyledon attachment, but never from the root apex. For compatible interactions, systemic colonization of plants by Foc, as indicated by fungal growth from the cut ends of plated tissues, occurred 10-15 days after inoculation. Afterwards race 5 colonized the plant axis faster than race 0. Thus, while growth of race 5 was discontinuous up to 15 days after inoculation and it took 20 days to fully colonize the plant axis, growth of race 0 was discontinuous up to 20 days after inoculation and it took 30 days to colonize the plant axis entirely. For the incompatible interactions, Foc never grew from the cut ends of plated tissues, indicating that systemic colonization did not occur.

Histology

In the histological examination no fungal structures were found in root and stem tissues from the incompatible interactions. In the compatible interactions, thin, unbranched hyphae were first observed in the xylem vessels of the root and lowermost stem internodes 15 days after inoculation with race 5. At this time, few xylem bundles were infected and these had a low intensity of colonization. Incidence and intensity of colonization of xylem vessels in the root, and first and second stem internodes increased by 20 days after inoculation, and occurred to a lower degree in the fifth stem internode. Fungal structures in the xylem of the root and lowermost stem sections were abundant, including numerous microconidia (Fig. 1A), chlamydospores, and thick, irregular and branched hyphae which occluded the xylem vessels (Fig. 1B). Thin hyphae were found to invade adjacent vessel elements through pits in the lateral walls (Fig. 1C). Fungal structures formed in xylem vessels infected by race 0 were similar to those found for race 5, although they developed later. Thin hyphae were observed by 20 days after inoculation, occurring with a low intensity of colonization in a small proportion of xylem bundles of the first and second stem internodes. By 40 days after inoculation, incidence and intensity of xylem colonization were much higher in the root and lowermost stem internodes, and to a lesser extent in the fifth stem internode. At this time numerous microconidia and chlamydospores were formed in the vessels.

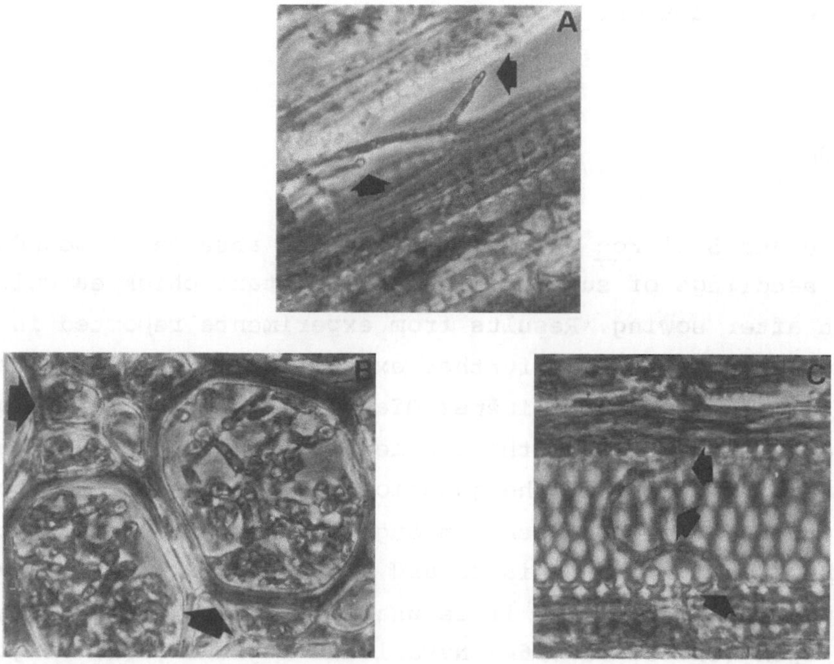

Fig. 1. Growth of Fusarium oxysporum f. sp. ciceri race 5 in xylem vessels of susceptible chickpea cultivar JG 62. (A) Phialides and microconidia (x550). (B) Mycelia in mature and immature vessels (x550). (C) Invasion of an adjacent vessel element by hyphae through pits in lateral walls (x550).

Histological modifications in compatible interactions consisted of cavities which formed mainly between phloem and xylem and xylem and medula, and occasionally between phloem and cortical parenchyma. These cavities appeared by 20 days after inoculation irrespective of the race of the pathogen, and they were larger and more frequent in the first and second stem internodes. In plants severely affected, i.e. by 20 and 40 days after inoculation with races 5 and 0, respectively, walls of parenchyma cells in the medula degenerated. Additional histological modifications in cultivar JG 62 infected by race 5 consisted of anomalous cellular proliferation in the vascular cambium and formation of new vessel elements adjacent to infected vessels. These modifications occurred in the first and second stem internodes but not in the root or fifth stem internode and were not found in any other compatible interaction.

DISCUSSION

Races 0 and 5 of Foc gain ingress in the absence of wounds in young seedlings of susceptible and resistant chickpea cultivars soon after sowing. Results from experiments reported in this paper as well as from further experiments (Basallote-Ureba 1987; Basallote-Ureba and Jiménez-Díaz 1987) suggest that ingress takes place mainly through the cotyledons and the zones of the epicotyl and hypocotyl at the junction of, or close to the cotyledons, and to a lesser extent through zones of the main root except the root apex. In this regard, Foc behaves like F. oxysporum f. sp. pisi except that it is unable to penetrate through the root apex (Chi et al 1964; Nyvall and Haglund 1972; Virgin and Walker 1940) and it appears to differ from other formae speciales of F. oxysporum which penetrate the host plant mainly through the root apex (Anderson and Walker 1935; Nelson 1981; Peterson and Pound 1960).

After penetration, vascular colonization of susceptible cultivars by races 0 and 5 of Foc seems to follow the same pattern of most wilt diseases (Beckman 1987; Nelson 1981). Advance of the pathogen along the plant axis occurs by means of hyphal growth and microconidia carried in the xylem vessels, and systemic colonization appears to proceed slowly since it took some 15-20 days after sowing to form fungal structures in the xylem. Fungal growth was maximum in the root and first stem internode, which correspond to penetration sites, and decreased acropetally along the plant axis. Increased vascular colonization of upper stem internodes may occur by further infection of xylem bundles in the underground tissues as well as by lateral colonization of adjacent vessels.

Race 5, which is much more virulent to cultivar P 2245 than race 0, colonized xylem tissues of this cultivar more rapidly and extensively than race 0. Thus, race 5 fully colonized the xylem tissues by 20 days after sowing, while race 0 barely showed the first indications of vascular colonization at that time and it took 20 more days to reach an incidence and intensity of colonization comparable to that of race 5. For both races, severe symptoms developed when colonization of xylem vessels in the root and lower stem internodes was intense, although the pathogen was spread along the stem. Intense colonization of the root and lower stem by the pathogen resulted in occlusion of xylem vessels by fungal structures. We did not observe formation of tyloses or gels which additionally might be involved in xylem blockage. However, root and stem sections may have lost gels during histological preparation, since recent observations (Basallote-Ureba and Jiménez-Díaz unpublished) of fresh sections stained with ruthenium red indicate gel formation.

Symptomless reactions in all incompatible interactions were associated with lack of vascular invasion by the fungus as indicated by isolation and microscopic observations. However, the fungus seems to be able to superficially associate with the underground tissues of the plant from which it could be isolated, though to such a minor extent that no fungal structures were observed microscopically. Thus, it appears that in chickpea, mechanisms of resistance to Foc may operate in the determinative phase I (Beckman 1987), as suggested by the inhibition of spore germination and mycelial growth by seed exudates (Haware and Nene 1984; Satyaprasad and Ramarao 1983).

ACKNOWLEDGMENTS

This research was supported by grants C2-0238/81 and PA85-0367 from Comision Asesora de Investigación Científica y Técnica, Ministerio de Educación y Ciencia.

REFERENCES

Anderson ME and Walker JC (1935) Histological studies of Wisconsin Hollander and Wisconsin Ballhead cabbage in relation to resistance to yellows. J Agric Res 50:823-836

Ayyar VR and Iyer RR (1936) A preliminary note on the mode of inheritance of reaction to wilt in *Cicer arietinum*. Proc Indian Acad Sci B 3:438-443

Basallote-Ureba MJ (1987) Relaciones huesped-parásito en la marchitez del garbanzo causada por *Fusarium oxysporum* f. sp. *ciceri*, PhD Thesis. Universidad de Córdoba, Córdoba

Basallote-Ureba MJ and Jiménez-Díaz RM (1987) Infection courts on chickpeas for races 0 and 5 of *Fusarium oxysporum* f. sp. *ciceri*. In: SEF (ed) Proc 7th Congr Un Phytopath Mediterr 20-26 Sept 1987. Dirección General de Investigación y Extensión Agrarias, Sevilla, pp 123-124

Beckman CH (1987) The nature of wilt diseases of plants. APS Press, St Paul, Minnesota

Cabrera de la Colina J (1986) Variación patogénica en *Fusarium oxysporum* Schlecht. emend. Snyder & Hansen y *Fusarium solani* (Mart.) Sacc. emend. Snyder & Hansen agentes de la Marchitez y Podredumbre de raíz del garbanzo en Andalucía, PhD Thesis. Universidad de Córdoba, Córdoba

Cabrera de la Colina J, Trapero Casas A and Jiménez-Díaz RM (1985) Races of *Fusarium oxysporum* f. sp. *ciceri* in Andalucía, southern Spain. Int Chickpea Newsletter 13:24-26

Chi CC, Childers WR and Hanson EW (1964) Penetration and subsequent development of three *Fusarium* species in alfalfa and red clover. Phytopathology 54:434-437

González Torres R (1985) Densidad de inóculo y estructura de virulencia de poblaciones de *Fusarium oxysporum* Schlecht. y *Fusarium solani* (Mart.) Appl. et Wr. que infectan garbanzo en Andalucía Occidental, PhD Thesis. Universidad de Córdoba, Córdoba

Haware MP and Nene YL (1980) Sources of resistance to wilt and root rots of chickpea. Int Chickpea Newsletter 3:11-12

Haware MP and Nene YL (1982) Races of *Fusarium oxysporum* f. sp. *ciceri*. Plant Disease 66:809-810

Haware MP and Nene YL (1984) The role of chickpea root exudates in resistance to Fusarium wilt. Int Chickpea Newsletter 10: 12-13

Jensen WA (1962) Botanical histochemistry. W.H. Freeman and Co, San Francisco

Johansen DA (1940) Plant microtechnique. Mc Graw Hill Book Co, New York

Nelson PE (1981) Life cycle and epidemiology of *Fusarium oxysporum*. In: Mace ME, Bell AA and Beckman CH (eds) Fungal wilt diseases of plants. Academic Press, New York, pp 51-80

Nene YL and Haware MP (1980) Screening chickpea for resistance to wilt. Plant Disease 64:379-380

Nene YL and Reddy MV (1987) Chickpea diseases and their control. In: Saxena MC and Singh KB (eds) The Chickpea. CAB International, Wallingford, pp 233-270

Nyvall RF and Haglund WA (1972) Sites of infection of *Fusarium oxysporum* f. sp. *pisi* race 5 on peas. Phytopathology 62:1419-

1424
Peterson JL and Pound GS (1960) Studies on resistance in radish to *Fusarium oxysporum* f. *conglutinans*. Phytopathology 50:807-816

Satyaprasad K and Ramarao P (1983) Effect of chickpea exudates on *Fusarium oxysporum* f. sp. *ciceri*. Indian Phytopathology 36:77-81

Trapero Casas A and Jiménez-Díaz RM (1985) Fungal wilt and root rot diseases of chickpea in southern Spain. Phytopathology 75:1146-1151

Virgin WJ and Walker JC (1940) Relation of the near-wilt fungus to the pea plant. J Agric Res 60:241-248

CURRENT STATUS OF VERTICILLIUM WILT OF COTTON IN SOUTHERN SPAIN: PATHOGEN VARIATION AND POPULATION IN SOIL

M.A. Blanco López[1], J. Bejarano Alcazar[1,2], J.M. Malero-Vera[1,2] and R.M. Jiménez-Díaz[1]
Departamento de Agronomia
Universidad de Córdoba
Apartado 3048
14080 Córdoba
Spain

INTRODUCTION

Aproximately 80% of the total national acreage of cotton (*Gossypium hirsutum* L.) is grown under irrigation in the Guadalquivir Valley of Andalucía, southern Spain. In this region, cotton has usually been grown in rotation with crops such as corn, sunflower, sugar beet and wheat. However, high yields and income obtained by farmers with cotton crops as compared to other crops, have led to shorten rotations and intensified cotton cultivation. This trend has been associated with increased attacks by pests and diseases, of which Verticillium wilt (VW) induced by *Verticillium dahliae* Kleb, is the most important. Recently, VW epidemics have become highly severe in the lower Guadalquivir Valley (Marismas) as a result of high inoculum densities (ID) in the soil and/or the existence of a defoliation-inducing pathotype. These severe epidemics are limiting the use of cotton as a crop in many fields at Marismas.

This paper presents an overview of our studies on VW of cotton in Andalucía, including a) importance and distribution of the

2. Departamento de Protección Vegetal
DGIEA, Junta de Andalucía
Apartado 240
Spain

NATO ASI Series, Vol. H28
Vascular Wilt Diseases of Plants
Edited by E.C. Tjamos and C. Beckman

disease, b) the variation in virulence, and ID in soil, of the pathogen, c) the epidemic development of VW with different ID of defoliating and non-defoliating pathotypes of V. dahliae, and d) the efficacy of soil solarization to control VW under different conditions.

IMPORTANCE AND DISTRIBUTION OF THE DISEASE

Systematic disease surveys were carried out in the Guadalquivir Valley in September, at the time of boll opening, in 1981-1983 and 1985. A total of 139 fields were inspected, including 21, 20, 44 and 54 in 1981, 1982, 1983 and 1985, respectively. Disease incidence (DI) was assessed by counting the number of plants with foliar symptoms (FS) in three or six groups of 20 consecutive plants which were randomly chosen in each field. Disease severity was assessed for each plant on a 1-5 scale according to the percentage of affected foliage in acropetal progression (1 = 0%, 2 = 1-33%, 3 = 34-66%, 4 = 67-100%, 5 = dead plant). Stem and petioles sampled from affected plants were used for isolation of the pathogen in the laboratory.

In 1985, ID of V. dahliae was assessed in soil of 57 cotton fields, of wich 54 had been inspected for disease incidence and severity. For each field, 30 soil samples were taken after sowing, in May, from the upper 20-cm-soil. Samples from each field were mixed and ID was determined in 0.5 g aliquots with the Andersen sampler (Butterfield and DeVay 1977).

Incidence and severity of the disease (Table 1) were fairly similar the 4 years of the surveys, with a disease prevalence varying from 74 to 82% of the inspected fields. Distribution of VW along the Valley was uneven, with more fields being affected in the Upper and Lower Vàlley than in the Central Valley. DI values were more frequently less than 25% with an average of 19.2%, and mean disease severity was 2.9 (Table 1).

In 1985, V. dahliae was found in soil of 35.1% of the sampled fields, with ID values ranging from 2 to 132 propagules per gram of dry soil (ppg) (Table 2).

Table 1. Incidence and severity of Verticillium wilt of cotton in areas of the Guadalquivir Valley

Area	Affected field(%)	Affected fields (%) with incidence (%)				Mean disease	
		<25	25-50	50-75	75-100	Incidence(%)	Severity(a)
Upper	88.2	82.2	13.3	4.4	0.0	15.1	2.8
Central	66.0	71.4	22.9	5.7	0.0	19.0	2.8
Lower	80.0	71.4	0.0	17.9	10.7	26.1	3.2
Total	77.7	75.9	13.0	8.3	2.8	19.2	2.9

(a) Based on a 1-5 scale (1=no symptoms, 5=dead plants)

Table 2. Frequency and inoculum density of Verticillium dahliae in areas of the Guadalquivir Valley in 1985

Area	Fields		Inoculum density (Nº of propagules/g soil)	
	Sampled(Nº)	Verticillium detected (%)	Mean	Range
Upper	16	37.5	6.4	2.0-10.0
Central	23	34.8	3.4	2.0-6.5
Lower	18	33.3	37.1	2.0-132.0
Total	57	35.1	14.4	2.0-132.0

Fields with ID values higher than 10 ppg occurred only in the Lower Valley. Fields affected by VW, where V. dahliae was not detected in soil, had DI values lower than 30%.

VIRULENCE OF ISOLATES

Virulence of 14 V. dahliae isolates to cotton cultivars Coker 310, PI 70-110, Acala 4-42, Acala SJ2, Acala SJ5 and Acala SJC-1, was investigated in a series of five experiments. All experiments were carried out in a growth chamber under environmental conditions repeatable and suitable for disease development. Three-week-old plants were inoculated by injecting a droplet of inoculum suspension ($3x10^6$ conidia/ml) at two opposite sites of the epicotyl. Disease reaction was assessed by the severity of symptoms on the 1-5 scale at 4-5 weeks after inoculation. Experiments were designed as randomized complete bloks with 14-20 replicated plants.

Results were consistent over the experiments and some of them are shown in Table 3.

Table 3. Severity of disease reaction of cotton cultivars to isolates of Verticillium dahliae (x)

Isolate	Cultivar				Mean (y)
	AC SJC-1	AC SJ2	AC SJ5	P1 70-110	
V4	2.2	2.2	2.1	2.8	2.3 a
V99	3.9	3.4	3.7	4.3	3.8 b
V96	4.6	4.4	4.7	4.9	4.7 c
V101	4.7	4.9	4.5	4.9	4.8 c
V102	4.2	4.4	4.5	4.9	4.7 c
V103	4.7	4.4	4.5	5.0	4.7 c

(x) Assessed on a 1-5 scale (1=no symptoms, 5=dead plant) 4 weeks after inoculation.
(y) Isolate x cultivar interaction was not statistically significant. Means followed by same letter are not statistically different according to the t Student test.

Disease reaction varied with isolates of the pathogen, and there was not statistically significant isolate x cultivar interaction. Most isolates, as represented by V4, induce mild foliar symptoms restricted to the lower portion of the plant. Isolate V99 induces severe foliar symptoms without defoliation, and death of the plant with incidence varying upon cultivars. A third group of isolates, represented by V96, 101, 102 and 103, induce stunting, severe foliar symptoms, defoliation, and death of the plant. All defoliating isolates were obtained from the lower Guadalquivir Valley, while all isolates obtained from VW-affected plants in the Central and Upper Valley are mildly or highly virulent, but non-defoliating. All defoliating isolates form elongated microsclerotia in water agar (WA), have an optimum temperature for mycelial growth on potato-dextrose agar (PDA) near 27 C, and grow on PDA supplemented with 0.001 M sanguinarine nitrate. Thus, these defoliating isolates appear to be similar to the T-1 type isolates occurring in California. In contrast, isolates inducing mild or severe foliar symptoms without defoliation form round microsclerotia in WA, have an optimum temperature for mycelial growth on PDA at, or lower than, 24 C, and have a restricted growth on PDA supplemented with sanguinarine nitrate (Schnathorst 1973). These non-defoliating isolates are considered similar to the SS-4 type isolates infecting cotton in California.

INFLUENCE OF INOCULUM DENSITY AND VIRULENCE OF ISOLATES ON DISEASE DEVELOPMENT

Five cotton fields with different ID in 1985, and designated A-E, were selected for this study. In 1986, ID was assessed in a 900-m^2 plot at each field sown to susceptible cultivar Coker 310. At each plot, disease progress was determined at 7-14 day-intervals from mid-June to mid-September, by the incidence of FS in 50 consecutive plants at each of six adjacent rows. Maximum and minimum daily temperature were recorded for each field to calculate physiological time (degree-day, with a threshold of 11.9 C). Mass isolates of *V. dahliae* were obtained from affected plants in each plot and single-spored. Single-spore isolates were characterized as defoliating or no-defoliating types by means of virulence to cotton cultivars and morphological and physiological observations.

In 1987, three plots were sown to cultivar Coker 310 at fields A (75 m^2), B (40 m^2) and D (900 m^2) and ID and disease progress was determined as for 1986 by the incidence of FS in 75-100 consecutive plants at each of three or four adjacent rows.

All isolates from fields A, B and C were of the defoliating type. All isolates from field D and E were of the non-defoliating type, except for one from field E that was characterized as defoliating.

Disease progress over physiological time was fairly adjusted to linear relationship. Regardless of ID values, rates of disease progress in fields A, B and C infested with the defoliating pathotype were higher than those rates obtained for fields D and E infested mainly with the non-defoliating pathotype (Tables 4, 5).

Table 4. Disease incidence-inoculum density relationship in cotton fields infested with pathotypes of Verticillium dahliae in 1986

Field	Inoculum density (Nº propagules/ g soil)	Pathotype(a)	Disease progress equation	R^2	Final DI (%)
A	44.0	D	Y=0.103x-10.0	0.92	99.3
B	24.0	D	Y=0.100x-1.5	0.95	97.0
C	5.5	D	Y=0.084x-45.8	0.94	83.3
D	34.0	ND	Y=0.064x-24.1	0.99	93.7
E	9.0	ND	Y=0.005x-30.5	0.94	65.3

(a) D = Defoliating, ND = Non-defoliating.

Rates of disease progress and final incidence of FS appear to be positively correlated with ID values for fields A-E in 1986, and for plots B1-B3 and field D in 1987 (Tables 4, 5). This relationship did not hold for plots A1-A3 in 1987. Apparently, ID values of the defoliating pathotype in these plots, ranging 44.5-75.5 ppg, were in the threshold to determine similarly high rates ca. 0.15 and a final DI of 100% in all plots.

Table 5. Disease incidence-inoculum density-relationship in cotton fields infested with pathotypes of <u>Verticillium</u> <u>dahliae</u> in 1987

Field	Inoculum density (Nº propagules/ g soil)	Pathotype(a)	Disease progress equation	R^2	Final DI (%)
A1	75.5	D	Y=0.147x-58.0	0.98	100
A2	55.5	D	Y=0.15x-59.2	0.99	100
A3	44.5	D	Y=0.154x-73.5	0.98	100
B1	10.0	D	Y=0.108x-21.6	0.99	98.7
B2	8.0	D	Y=0.09x-22.8	0.91	96.0
B3	2.0	D	Y=0.067x-26.2	0.96	68.4
D	27.5	ND	Y=0.056x-27.2	0.97	67.7

(a) D = Defoliating, ND = Non-defoliating

EFFICACY OF SOLARIZATION TO CONTROL THE DISEASE IN ANDALUCIA

In 1985, a preliminary solarization trial was conducted in a loamy soil infested with the non-defoliating pathotype near Cordoba in the Central Valley. An irrigated plot, 12x33 m^2, was covered (solarized), with transparent polyethylene sheets (37 µm thick) from July 22-September 22. An adjacent non-solarized, unirrigated plot of the same size used as control. Inoculum density, assessed just before solarization, was 9.5 ppg in the solarized plot and 25.5 ppg in the non-solarized control. After solarization, ID values were 0 ppg in the solarized plot and 8.5 ppg in the control.

In 1986, a solarization experiment was carried out in two fields (A and B) with very clay soil infested with the defoliating pathotype at the Lower Guadalquivir Valley. Six plots 7 x 10 m^2 for field A, and nine plots 3.8 x 10 m^2 for field B, were solarized, respectively, with 37 µm or 25 µm-thick sheets of transparent polyethylene. Equal numbers of untreated, irrigated plots served as controls. Plots were solarized from July 9 -

August 31. Inoculum density in solarized and non-solarized plots was assessed just before and after solarization. Solarized and non-solarized plots were sown to cotton cultivar Coker 310 in late March or early April, 1987. Inoculum density in these plots was assessed at planting time, and incidence of FS was determined in late July and early September (Table 6). The influence of solarization on seed yield was determined in field B, for which plots were harvested by hand-picking.

Table 6. Control of Verticillium wilt of cotton by soil solarization in 1987

Field	Planting date	Solarization treatment	Disease incidence (%) at July 28	Disease incidence (%) at September 9
A	April 13	Un-solarized	90.5	99.5
		Solarized	5.0	22.3
B	March 24	Un-solarized	60.0	89.6
		Solarized	4.5	6.7

In 1986, ID in plots at fields A and B averaged, respectively, 21.3 and 2.1 ppg before solarization. In contrast, ID was at negligeable levels in solarized plots, both just after solarization and at planting time in 1987. ID values in untreated plots at fields A and B averaged 51.3 ppg and 6.1 ppg, respectively, when assessed at planting time in 1987.

In untreated plots, final incidence of FS was high for both fields (Table 6), but severe disease occurred earlier in the heavily infested field A. In contrast, incidence of FS in solarized plots was low for both field in the earlier assessment, and increased to some extent only in field A at the later assessment (Table 6). Such late increase in DI had probably a rather small influence on final yield. Seed yield in solarized plots at field B increased by an average of 230% as compared to untreated controls.

CONCLUSIONS

Our results show that VW of cotton occurs widespread along the Guadalquivir Valley of Andalucia although remarkable differences exist among areas of the Valley. Thus, in the Central and Upper Valley attacks occur with moderate disease incidence and severity, associated with the existence of a non-defoliating pathotype of V. dahliae at ID values less than 10 ppg. In contrast, in the Lower Valley, although disease prevalence is similar to that in the Central and Upper Valley, severe disease incidence and severity occur in some fields as a difference from the other areas of the Valley. Severe disease incidence and severity in the Lower Valley are associated either with a high ID values of the non-defoliating pathotype, or with low to high ID values of a defoliating pathotype which is restricted to this area. Pathogenicity and morphological and physiological characteristics indicate that defoliating and non-defoliating pathotypes in Andalucia are similar to the T-1 and SS-4 type isolates of the pathogen in California respectively.

Disease progress was linear with physiological time. While ID values ranging 2.0-10.0 ppg of the defoliating pathotype caused a final DI in the range of 68.4-98.7%, 27.5-34.0 ppg of the non-defoliating pathotype were required to attain similar DI. The existence of high ID in soil and of the defoliating pathotype may be limiting factors for cotton production in the Lower Valley. However, soil solarization in this area has proven successful in reducing the ID in soil, delaying the increase of DI and increasing seed yield. Thus, solarization may be a suitable control measure for VW in Andalucia.

AKNOWLEDGMENTS

This research was supported by grants INIA-7637 from Instituto Nacional de Investigaciones Agrarias, Ministerio de Agricultura, and CCA 83059 from the Spanish American Joint Committee for Scientific and Technological Cooperation. We are grateful to Mr. J. Cobos for technical assistance.

REFERENCES

Butterfield EJ and DeVay JE (1977) Reassessment of soil assays for Verticillium dahliae. Phytopathology 67:1073-1078

Schnathorst WC (1973) Additional strains of Verticillium dahliae from cotton in California. In: Ariz Natt Cotton Counc (ed) Proc Beltwide Cotton Prod Res Conf, Phoenix, pp 22-23

A LEAF INJECTION TECHNIQUE FOR THE ENHANCEMENT OF LOW POPULATIONS OF CLAVIBACTER MICHIGANENSIS SUBSP. SEPEDONICUS

A.S. Alivizatos
Benaki Phytopathological Institute
145 61 Kifissia, Athens
Greece

INTRODUCTION

Clavibacter michiganensis subsp. sepedonicus (Cms) Davis et al causes ring rot of potatoes (Solanum tuberosum L.) and is responsible for financial losses resulted from yield and storage losses in ware and seed potatoes and disqualification of seeds crops (Nelson and Torfason 1974; Easton 1979; Sletten 1985; Lelliott 1988). Other plants reported as natural hosts of Cms include some weed species of Solanum (including S. dulcamara) (Lelliott 1988) and sugar beet (Bugbee et al 1987).

Infected seed potatoes are the most important means of spreading Cms over long distances, although sugar beet seed may play a role in distributing the pathogen (Bugbee and Gudmestad 1988).

The infection of tubers may be obvious or remain latent. Tubers with typical symptoms allow visual diagnosis and isolation of the pathogen from the discoloured vascular tissue in most cases. However, latent infections can be detected by appropriate methods. The aubergine inoculation test (Lelliott and Sellar 1976) is quite sensitive and reliable, but it takes 6-40 days to complete and may give false positive reaction (due to other organisms present in the potato extract) (Olsson 1976) or it may give latent infection of egg-plants. An immunofluorescence (IF) test (De Boer and Copeman 1980) is widely used as a rapid screen method, but it can give false positive reaction and is not expected to detect very low levels of Cms. A recently published method of the European Communities (EUR 11288 1987) combines

NATO ASI Series, Vol. H28
Vascular Wilt Diseases of Plants

the IF test (as a screening test) with the egg-plant test (confirmatory test) and further requires the isolation and identification of Cms.

Because of the above mentioned limitations of the aubergine test and the fact that egg-plants become more resistant by aging (Alivizatos 1983) the EEC method still offers little help for a fast diagnosis of low levels of Cms in the potato extract.

To fasten the diagnosis, we studied the multiplication of low populations of Cms (as pure culture or in a potato tuber extract) in the leaflets of potato and tomato plants. Furthermore the effect of saprophytic bacteria (present in the potato extract) and of various inhibitors on growth of Cms were also examined.

MATERIALS AND METHODS

Bacterial cultures

Isolate BCPPB 780 (Benaki Collection of Plant Pathogenic Bacteria, Kifissia, Athens, Greece) of Cms (isolated in 1982 from diseased potato tubers kindly sent by Prof. Manzer, Univ. of Maine, USA for an EEC experiment), grown on Yeast extract-glucose-mineral salts agar (YGM) medium (Yeast extract 0.2%, glucose 0.25%, K_2HPO_4 0.025%, KH_2PO_4 0.025%, $MgSO_4.7H_2O$ 0.01%, $MnSO_4.H_2O$ 0.0015%, NaCl 0.005%, $FeSO_4.7H_2O$ 0.0005%, Bacto-Difco agar 1.5% w/v) at 22±1 C, was used throughout this study. Bacterial suspension was prepared in sterile 0.05M phosphate buffered saline (PBS) pH 7.0 (EUR 11288 1987) from a 24 hours slope culture on YGM.

"Saprophytic" bacterial isolates were obtained from an extract of healthy foundation potato tubers (S. tuberosum L. cv. Kennebec), 10-fold serially diluted, plated on YGM (containing 200 µg/ml cycloheximide) and incubated at 25-26 C. The various isolates were selected on the basis of the difference in their cultural characteristics only. Potato extract was obtained as previously described (EUR 11288 1987).

All cultures were maintained on YGM medium titrated and assayed for colony forming units (cfu) per ml as described elsewhere (Alivizatos and Markham 1986).

Test plants and inoculation method

Tests plants were potato (S. tuberosum L. cv. Kennebec) and tomato (Lycopersicon esculentum Mill. cv. Ace 55 VF) plants with 8 permanent leaves grown at 22 C under 16 hours photoperiod.

Test plants were inoculated with a suspension of Cms in 0.05 M sterile PBS containing $2x10^3$ cfu/ml or with a mixture of Cms suspension and a tissue extract obtained (EUR 11288 1987) from healthy potato tubers (cv. Kennebec). This mixture contained Cms at a concentration of $2x10^3$ cfu/ml and total "saprophytic" bacterial population $1x10^6$ cfu/ml. Inocula were introduced in a 10 ml plastic syringe fitted with a 27 G hypodermic needle and injected in the leaflets of the 4-5 leaf at the junctions of the veins until the mesophyl became infiltrated. Controls were plants similarly injected with sterile 0.05 M PBS.

Injected leaves were washed with sterile 0.05 M PBS to remove possible epiphytic contamination by Cms and "saprophytes" and kept at 22±1 C under 16 hours photoperiod. The margins of the injected areas were marked with a black textmark.

Assay of bacteria in plants

Twenty four hours (24h) after injection and then at various time intervals, leaf disks, 6 mm in diameter, were cut aseptically from the marked areas of potato and tomato leaflets and immediately transferred in 1 ml of sterile 0.05 M PBS. The disks were further ground in a sterile mortar and the extract was serially 10-fold diluted in 0.05 M PBS and assayed for cfu/disk (Alivizatos and Markham 1986). Plates were incubated at 22±1 C and examined after 6 days.

Bacteria isolated from leaflets injected with pure culture of Cms, were identified by cultural, biochemical, serological (IF test) and pathogenicity tests (EUR 11288 1987). Pathogenicity was performed on tomato seedlings (with 2-3 permanent leaves) by picking up individual colonies with sterile toothpicks and pricking their stem above the 1st leaf. Five plants were inoculated with each colony. Pathogenicity of the bacteria isolated from the potato tissue extract (before the addition of Cms) was determined as above, by performing an HR test (Klement 1963) and by their ability to rot potato slices. Inoculated plants were maintained at 22±1 C.

Small pieces (1-2 mm) cut aseptically from the injected leaves or from inoculated tomato stems were transferred on a glass slide in sterile 0.05 M PBS and examined by phase contrast microscopy for the presence of bacteria.

Susceptibility of Cms to inhibitors

The effect of the "saprophytic" bacteria isolated from the potato extract, on growth of Cms was tested by inoculating (with the saprophytes) YGM medium in glass plates (4 isolates/plate), incubating at 25-26 C for 2 days, killing the bacteria following exposure to chloroform vapours and applying on plates a soft agar (0.8% w/v Bacto-Difco agar in phosphate buffer[1]) layer (3 ml/plate) containing Cms at a concentration of $1x10^7$ cfu/ml. Plates were incubated at 22±1 C and examined after 5 days for the presence of inhibition zones.

The effect of various concentrations of ammonium chloride, ammonium nitrate, bacitracin, berberine, chloramphenicol, erythromycin, kanamycin, nystatin, penicillin, polymyxin-B, potassium dichromate, sodium nitrate, streptomycin, tergitol, tetracycline-HCl, thallium nitrate and tyrotricin on growth of Cms was tested on solid YGM medium using the impregnated paper disk method (Alivizatos 1987). The inhibition zones were recorded after 5 days incubation.

Some positive inhibitors were tested in YGM medium to determine inhibition of saprophytic bacteria.

---

[1]phosphate buffer (g/l) : $Na_2HPO_4.12H_2O$ 4.3g, $NaH_2PO_4.2H_2O$ 1.2g, H_2O 1l.

RESULTS

Multiplication of Cms in plants

Cms multiplied both in potato and tomato leaflets when injected as pure culture (Fig. 1). The titre of the bacterium in

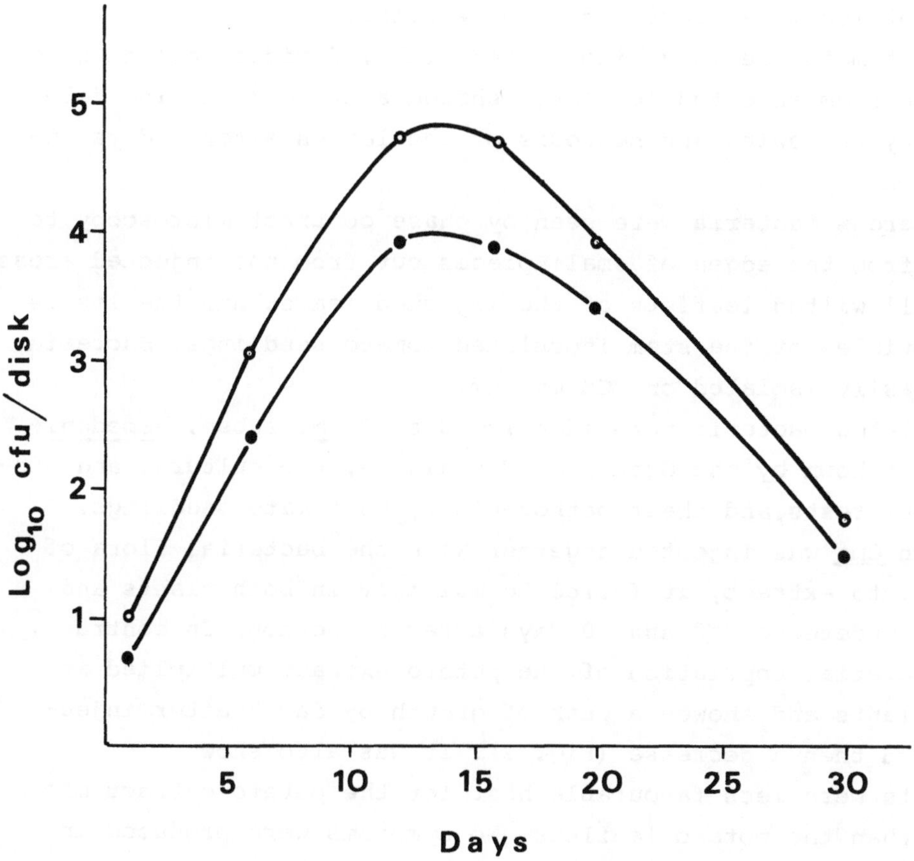

Fig. 1. Multiplication of Clavibacter michiganensis subsp. sepedonicus in the leaflets of tomato (o) and potato (●), as measured by the number of cfu per leaf disk.

the leaflets of potato and tomato was 5 cfu/disk and 10 cfu/disk respectively after 24 hours and then increased to $8x10^3$ cfu/disk and $5x10^4$ cfu/disk after 12 days. This titre was maintained until day 16 and then showed a decrease to approximately 10^2 cfu/

disk by day 30 (Fig. 1). No other bacteria were recovered from the plants. Fig. 1 shows that Cms was grown in tomato better than in potato since it gave a higher maximum titre in tomato and the cell doubling time during the exponential phase was 45 min in tomato and 50 min in potato.

One month after injection the inoculated leaflets of tomato showed wilting and yellowing. After 2 months they died while adjacent leaflets started to show wilting.

All tomato seedlings inoculated with individual colonies recovered from injected leaflets, showed a typical wilting followed by yellowing and necrosis of the leaves after 5 days incubation.

Mumerous bacteria were seen by phase contrast microscopy to exude from the edges of small pieces cut from the injected areas, from all wilted leaflets of the injected leaves and the leaves and petioles of the stem inoculated tomato seedlings. Bacteria were easily isolated on YGM medium.

Isolated bacteria were identified as C. m. subsp. sepedonicus, as shown by the Gram and IF staining, the cultural and biochemical tests and their pathogenicity on tomato seedlings.

When Cms was injected together with the bacterial flora of the potato extract, it failed to multiply in both plants and it was undetected 10 and 20 days after injection. In contrast, the bacterial population of the potato extract multiplied in both plants and showed a peak of growth by day 7 after injection and then a decrease (Fig. 2). It was also show leaflets were less favourable host for the potato extract microflora than the potato leaflets. No symptoms were produced in the inoculated leaflets even after 3 months.

All six strains of bacteria isolated from the potato extract were not pathogenic to tomato and gave negative HR or rot of potato slices.

Effect of the potato extract bacteria on Cms

Of the six potato isolates, one inhibited in vitro the growth of Cms and produced an inhibition zone of 5 mm. This isolate which was a Gram positive, non-spore forming motile rod, was isolated also from the leaflets of potato and tomato plants injected with the potato extract and comprised the predominant isolate.

Susceptibility of Cms to inhibitors

Bacitracin and tetracycline-HCl inhibited the growth of Cms at very low quantities (< 0.025 µg). Very effective were also erythromycin, tergitol, streptomycin (0.025 < i < 0.25 µg), followed by tyrotricin (0.125 < i < 1.25 µg) polymyxin-B sulphate, kanamycin, chloramphenicol (0.25 < i < 2.5 µg) and thallium nitrate (0.4 < i < 4 µg). Penicillin, berberine and potassium dichromate were less effective (25 < i < 250 µg) as was nystatin (i > 250 µg). The least effective were ammonium chloride (4.5 mg), sodium nitrate (5 mg) and ammonium nitrate (i > 5 mg). Of the last seven inhibitors polymyxin-B at 0.25 µg/ml suppressed the bacterial population of potato extract by 25%, berberine at 50 µg/ml by 50% and potassium dichromate at 25 µg/ml by 95%. These concentrations of polymyxin-B and berberine were not inhibitory to Cms but that of potassium dichromate was slightly inhibitory. These inhibitors as well as penicillin, nystatin and the three inorganic salts are to be introduced in a growth medium to test their effect on Cms and saprophytes.

DISCUSSION

Both tomato and potato leaflets proved to be suitable for the multiplication of Cms, the titre of which was increased by 3-3.5 long units within 12 days, when injected as a pure culture. Meanwhile tomato was found to be superior over potato, since Cms showed in tomato leaflets higher titre and shorter doubling time and the population of the potato extract microflora was suppressed in tomato compared to potato. However eggplant is the indicator plant recently used to perform a pathogenicity test. The factors affecting the sensitivity of the

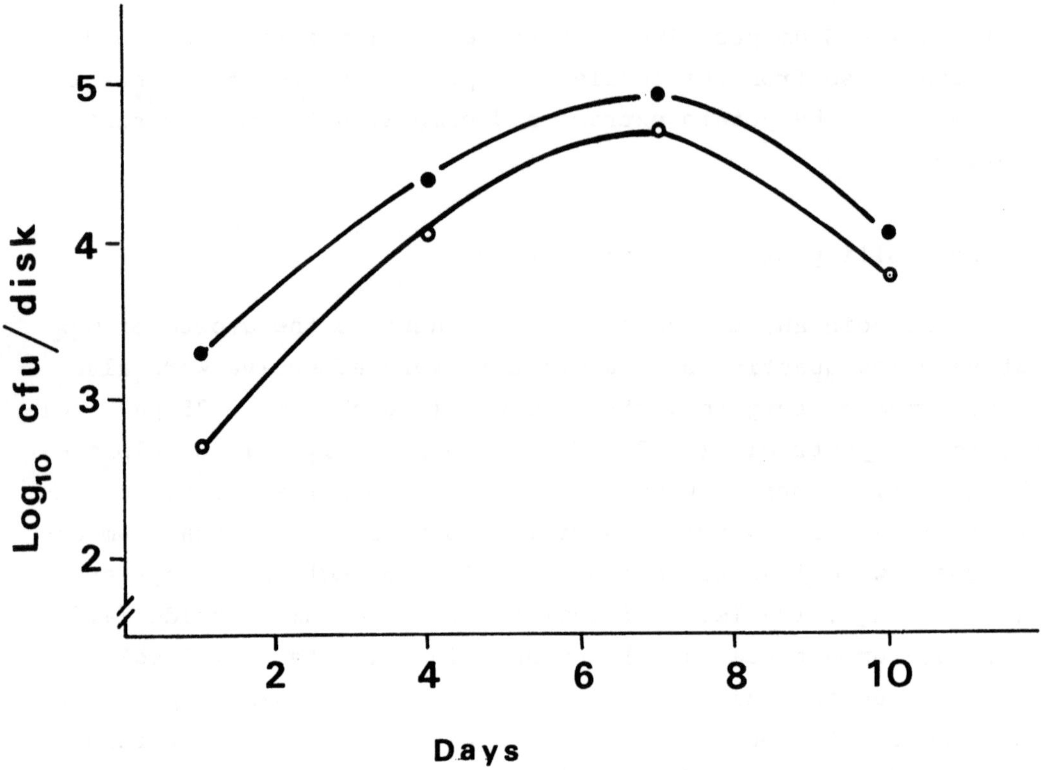

Fig. 2 Multiplication of the potato extract bacterial microflora in the leaflets of tomato (o) and potato (●), as measured by the total number of cfu per leaf disk.

egg-plant test (EP. test) (growth stage of plants, incubation conditions, virulence and inoculum concentration of viable Cms) have been recently discussed (Janse and Van Vaerenbergh 1987). Some of these factors can be controlled but others may create problems in the diagnosis.

The failure of Cms to multiply in the leaflets of potato and tomato to which it was introduced together with potato tuber microflora, revealed the presence of bacteria which inhibited the growth of Cms. This finding consists a new problem, directly affecting the survival of Cms and symptom expression in the egg-plant test. It is therefore possible that positive sample extracts can give falsely negative result in the EP test.

This problem could be overcome by eliminating the population of potato microflora and thus giving a chance to Cms to multiply. In this study we found some interesting inhibitors the use of which may help the formulation of a semi- or selective medium for Cms.

REFERENCES

Alivizatos AS (1983) The use of egg-plant test for the detection of *Corynebacterium sepedonicum* in latently infected pototo tubers. In Abstracts, 1st Hellenic Congress on Plant Diseases and Pests, pp 30-31. Hellenic Phytopathological Society, Athens

Alivizatos AS (1987) Comparison of the antibiotic sensitivity in vitro of *Spiroplasma kunkelii* and *Spiroplasma citri*. Annls Inst Phytopath Benaki 15:121-127

Alivizatos AS and Markham PG (1986) Multiplication of corn stunt spiroplasma in *Dalbulus maidis* and transmission *in vitro*, following injection. Ann appl Biol 108:545-554

Bugbee WM and Gudmestad NC (1988) The recovery of *Corynebacterium sepedonicum* from sugar beet seed. Phytopathology 78:205-208

Bugbee WM, Gudmestad NC, Secor GA and Notle P (1987) Sugar beet as a symptomless host for *Corynebacterium sepedonicum*. Phytopathology 77:765-770

Davis MJ, Gillasp AG, Vidaver AK and Harris RW (1984) *Clavibacter*: a new genus containing some phytopathogenic coryneform bacteria, including *Clavibacter xyli* subsp. *xyli* sp. nov., subsp. nov., and *Clavibacter xyli* subsp. *cynodontis* subsp. nov., pathogens that cause ratoon stunting disease of sugarcane and Bermudagrass stunting disease. Int J Syst Bacteriol 34:107-117

De Boer SH and Copeman RJ (1980) Bacterial ring rot testing with the indirect fluorescent antibody staining procedure. Am Pot J 57:457-465

Easton GD (1979) The biology and epidemiology of potato ring rot. Am Pot J 56:459-460

EUR 11288 (1987) Scheme for the detection and diagnosis of the ring rot bacterium *Corynebacterium sepedonicum* in batches of potato tubers, p 21. Commission of the European Communities. Brussels-Luxembourg

Janse JD and Van Vaerenbergh J (1987) Interpretation of the EC method for the detection of latent *Corynebacterium sepedonicum* infections in potato. Bulletin OEPP/EPPO Bulletin 17: 1-10

Klement Z (1963) Rapid detection of the pathogenicity of phytopathogenic pseudomonads. Nature 199:299-300

Lelliott RA (1988) *Corynebacterium sepedonicum* (Spieckermann and Kotthoff) Skaptason and Burkholder. In European Handbook of Plant Diseases (Smith IM, Dunez J, Lelliott RA, Phillips DH and Archer SA Eds) pp 184-186. Blackwell Scientific Publications, Oxford U.K.

Lelliott RA and Sellar PW (1976) The detection of latent ring rot Corynebacterium sepedonicum in potato stocks. Bulletin OEPP/EPPO Bulletin 6:101-106

Nelson GA and Torfasson WE (1974) Associative effects of leaf roll and ring rot on disease expression and yield of potatoes. Am Pot J 51:12-15

Olsson K (1976) Experience of ring rot caused by Corynebacterium sepedonicum in Sweden. Particularly detection of the disease in its latent form. Bulletin OEPP/EPPO Bulletin 6:209-219

Sletten A (1985) The effect of Corynebacterium sepedonicum on symptoms and yield of four potato cultivars. Potato Research 28:27-33

SOURCES AND PATHWAYS OF CONTAMINATION OF POTATOES BY SOFT ROT ERWINIAS IN SCOTLAND

M.C.M. Pérombelon and L.J. Hyman
Scottish Crop Research Institute
Invergowrie
Dundee DD2 5DA
Scotland

INTRODUCTION

Erwinia carotovora subsp. atroseptica (Eca) causes potato blackleg in the field in temperate climates and together with E. carotovora subsp. carotovora (Ecc) is commonly associated with tuber soft rot in the stored crop. Blackleg is a seed-borne disease and its development is affected by several factors, of which environmental conditions, especially wet soils, which favour growth of the bacteria, and the level of contamination of the seed tubers by Eca are the most important.

Until fairly recently, it was commonly accepted that the source of inoculum for blackleg from one potato generation to the next was either the seed tubers from blackleg-affected plants or the bacteria surviving freely in the soil. By the mid sixties it was generally believed that survival of the bacteria in soil was limited and was clearly less than the time lapse (c. 3-5 years) from one potato crop to another in the average crop rotation system (Pérombelon and Kelman 1980). The concept that the pathogen was transmitted in latently infected progeny tubers from diseased plants led to the development of seed certification schemes, based on crop inspection and roguing out diseased plants and their progeny tubers, to produce cleaner crops, a process which had been used successfully to control many viral diseases. However, these measures have patently failed to have a significant impact on blackleg control. It is common knowledge that seed from blackleg-free crops can produce blackleg

NATO ASI Series, Vol. H28
Vascular Wilt Diseases of Plants
Edited by E.C. Tjamos and C. Beckman

plants and the converse is also true. Furthermore, no explanation was proposed regarding the source of Ecc which typically does not cause blackleg in temperate regions although it is frequently present on tubers.

Several sources of erwinias and methods of dispersal and spread within and between crops have now been identified and their relative importance to tuber contamination is discussed in this paper.

MOTHER TUBERS

Large scale surveys in Scotland and elsewhere have shown that a high proportion of tubers in commercial stocks regardless of the presence or absence of blackleg in the parent crops were contaminated with both Ecc and Eca (Pérombelon and Kelman 1980). The bacteria are usually located on the surface where they tend to die out in storage under dry conditions and in lenticels where they persist in undiminished numbers for a long time. They are rarely present in the vascular ring except in tubers taken from blackleg affected plants (Pérombelon 1973). Therefore, a high proportion of seed tubers at planting are likely to be contaminated by both Ecc and Eca and hence could be a major source of contamination for the progeny tubers (Pérombelon 1974). Progeny tuber contamination occurs after rotting of the contaminated mother tubers, usually in July-August, when the bacteria are liberated in the soil and are transmitted to the tubers by soil water. Erwinias are moved readily by soil water vertically and horizontally over a distance of several meters in soil (Pérombelon 1976). Breakdown of mother tubers depends on several factors; erwinia contamination level, cultivar tuber resistance to rotting and especially soil water status. In dry soils, fewer mother tubers will rot and there is little or no transmission of the bacteria to progeny tubers because of discontinuity of water film around soil particles (Elphinstone and Pérombelon 1986a).

Tuber contamination level can vary during the growing season depending on soil moisture level and the condition of the tuber lenticels (Pérombelon 1976). In wet soils, tuber water potential is high and lenticels tend to open (proliferate) allowing entry of the bacteria and in dry soils, tuber water potential is low and lenticels tend to close (suberise).

The concept that the contaminated mother tuber and not the blackleg plant is an important source of contamination of progeny tubers by both erwinias has been a major contributing factor in attempts to produce erwinia-free seed tubers from stem cuttings or micropropagated plants. However, surveys carried out on commercial farms in Scotland showed that a high proportion of stem cutting-derived stocks were contaminated after only the first multiplication cycle and most were extensively contaminated after the third multiplication cycle (Pérombelon et al 1980). Once contamination has occurred, at however a low level, it can spread throughout stocks by the rotting mother-daughter tuber pathway even in the absence of further contamination.

These findings clearly imply that sources of the bacteria other than the mother tubers are also implicated in tuber contamination.

AIR-BORNE ERWINIAS

The role of insects as vectors of soft rot erwinias has long been known (Chiu et al 1958). Several insect species trapped near potato and vegetable refuse dumps in Scotland were found to be frequently contaminated, more often with Ecc than Eca (Harrison et al 1977). In Colorado insects have been shown to transmit marker strains of erwinias from potato dumps to nearby potato crops (Molina and Harrison 1980).

Aerial transmission by wind-blown rain has long been implicated in the spread of several plant bacterial diseases; but the recognition that long distance dispersal of soft rot erwinias can occur in aerosols is a relatively recent development (Graham et al 1977, 1979; Pérombelon et al 1979; Quinn et al 1980). Aerosols containing erwinias (c.10^8 cells/ha) are generated by rain impaction on diseased stems or on any other contaminated substrate such as leaves and surface water and when potato stems are pulverised before harvest as is commonly done in Scotland. When air was sampled in Scotland even with no potato crop visible, both Ecc and Eca were readily detected, albeit in low numbers, more often on rainy than dry days and not before August. Although survival of the bacteria captured on microthreads was poor under conditions similar to those prevailing at harvest time in Scotland (September-October), 50% remained viable for 5-10 min, which would be long enough for them to be blown for several hundred meters. Erwinias in aerosols can be washed down by rain to contaminate leaves and soil.

POTATO LEAVES

Contamination mostly by Ecc of leaves of potato plants and of other crops, weeds and trees in Scotland was detected by an enrichment procedure mostly after August when rain was more frequent and air humidity greater (Pérombelon unpublished). Field and controlled environment studies indicated that survival and miltiplication of Ecc and Eca on potato leaves occurred only as long as the leaf surface remained wet but bacterial numbers were $< 10^2$ cells per cm^2 leaf area. Low erwinia numbers could be attributed to competition with phylloplane microflora and low levels of nutrients on leaf surfaces (Pérombelon 1979). Therefore, it is unlikely that leaf-borne erwinias could be transmitted directly by rain from leaf surfaces in numbers high enough to have a significant impact on progeny tuber contamination. However, bacterial multiplication can occur in the rotting leaf debris on the soil surface to reach 10^4 cells/g debris under wet conditions in September and can be washed down by rain into the soil to contaminate the progeny tubers (Elphinstone and Pérombelon 1986a).

SOIL

Since the first studies on soft rot erwinias early this century, longevity of the bacteria in soil has been and still remains a controversial subject. Survival of Ecc and Eca inoculated in field soil was more affected by temperature than soil moisture content, longevity decreased gradually over a period measured in months and weeks with increasing temperature but fell sharply within a few days at temperatures > 20 C, Ecc surviving marginally better than Eca. Erwinias, as many plant pathogenic bacteria adapted to grow in nutrient rich plant tissues, cannot gear down their metabolic rate to a sufficiently low level especially at high temperatures to compete successfully in the relatively nutrient poor soil environment (Pérombelon and Hyman 1988).

When the presence of erwinias in potato fields were monitored soon after harvest by monthly sampling over several years, contamination level fell sharply by the following spring and was subsequently rarely present except for occasional short-lived increases in winter and as long as potatoes were not grown. This apparent persistence of the bacteria, usually Ecc, in low numbers also found in other countries led Stanghellini (1982) to conclude that the bacteria are true soil-borne organisms surviving in the rhizosphere of weeds and crops. In contrast, Pérombelon and Hyman (1988) suggested that regardless of whether they are endemic in soil or not, erwinias are repeatedly introduced from a wide range of sources such as rain-washed contaminated aerosols, irrigation water, from the subsoil since field drain water is frequently contaminated, and host crops in short rotation systems. Longevity in soil may be prolonged when the bacteria are protected in crop debris until complete decomposition and in volunteeers.

WATER

Less than a decade ago little was known of the presence of soft rot erwinias in natural water sources although several reports suggested that they probably had been repeatedly isolated but rarely fully characterised. Therefore, it was not unexpected to detect erwinias in water when they were specifically looked for, but what was unexpected was their widespread distribution in surface water. A high proportion of random water samples from lochs, streams and rivers were found to be contaminated by erwinias, mostly Ecc by McCarter-Zorner et al (1984) in Scotland and Colorado, USA and by Gudmestad and Secor (1983) in North Dakota, USA. The bacteria were also commonly found in river and irrigation canal water in the Netherlands, France, Spain, Israel and tropical Peru (Pérombelon unpublished). The numbers varied considerably within and between countries, from 0 to > 10^6 cells/liter, probably reflecting the degree of pollution present. Water from deep wells was usually erwinia free.

Pérombelon and Hyman (1987) on the basis of their studies on the ecology of erwinias in river water in eastern Scotland suggested that the erwinias were probably of allochthonous origin in agreement with the generally held view that most bacteria in rivers are washed in from surrounding land, in drains and run off water. Rivers flowing through uninhabited moorland were found not to be contaminated at any time during the year but they became increasingly so when they flowed through cultivated, inhabited land, with greater erwinia numbers in late summer and early autumn than at other times.

MECHANICAL HANDLING EQUIPMENT

Mechanical grading has long been thought to cause spread of erwinias because there was more soft rot and blackleg in crops grown from graded than in those from ungraded tubers. The higher disease levels were attributed to the transmission of the pathogens from soft rotting to healthy tubers, either by direct contact or via bacterial slime left on the grader. Using placement rotting tubers infected with a marker strain of Ecc, Elphinstone and Pérombelon (1986b) showed that bacteria from one rotting tuber can spread to c. 17% of a total of 600 kg tubers during grading, although only the first 50 kg of tubers were heavily infected usually by > 10^4 cells/tuber. Therefore, whenever tubers are mechanically handled at planting time, during harvest, store filling and grading, widespread contamination is also likely to occur if rotting tubers are present. Most soft rotting tubers in stored potato stocks contain large numbers of erwinias, more often Eca than Ecc because of the low prevailing temperatures favour the former organism, even if the initial cause of rotting is of fungal origin such as late blight. In addition, survival of the bacteria after grading was related to the extent and nature of external damage incurred. Tuber wounding is unavoidable and the incidence and type of wounds depends on the design of the machinery and the operating speed and on cultivar susceptibility to wounding. The bacteria tended to die out within < 1 month in cold storage in shallow (< 1 mm) wounds but they can survive in large numbers for > 2 months in deeper wounds where they would be better protected from desiccation before wound healing can seal them in.

RELATIVE IMPORTANCE OF ERWINIA SOURCES

As assessment of the relative importance of the different sources and transmission pathways of erwinias is essential in the development of an integrated approach to disease control. Crops harvested early, before mid August, are less likely to be contaminated with air, insect and leaf-borne contamination than late harvested crops because widespread leaf contamination and conditions (wet weather) favouring rotting of fallen leaves tend to occur late. Moreover, contamination is mostly by Ecc which is not the causal agent of blackleg in temperate areas. In contrast, erwinias from mother tubers can contaminate progeny tubers at an earlier time in the growing season, as soon as conditions favour rotting of the mother tubers and spread of the bacteria in soil, usually in July and August in Scotland. The fact that the mother tuber can be a source of both Ecc and Eca further increases its importance relative to leaves in the contamination of tubers.

Water is not a major contributor to tuber contamination. Results obtained in Israel and Spain where irrigation water is often heavily contaminated, mostly by Ecc and frequently with $> 10^6$ cells/ml (Lumb and Pérombelon 1986) suggest that the risk of increased blackleg level is low and that progeny tuber contamination at harvest is not high. But water containing even low numbers of Eca can have serious consequences when used to irrigate high grade erwinia-free seed crop because once introduced, the bacteria can spread during stock multiplication.

Only contamination brought about during mechanical hadnling of tubers could possibly explain the spread of Eca in stocks derived from micropropagated plants especially after the third multiplication cycle when, because of the bulk, more mechanical handling is involved. As discussed before, Eca is rarely found in the environment, air, water and soil and on potato leaves and should be also rare on progeny tubers in crops derived from axenic plants. But when the few Eca-infested tubers rot in storage, the organism tends to predominate in the rot and it is therefore more likely than Ecc to spread within and between stocks during grading. Moreover, because large numbers of erwinias are present in rotting tuber tissue, tuber contamination level can be much higher than that occurring before harvest in the field.

REFERENCES

Chiu WF, Yuen CS and Wu CA (1958) On the overwintering and dissemination of soft rot organisms, Erwinia aroideae (Towns) Holland. Acta Phytopath Sin 4:8-15 (in Chinese).

Elphinstone JG and Pérombelon MCM (1986a) Contamination of progeny tubers of potato plants by seed- and leaf-borne Erwinia carotovora. Pot Res 29:77-93

Elphinstone JG and Pérombelon MCM (1986b) Contamination of potatoes by Erwinia carotovora during grading. Plant Pathology 35:25-33

Graham DC, Quinn CE and Bradley LF (1977) Quantitative studies on the generation of aerosols of Erwinia carotovora var. atroseptica by simulated raindrop impaction on blackleg-infected potato strains. J Appl Bacteriol 43:412-424

Graham DC, Quinn CE, Sells IA and Harrison MD (1979) Survival of strains of soft rot coliform bacteria on microthreads exposed in the laboratory and in the open air. J Appl Bacteriol 46:

367-376

Gudmestad NC and Secor GD (1983) The bionomics of Erwinia carotovora in North Dakota. Am Potato J 60:759-771

Harrison MD, Quinn CE, Sells A and Graham DC (1977) Waste potato dumps as sources of insects contaminated with soft rot coliform bacteria in relation to recontamination of pathogen-free potato stocks. Pot Res 20:37-52

Lumb VM, Pérombelon MCM and Zutra D (1986) Studies of a wilt disease of the potato plant in Israel caused by Erwinia chrysanthemi. Plant Pathology 35:196-202

Molina JJ and Harrison MD (1980) The role of Erwinia carotovora in the epidemiology of potato blackleg. II. The effect of soil temperature on pathogen activity. Amer Pot J 57:351-363

McCarter-Zorner NJ, Franc GD and Harrison MD (1984) Soft rot Erwinia bacteria in surface and underground waters in southern Scotland and in Colorado, United States. J Appl Bacteriol 57:95-105

Pérombelon MCM (1972) The extent and survival of contamination of potato stocks in Scotland by Erwinia carotovora var. carotovora and E. carotovora var. atroseptica. Ann Appl Biol 71: 111-117

Pérombelon MCM (1973) Sites of contamination and numbers of Erwinia carotovora present in stored seed potato stocks in Scotland. Ann Appl Biol 74:95-65

Pérombelon MCM (1974) The role of the seed tuber in the contamination by Erwinia carotovora of potato crops in Scotland. Pot Res 17:187-189

Pérombelon MCM (1976) Effects of environmental factors during the growing season on the level of potato tuber contamination by Erwinia carotovora. Phytopath Z 85:97-116

Pérombelon MCM (1979) Contamination of potato crops by airborne Erwinia carotovora. Proc Int Conf Plant Path Bact 4th, Angers 1978. pp 63-65. ed Stat Path Veg Phytobacteriol Angers, France INRA, p 978

Pérombelon MCM and Hyman LJ (1987) Frequency of Erwinia carotovora in the Alyth Burn in eastern Scotland and the sources of the bacterium. J Appl Bacteriol 63:281-291

Pérombelon MCM and Hyman LJ (1988) Survival of soft rot coliforms Erwinia carotovora subsp. carotovora and E. carotovora subsp. atroseptica in soil in Scotland. J Appl Bacteriol (in Press)

Pérombelon MCM and Kelman H (1980) Ecology of the soft rot erwinias. Ann Rev Phytopathol 18:361-387

Pérombelon MCM, Fox RH and Lowe RH (1979) Dispersion of Erwinia carotovora in aerosols produced by the pulverisation of potato haulm prior to harvest. Phytopath Z 94:249-260

Quinn CE, Sells IA and Graham DC (1980) Soft rot Erwinia bacteria in the atmospheric bacterial aerosol. J Appl Bacteriol 49:175-181

Stanghellini ME (1982) Soft rotting bacteria in the rhizosphere. In Phytopathologenic Prokaryotes Vol 1, eds pp 249-261. Academic Press, New York

RECOGNITION AND RESPONSE BETWEEN HOST AND PARASITE AS DETERMINANTS IN RESISTANCE AND DISEASE DEVELOPMENT

C.H. Beckman
Department of Plant Sciences
University of Rhode Island
Kingston, RI 02881
USA

INTRODUCTION

The entire metabolic organization that characterizes life forms is dependent upon the capacity to detect ambient conditions, or changes in ambient conditions, and respond appropriately. This capacity is readily apparent in interactions between vascular parasites and potential hosts. Soil-borne, fungal, root parasites, for example produce thick-walled, resting structures that can survive in a dormant state for long periods of time. As plant roots ramify through the soil, root cap cells are sloughed off and root exudates leach out of these cells and from young roots in a gradient that may extend a millimeter or two from the root surface (Griffin 1969). The resting spores are able to detect the presence of these nutrients (and probably the gradient) and respond by rapid germination (Griffin 1981), growth toward the source, and colonization of the root rhizosphere, rhizoplane, and to varying degrees, cells of the epidermis, cortex (Gerik 1984; Gerik and Huisman 1985) and, sometimes, vascular tissues. These precesses, and the invasion of the vascular elements is apparently enhanced by senescence (Hepple 1963) or nematode infestation of roots (Bergeson et al 1970; Conroy et al 1972). When tissues have been invaded for some time and nutrients, especially carbohydrates, become depleted, cells of the parasites are rapidly converted into new, thick-walled, resting structures (Meyers and Cook 1972) to await the arrival of a new nutrient source. Thus, these parasites possess a remarkable system of re-

NATO ASI Series, Vol. H28
Vascular Wilt Diseases of Plants
Edited by E. C. Tjamos and C. Beckman

cognition and response that permits them to take advantage of sometimes fleeting nutrient sources.

The host, also, has a keen capacity to sense its environment. Epidermal and cortical cells have been shown to respond to the presence of many organisms, including Fusarium or Verticillium, apparently on contact, with the rapid synthesis and deposition of callose on abutting cell walls at specific sites of attempted penetration. These deposits frequently prevent the infection of epidermal and cortical cells of hops (Talboys 1958), and other plants (Griffiths 1971), including tomato (Bishop and Cooper 1983). They are, in themselves, highly impermeable to aqueous solutes (Heslop-Harrison 1966), and become permeated with lignin-like substances that make them highly resistant to deterioration, both physically and chemically (Vance and Sherwood 1977).

Similar capacities of host and parasite are played out within the vascular tissues of plants. When Fusarium spores are introduced into the vascular system of tomato plants through severed roots or stems, the spores germinate within a few hours. They then grow toward adjacent pits in the vessel walls, apparently following a nutrient gradient toward the vascular parenchyma cells that lie beyond the pits (Fig. 1A). When contacts with pits are made, the hyphae enlarge (Fig. 1B) and form bulbous tips that often fill the pit cavity (Fig. 1C). Here they apparently produce hydrolytic enzymes, presumably by inductive processes (Cooper and Wood 1975), that erode the pit membranes (Fig. 1B) and form penetration pegs (Fig. 1C, arrow) by which cell invasion is attempted.

In the meantime the host cells respond, whithin 8-12 hours after inoculation, by depositing callose as papillae that surround the penetrating hyphal tips (Fig. 1D, arrows), or as extended, (Fig. 1D) often layered (Fig. 2A) or marbled (Fig. 2B) wall appositions (Beckman et al 1982; Mueller and Beckman 1988). These deposits generally prevent infection of the cells and the protoplasts remain healthy (Fig. 2A,C). A somewhat smaller proportion of cells respond to attempted infections by a response in which the protoplasts become disorganized and stain strongly with Osmium tetroxide, thus becoming electron dense (Fig. 2C).

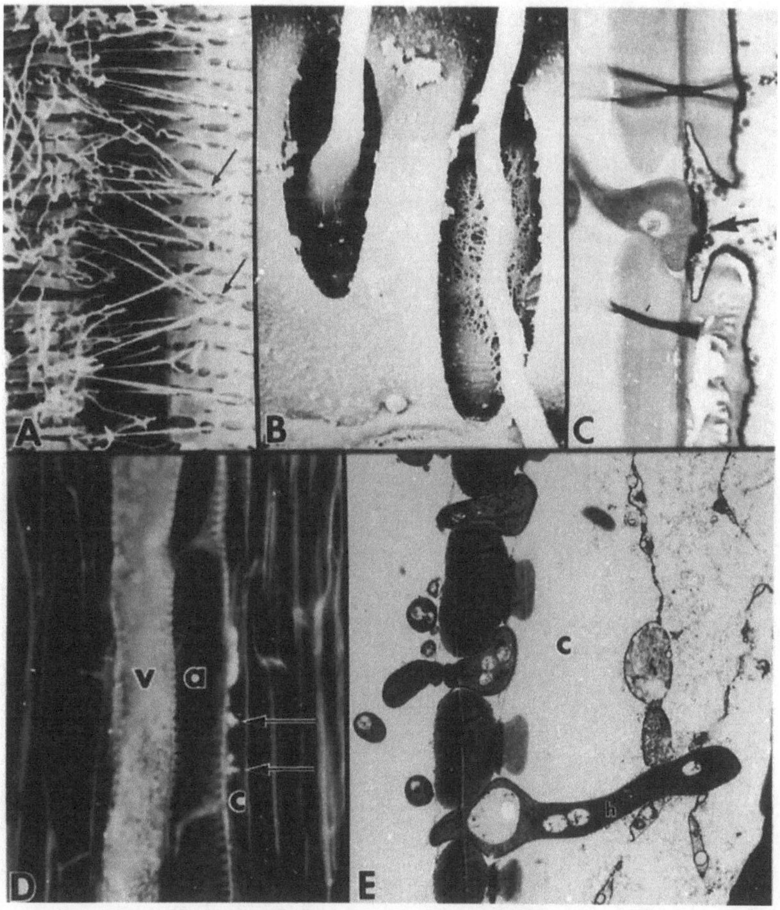

Fig. 1A-E. Light and transmission electron micrographs of longitudinal sections of the vascular tissues infected with Fusarium oxysporum showing A) an infected vessel in banana (far left) and an adjacent vessel across which hyphae have grown, apparently chemotropically, toward pits in the vessel wall (arrows) and the contact parenchyma cells that lie immediately beyond them, B) hyphae, after a day of incubation) that have made contact with, adhered to, and begun eroding a pit membrane in banana (note swollen condition of hyphae C) a hyphal tip that has, after 3 days, formed a large bulbous structure that has adhered tightly to a pit membrane and has produced an infection peg (arrow) that has been unsuccessful in penetrating the pit membrane in tomato. The pit membrane is strongly osmiophilic and apparently lignified or suberized, D) an infected vessel (V) and adjacent vessel (a) across which hyphae have grown to attack a contact parenchyma cell in which callose deposits in the form of papillae (arrow) and an apposition layer have been deposited and E) a contact parenchyma cell (c) that has, at 24 hours after infection, been successfully invaded by a fungal hypha (h) and the protoplast of which has become totally disorganized and disintegrated.

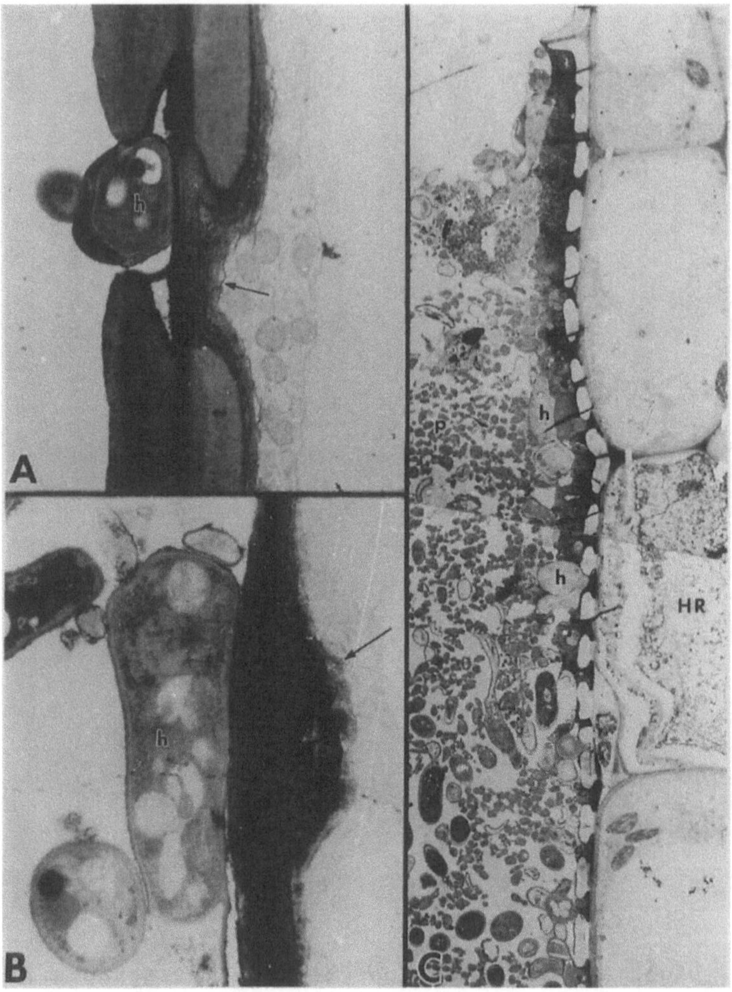

Fig. 2A-C. Transmission electron micrographs of longitudinal sections of vessels and contact cells of tomato infected with *Fusarium oxysporum* f.sp. *lycopersici*, race 1, showing A) a fungal hypha (h) that is tightly appressed to the pit membrane within the pit cavity of a secondary xylem vessel wall and a multilayered, callose-containing deposit (arrow) within the pit region, but also along the wall abutting the infected vessel, B) a marbled-appearing deposit of callose (arrow) in a contact cell at the immediate site of adherence of an enlarged hyphal tip (h) of the parasite that is closely appressed to it, and C) a heavily infected vessel showing hyphae (h) and small tracer particles (p) and four contact parenchyma cells, three of which have deposited callose (not visible at this magnification) and remain healthy, and one of which shows a hypersensitive reaction (HR) and is highly osmiophilic and dead, but not invaded by the fungus.

Such cells are rarely invaded. When they are, the fungus remains confined within them. This response, which apparently corresponds to that reported by Tjamos and Smith (1975), has therefore been designated a Hypersensitive Reaction (HR).

A study was next undertaken to compare the responses of vascular parenchyma cells in a single-dominant-gene resistant tomato cultivar (VF-11) and its susceptible near-isoline, Improved Pearson (IP). It was quickly apparent that some cells of both cultivars were successful in resisting penetration and by similar means. The rate of success, however, appeared to differ (Beckman et al 1982). A subsequent experiment was set up in which the responses in these cultivars could be compared in quantitative terms (Beckman et al 1988). Spores and red tracer particles (Beckman et al 1962) were introduced into the vascular system by severing 4-5 leaf-old plants at the base of the hypocotyl, immersing the cuttings in inoculum, and allowing transpirational uptake for 30 minutes. The cuttings were then transferred to tap water and held in an environment chamber at 27 ± 1 C until harvested. Cuttings were removed periodically and thick, longitudinal sections were cut with a razor blade and cutting guide (Beckman et al 1962). The sections were embedded and sectioned longitudinally so that vascular parenchyma cells, to a depth of 3 cells from infected vessels, could be examined by means of electron microscopy (Mueller and Beckman 1988; Beckman et al 1988). These cells were designated and recorded as <u>contact cells</u>, those in immediate contact with a vessel, and <u>first</u> (1^0) and second (2^0) <u>adjacent</u> cells.

The resulting data showed that the parenchyma cells associated with protoxylem tissues (i.e., those derived from the apical meristem) were highly resistant (95-97%) in both resistant and susceptible cultivars, respectively. The data obtained from secondary xylem tissues (i.e., those derived from cambial activity) differed greatly in the resistant and susceptible cultivars. The data for secondary xylem tissues are presented diagrammatically in Fig. 3A, B in which it can be seen that, although similar responses occurred in both cultivars, the rate of successful defense was much higher in the resistant than in the susceptible cultivar (See legend). Conversely, the rate of successful penetration and colonization of cells was much higher in the susceptible host (which has the "I" gene in a homozygous recessive "ii" condition) than in the resistant cultivar (which, in this case, has the "I" gene in the homozygous dominant "II" condition).

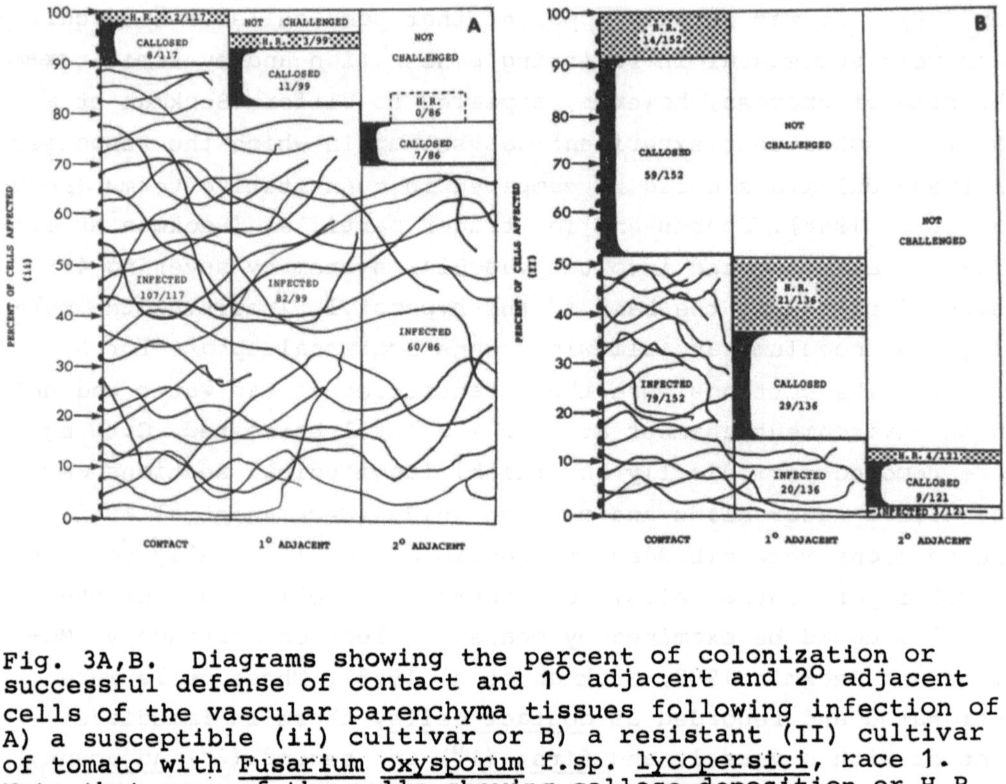

Fig. 3A,B. Diagrams showing the percent of colonization or successful defense of contact and 1° adjacent and 2° adjacent cells of the vascular parenchyma tissues following infection of A) a susceptible (ii) cultivar or B) a resistant (II) cultivar of tomato with Fusarium oxysporum f.sp. lycopersici, race 1. Note that none of the cells showing callose deposition or H.R. were infected and that such cells protected next adjacent cells from challenge by the fungus. The fractions represent the number of cells that showed a particular response or condition divided by the total number of cells counted in each of the cell layers.

These data are integrated into the overall model of host-parasite interactions (Fig. 4) within a time frame of 0-3 days (t_{0-3}) and that portion of the vascular system that was directly infested at the time of inoculum uptake, i.e., Space zero (S_0). A quotient for infection (Q_i) and the inverse quotient for defense (Q_d) were also calculated to begin the process of modeling colonization by the parasite or defense by the host in mathematical terms. These quotients were calculated by dividing the number of cells that had become infected by the total number of cells observed in the 3-cell layers (Q_i) or by dividing the total number of cells that remained healthy by the total number of cells observed in the 3-cell layers (Q_d).

Fig. 4 illustrates, first, the relative capacity of cells of plants, that have different levels of resistance, to mount a defense in depth. Secondly, it illustrates that the relative success or failure of that defense becomes extended in time and space as the infection progresses or becomes static as the infection fails to progress, through a succession of cells. The net effect, in the case of the two cultivars employed in this study, was that the pathogen colonized 4 times as many cells in the susceptible host as in the resistant host within the three cell layers observed. At that point the infection in the resistant cultivar would be essentially checked, while that in the susceptible cultivar would continue to advance. Furthermore, if the third or fourth, or fifth cell encountered in a susceptible host were another vessel and that vessel were not already infected, an entirely new infection would be established in which the processes of colonization and defense would begin anew.

These results raise some intersting and pertinent questions about the basic mechanisms by which a host plant recognizes and responds to the challenge of a foreign presence. Since the recognition and response processes are non-specific with respect to the challenging organism (Beckman et al 1982), it is apparent that the challenging organism, in order to be a successful colonizer and pathogen, must somehow by-pass or inhibit the processes of recognition and/or response.

It is suggested, on the basis of the immunological studies of DeVay and his colleagues (1967, 1981), that the basis for the successful colonization of specific host cultivars by specific races of a parasite lies in the degree of commonality within the molecular structure of their respective cell walls and/or plasma membranes. If that commonality is small, the differences are readily detected and responses are initiated quickly and strongly.

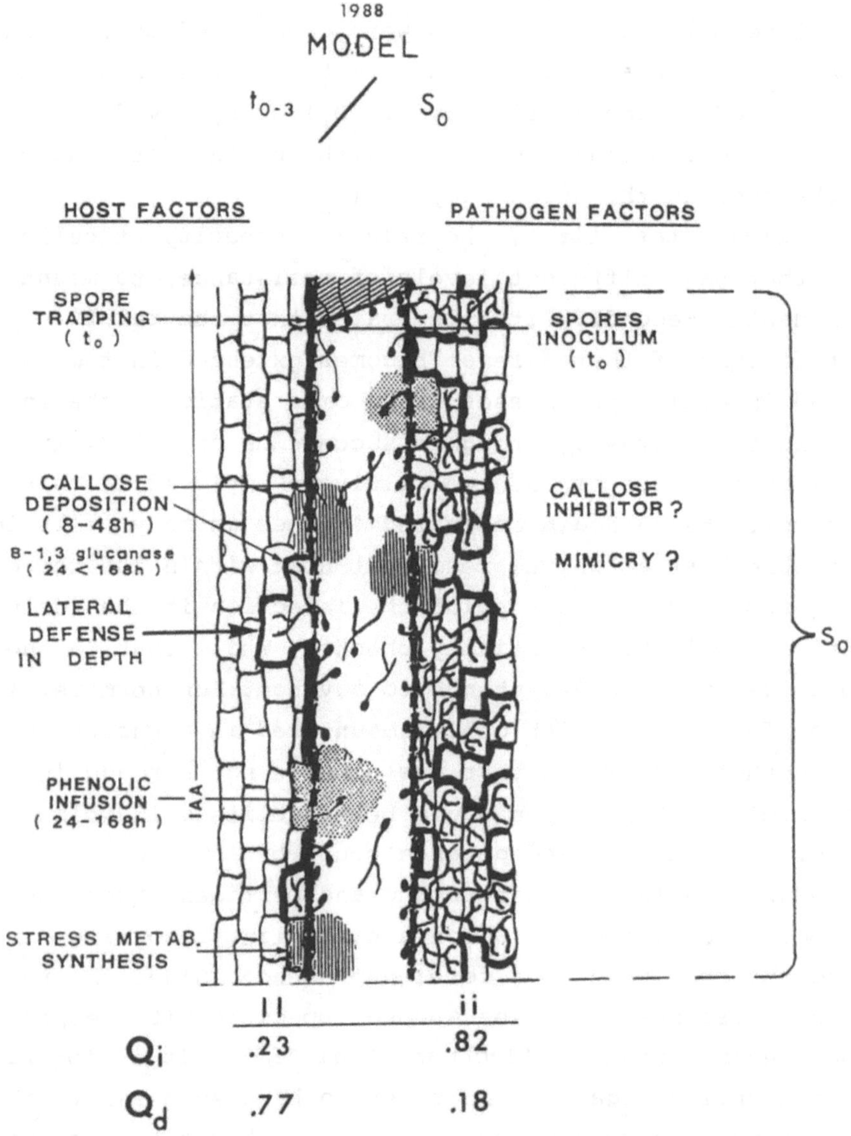

Fig. 4. Model of the colonization by *Fusarium oxysporum* f.sp. *lycopersici*, race 1 during the first 3 days of incubation (t_{0-3}) from an initially infected (S_0), secondary xylem vessel into the xylem parenchyma tissues of a single-dominantly resistant (II) cultivar (left) or a susceptible (ii) cultivar (right). Shown also are host factors (left) and pathogen factors (right) that appear to contribute to the outcome. Quotients based on the proportion of cells successfully invaded by the parasite (Q_i) or successfully defended by the host (Q_d) are also given for the resistant (II) and susceptible (ii) hosts.

If the commonality is great, i.e., if the surface molecules of the parasite closely mimic those of the host cells (Damian 1979), recognition of a "foreign" presence does not occur, or occurs more sporadically or at a lesser rate. Such differences, determined within hours or possibly minutes after contact between the cells of host and parasite, could ultimately determine, through a chain of events, whether the metabolic potential of the host or that of the parasite will predominate within a particular time and space frame and, ultimately, whether the host will successfully sealed off the parasite and neutralize the infection, or whether the parasite will avoid, prevent, or overcome these responses and continue to colonize the host in ever broadening time and space frames.

REFERENCES

Beckman CH (1987) The Nature of Wilt Diseases of Plants. 175 pp. APS Press, St Paul, MN, U.S.A.
Beckman CH, Halmos S and Mace ME (1962) The interaction of host, pathogen, and soil temperature in relation to susceptibility to Fusarium wilt of bananas. Phytopathology 52:134-140
Beckman CH, Mueller WC, Tessier BJ and Harrison NA (1982) Recognition and callose deposition in response to vascular infection in fusarium wilt-resistant or susceptible tomato plants. Physiol Plant Pathol 20:1-10
Beckman CH, Verdier PA and Mueller WC (1988) A system of defense in depth provided by vascular parenchyma cells in response to vascular infection with Fusarium oxysporum f.sp. lycopersici, race 1. Physiol Mole Plant Pathol (in Press)
Bergeson GB, Van Gundy SD and Thomason IJ (1970) Effect of Meloidogyne javanica on rhizosphere microflora and Fusarium wilt of tomato. Phytopathology 60:1245-1249
Conroy JJ, Green RJ, Jr and Ferris JM (1972) Interaction of Verticillium albo-atrum and the root lesion nematode, Pratylenchus penetrans, in tomato roots at controlled inoculum densities. Phytopathology 62:362-366
Damian RT (1979) Molecular mimicry in biological adaptation. Pages 103-126 in: Host-parasite Interfaces. Nickol BB (ed) Academic Press, New York
DeVay JE, Schnathorst WC and Foda MS (1967) Common antigens and host-parasite interactions. In: The Dynamic Role of Molecular Constituents in Plant-Parasite Interaction. Microcha CJ and Uritani I (eds). American Phytopathological Society, St Paul, Minnesota, U.S.A.
DeVay JE, Wakeman RJ, Kavanaugh JA and Charudattan R (1981) The tissue and cellular location of a major cross-reactive antigen

shared by cotton and soil-borne fungal parasites. Physiol Plant Pathol 18:59-66
Gerik JS (1984) Fungal colonization of the cotton root system with emphasis on Verticillium dahliae. PhD Thesis, University of California, Berkeley
Gerik JS and Huisman OC (1985) Mode of colonization of roots by Verticillium and Fusarium. Pages 80-83 in: Ecology and Management of Soilborne Plant Pathogens. CA Parker, AD Rovira, KJ Moore, PTW Wong and JS Kollmorgen (eds). American Phytopathological Society, St Paul, MN
Griffin GJ (1969) Fusarium oxysporum and Aspergillus flavus spore germination in the rhizosphere of peanut. Phytopathology 59: 1214-1218
Griffin GJ (1981) Physiology of conidium and chlamydospore germination in Fusarium. Pages 331-339 in: Fusarium: Diseases, Biology and Taxonomy. PE Nelson, TA Toussoun and RJ Cook (eds). Pennsylvania State University Press, University Park
Griffiths DA (1971) The development of lignitubers in roots after infection by Verticillium dahliae Kleb. Can J Microbiol 17: 441-444
Hepple S (1963) Infection of pea plants by Fusarium oxysporum f.sp. pisi in naturally infested soil. Trans Brit Mycol Soc 46:585-594
Heslop-Harrison J (1966) Cytoplasmic continuities during spore formation in flowering plants. Endeavor 25:65-72
Meyers JA and Cook RJ (1972) Induction of chlamydospore formation in Fusarium solani by abrupt removal of the organic carbohydrate substrate. Phytopathology 62:1148-1153
Mueller WC and Beckman CH (1988) Correlated light and electron microscope studies of callose deposits in vascular parenchyma cells of tomato plants inoculated with Fusarium oxysporum f. sp. lycopersici. Physiol Mole Plant Pathol (in Press)
Talboys PW (1958) Some mechanisms contributing to Verticillium-resistance in the Hop roots. Trans Brit Mycol Soc 41:227-241
Tjamos EC and Smith IM (1975) The expression of resistance to Verticillium albo-atrum in monogenically resistant tomato varieties. Physiol Plant Pathol 6:215-225
Vance CP and Sherwood RT (1977) Lignified papilla formation as a mechanism for protection in reed canarygrass. Physiol Plant Pathol 10:247-256

SECONDARY METABOLITES PRODUCED IN RESISTANT AND SUSCEPTIBLE HOST PLANTS IN RESPONSE TO FUNGAL VASCULAR INFECTION

M.E. Mace
Cotton Pathology Research Unit
Southern Crops Research Laboratory
USDA, ARS
College Station, Texas 77841
USA

INTRODUCTION

Resistance is a relative term used to describe the ability of a plant to prevent, restrict, or retard the penetration and development of pathogens in host tissue. The biochemical events involved with resistance are invariably invoked in "susceptible" as well as in resistant plants in response to transient biotrophs, e.g., Verticillium and Fusarium wilt fungi. Thus, from a biochemical perspective susceptible plants also resist infection, but the speed or intensity of the response is inadequate, and the pathogen is able to progressively colonize the plant. Therefore, the degree of resistance depends on the speed and intensity of the defensive response of the host relative to the speed of infection by the pathogen (Mace 1978).

Host plant resistance is the major method used to control fungal wilt diseases. Thus, many attempts have been made to determine the mechanisms of resistance with the hope that such knowledge would facilitate the development and use of resistant cultivars. Although this objective has not been achieved for most wilt diseases, several general concepts of the nature and mechanisms of resistance are emerging. This review paper will evaluate the function of secondary compounds in susceptible and resistant responses to fungal wilt diseases. Special emphasis will be placed on the phytoalexins associated with Verticillium wilt of cotton.

NATO ASI Series, Vol. H28
Vascular Wilt Diseases of Plants
Edited by E. C. Tjamos and C. Beckman

SECONDARY COMPOUNDS

Many secondary compounds produced by host plants are toxic to fungi. Antibiotic compounds produced by plants frequently have been implicated in resistance responses. Resistance to fungal wilt diseases is composed of multiple biochemical factors. Thus, any putative factor should be viewed from several perspectives if its role in the resistance complex is to be fully understood.

Antibiotic secondary compounds produced by plants can be divided into two major groups: constitutive antibiotics and phytoalexins. Constitutive compounds are synthesized and stored in living cells for extended periods of time. Wound-induced antibiotics may be regarded as types of constitutive compounds that have undergone wound-induced chemical change. Phytoalexins are synthesized de novo after the plant undergoes stress. Constitutive compounds frequently undergo quantitative and qualitative changes after infection of the host plant; thus, they share some of the characteristics of the phytoalexins, and will be considered along with the phytoalexins.

Constitutive Compounds

Infection has variable effects on the concentration of constitutive antibiotics. Tomatine increases similarly in resistant and susceptible tomato plants inoculated with F. oxysporum (Langcake et al 1972; McCance and Drysdale 1975), but in wounded roots tomatine content increases most rapidly in resistant cultivars. Flavanol concentrations in cotton stems also increase more rapidly and to a greater extent in resistant than in susceptible cultivars infected with V. dahliae (Bell and Stipanovic 1978; Mace et al 1978). Flavanols also increase in young, but not old, cotton leaves in response to the fungus (Howell et al 1976).

Except for flavanols in cotton, little is known about cellular localization of constitutive antibiotics. Tomatine occurs in extracts of xylem tissues but not in xylem exudates from severed stem stubs (Langcake et al 1972; McCance and Drysdale

1975). Thus, it may be formed apart from the vessel. Less than 5% of the flavanols from cotton were recovered from xylem vessels eluted with ethanol (Bell and Stipanovic 1978), and histochemical studies showed that most of the induced synthesis was in the xylem ray cells apart from the vessels (Mace et al 1978). Thus, flavanols and possibly other constitutive antibiotics may be more important for restricting the fungus to vessels than for restricting its growth within the vessel system.

Flavanols in cotton roots occur in the root cap, endodermis, hypodermis, and occasional scattered cells of the xylem parenchyma (Mace and Howell 1974). Mueller and Beckman (1976) and Mueller and Greenwood (1978) studied the ultrastructure of flavanol-storing cells in the endodermis. Flavanols first appear 1-2 cm back of root tips in numerous small vacuoles in cells. Later, these vacuoles coalesce into one or a few vacuoles. The cytoplasm and organelles of flavanol-storing cells are similar to adjoining cells free of flavanols. Thus, the intracellular origin of flavanols in uncertain.

Many monomeric polyphenols, such as caffeic acid and chlorogenic acid, have limited toxicity to wilt fungi, but their oxidation products are antibiotics (Patil et al 1964) and enzyme denaturants (Beckman et al 1974). In these cases, the phenols stored in vacuoles (Mueller and Beckman 1974) must be mixed with phenol oxidase enzymes, such as peroxidase, tyrosinase, or laccase (Mace et al 1972; Mace 1963; Mace and Wilson 1964) to form the most active components.

Xylem vessels infected by wilt fungi characteristically become filled with gums that also impregnate vessel walls and surrounding vessels, and the vessels turn brown in color. Streaks of vascular browning are often shorter, more intense, and more scattered in immune and resistant than in susceptible plants (Smalley and Riker, 1962). Most gums exuded from paratracheal cells into xylem vessels and surrounding tissue contain polysaccharides and phenolic compounds that probably contribute to the browning (Mace et al 1972; Davis et al 1953). For example, vascular browning is due to oxidation products of dopamine in banana (Mace 1963; Mace and Wilson 1964) and of catechin in cotton (Bell and Howell 1971). Oxidation products of dihydroxy phenols also apparently cause browning in elm (Gagnon 1967) and tomato (Mace et al 1972). The specific compounds involved have not been identified, although (+)-catechin and caffeic acid derivatives, respectively are likely involved.

Phenol concentrations before and after infection have been correlated with resistance. Verticillium-resistant tobacco cultivars (Sheppard and Peterson 1976) and potato cultivars (McLean et al 1961) have the greatest amounts of the dihydroxyphenol chlorogenic acid. The decrease of resistance to Verticillium wilt with age in potato also is associated with a similar decrease in the chlorogenic acid content of the roots (Patil et al 1964). In strawberry (Okasha et al 1968) and tomato (Matta et al 1969), phenol contents of healthy cultivars are similar, but greater concentrations of phenols are formed in resistant than susceptible cultivars soon after inoculation with V. dahliae or F. oxysporum, respectively. Resistance to Fusarium wilt induced in tomato by catechol (Retig and Chet 1974) or by quinic acid and phenylalanine (Carrasco et al 1978) also is associated with increased phenolic contents.

Dopamine is the major substrate for vascular browning in Fusarium-infected banana roots (Mace 1963). Indoleacetic acid (IAA), produced by the banana wilt pathogen Fusarium oxysporum f. sp. cubense, stimulates tylosis, a key resistance factor in banana roots (Beckman et al 1962). Furthermore, dopamine inhibits the IAA oxidase of banana roots (Mace and Solit 1966). Consequently, dopamine may enhance tylosis and resistance to Fusarium wilt by protecting IAA from enzymatic oxidation.

Phytoalexins

Phytoalexins elicited by wilt fungi have been studied in alfalfa, cotton, elm, and tomato. All but alfalfa, form sesquiterpenoid phytoalexins. Alfalfa produces pterocarpans, sativan and medicarpin (Khan and Milton 1975).

The major terpenoid aldehyde phytoalexins in cotton are hemigossypol, and hemigossypol-6-methyl ether, (Bell et al 1975). Their precursors, desoxyhemigossypol and desoxyhemigossypol-6-

methyl ether, are also prominent, especially 12-48 hours after infection (Stipanovic et al 1975). Only small amounts of gossypol and its 6-methyl and 6,6'-dimethyl ethers are formed as phytoalexins in the xylem (Mace et al 1976), although these are the major constitutive terpenoids in the epidermis of roots (Mace et al 1974) and in lysigenous pigment glands of the phloem (Bell et al 1978).

Mansonone E and mansonone F were identified as phytoalexins in Ulmus hollandica (Overeem and Elgersma 1970; Elgersma and Overeem 1971). The predominant terpenoids in the taxonomic section Madocarpus (including U. hollandica) are the aldehydes, 7-hydroxycadalenal and 7-hydroxy-3-methoxycadalenal; other prominent compounds are 7-hydroxycalamenene, 7-hydroxycadalene, and mansonone C (Rowe et al 1972; Krishnamoorthy and Thomson 1971). Both these results and ours with cotton (Mace et al 1985) suggest that terpenoid aldehydes, mansonones and their precursors probably are formed as phytoalexins in xylem vessels of resistant elm.

Rishitin is the only phytoalexin identified from tomato, although several others are formed, especially by a resistant variety in response to V. albo-atrum (Tjamos and Smith 1974). More than 20 other phytoalexins have been identified from other solanceous hosts of wilt fungi such as potato, pepper, eggplant, and tobacco (Stoessl et al 1977), but wilt fungi were not used to induce their synthesis.

Phytoalexins generally inhibit mycelial growth, conidial germination, and sporulation. Rishitin had an ED_{50} value between 30-100 µg/ml for inhibition of conidial germination of V. albo-atrum, whereas the ED_{50} for inhibition of germ tube elongation was 3 µg/ml (McCance and Drysdale 1975). Corresponding values for F. oxysporum f. sp. lycopersici were 50-100 µg/ml and 8 µg/ml, respectively. Rishitin also may inhibit tylosis in Fusarium wilt of tomato (Harrison and Beckman 1987). Phytoalexins from alfalfa also were slightly more toxic to mycelial elongation than to conidial germination of V. albo-atrum (Khan and Milton 1975). More importantly, virulent strains of Verticillium were much less sensitive to sativan and medicarpin than avirulent strains.

Hemigossypol (HG), methoxyhemigossypol (MHG), desoxyhemigossypol (dHG) and desoxymethoxyhemigossypol (dMHG) are the major phytoalexins elicited in V. dahliae-infected, wilt-resistant Seabrook Sea Island (SBSI) cotton. The terpenoids dHG, HG, dMHG, and MHG at 25 C killed V. dahliae conidia after 18-40 hours at 10, 45, 25, and 60 µg/ml, respectively; and mycelia after 48 hours at 15, 35, 25, and 45 µg/ml, respectively (Mace et al 1985). Dimethylsulfoxide at 2 to 5% and/or increase of the pH above that of infected stem xylem (pH 6.3) was required to solubilize HG, MHG, and dMHG at inhibitory concentrations. Only dHG was sufficiently water soluble at pH 6.3 to account for death of V. dahliae conidia and mycelia at most sites in the stem stele of SBSI cotton 10 days after inoculation (Mace 1978). The dHG in the stem stele was well in excess of fungicidal concentrations. Recent studies (Mace unpublished) indicate no significant differences between the sensitivity of V. dahliae isolates of high and low virulence to cotton (Puhalla 1978) to the four major terpenoid phytoalexins.

The terpenoid aldehyde phytoalexins of cotton are formed primarily by paratracheal parenchyma cells and are exuded into xylem vessels where they are absorbed by fungal hyphae and conidia, cell walls, and tyloses (Mace et al 1976). Over one-half of the induced terpenoid aldehydes were recovered in ethanolic eluates of vessels (Bell 1969) compared to less than 5% of the induced flavanols (Bell and Stipanovic 1978). This shows, at least in cotton, that the terpenoid phytoalexins are the main secondary compounds exuded into vessels.

Phytoalexins in various plants are apparently exuded into gums, gels, or tylose walls of infected xylem vessels. These structures may block phytoalexin movement in the vascular system (Mace et al 1976; Mace 1978) and permit its accumulation in regions of infection. Xylem sap exuded from cut stems or aqueous xylem eluates collected under vacuum generally contain little or no phytoalexin, presumably because it occurs only in vessels that are occluded as a result of infection (McCance and Drysdale 1975). Phytoalexins have only been obtained from vessels by eluting, under vacuum, with ethanol or acetone, solvents that dehydrate and shrink gels and gums (Bell 1967; Stromberg and Corden 1977). These facts need careful consideration in further studies of phytoalexins in the xylem of other host plants.

Both the rate and magnitude of phytoalexin synthesis have been consistently correlated with the resistance of cultivars to wilt diseases. Differences in phytoalexin synthesis between susceptible and resistant cultivars have been particularly striking in the xylem tissue of the root and stem where critical differences in the speed and extent of fungal development also are observed.

Stem tissues of resistant 'Marvis Kabul' alfalfa accumulate twice as much medicarpin and sativan as the susceptible 'Dupuits' by 48 hours after inoculation with V. albo-atrum (Khan and Milton 1978). This difference occurs regardless of the virulence of the fungal strain, although avirulent strains induce much higher levels of phytoalexins than virulent strains in both cultivars (Khan and Milton 1975). Leaf tissues produce more phytoalexins than stems, but amounts produced by leaves of resistant and susceptible cultivars are similar.

More rapid synthesis of phytoalexins occurs in xylem tissue of resistant than in susceptible tomato cultivars affected by either Verticillium or Fusarium wilt (Khatib et al 1974; McCance and Drysdale 1975; Stromberg and Corden 1977). Uniform inoculation of stem segments with conidia by application of vacuum has given greater and more consistent differences in the response of resistant and susceptible cultivars to V. albo-atrum (Tjamos and Smith 1974) and to F. oxysporum (McCance and Drysdale 1975) than root inoculations of whole plants. When plants are inoculated with F. oxysporum through severed roots, both root and stem xylem of resistant cultivars contain about twice as much rishitin as susceptible cultivars at 24 and 48 hours after inoculation. By 72 hours, concentrations are similar in the cultivars (McCance and Drysdale 1975). The generation time of F. oxysporum is only 18-30 hours, depending on temperature and cultivars. Thus, even at 72 hours the host could be responding to three generations of this fungus. The more rapid growth, sporulation, and distribution of the fungus in the susceptible than in the resistant plant eventually would cause phytoalexin concentrations in the susceptible plant to reach and exceed those in the resistant plant.

Root and fruits from tomato cultivars with monogenic resistance to wilts also may show greater rishitin concentrations than those from susceptible cultivars (Khatib et al 1974; Tjamos and Smith 1975). The rishitin in roots accumulated in necrotic tissue resulting from a hypersensitive response. Tjamos and Smith (1975) thought that this response greatly reduced the number of successful penetrations by V. albo-atrum into xylem vessels in resistant compared to susceptible cultivars. When inoculated with conidia of V. albo-atrum, fruit from a resistant cultivar accumulated 22.2 µg rishitin/g tissue compared to only 1.4 µg/g in fruit from a susceptible cultivar (Tjamos and Smith 1974).

Relationships between resistance and phytoalexin synthesis have been studied extensively in cotton. Cultivars of Gossypium barbadense, resistant to Verticillium wilt produce phytoalexins 24-48 hours sooner in response to V. dahliae infection than susceptible or tolerant G. hirsutum cultivars (Bell 1969; Zaki et al 1972b; Mace 1978). Tolerant cultivars of G. hirsutum form phytoalexins more rapidly than susceptible cultivars in response to avirulent or moderately virulent, but not highly virulent, strains of V. dahliae (Bell 1969; Zaki et al 1972b).

Wilhelm et al (1970, 1974) described the localization of V. albo-atrum (sic) (V. dahliae) in the lower four to five internodes of Verticillium wilt-resistant G. barbadense cottons. When the G. barbadense cultivar Seabrook Sea Island (SBSI) was infected early in the growing season, Verticillium was isolated only from the older or inner xylem of lower internodes in mature plants. The pathogen often colonized a few lower leaves and induced localized wilt, but most of the mature plant showed no wilt symptoms and was generally free of Verticillium. Comparable Gossypium hirsutum plants failed to restrict the pathogen to the lower stem, and systemic spread of the pathogen and severe wilt ensued.

The combined effects of rapid tylosis and TA synthesis are apparently needed for the resistance of SBSI to Verticillium wilt (Mace 1978; Mace et al 1985). Rapid occlusion of vessels at primary infection sites in SBSI apparently prevents systemic distribution of secondary conidia and allows subsequent accumulation of terpenoid phytoalexins within the localized primary infection sites. Once V. dahliae is effectively localized in wilt-resistant SBSI, formation of new, non-infected xylem tissue apparently enables SBSI to compensate for vessels occluded by responses to the primary infections. Occlusion of vessels in the G. hirsutum cultivar Rowden is delayed and allows many secondary conidia to escape containment at primary infections sites and thereby to escape exposure to the localized accumulation of terpenoid phytoalexins. Because of extensive systemic distribution of secondary conidia in Rowden, infection of newly formed xylem vessels generally kept pace with their formation, and severe wilt disease developed.

CONCLUSIONS

Induced resistance to fungal wilt diseases usually results in a marked reduction of the extent of secondary colonization by the fungus within the vascular tissue. This resistance appears to depend largely on anatomical trapping of the fungus followed by rapid accumulation of antibiotics around the fungal cells. Without these suppressive chemicals, the fungus apparently digests or penetrates through the anatomical barrier. Consequently, rapid formation of coordinated anatomical and antibiotic mechanisms is necessary for successful resistance.

There is a need for detailed histochemical studies in evaluating the role of host metabolites in resistance. When these are used with normal histological and biochemical studies, a better understanding of resistance mechanisms in possible. Detailed histochemical studies (Mace and Howell 1974) have shown that the endodermis of cotton contains catechin and its proanthocyanidin derivatives, which may offer resistance to fungal penetration, whereas the compounds exuded into xylem vessels to contain secondary colonization of wilt fungi are mostly sesquiterpenoid aldehydes (Mace et al 1976; Mace et al 1985). Moreover, histochemical studies show that the terpenoid aldehydes are absorbed and accumulated by fungal hyphae and tyloses at a proper time to be effective in resistance (Mace 1978). Detailed histochemical studies are needed to clarify the time-space relationships in other wilt diseases.

REFERENCES

Beckman CH, Halmos S and Mace ME (1962) The interaction of host pathogen, and soil temperature in relation to susceptibility to Fusarium wilt of banana. Phytopathology 52:134-140

Beckman CH, Mueller WC and Mace ME (1974) The stabilization of artificial and natural cell wall membranes by phenolic infusion and its relation to wilt disease resistance. Phytopathology 64:1214-1220

Bell AA (1969) Phytoalexin production and Verticillium wilt resistance. Phytopathology 59:1119-1127

Bell AA and Howell CR (1971) Comparative physiology of Verticillium albo-atrum isolates. II. Induction of phytoalexins in cotton. In: Proc Beltwide Cotton Prod Res Conf National Cotton Council, Memphis, p 84

Bell AA and Stipanovic RD (1978) Biochemistry of disease and pest resistance in cotton. Mycopathologia 65:91-106

Bell AA, Stipanovic RD, Howell CR and Fryxell PA (1975) Antimicrobial terpenoids of Gossypium: hemigossypol, 6-methoxyhemigossypol and 6-deoxyhemigossypol. Phytochemistry 14:225-231

Bell AA, Stipanovic RD, O´Brien DH and Fryxell PA (1978) Sesquiterpenoid aldehyde quinones and derivatives in pigment glands of Gossypium. Phytochemistry 17:1297-1305

Carrasco A, Boudet AM and Marigo G (1978) Enhanced resistance of tomato plants to Fusarium by controlled stimulation of their natural phenolic production. Physiol Plant Pathol 12:225-232

Davis D and Dimond AE (1953) Inducing disease resistance with plant growth-regulators. Phytopathology 43:137-140

Elgersma DM and Overeem JC (1971) The relation of mansonones to resistance against Dutch elm disease and their accumulation, as induced by several agents. Neth J Plant Pathol 77:168-174

Gagnon C (1967) Polyphenols and discolouration in the elm disease investigated by histochemical techniques. Can J Bot 45:2119-2124

Harrison NA and Beckman CH (1987) Growth inhibitors associated with Fusarium wilt of tomato. Physiol and Mole Plant Pathol 30:401-420

Howell CR, Bell AA and Stipanovic RD (1976) Effect of aging on flavonoid content and resistance of cotton leaves to Verticillium wilt. Physiol Plant Pathol 8:181-188

Khan FZ and Milton JM (1975) Phytoalexin production by lucerne (Medicago sativa L.) in response to infection by Verticillium. Physiol Plant Pathol 7:179-187

Khatib AEl, Aramouni A, Hassan A and Ravise A (1974) Accumulation of phenolic substances and phytoalexins in varieties of tomato cultivated in infested soils in Lebanon. Phytopathol Z 81: 23-37

Krishnamoorthy V and Thomson RH (1971) Mansonone C in elm wood. Phytochemistry 10:1669-1670

Langcake P, Drysdale RB and Smith H (1972) Postinfectional production of an inhibitor of Fusarium oxysporum f. sp. lycopersici by tomato plants. Physiol Plant Pathol 2:17-25

Mace ME (1963) Histochemical localization of phenols in healthy and diseased banana roots. Physiol Plant 16:915-925

Mace ME (1978) Contribution of tyloses and terpenoid aldehyde phytoalexins to Verticillium wilt resistance in cotton. Physiol Plant Pathol 12:1-11

Mace ME and Howell CR (1974) Histochemistry and identification of condensed tannin precursors in roots of cotton seedlings. Can J Bot 52:2423-2426

Mace ME and Solit E (1966) Interactions of 3-indoleacetic acid and 3-hydroxytyramine in Fusarium wilt of banana. Phytopathology 56:245-247

Mace ME and Wilson EM (1964) Phenol oxidases and their relation to vascular browning in Fusarium-invaded banana roots. Phytopathology 54:840-842

Mace ME, Bell AA and Beckman CH (1976) Histochemistry and identification of disease-induced terpenoid aldehydes in Verticillium wilt-resistant and -susceptible cottons. Can J Bot 54: 2095-2099

Mace ME, Bell AA and Stipanovic RD (1978) Histochemistry and identification of flavanols in Verticillium wilt-resistant and -susceptible cottons. Physiol Plant Pathol 13:143-149

Mace ME, Stipanovic RD and Bell AA (1985) Toxicity and role of terpenoid phytoalexins in Verticillium wilt resistance in cotton. Physiol Plant Pathol 26:209-218

Mace ME, Veech JA and Beckman CH (1972) Fusarium wilt of susceptible and resistant tomato isolines: histochemistry of vascular browning. Phytopathology 62:651-654

Matta A, Gentile I and Giai I (1969) Accumulation of phenols in tomato plants infected by different forms of Fusarium oxysporum. Phytopathology 59:512-513

McCance DJ and Drysdale RB (1975) Production of tomatine and rishitin in tomato plants inoculated with Fusarium oxysporum f. sp. lycopersici. Physiol Plant Pathol 7:221-230

McLean JG, LeTourneau DJ and Guthrie JW (1961) Relation of histochemical tests for phenols to Verticillium wilt resistance in potatoes. Phytopathology 51:84-89

Mueller WC and Beckman CH (1974) Ultrastructure of the phenol-storing cells in the roots of banana. Physiol Plant Pathol 4:187-190

Mueller WC and Beckman CH (1976) Ultrastructure and development of phenolic-storing cells in cotton roots. Can J Bot 54:2074-2082

Mueller WC and Greenwood AD (1978) The ultrastructure of phenolic-storing cells fixed with caffeine. J Exp Bot 29:757-764

Okasha KA, Ryugo K and Bringhurst RS (1968) Inhibition of growth of Verticillium albo-atrum sporelings by tannins and polyphenols from infected crowns of Verticillium-resistance and -susceptible strawberry cultivars. Phytopathology 58:1118-1122

Overeem JC and Elgersma DM (1970) Accumulation of mansones E and F in Ulmus hollandica infected with Ceratocystis ulmi. Phytochemistry 9:1949-1952

Patil SS, Powelson RL and Young RA (1964) Relation of chlorgogenic acid and free phenols in potato roots in infection by Verticillium albo-atrum. Phytopathology 54:531-535

Puhalla JE (1979) Classification of isolates of Verticillium dahliae based on heterokaryon incompatibility. Phytopathology 69:1186-1189

Retig N and Chet I (1974) Catechol-induced resistance of tomato plants to Fusarium wilt. Physiol Plant Pathol 4:469-475

Rowe JW, Seikel MH, Roy DN and Jorgensen E (1972) Chemotaxonomy of Ulmus. Phytochemistry 11:2513-2517

Sheppard JW and Peterson JF (1976) Chlorogenic acid and Verticillium wilt of tobacco. Can J Plant Sci 56:157-160

Smalley EB and Riker AJ (1962) Reactions of several tropical members of the Ulmaceae to inoculation with Ceratocystis ulmi. Phytopathology 52:28

Stipanovic RD, Bell AA and Howell CR (1975) Naphthofuran precursors of sesquiterpenoid aldehydes in diseased Gossypium. Phytochemistry 14:1809-1811

Stoessl A, Ward EWB and Stothers JB (1977) Biosynthetic relationships of sesquiterpenoidal stress compounds from the Solanaceae In Hedin PA (ed) Host Plant Resistance to Pests. American Chemical Society, Washington, p 61

Stromberg EL and Corden ME (1977) Fungitoxicity of xylem extracts from tomato plants resistant or susceptible to Fusarium wilt. Phytopathology 67:693-697

Tjamos EC and Smith IM (1974) The role of phytoalexins in the resistance of tomato to Verticillium wilt. Physiol Plant Pathol 4:249-259

Tjamos EC and Smith IM (1975) The expression of resistance to Verticillium albo-atrum in monogenically resistant tomato varieties. Physiol Plant Pathol 6:215-225

Wilhelm S, Sagan JE and Tietz H (1970) Seabrook (Gossypium barbadense) x Rex (Gossypium hirsutum) crosses give Verticillium wilt-resistant, Upland type all fertile offspring In: Proc Beltwide Cotton Prod Res Conf National Cotton Council, Memphis, p 70

Wilhelm S, Sagan JE and Tietz H (1974) Resistance to Verticillium wilt in cotton: sources techniques of identification, inheritance trends, and the resistance potential of multiline cultivars. Phytopathology 54:924-931

Zaki AI, Keen NT, Sims JJ and Erwin DC (1972a) Vergosin and hemigossypol, antifungal compounds produced in cotton plants inoculated with Verticillium albo-atrum. Phytopathology 62:1398-1401

Zaki AI, Keen NT and Erwin DC (1972b) Implication of vergosin and hemigossypol in the resistance of cotton to Verticillium albo-atrum. Phytopathology 62:1402-1406

INDUCED RESISTANCE TO FUSARIUM WILT DISEASES

A. Matta
Istituto di Patologia vegetale
Università di Torino
Via P. Giuria 15
10126 Torino
Italy

INTRODUCTION

Combined inoculation of different pathogens can confer a pathological picture drastically different from the syndrome singularly evoked by each of two pathogens or, if not, accelerate or delay the severity of one disease or another. We are here expected to deal with the situation in which a first microorganism avirulent on a given host and deprived of antagonistic properties in vitro (the inducer) decreases the severity of the disease caused by a vascular pathogen (the challenger).

This phenomenon, first observed within vascular diseases more than 40 years ago (Borlaug 1945; in Davis 1967) and since then reported many times for both bacterial and fungal vascular pathogens, has been often indicated as "cross-protection", a term borrowed from plant virology where its use is limited to indicate interactions between closely related strains of a virus. In fungal plant diseases in general it has been loosely used for non specific phenomena that should be better referred to as "induced resistance". As regards vascular diseases the uncomplete knowledge of the involved mechanisms makes critical a definitive choice: while induced resistance does not entail specificity, cross-protection has a more generic meaning and does not entail any particular mechanism. As they are not contrasting, I think it is possible to use, both provisionally, preferring the second when plant mediated effects are not clearly evident.

Experiments of cross-protection in fungal vascular diseases

NATO ASI Series, Vol. H28
Vascular Wilt Diseases of Plants
Edited by E. C. Tjamos and C. Beckman

are generally carried out in laboratory with rather simple techniques, consisting of an initial root, stem or cutting inoculation of plants, generally grown in pots, with conidial suspension of avirulent microorganisms and subsequent inoculation with the pathogen or, more rarely, of simultaneous inoculation of inducer and challenger. Such experimental procedures bring about some difficulties in:

1) mantaining gnotobiotic conditions, except when using small plants in test tubes that can give misleading responses;
2) distinguishing between the effects on the plant of inducer and challenger;
3) appreciating parenchymatic and vascular colonization separately;
4) avoiding in some cases interferences due to wounding;
5) evaluating the importance of competition for the sites or the substrates.

For such reasons and for the complexity of wilt pathogenesis, the research on cross protection with vascular diseases is much less handy and scientifically rewarding than with plant-parasite systems that use slices of uniform tissues or pod halves.

An impulse to extend our knowledge of the phenomenon and the involved mechanisms comes however from the following considerations:

1) under natural conditions vascular fungi can extensively colonize uncongenial host plants (Armstrong and Armstrong 1948; Waite and Dunlop 1953);
2) Fusarium-suppressive properties are conferred to agricultural soils by non-pathogenic populations of Fusarium, especially Fusarium oxysporum (Louvet et al 1981) and parasitic competition (possibly a plant mediated type of protection) has to be taken into account as a mechanism of suppression (Schneider 1984);
3) the applicability of cross-protection against vascular diseases in the field has been shown in some cases (Shimotsuma et al 1971; Matta and Garibaldi 1977; Ogawa and Komada 1982).

A great deal of research on cross protection in fungal vascular diseases concern Fusarium wilts. Unfortunately, owing to the great variability of experimental conditions, their results are not always easily comparable. However they place at our disposal enough data to discuss the phenomenon and attempt a few suggestions about its nature and mechanisms.

ASPECIFICITY OF INDUCTION AND PROTECTION, COLONIZATION

The inducers most commonly used against Fusarium wilt diseases are non-virulent races or non-pathogenic formae speciales of F. oxysporum, but saprophytic isolates of F. oxysporum, and other species of fungi (F. solani, Verticillium spp., Cephalosporium spp., Helminthosporium carbonum, etc.) have also been proved effective (Table 1).

Different non-pathogenic forms of F. oxysporum can be active against the same disease: in the particular case of Fusarium wilt of tomato protection has been obtained among others with ff. spp. dianthi, callistephi, melonis, lini, niveum, batatas and vasinfectum (Matta 1966; Davis 1967). An exchange of rôles between inducer and challenger is not unusual, i.e.: F. oxysporum f. sp. lycopersici protects melon plants from f. sp. melonis and f. sp. melonis tomato plants from F. oxysporum f. sp. lycopersici (Table 1); Verticillium can give protection against Fusarium wilts of watermelon and tomato (Matta 1966; Shimotsuma et al 1971) and F. oxysporum against Verticillium wilt of tomato (Matta and Garibaldi 1977; Tigchelaar and Dick 1975). Thus, induction as well as the resulting protective effect appear to be largely aspecific. However, not all fungi possess the ability to induce protection and many species show great variability in this character at a subspecific level.

Saprophytes with antagonistic properties such as Penicillium sp. and Trichoderma viride, parasites unrelated to F. oxysporum such as Phialophora cinerescens or Colletotrichum musae and also forms of F. solani do not, for instance, protect tomato plants from F. oxysporum f. sp. lycopersici (Matta 1967; Homma and Ohata 1977).

Variability in activity among isolates of a given form may be even higher than among ff. spp., and some saprophytic isolates

Table 1. Examples of cross-protection against Fusarium wilt diseases

Pathogen (F.o.ff.spp.)	Inducers (°)	References
batatas	F.solani f.sp. batatas, F.o. (a.i.)	McClure 1951; Bega 1954; Ogawa and Komada 1982
cucumerinum	F.o. (ff.spp., TNV, Colletotrichum lagenarium)	Gessler and Kuč 1982
dianthi	F.o. dianthi (a.r.)	Garibaldi (pers. comm.)
elaeidis	F.o. elaeidis (a.i.)	Taquet et al 1985
lini	F.o. lini (a.r.)	Borlaugh 1945
lycopersici	F.o. (ff.spp.) Verticillium spp.	Matta 1966; Davis 1968; Langton 1969; Homma and Ohata 1977; Vanachter et al 1978
lycopersici	Cephalosporium sp. Chaethomium sp.	Phillips et al 1967; Verma and Allison 1970
melonis	F.o. melonis (a.r.) F.oxysporum f. sp. lycopersici	Mas 1967; Maraite 1982
pisi	F.solani	Perry 1959
niveum	Helminthosporium carbonum, F.oxysporum f.sp. lycopersici, V. albo-atrum	Shimotsuma et al 1972
udum	F.o. (ff.spp.)	Kaiser and Sen Gupta 1969
vasinfectum	Cephalosporium maydis, F.o. vasinfectum (a.r.)	Sabet et al 1966; Hillocks 1986
different forms	F.o. (ff.spp.)	Davis 1967

(°) F. oxysporum (F.o), avirulent isolates (a.i.), avirulent races (a.r.), formae speciales (ff.spp.).

of F. oxysporum from soil can be as such or more active than non-pathogenic forms. For instance avirulent isolates of F. oxysporum but not heterologous ff. spp. perform as inducers in sweet potato (Ogawa and Komada 1982).

Plants can be symptomless carriers of Fusaria of the section elegans that can colonize the roots and, to a limited extent, the vascular system of aerial parts (Armstrong and Armstrong 1948; Waite and Dunlop 1953). Other fungi, related or not to the genus Fusarium, can also colonize uncongenial host plants in nature (Sabet et al 1966; Verma and Allison 1970). Although the phenomenon is apparently quite common, data about frequence and quantitative variations of vascular non-pathogens in plant tissues are scanty.

The behaviour of homologous forms of F. oxysporum in resistant varieties is better known (Mace and Veech 1970; Elgersma et al 1972) but it is unfortunate that these forms have not been investigated in relation to induced resistance.

Conidia of non-pathogenic ff. spp. and saprophytic isolates of F. oxysporum from soil, inoculated on the roots of tomato seedlings grown in gnotobiotic conditions, germinate and, like the pathogen does, penetrate in the epidermal cells of the primary differentiation zone and induce extensive production of lignitubers (unpublished). Also in the xylem vessels the behaviour of a non-pathogenic f. sp. does not apparently differ in tomato from that of F. oxysporum f. sp. lycopersici for a period of 20 days after inoculation (Homma et al 1978). By contrast, conidia of species (Colletotrichum sp., Trichoderma sp., Penicillium sp., etc.) that are ineffective in cross-protection do not germinate.

Is the ability of non-pathogenic fungi to protect correlated with their ability to colonize the xylem? Years ago I thought it was reasonable to give to this question a positive answer and suggested the use in cross-protection of fungi-capable to colonize extensively the vascular system without causing disease symptoms (Matta 1971). This suggestion has not been experimentally substantiated so far. Helminthosporium carbonum, a foliar parasite probably unable to invade the xylem, appeared on the contrary more active than vascular fungi such as F. oxysporum f. sp. lycopersici and V. albo-atrum against F.o. f.sp. niveum in watermelon (Shimotsuma et al 1971) and investigations carried out in this laboratory showed that the degree of colonization of vascular bundles of tomato stem by different forms of F. oxysporum is not consistently correlated with their protective activity (Fig. 1).

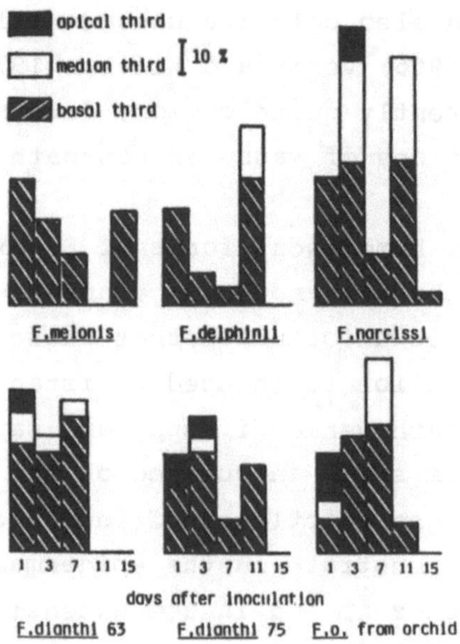

Fig. 1. Percentage of vascular bundles colonized with different ff. spp. of F. oxysporum, in the basal, median and apical third of "Bonny Best" tomato stems, at different times after inoculation. Protective activity was high for F.o. f.sp. melonis,delphinii, narcissi, low for the others.

LOCALIZATION OF THE PROTECTION

The effect of cross protection appears to be generally localized in the surroundings of the tissues inoculated with the inducer. There are however exceptions that deserve careful consideration: non-pathogenic isolates of F. oxysporum protect from f. sp. batatas sprouts of sweet potato up to 10 cm from the cut end (Ogawa and Komada 1986) and slight systemic protection of melon plants against F. oxysporum f. sp. melonis has been demonstrated with the split root technique (Maraite 1982). Resistance to Fusarium wilt of cucumber (F. oxysporum f. sp. cucumerinum) can be also induced by leaf infection with Colletotrichum lagenarium or TNV (Gessler and Kuč 1982). That some systemic protection is possible in vascular diseases is stressed by results of experiments on Verticillium wilt of cotton: systemically formed inhibitors appear to account for the protection against stem inoculation of the pathogen given by root inoculation with a mild strain of V. albo-atrum (Schnathorst 1970) and good protection is obtained by separately applying inducer (Thielaviopsis basicola) and pathogen to opposite halves of split roots (Mathre et al 1967). To get a more adequate picture of the degree of translocation of the defensive effect in vascular diseases new, innovative investigations are needed, especially on the extent of the sites conditioned by single infections on the roots. In diseases caused by soil-borne inocula eventual long distance translocation of protecting factors from the roots to the aerial parts, although theoretically interesting, is practically less important indeed than effects at short distance on the roots. By contrast, the possibility shown by Tjamos (1982) for Verticillium wilt of cucumber of protecting the roots by treating the aerial parts by application of high molecular weight fungal metabolites, would be of paramount practical importance.

PERSISTENCE

The duration of the interval of time between induction and challenge required for protection depends on the host-parasite system and the experimental conditions. In experiments carried out by Davis (1967) in test tubes, with conditions highly favourable to the inducers, no difference was found in the extent of protection in the range of 1-15 days intervals. Prolonged immersion of the roots or the basal end of cuttings of tomato plants in mixed inocula of F. oxysporum f. sp. pisi and F. oxysporum f. sp. lycopersici were quite effective in protecting against Fusarium wilt (Langton 1969). On the contrary when roots were inoculated with inducer and challenger separately, protection was very low at 0 hours, maximal at 24-48 hours intervals (Matta 1966 and Fig. 2).

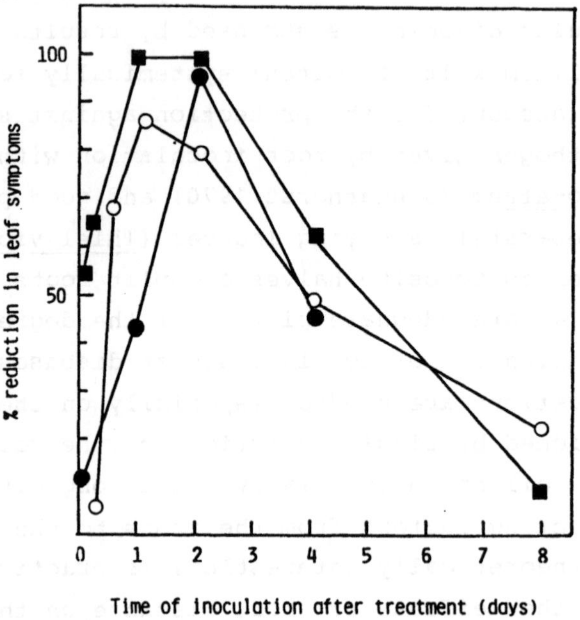

Fig. 2. Percentage reduction in Fusarium wilt symptoms in leaves of tomato plants whose roots have been immersed in hot water at 51 C for 30 sec (■), exposed at chloroform vapors for 3 min (●) or which have been inoculated with F. oxysporum f. sp. dianthi (o), in relation to the time between treatment and challenge inoculation with F. oxysporum f. sp. lycopersici.

The necessity of separating by one or a few days the first inoculation with the non-pathogen and the second challenge inoculation has been repeatedly indicated for different Fusarium wilts (McClure 1951; Sabet et al 1966; Phillips et al 1967; Verma and Allison 1970; Ogawa and Komada 1982; Hillocks 1986). Also in Verticillium wilts variable intervals of time between the two inoculations are required for protection (Schnathorst and Mathre 1966; Melouk and Horne 1975; Price and Sackston 1984).

PROTECTION AS A PLANT MEDIATED EFFECT

Absence of antagonism between inducer and challenger in vitro, and their ability to exchange rôles differentiate cross-protection from other types of biological control and are generally considered as proof of indirect mechanisms mediated by the plant. Moreover the fact that protection is not induced by dead or inactive Fusarium conidia (Homma and Ohata 1986) and is not prevented by paraffin coating (Ogawa and Komada 1986) rules out among possible mechanisms passive occlusion of wounds and vessels. However fungus-fungus competition in plant tissues can be definitively excluded only in the cases, apparently exceptional, of systemic protection. In all the other cases the possibility of direct interaction between inducer and challenger should not be overlooked.

The need of an interval of a few days between induction and challenge inoculation for the expression of maximum activity suggests that protection is conditioned by prolonged fungus-plant interaction and regulated by the consequent metabolic changes of the plant. It remains to be disproved however, that time is not necessary to allow the first fungus to acquire or strenghten its ability to compete with the second for sites and nutrients.

The finding that inhibition of the inducer by benomyl followed by inoculation with a benomyl-resistant strain of the pathogen does not affect the decrease of Fusarium wilt of melon confirms that induced resistance is probably a major component in the observed delay of symptoms (Maraite 1982). Data about the influence of metabolic inhibitors on the efficiency of protection might also shed light on the matter.

The role of injury

Heterologous forms or isolates from soil of F. oxysporum are known to produce discolouration and necrotic lesions on the roots, browning of vascular bundles and necrotic lesions at the end of cuttings. The relevance of the phenomenon is such that it cannot be disregarded and an evalutation of the effect on Fusarium wilt of abiotic treatments having cell-damaging effects might be helpful for understanding the rôle of cell injury in cross-protection.

Since 1931 Linford observed delay of Fusarium wilt in pea plants drastically pruned before inoculation. Keyworth and Dimond reported later (1952) that by injuring the roots by hot water or phytotoxic compounds before inoculation with the pathogen, the severity of Fusarium wilt of tomato was markedly reduced. Tests involving 272 chemicals showed that severity of Fusarium wilt and root injury were negatively correlated to a highly significant degree (Biehn and Dimond 1971).

More recently treatments of tomato roots with hot water or chloroform vapors have given maximum protection at exposure times that cause irreversible damage to part of the root system (apex and cortical layers of the rootlets) without affecting the survival of the main root, lateral roots, and proximal portion of the rootlets. Like inoculation with heterologous ff.spp., abiotic treatments of the roots delay expression of symptoms and colonization of stem vascular bundles. At the same time, they are ineffective in protecting the stem systemically and exhibit maximum activity when preceeding the inoculation with the pathogen by 24-48 hours (Fig. 2) (Anchisi et al 1985; Perucca and Matta 1986).

Heat and chloroform treatments and a non-pathogenic form of F. oxysporum cause, in tomato plants, transitory increases in soluble phenols content, polyphenoloxidase and peroxidase, chitinase and 1,3-β-glucanase activity and in production of ethylene (Ferraris et al 1987; Abbattista et al 1988; Bovio et al 1988; Matta et al 1988 in press). Phenols, and activity of peroxidase and polyphenoloxidase increase more in leaves than in other parts. Also 1,3-β-glucanase and chitinase are systemically increased following the treatments, the first in leaves more than in roots, the second in roots more than in leaves. Increased production of ethylene is exhibited by roots and stem only.

Although the abiotic stresses are generally more active than the fungus in stimulating the tested enzyme activities, variations with time of the enzymes, phenols and ethylene show the same trend with peaks at day 3-4 after the treatments. This suggests that a common physiological syndrome associated with and possibly generated by cell injury in the roots is evoked by application of biotic and abiotic stresses.

There seems little reason to doubt that the protection against F. oxysporum f. sp. lycopersici due to treatment before inoculation with heat and chloroform depends on plant mediated mechanisms. The profound analogies between the effect of abiotic and biotic stresses also suggest that in cross protection with fungi, cell injury and the associated transitory changes of plant metabolism play an important rôle.

A mechanism of protection due to the passive presence of muribund cells preventing penetration and transport to the main roots of conidia can be excluded on the basis that injured, senescent and necrotic cells are readily invaded by (Bishop and Cooper 1984) and generally considered a port of entry for Fusaria. In fact the damaged roots can be removed after the inducing inoculation (Davis 1967) or after abiotic treatments (Anchisi et al 1985) without reducing protection.

An alternative possibility is that resistance to a second invader operates in the xylem where it finds conditions unfavourable to growth and/or secondary distribution in form of vascular occlusions and fungitoxic compounds or a higher response potential.

Vascular occlusion

Localization of vascular infections by callose deposition, gel plugs and/or tyloses is a basic mechanism of resistance in numerous plant species (Beckman and Talboys 1981).

High rates of callose deposition can occur in vessel contact cells of tomato plants, not only in response to non-virulent races of F. oxysporum f. sp. lycopersici but also to root flora (Beckman et al 1982). The precocity of this event stresses its possible importance and the need of studies directed to its better understanding in relation to the cross-protection mechanisms.

Presence of vascular occlusions induced by the avirulent fungus in advance of the virulent one has been suggested as a possible mechanism of induced resistance. Gummosis and/or tyloses as a consequence of inoculation with non-pathogenic or non-virulent fungi have been detected in a number of plants (Moreau et al 1978; Homma et al 1979; Ogawa and Komada 1982; Hillocks 1986). Phillips et al (1967) considered that vascular occlusion preventing extensive spread of the pathogen in the plant and invasion of vessels formed after challenge was involved in the retardation of Fusarium wilt symptoms caused by a Cephalosporium species in tomato. Hillocks (1986) found that a slow growing non-virulent strain of F. oxysporum f. sp. vasinfectum was as much effective as a virulent strain in inducing vascular occlusion in cotton while inoculation of both strains resulted in a higher percentage of occluded vessels 24 hours after the challenge inoculation. Investigations on behaviour and effects of F. oxysporum f. sp. lycopersici, F. oxysporum f. sp. melonis, and three isolates from soil of F. oxysporum in tomato xylem show that occlusions are present after 4 and 7 days in a higher percent of vessels in cuttings inoculated with the non-pathogens than in those inoculated with the pathogen, but correlation in ability to protect and ability to induce vascular occlusions is scarce (Table 2). On the other hand Davis (1967) expressed the view that vascular occlusion was not involved in cross protection between forms of F. oxysporum and we have not found tyloses and gummosis in tomato plants after physical and chemical treatments that increase resistance to F. oxysporum f. sp. lycopersici.

Antifungal compounds

The participation of antifungal compounds in the resistance induced to vascular diseases seems to be possible in theory and, indeed, has been often postulated, but direct experimental evidence of a cause and effect relationship is meager and generally residing in scarcely consistent results.

Indirect evidence could be the accumulation of unspecified induced antifungal compounds (inhibitins sensu Ingham (1973) and phytoalexins) following infection with non-pathogens and abiotic treatments that induce resistance.

Table 2. Percent of xylem vessels with mycelium, tyloses and gummosis in trans sections of stem of tomato cuttings, 4 (among brackets) and 7 days after inoculation with conidial suspensions (10^6 ml^{-1}) of ff.spp. (lycopersici, melonis) and isolates from soil (5A, 6/5, 11V) of F. oxysporum. Protective activity is indicated: high (++), moderate (+) and low (+-)

		A) Mycelium	B) Tyloses	C) Gummosis	D) B+C
lycopersici		(45) 44	(0) 2	(7) 10	(7) 12
melonis	(++)	(24) 25	(1) 5	(10) 18	(11) 23
5A	(++)	(17) 17	(0) 12	(18) 19	(18) 31
6/5	(+)	(15) 14	(1) 7	(12) 7	(13) 13
11V	(+-)	(11) 8	(0) 5	(18) 11	(18) 16

The best known class of inhibitins that have been found to accumulate as a consequence of infection with non-pathogenic forms of F. oxysporum are phenols and their derivatives.

The accumulation of soluble phenols caused by inoculation of non-pathogens is accompanied by increased peroxidase and polyphenoloxidase activity in tomato plants (Ferraris et al 1987; Abbattista Gentile et al 1988) and discolouration of the colonized xylem (Homma et al 1978). It is generally admitted that 1) the transient intermediates of phenol oxidation are relatively good fungitoxicants (Bell and Mace 1981); 2) wall infusion with polymerized phenols provides a barrier against longitudinal and lateral advance of colonization; 3) resistance to vascular diseases can be increased by exogenous phenols and phenolic precursors (Retig and Chet 1974; Carrasco et al 1978). This leads to the suggestion that the overall activation of the phenol metabolism is one of the mechanisms of induced resistance. Unfortunately a primary direct role of phenols in the induced resistance is brought into question by the fact that increase of phenols and phenoloxidizing enzymes occurs not only in the roots but also in the aerial parts of cross protected tomato plants whereas protection is localized in the roots (Ferraris et al 1987; Abbattista Gentile et al 1988).

It would be reasonable to assume that the resistance induced by biotic and abiotic stresses is associated with accumulation of typical stress compounds such as the phytoalexins. Unfortunately, although different phytoalexins have been detected in relation to vascular infection (see: Bell and Mace 1981) very little is known about their presence and function in cross protection. Davis (1967) stated that the acquisition of resistance in as short an interval as 1 day reduces the probability that phytoalexins are involved in cross protection. However it is known at present that accumulation of phytoalexins can already be evident a few hours after inoculation. Indeed, substantial amounts of phytoalexins are induced by mildly virulent strains of V. albo-atrum, increasing resistance in cotton to highly virulent forms of the fungus and the interval required for protection (Bell and Presley 1969). We have not found terpenoid phytoalexins in roots of preinoculated tomato plants but this does not exclude the possibility that they are localized in undetected amounts in the xylem. The need of more investigation on this problem is evident.

The role of ethylene

As mentioned earlier, increased ethylene production in tomato follows root treatments with heat and chloroform and inoculation with non-pathogens that induce resistance to Fusarium wilt (Bovio et al 1988). The increase of ethylene is a generic expression of stress the aspecificity of which well matches, in this case,the aspecificity of induced resistance. The question arises: what, if any, is the actual role of ethylene in the expression of cross-protection? Treatments with ethylene or ethylene releasing compounds, provided they are applied before inoculation, increase resistance of tomato plants to vascular diseases (Orion and Hoestra 1974; Pegg 1976; Bovio et al 1988). A transitory increase of resistance to vascular diseases, namely to Fusarium wilt of tomato, is also determined by different other growth regulators and herbicides. Root treatments with naphthylacetic acid and trifluralin cause consistent increases of ethylene production in the roots, and to a lesser extent in the stem of tomato plants, for a period of 4-5 days corresponding to those caused by biotic or abiotic stresses (Abbattista Gentile and Bovio 1986; Bovio et al 1988). The analogous stimulation of ethylene emission by growth regulators, herbicides, biotic and abiotic stresses,all sharing the ability to induce resistance to Fusarium wilt of tomato, and the capacity of ethylene itself to increase resistance and, at the same time, vessel occlusion (Van der Molen et al 1983), polyphenoloxidase, peroxidase, phenylalanine ammonia lyase activity and the activity of fungal cell wall degrading enzymes can be converging lines of evidence for a regulatory rôle of this hormone in induced resistance.

Activation of the response potential

When tomato plants previously subjected to biotic and abiotic stresses are inoculated with F. oxysporum f. sp. lycopersici, the phenolic content and the phenoloxidase and glycosidase activities show further increases over those caused by the previous stresses alone. In absolute values the increases are markedly higher than those caused by the pathogen in the untreated controls. Thus it would appear that the infection has an amplifying effect on stress-activated plant responses and that inoculation with non-pathogens or abiotic stresses condition the plant in such a way as to increase its potential to respond more actively to the second inoculation. Unfortunately a relationship between these responses and the build up of defensive factors awaits to be demonstrated.

FUNGAL METABOLITES OR CELL-FREE COMPONENTS ACTIVE IN INDUCING RESISTANCE TO FUSARIUM WILTS

Elicitors of resistance have been looked for in culture filtrates, germination fluids and cellular components. Resistance to Fusarium wilt of melon and sweet potato has been induced by undiluted or concentrated culture filtrates or germination fluids of non-pathogenic Fusaria (Mas and Molot 1974; Ogawa and Komada 1986).

Culture filtrates of virulent and avirulent races of F. oxysporum f. sp. melonis contain fusaric acid and this toxin induces resistance in melon plants, but ability to give protection and production of fusaric acid are not correlated (Molot and Mas 1974). Previous treatments with fusaric acid desensitize tomato cuttings to fusaric acid (Jost 1965). A desensitization process could increase tolerance to the pathogen toxins. However as reduction of symptom severity is correlated with marked reduction of the pathogen population in the plant (Phillips et al 1967; Langton 1969; Meyer and Maraite 1972; Molot et al 1979; Hillocks 1986) increased tolerance appears to be an improbable mechanism of cross-protection.

Cell wall of F. oxysporum contains 1,4-1,6-β glucans eliciting phytoalexin production (Anderson 1980) but the involvement of such kinds of elicitors in the induction of resistance to Fusarium wilts is far from being proved. Autoclaved conidia of F. oxysporum are generally reported as being inactive but some resistance is induced in melon (Maraite 1982) and tomato (Matta unpublished data) by cell-free preparations of F. oxysporum f. sp. melonis. Resistance to Fusarium wilt in tomato is also induced by cell wall fragments of Phytophthora capsici whose inducing properties, being lost after delipidation, apparently depend on their unsaturated fatty acid content (Morone et al 1988). Some encouragement to persist in the study of the elicitors of resistance in F. oxysporum comes from the findings that high molecular weight components in the culture filtrates and in the wall of V. albo-atrum elicit resistance to Verticillium wilt in cucumber (Tjamos 1982) and rishitin accumulation in tomato (Woodward and Pegg 1986), respectively.

CONCLUSIONS

In conclusion a certain degree of ability to establish a physiological relationship and interact with the plant appears to be a necessary prerequisite of non-pathogenic fungi for inducing resistance to Fusarium wilt diseases. However induced resistance is not related to the extent of xylem colonization. Rather, it appears to depend on cellular damage induced in roots or stem by the inoculum of the inducers. A transitory state of resistance would be associated with morphological and physiological alterations similar to those caused by physical and chemical stresses and taking place few days following induction.

Absence of obstacles to the normal development of the pathogen once it has succeeded in spreading ahead of the locally protected tissues, and production of new unprotected roots might explain the limited persistence of the effect.

Defense triggered by previous infections can be considered a case of biological control. With few exceptions however, the short persistence of the effect makes extremely critical its practical application.

The situation appears to be different in Fusarium suppressive soils where natural populations of non pathogenic Fusaria predominate over the pathogens and, being thoroughly diffused in the soil, might then colonize new roots as soon as they develop. The increasing evidence that the most active isolates of Fusarium in inducing suppressive properties are efficient colonizers of plant roots (Schneider 1984; Cugudda and Garibaldi 1988) suggests that root mediated effects are more important than saprophytic antagonism in soil. In the case of celery, parasitic competition appears to be the primary mechanism of Fusarium wilt suppression (Schneider 1984).

A possible involvement of induced resistance in Fusarium suppressive soils is also suported by evidence that: 1) F. oxysporum isolates associated with suppressive soils are active in cross-protecting and conversely non-pathogenic forms of F. oxysporum commonly employed in cross-protection can be Fusarium suppressive; 2) suppressive isolates as well as non-pathogens and abiotic treatments that induce resistance cause cellular damage on roots (Tamietti personal communication). On the other hand a participation of induced resistance in Fusarium-suppressive soils is denied by the fact that soils suppressive to F. oxysporum are permissive to Verticillium whereas induced resistance can be effective against both.

Whether similar mechanisms operate in Fusarium suppressive soil and in cross-protection will be more easily ascertained by avoiding separation of research fields and comparing the effect of different methodologies.

REFERENCES

Abbattista Gentile I and Bovio M (1986) Fusarium wilt severity and ethylene evolution in tomato plants after treatment with trifluralin and naphthylacetic acid. PflKrankh 93:624-631

Abbattista Gentile I, Ferraris L and Matta A (1988) Variations of phenol-oxidase activities as a consequence of stresses that induce resistance to Fusarium wilt of tomato. Phytopathol Z 122:45-53

Anchisi M, Gennari M and Matta A (1985) Retardation of Fusarium wilt symptoms in tomato by pre- and post-inoculation treatments of the roots and aerial parts of the host in hot water. Physiol Plant Pathol 26:175-183

Anderson AJ (1980) Studies on the structure and elicitor activity of fungal glucans. Can J Bot 58:2343-2348

Armstrong GM and Armstrong JK (1948) Nonsusceptible hosts as carriers of Fusaria. Phytopathology 38:808-826

Beckman CH, Mueller WC, Tessier BJ and Harrison NA (1982) Recognition and callose deposition in response to vascular infection in Fusarium wilt-resistant or susceptible tomato plants. Physiol Plant Pathol 20:1-10

Bega R (1954) Biological control of Fusarium wilt of sweet potato. Phytopathology 44:482

Bell AA and Mace ME (1981) Biochemistry and physiology of resistance. In: Mace ME, Bell AA and Beckman CH (eds) Fungal wilt diseases of plants. Academic Press 431-486

Bell AA and Presley JT (1969) Heat-inhibited or heat-killed conidia of Verticillium albo-atrum induce disease resistance and phytoalexin synthesis in cotton. Phytopathology 59:1147-1151

Biehn WL and Dimond AE (1971) Correlation of chemically induced root injury with a reduction of Fusarium wilt of tomato. Phytopathology 61:127

Bishop CD and Cooper RM (1984) Ultrastructure of vascular colonization by fungal wilt pathogens. II. Invasion of resistant cultivars. Physiol Plant Path 24:277-289

Borlaug NE (1945) Variation and variability of Fusarium lini. Tech Bull Minn agric Exp Sta 168:pp40

Bovio M, Abbattista Gentile I and Matta A (1988) Release of ethylene following abiotic and biotic stimuli that induce resistance to Fusarium wilt of tomato. Phytopath medit 26:177-182

Carrasco A, Boudet AM and Marigo G (1978) Enhanced resistance of tomato plants to Fusarium by controlled stimulation of

their natural phenolic production. Physiol Plant Path 12:225-232
Cugudda L and Garibaldi A (1988) Ruolo della competizione per il ferro e per il glucoso e della premunità nella repressività verso Fusarium oxysporum f. sp. dianthi associata a Fusaria saprofiti. Riv Pat veg (in Press)
Davis D (1967) Cross-protection in Fusarium wilt diseases. Phytopathology 57:311-314
Davis D (1968) Partial control of Fusarium wilt of tomato by formae of Fusarium oxysporum. Phytopathology 58:121-122
Elgersma DM, Mac Hardy WE and Beckman CH (1972) Growth and distribution of Fusarium oxysporum f. sp. lycopersici in near isogenic lines of tomato resistant or susceptible to wilt. Phytopathology 62:1232-1237
Ferraris L, Abbattista Gentile I and Matta A (1987) Variations of phenols concentration as a consequence of stresses that induce resistance to Fusarium wilt of tomato. PflKrankh 94: 624-629
Gessler C and Kuč J (1982) Induction of resistance to Fusarium wilt in cucumber by root and foliar pathogens. Phytopathology 72:1439-1441
Grinstein A, Lisker N, Katan J and Eshel Y (1984) Herbicides-induced resistance to plant wilt diseases. Physiol Plant Pathol 24:347-356
Hillocks RJ (1986) Cross protection between strains of Fusarium oxysporum f. sp. vasinfectum and its effect on vascular resistance mechanisms. J Phytopath 117:216-225
Homma Y, Ishii M and Ohata K (1978) Behaviour of tomato Fusarium wilt pathogen and wilt-suppressive prior-inoculants in xylem tissues of tomato plants and some enzyme-histochemical changes of infected tissues (in Japanese). Bull Shik Agric Exp Sta 31:71-86
Homma Y and Ohata K (1977) Suppression of Fusarium wilt symptoms in tomato by prior-inoculation of other formae of Fusarium oxysporum and F. solani (in Japanese). Bull Shik Agric Exp Sta 30:129-134
Ingham JL (1973) Disease resistance in higher plants. The concept of pre-infectional resistance. Phytopath Z 78:314-355
Jost J-P (1965) Contribution à l'étude de la résistance toxicologique des végétaux à l'acide fusarique. Phytopath Z 54(4): 338-378
Kaiser SAKM and Sen Gupta PK (1969) Cross protection against wilt disease caused by Fusarium oxysporum f. sp. undum in pigeon pea. Indian J mycol Res 7:38-39
Keyworth WG and Dimond AE (1952) Root injury as a factor in the assessment of chemotherapeutants. Phytopathology 42:311-315
Langton FA (1969) Interactions of the tomato with two formae speciales of Fusarium oxysporum. Ann appl Biol 62:413-427
Linford MB (1931) Studies of pathogenesis and resistance in pea wilt caused by Fusarium orthoceras var. pisi. Phytopathology 21:797-826
Louvet J, Alabouvette C and Rouxel F (1981) Microbiological suppressiveness of some soils to Fusarium wilts. In: Nelson PE, Toussoun TA and Cook RJ (eds) Fusarium: diseases, biology and taxonomy. State University Press, Univ Park 261-275
Mace ME and Veech JA (1970) Uptake and distribution of Fusarium

spores in Fusarium wilt-susceptible or resistant tomatoes. Phytopathology 60:1302
Maraite H (1982) Induced resistance in muskmelon wilt: facts and limitations. In: Wood RKS (ed) Active defence mechanism in plants. Plenum Press 339-340
Mas PM (1967) Protection du melon contre la fusariose par infection préalable de la plantule avec d'autres souches de Fusarium. Acad Agric de France 1034-1040
Mas PM and Molot PM (1974) Atténuation de la sensibilité du melon (Cucumis melo) au Fusarium oxysporum Schl. f. sp. melonis Sn. et Hans. I. Rôle des filtras de milieux de germination de spores. Ann Phytopathol 6:237-244
Mathre DE, Garber RH and Ravenscroft AV (1967) Effect of Thielaviopsis basicola root rot of cotton on incidence and severity of Verticillium wilt. Phytopathology 57:604-607
Matta A (1966) Effetto immunizzante di alcuni micromiceti verso le infezioni di Fusarium oxysporum f. sp. lycopersici su pomodoro. Ann Fac Sc Agr Univ St Torino 3:85-98
Matta A (1971) Microbial penetration and immunization of uncongenial host plants. Ann Rev Phytopath 9:387-410
Matta A, Abbattista Gentile I and Ferraris L (1988) Stimulation of 1,3-β-glucanase and chitinase by stresses that induce resistance to Fusarium wilt in tomato. Phytopath medit (in Press)
Matta A and Garibaldi A (1977) Control of Verticillium wilt of tomato by preinoculation with avirulent fungi. Neth J Plant Path 83 (supp 1):457-462
McClure TT (1951) Fusarium foot rot of sweet-potato sprouts. Phytopathology 41:72-77
Melouk HA and Horner CE (1975) Cross protection in mints by Verticillium nigrescens against V. dahliae. Phytopathology 65: 767-769
Meyer JA and Maraite H (1972) Multiple infection and symptom mitigation in vascular wilt diseases. In: Wood RKS, Ballio A and Graniti A (eds) Phytotoxins in plant diseases. Academic Press 469-470
Molot PM and Mas PM (1974) Atténuation de la sensibilité du melon (Cucumis melo) au Fusarium oxysporum Schl. f. sp. melonis Sn. et Hans. II. Rôle de l'acide fusarique. Ann Phytopathol 6:245-253
Molot PM, Mas MP and Ferrière H (1979) Mise au point d'une technique de dosage biologique du Fusarium oxysporum f. sp. melonis dans les tissus de melon. Application à la prémunition. Ann Phytopathol 11:209-222
Moreau M, Catesson A, Peresse M and Czaninski Y (1978) Dynamique comparée des réactions cytologiques du xylème de l'oeillet en présence des parasites vasculaires. Phytopath Z 96:289-306
Morone C, Regnani C and Matta A (1988) Necrogenous Phytophthora capsici hyphal wall components in relation to induced resistance to Fusarium oxysporum f. sp. lycopersici in tomato plants. NATO Adv Res Workshop on Phytotoxins and Plant Pathogens. Capri May 30-June 3 1988
Ogawa K and Komada H (1986) Induction of systemic resistance against Fusarium wilt of sweet potato. Ann Phytopath Soc Japan 52:15-21
Orion D and Hoestra H (1974) Effect of root knot nematodes and

ethrel on Fusarium wilt of tomatoes. Neth J Plant Pathol 80: 28-36

Pegg GF (1976) The response of ethylene-treated tomato plants to infection by Verticillium albo-atrum. Physiol Plant Pathol 9:215-226

Perry DA (1959) Studies on the mechanism underlying the reduction of pea wilt by Fusarium solani var. pisi. Trans Brit mycol Soc 42:388-396

Perucca M and Matta A (1986) Induzione di uno stato transitorio di resistenza alla tracheofusariosi del pomodoro (Fusarium oxysporum f. sp. lycopersici) mediante stress abiotici diversi. Riv Pat Veg 22:116-124

Phillips DV, Leben C and Allison CC (1967) A mechanism for the reduction of Fusarium wilt by a Cephalosporium species. Phytopathology 57:916-919

Price D and Sackston WE (1983) Cross protection in sunflower against Verticillium wilt and downy mildew. Can J Plant Path 5: 210

Retig N and Chet I (1974) Cathecol-induced resistance of tomato plants to Fusarium wilt. Physiol Plant Path 4:469-475

Sabet KA, Samra AS and Mansour IS (1966) Interaction between Fusarium oxysporum f. vasinfectum and Cephalosporium maydis on cotton and maize. Ann Appl Biol 58:93-101

Schnathorst WC (1970) Evidence for involvement of a systemic factor in the cross-protection response in cotton plants infected with Verticillium wilt. Phytopathology 60:1537

Schnathorst WC and Mathre DE (1966) Cross-protection in cotton with strains of Verticillium albo-atrum. Phytopathology 56-1204-1209

Schneider RW (1984) Effects of nonpathogenic strains of Fusarium oxysporum on celery root infection by F. oxysporum f. sp.apii and a novel use of the Lineweaver-Burke double reciprocal plot technique. Phytopathology 74:646-653

Shimotsuma M, Kuč J and Jones CM (1972) The effects of prior inoculations with non-pathogenic fungi on Fusarium wilt of watermelon. Hort Science 7:72-73

Taquet B, Ravisé A, Renard JL and Kunesch G (1985) Modulation des réactions de défense du palmier a huile contre le Fusarium oxysporum f. sp. elaeidis. Applications: Prémunition et stimulation chimique. Phytopath Z 112:298-314

Tigchelaar EC and Dick JB (1975) Induced resistance from simultaneous inoculation of tomato with Fusarium oxysporum Sacc. and Verticillium albo-atrum Reinke & Berth. Hort Science 10: 623-624

Tjamos EC (1982) Systemic induction of protection to Verticillium wilt in cucumbers. In: Wood RKS (ed) Active defense mechanisms in plants. Plenum p 360

Vanachter A, Van Wambeke E and Van Assche C (1978) Influence of pre-inoculation treatments against Fusarium oxysporum f. sp. lycopersici on disease development in tomato plants. Med Fac Landbouww Rkjksuniv, Gent 43/2:873-880

Van der Molen GE, Labavitch JM, Strand LL and DeVay JE (1983) Pathogen-induced vascular gels: ethylene as a host intermediate. Physiologia Plantarum 59:573-580

Verma PS and Allison CC (1970) Possible modification of susceptibility of tomato to Fusarium wilt by a Chaetomium sp. Phy-

topathology 60:1318
Waite BH and Dunlap VC (1953) Preliminary host range studies with *Fusarium oxysporum* f.sp.*cubense*. Plant Disease Reporter 37:79-80
Woodward S and Pegg GF (1986) Rishitin accumulation elicited in resistant and susceptible isolines of tomato by mycelial extracts and filtrates from cultures of *Verticillium albo-atrum*. Physiol Mol Plant Pathol 29:337-347

PHYSIOLOGICAL AND BIOCHEMICAL MECHANISMS IN HOST RESISTANCE AND SUSCEPTIBILITY TO WILT PATHOGENS

J.E. DeVay
Department of Plant Pathology
University of California
Davis, CA 95616
USA

INTRODUCTION

The literature on plant host-pathogen interactions which contribute to symptom development in vascular wilt diseases is extensive, controversial and comprises some elegant and some not-so elegant studies. Certain controversies revolve around the possible roles of wilt toxins, the importance of mechanical plugging of xylem elements, the involvement of cell wall degrading enzymes, the effects and origin of plant growth regulating substances that appear to influence the wilting of plants, and the mechanisms of susceptibility and resistance to pathogens and disease development. Another poorly understood process in wilt diseases involves recognition phenomena between host and pathogen. The more we know about the various processes that contribute to a compatible or incompatible association of a plant host-pathogen relationship, the more we can appreciate the many subtle transitions and the sequential interdependence of physiological interactions between host and parasite that lead to disease development. A reading of the many excellent monographs, reviews, and symposium articles that have been written on various topics concerning wilt diseases of plants, hardly leaves much more to be said (Beckman 1964 and 1967, Duniway 1973, Mace et al 1981); yet there are concepts and some new information to be reviewed that may contribute to a better understanding of wilt diseases. The object of this essay is to focus on certain phenomena which may involve recognition between host and pathogen and the apparent

NATO ASI Series, Vol. H28
Vascular Wilt Diseases of Plants
Edited by E.C. Tjamos and C. Beckman

cause and effect relationships among ethylene, induction of polygalacturonase activity, vascular gelation and the resulting changes in plant-water relationships that contribute to vascular wilt symptoms.

CROSS - REACTIVE COMPATIBILITY FACTORS (CRCF)

The presence of antigenically similar substances unique to certain plant host-pathogen combinations was recognized over 50 years ago in pioneering studies by Fedotova (1935). She and her colleagues at the Plant Breeding Institute in Leningrad have used this form of molecular mimicry between host and pathogen to predict by serodiagnosis, varietal susceptibility of crop plants to certain pathogens and diseases. Simply stated, the more intense the serological relationship, the greater the likelihood of a compatible association between the host and pathogen and disease development. These serological relationships have been used for the artificial selection of disease resistant crops in the Soviet Union (Fedotova 1935; Gromova et al 1975) and for selection of Fusarium-resistant varieties of watermelons in Egypt (Abd-El-Rehim 1971). Subsequent studies, beginning with the research of Doubly et al (1960) on flax rust, have verified the presence of cross-reactive antigens (common antigens) shared by host and pathogen in a wide range of plant diseases with reference to their specific interactions (DeVay and Adler 1976; Dazzo 1980; Damian 1979; Heide and Smedegaard-Petersen 1985; Purkayastha and Ghosal 1985). These host-pathogen interactions include local and systemic infections and diseases affecting roots, stems and foliar tissues. However, a test of the hypothesis that the antigenic similarity between host and pathogen is related to the susceptibility of the host to disease has failed in several studies (DeVay and Adler 1976; Charudattan and DeVay 1972; Heide and Smedegaard-Petersen 1985). In some comparisons of host and pathogen, cross-reactive antigens were not found (DeVay et al 1972); in others, the antigenic relationships were specific for the plant species but did not distinguish between susceptible and resistant cultivars within the species; moreover, in some studies the presence of CRCF in races of a pathogen or in near-isogenic hosts with single gene differences for virulence or disease resistance, respectively, were not found (Heide and Smedegaard-Petersen 1985). Yet in some studies, such as those on coffee rust (Alba et al 1983) and late blight of potato (Alba and DeVay 1985), races of the pathogens were distinguished. More often than not in serological comparisons, multiple cross-reactive antigens are apparent between host and pathogen.

In studies on animal pathogens some of these multiple cross-reactive antigens have been sorted out and only one or a very few have any importance in host-pathogen compatibility (Damian 1979). The problems or differences reported among studies on plant pathogens and hosts may in some cases, involve a failure to detect disease specific antigens either through differences in methodology or an insufficient concentration of antigens used for immunization of experimental animals. For example, in Fusarium and Verticillium wilts of cotton, resistance to disease is mainly expressed during colonization of the vascular system by the pathogens (DeVay 1986): whereas colonization of the root cortex is common to both virulent and avirulent isolates of _Fusarium oxysporum_ f. sp. _vasinfectum_ (_FOV_) and _Verticillium dahliae_ (_VD_) (Charudattan and DeVay 1972). From these studies it became apparent that the sharing of cross-reactive antigens between host and parasite was indicative of a basic cell compatibility but not necessarily for disease resistance or for susceptibility. Disease development is commonly dependent on a whole range of additional factors such as the nutrition-inhibition environment for the parasite's growth and development and the production of phytoalexins by host plants. The concept of a basic cell compatibility based on the presence of cross-reactive substances between host and parasite, the absence of which precluded further development of a parasite or pathogen, was discussed by DeVay and Adler (1976). This concept excluded perthophytic pathogens whose establishment in host plants was dependent on the killing of tissues in advance of cell infection rather than on a live cell-to-cell encounter.

Some confusion has arisen over the use of terminology such

as cross-reactive antigens or common antigens; these terms suggest that plants have an antigen-antibody system similar to vertebrate animals. However, there is no evidence to support the view that immunoglobulins function or are even present in plants as they are in the humoral system of vertebrate animals. These terms refer only to the antigenicity of substances when injected into experimental animals. A more appropriate and less confusing terminology for these cross-reactive substances may be cross-reactive compatibility factors (CRCF) as used by Goodwin (1987).

In regard to vascular wilt pathogens, both VD and FOV share CRCF with cotton (*Gossypium hirsutum* L.)(Charudattan and DeVay 1972; DeVay et al 1981). Other pathogens of cotton such as the root knot nematode, *Meloidogyne incognita*, *Thielaviopsis basicola* which causes black root rot, and the bacterial blight pathogen, *Xanthomonas campestris* pv. *malvacearum*, all share CRCF with cotton (DeVay et al 1964 and 1981; McClure et al 1973). Non-pathogens of cotton such as *F. moniliforme* do not usually share these CRCF with cotton (Charudattan and DeVay 1972). However, in studies of wilt fungi, Charudattan and DeVay (1972) found that the presence of CRCF was not always reflective of disease susceptibility; the presence of these factors was more associated with the colonization and invasion of root tissues by these parasites. They compared a highly virulent parent isolate of FOV with a series of nonpathogenic mutants induced by ultra-violet irradiation for CRCF, for colonization of cotton roots, and for virulence in cotton seedlings. All of the nonpathogenic mutants still retained both the CRCF with cotton roots as well as their parasitic habit on the roots. Moreover, extracts of roots and shoots of cotton cultivars, including those resistant (cv. Auburn 56) or susceptible (cv. Rowden) to Fusarium wilt, also were strongly reactive in serological tests with FOV and the roots were colonized by this pathogen. In agar-gel double-diffusion tests, a single precipitin band was formed when antiserum to cotton antigens was reacted with extract from the *Fusarium* cultures. When isolated and purified (Charudattan and DeVay 1981), the cross-reactive substance formed a single band in polyacrylamide gel electrophoresis and a single precipitin band in gel diffusion tests showing identity with the substance in the crude culture extracts. The cross-reactive substance from the Fusarium cultures also showed identity in the gel diffusion tests with the CRCF in extracts from T. basicola. The homogeneity of the purified substance from Fusarium cultures was indicated by a single sedimentation peak in ultracentrifugation, and by a single band in both polyacrylamide gel electrophoresis and agar gel immunoelectrophoresis; the CRCF consisted of carbohydrate (75%), protein (9%) and phosphorus (16%). Immunization of rabbits with the purified CRCF resulted in an antiserum which formed a single precipitin band with fungal and plant extracts that showed identity to the CRCF in the agar-gel diffusion tests.

The cellular location of the CRCF in cotton root tissues was indicated by the staining of root sections with antibodies labeled with fluorescein isothiocyanate (FITC) (DeVay et al 1981). In cross-sections of roots, cut just below the root hair zone, the CRCF was present mainly around xylem elements, the endodermis and epidermal cells and was present throughout the cortex tissue. Protoplasts prepared from young root tissues reacted with antiserum for FOV labeled with FITC, indicating the presence of CRCF in cell membranes and newly formed cell walls. Staining of conidia and hyphae of FOV with FITC-antibodies to cotton indicated that the CRCF was mainly present in hyphal tips and in patch-like areas on conidia (DeVay et al 1981). In view of the studies that have been made on the association of CRCF between host and parasite, the question remains, what is their role or possible significance?

Several studies have provided evidence that the critical time for possible involvement of CRCF in plant host-parasite compatibility relationships occurs during the initial or early cellular interactions of host and parasite (DeVay and Adler 1976; Charudattan and DeVay 1972), a time when an exchange of metabolites may occur (Andrews 1975). The host's response may involve the production of phytoalexins, a hypersensitive reaction or a passive noninhibitory reaction to the invading parasite or pathogen conditioned by the CRCF. Providing the nutrition-inhibition environment of the host may favor the development and establishment of the pathogen, disease development may follow.

At least two possible mechanisms have been visualized for

CRCF; however, no cause and effect relationship has yet been defined. During an incompatible reaction of potato and Phytophthora infestans, a disruption of lysosomes in parasitized leaf cells occurs releasing acid phosphatase and other hydrolases compartmentalized in lysosomes; rapid cell death results. In compatible reactions of host and pathogen, lysosome breakdown also occurs, but at a greatly reduced level (Jordan and DeVay, unpublished data); these results confirm those of earlier studies which suggested a role of lysosomal enzymes in pathogenicity (Pitt and Coombes 1969; Pitt and Golpin 1973). The possible role of CRCF in reducing the hypersensitive reaction in incompatible reactions of potato and P. infestans was shown by Alba and DeVay (1985) and by Jordan and DeVay (unpublished data). A second mechanism by which CRCF may affect host-parasite compatibility is the suppression of phytoalexin production, a phenomenon that may be associated with compatible interactions of potato and P. infestans; however, no direct data on phytoalexin accumulation in potato leaf tissues has been obtained to support this assumption. This idea is from the excellent studies of Oku et al (1987) who demonstrated that several species of plant pathogenic fungi belonging to the genus Mycosphaerella secrete in the spore germination fluid, low molecular weight peptides which suppress the production of phytoalexins in host plants. Three suppressors were isolated from M. pinodes, M. melonis, and M. ligulicola, each specific to the legume hosts of each pathogen. The suppressors of phytoalexins also compromised the defense mechanisms of host plants to nonpathogens. For example, M. pinodes was pathogenic on five out of 12 legume hosts tested; when the various legumes were first treated with F5, the phytoalexin inhibitor from M. pinodes, and then inoculated with an Alternaria species not pathogenic on these legume hosts, only those legumes susceptible to M. pinodes were colonized by the Alternaria species. Their study showed that the biological activity of these phytoalexin inhibitors was host specific. In further studies, Oku et al (1987) found that the mode of action of F5 was related to an inhibition of synthesis of enzymes involved in pisatin production. Amond the Mycosphaerella species tested, spore germination fluids also contained elicitors of phytoalexin production; however, in compatible host-pathogen interactions, the suppressors negated the activity of the elicitors.

PATHOGENESIS IN WILT DISEASES

Colonization of the host

In wilt diseases caused by forma specialis of Fusarium oxysporum and VD, leaves showing wilt symptoms typical of the disease are also colonized or infected by the pathogen (Beckman 1987, DeVay 1986). It also is generally accepted that resistance to these wilt diseases is not expressed by the host plant until after invasion of the vascular system by the pathogen (Bugbee and Sapenfield 1968; Charudattan and DeVay 1972; Garber 1973). Although the colonization of the root cortex apparently does not elicit a defense reaction in the host, the effects of the infecting hyphae of VD and FOV can be quite different. In cotton, VD seldom causes root necrosis while in advanced infections caused by FOV, rotting of roots is common. The mechanisms of root infection of cottom by VD have been well described by Gerik and Huisman (1985). For FOV, Jeffers et al (1985) have shown that in both resistant and susceptible cotton plants, infection of seedlings is well established within one to two weeks after emergence of seedlings from the soil; levels of susceptibility to wilt are reflected in the rate and height of shoot infection. It appears that infection and colonization of tissues by both FOV and VD precedes symptom development (DeVay et al 1972;Jeffers et al 1985). Beckman (1987) has presented this concept very forcefully since it is a major basis for the discussion of mechanisms of disease development in the wilt diseases caused by Verticillium and Fusarium species.

Another point to be made in comparing the infection of cotton by VD and FOV, is that soil temperature at planting time has a major effect on plant infection by FOV (Jeffers et al 1985), but not on VD which infects plant roots at a steady rate throughout the active growth of the cotton plants (DeVay and Pullman 1985). What then are the causes of symptom development in vascular wilt diseases?

Toxigenic substances

The case for phytotoxins produced by vascular wilt pathogens has been thoroughly reviewed and analysed by Beckman (1987). He concluded "that the concept of toxins 'acting in advance' to cause the final wilting syndrome not only has turned out to be invalid but has distracted us from the main events that occur early in the interaction in the first time-space frames". From the foregoing discussion it seems reasonable to question the role of phytotoxins in symptom development, certainly at a distance from infecting hyphae. If a role does exist for phytotoxins, their effect would be in the immediate vicinity of the pathogen. When plants such as cotton are infected with VD, symptom development in leaves can usually be delayed by putting the plants in a moist chamber for periods long past the time when they would have been killed by disease if kept on a greenhouse bench. After removal of the plants from the moist chamber, disease symptoms characterized by wilting of limited sections of leaf lamina followed by chlorosis and necrosis, usually appear within 24 to 48 hours (DeVay et al 1972). Likewise, leaves from such plants that are colonized by VD when placed between moist towels do not develop symptoms of Verticillium wilt until removed from the towels; if their petioles are placed in water, little or no water uptake occurs and leaf symptoms typical of those on intact plants soon develop. If toxins are involved in vascular wilt diseases, their effects are probably not associated with damage to cell membranes since discs from diseased leaves showing incipient wilting and slight chlorosis will regreen and regain their turgidity if floated on water (DeVay et al 1972; Duniway 1973) indicating that early leaf symptoms are reversible. The literature on wilt diseases and the possible role of microbial toxins in symptom development is extensive but a persuasive case has yet to be made for an active role for toxins in vascular wilt diseases. Milestones in our understanding of what contributes to the development of disease symptoms came from the studies of Duniway (1971) who demonstrated convincingly that Fusarium wilt of tomato and

Verticillium wilt of cotton resulted from occluded xylem tissues and not from the direct action of toxins. Similarly, Van Alfen and Turner (1975) found that the action of a toxin produced by Corynebacterium insidiosum, the cause of bacterial wilt of alfalfa, was related to a physical interference of water transport through pit membranes rather that to membrane damage (Ries and Strobel 1972). Further investigation of another bacterial wilt disease, Pierce's disease of grape, for which a phytotoxin was originally reported as a determinant of disease, has shown that a toxin may not be involved since the symptoms of Pierce's disease are caused by occlusion of xylem elements and water stress (Goodwin et al 1988a).

Within the realm of phytotoxins, growth regulating substances and various enzymes could be included; however, a gray area exists between the origin and substantive contribution of these substances in the interaction of host and pathogen. Thus, the categorization of these substances, whether toxin or plant metabolite is difficult to assess. However, among these substances, the involvement of ethylene and pectic enzymes in vascular wilt development has been well substantiated (Beckman 1987; DeVay 1986; Pegg 1981; Cooper and Wood 1980; VanderMolen et al 1983).

ETHYLENE

Among plant growth regulating substances, ethylene plays a major role in development of vascular wilt diseases and in some cases, it contributes to mechanisms of disease resistance. Pegg (1981) has made an in-depth review of the involvement of ethylene and other growth regulators in plant diseases. A point to be made at the outset is that the production of ethylene by plant pathogens in vitro may not mimic the production of ethylene in vivo. Thus, drawing conclusions about pathogenicity or virulence from the in vitro production of ethylene may be misleading; in some situations there is a direct relationship (Wiese and DeVay 1970; Tzeng and DeVay 1985a), whereas in others it may be inverse (Chalutz and DeVay 1969).

The significance of ethylene in Fusarium wilt of tomato was first reported by Dimond and Waggoner (1953b); since then other studies have confirmed the importance of ethylene as a cause of epinasty, leaf abscission, and vascular gelation which are among the major plant responses to ethylene in vascular wilt diseases (Pegg 1981; Tzeng and DeVay 1985a; VanderMolen et al 1983; Wiese and DeVay 1970). The extensive studies by Pegg and co-workers (1981) on the involvement of ethylene in tomato diseases caused by V. albo-atrum and Pseudomonas solanacearum have characterized the key role of ethylene in these diseases.

Vascular gelation or the process by which gel-like substances form in the lumen of vessels soon after infection or chemical stimulation of xylem tissue has been extensively studied by Beckman and co-workers (1987). It is a general phenomenon and is a natural sequence of events in the senescence of plants. In wilt diseases, ethylene tends to increase shortly after infection as a pulse until about the twelfth day in both cotton (Tzeng and DeVay 1985a) and tomato (Gentile and Matta 1975; Pegg 1981); it is apparently the major metabolite which activates the polygalacturonase (PG) of host origin which in turn causes a melting of pectin in primary cell walls (Bennett and DellaPenna 1987). It is hypothesized that the melted cell wall and pectic substances of the primary cell wall and middle lamella move through pits in the lateral vessel walls and accumulate in the vessel lumina (Moreau et al 1978; VanderMolen et al 1977; VanderMolen et al 1983), where they solidify and plug the vessels. The origin of the pectic substances was indicated in a study by VanderMolen et al (1986) who compared the composition of the gel plugs from vessels in banana roots from both resistant and susceptible cultivars with the components of cell walls from stele and leaf blade tissues. Gels from both Fusarium-resistant and susceptible cultivars of banana were similar in composition; they contained no hydroxyproline which was present in the cell wall preparations. The gel plugs contained more uronic acids and phenolic acids than wall preparations but lacked cellulose which was a component of the wall preparations. The same neutral sugars were found in varying concentrations in both the gel plugs and cell wall material. Combined gas chromatography/mass spectroscopy confirmed the presence of terminal-arabinosyl, 1,4-linked xylosyl and 1,4-linked and 1,4,6-linked glucosyl residues in wall preparations and gels. Moreover, the relatively high uronic acid content of the gel plugs suggested the presence of pectins and their possible origin from primary cell walls.

The importance of ethylene in the development of vascular wilt diseases is not controversial nor is the view that many plant stress situations induce increased levels of ethylene production in plants (Ables 1973). It is unclear, however, if ethylene produced by pathogens has a significant role versus that produced by the host and if pectic enzymes of pathogen origin that induce ethylene production have any further role in vascular gelation. Mussell and Strand (1977) found that purified endo-PG from VD would induce wilt symptoms in cotton; Mussell et al (1982) later demonstrated that endo-PG was probably the primary inducer or increased ethylene levels in tomato. Studies by Cooper and Wood (1980) yielded convincing evidence that the activity of endo-pectin lyase was greatly elevated in tomato cuttings infected by *V. albo-atrum*, prior to the appearance of vascular gelation and wilt symptoms. They also found that partially purified endo-PG in amounts similar to those detected in diseased plants would induce vascular gelation and disease symptoms. In other studies, Misaghi et al (1978) showed that cross-sections of midribs or petioles of leaves from healthy cotton plants, when bathed in solutions of pectin methyl esterase at 100 µg/ml or methyl alcohol at 25 µM, would develop vascular discolouration and plugging of vessels similar to that found in leaves with symptoms of Verticillium wilt. However, Howell (1976) found that pectic enzyme-deficient mutants of VD would cause Verticillium wilt of cotton; his studies indicated that pectic enzymes of fungal origin were not required for wilt development. It appears that in addition to pectic enzymes, other fungal metabolites and substances may induce ethylene production by the plant host and that ethylene then activates a PG of host origin (Bennett and DellaPenna 1987) and possibly a hemicellulase that are major contributors to the considerable primary cell wall erosion which occurs in Fusarium-infected plants (Robb et al 1975; VanderMolen et al 1977).

In studies on the role of ethylene in vascular gelation,

VanderMolen et al (1983) tested the sensitivity of castor bean plant to ethylene. Potted plants and excised leaves were enclosed in chambers in separate gas-flow systems and exposed to either ethylene-free air or to 3 ppm ethylene in air at 20 C in the dark. After 48 hours about half the vessels in the excised leaves exposed to ethylene were occluded by gel while occlusion in control leaves was approximately 5%. For the intact plants exposed to ethylene, 40% of the vessels were occluded after 60 hours with only 3% blockage for those in air.

The perennial question about the relative contribution of host and pathogen to ethylene production is difficult to answer since during the interaction, both organisms may produce ethylene. However, in further studies on vascular gelation, VanderMolen et al (1983), through the use of aminooxyacetic acid (AOA), an inhibitor which blocks the formation of 1-aminocyclopropane carboxylic acid, the immediate biosynthetic precursor to ethylene in higher plants, found that production of ethylene by host tissue can cause vascular gelation in response to the presence of PG produced by Fusarium oxysporum f.sp. cubense (FOC). When petioles of castor bean leaves (Ricinus communis) were placed in solutions of 25 µM AOA with crude culture filtrates of FOC containing major activities in PL and β-1,4-xylanase (XYL), vascular gelation was greatly reduced compared to treatments without AOA. The AOA did not after PG activity, but when ACC was introduced with the culture filtrate and AOA, the inhibition of vascular gelation was reversed. It has been convincingly demonstrated in pear fruits that PG of pathogen origin in diseased tissue can be selectively inhibited while the PG of host origin remains active (Abu-Goukh and Labavitch 1983). The results of VanderMolen et al (1983) suggest that the melting of primary cell wall material which results in vascular plugging may be catalysed by host enzymes induced by host ethylene.

PLANT-WATER RELATIONS IN VASCULAR WILT DISEASES

Ludwig (1952) is credited with the first observation that

blockage and wilting in vascular wilt disease is associated with the accumulation of a "hyaline substance" in vessels. In later studies by Beckman (1964) and VanderMolen et al (1977), it was apparent that the formation of gels in vessels was a general response to infection by both host-specific and non host-specific pathogens. New insights concerning the significance of these gels in relation to disease development resulted from the research of Duniway (1971) who studied the water relations of Fusarium wilt in tomato; he compared the water potential of leaflets (ψ_l), the leaf solute potential (ψ_s), the diffusive resistance of leaves to water vapor loss and the resistance to water flow in the stems and roots of healthy plants with those infected with F. oxysporum f.sp. lycopersici (FOL). His results showed that leaves of diseased and healthy plants wilted at similar ψ_l values and that the wilting which occurred during disease development was caused by water stress. Solute potential of sap expressed from fully turgid leaves was higher from leaves of diseased plants than that expressed from leaves of healthy plants, indicating that diseased leaves have more solute per cell. His results suggest that the direct action of fungal toxins which could affect the permeability of cell membranes and cause excess electrolyte leakage are probably not involved in Fusarium wilt of tomato; instead, water stress phenomena are more probably the basis of symptom development.

Experiments by Duniway (1971) on the transpiration rates of healthy and Fusarium-infected plants confirmed the results of earlier studies (Dimond and Waggoner 1953a, Scheffer and Walker 1953; Ludwig 1952) that the rate of diseased plants was less than that of healthy plants. Thus, it is unlikely that excessive transpiration is the cause of wilting in diseased plants. The role of transpiration on wilting was also reflected in the diffusive resistance of leaves. The diffusive resistance of both healthy and Fusarium-infected leaves increased rapidly with increases in ψ_l. The diffusive resistance of leaves of diseased plants was as high or higher than the resistance of healthy leaves at a given ψ_l value. These results further indicated that no alteration had occurred in the transpiration of Fusarium-infected tomato plants that would cause wilting. Measurements of stomatal aperture showed that infection of tomato by FOL inhibited stomatal opening in light. These results comfirmed those of other studies (Dimond and Waggoner 1953a; Ludwig 1952) that reduced stomatal apertures are associated with Fusarium wilt of tomato. Measurements of resistance to water flow in roots and stems of healthy and Fusarium-inoculated tomato plants also indicated that the infected roots were not significantly different in resistance to water flow than healthy roots. However, the stems of diseased plants were several times more resistant to water flow than healthy stems, which was consistent with previous studies (Ludwig 1952). Research by Duniway (1973); Tzeng and DeVay (1985); and Tzeng et al (1985) on water relations in Verticillium wilt of cotton gave results which were similar to those obtained in the study on Fusarium wilt of tomato. Phenological studies of field grown cotton plants infected with VD, indicated that the growth of diseased plants ceased or was greatly reduced beginning about 2 weeks before the appearance of foliar symptoms (DeVay and Pullman 1985). At the time when leaf symptoms were first observed, the midday ψ_1 of plants with new symptoms was consistently lower than that of adjacent and apparently healthy plants and it remained lower for the remainder of the season for naturally infected field-grown plants (Fig. 1A). Somewhat similar results were obtained when field-grown plants were inoculated with different pathotypes of VD (Fig. 1B). Stomatal resistance which is related to stomatal aperture, varied in cotton plants, depending on the pathotype of VD infecting the plants. For field grown cotton plants infected with either the defoliating or non-defoliating pathotype, stomatal resistance of young leaves was consistently higher than that of healthy plants; however, for older leaves, stomatal resistance dropped to levels similar to control plants approximately 3 weeks after inoculation (Tzeng et al 1985). Concentrations of proline, a water stress metabolite in cotton, followed a pattern of increase and decrease in cotton sap similar to the changes in stomatal resistance (Tzeng and DeVay 1985). Additionally, in Verticillium wilt of cotton, it was clearly demonstrated that water loss and wilting are reversible (Duniway 1973; DeVay et al 1972). When discs were cut from wilted sections of diseased leaves and floated on water, the

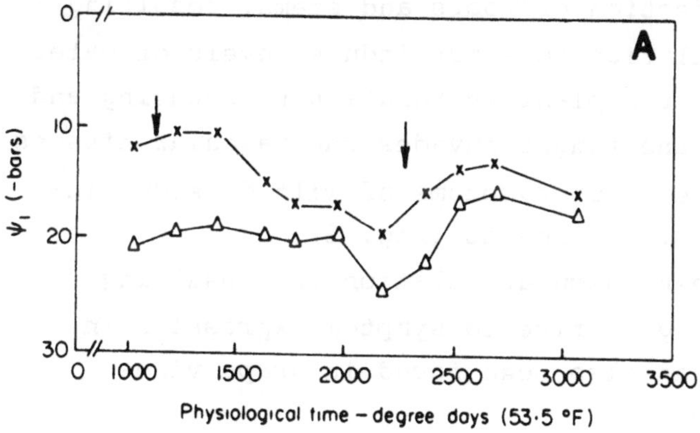

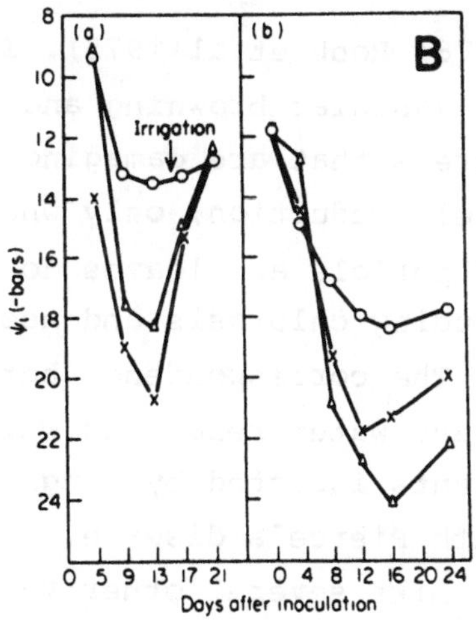

Fig. 1A. Midday leaf water potential (ψ_l) of healthy (X) and diseased (Δ) field-grown cotton plants (Gossypium hirsutum cv. Acala SJ-2), from the time when first foliar symptoms of Verticillium wilt were observed. Arrows indicate the times of field irrigation (after Tzeng et al 1985).

Fig. 1B. Midday leaf water potential (ψ_l) of field grown Acala SJ-2 cotton plants inoculated at first flowering time with defoliating (Δ) and nondefoliating (X) pathotypes of Verticillium dahliae. Control plants (O) were injected with sterile distilled water. (a) Davis, California; (b) Five Points, California (after Tzeng and DeVay 1985).

wilting was reversible. Also, placing infected Deltapine Smooth Leaf cotton plants, which were highly susceptible to Verticillium wilt, in a mist chamber 4 days after inoculation completely inhibited wilting. However, within 48 hours after the plants were removed from the mist, typical symptoms of Verticillium wilt developed. The close relation between plant water relations of cotton plants and the development of Verticillium wilt symptoms argues against a role for the direct action of a phytotoxin of fungal origin.

An important concept in Verticillium wilt of cotton and chrysanthemum is that vascular gelation of xylem in petiole and leaf veins is of crucial importance in the development of foliar symptoms (Hall et al 1975; McHardy et al 1976; Misaghi et al

1978; Robb et al 1975). Infection of roots and stems, resulting in vascular browning and gelation does not induce levels of water stress that are damaging to the plant in relation to stunting and yield reduction; only when the fungus invades the vascular tissues of petiole and leaves do the major symptoms of wilt develop, including chlorosis and necrosis of leaves (Fig. 2).

The correspondence between vascular gelation and resulting plant water relations which give rise to symptom expression in plants infected by fungi, has also been found in grape vines with Pierce´s disease.

Like several other vascular wilt diseases, a toxin of pathogen origin was reported to be the main cause of symptoms in Pierce´s disease which is caused by the bacterium, Xylella fastidiosa. This turned out not to be the case. Studies of water relations of grapevines, both healthy and diseased, indicated that in diseased plants, stomatal resistance increased, transpiration rate decreased, and the rate of photosynthesis decreased compared with healthy vines (Goodwin et al 1988a and 1988b). Associated with these changes in plant water relations of vines with Pierce´s disease, was a blockage of the xylem in petioles and leaves that caused leaf symptoms that closely resembled symptoms development in cotton leaves affected by Verticillium wilt.

CONCLUSION

Physiological and biochemical mechanisms of disease development in several vascular wilt diseases of plants are reasonalbly well understood based on the research and confirmatory results from various laboratories; however, the role of phytotoxins is still debateable. Much is known about possible mechanisms of resistance to vascular wilt diseases and to wilt pathogens in different crop plants. The excellent studies by Mace and co-workers (1981) have shed much light on the roles of phytoalexins, tyloses, and tissues changes in plants responding to infections by vascular pathogens; however, an understanding of the mechanisms of disease resistance, especially as affected by tempera-

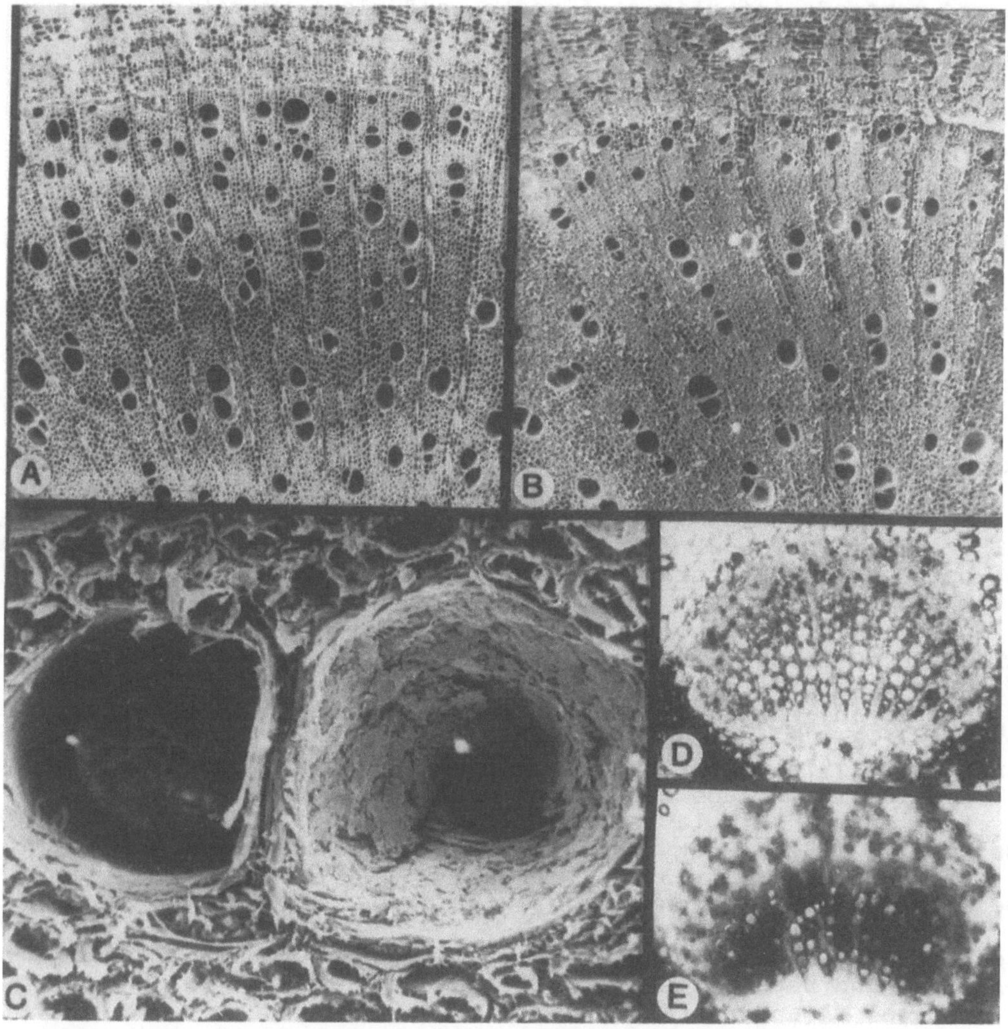

Fig. 2. Comparison of healthy and Verticillium dahliae-infected root and leaf tissues of cotton plants (Gossypium hirsutum) for vascular gelation. (A,B,C) Scanning electron micrographs of tap root sections, at 5 cm below the soil line. (A) is representative of roots of plants in solarized soil, free of V. dahliae; (B) is from plants in unsolarized soil, with advanced symptoms of Verticillium wilt; (C) is a closeup of vessel elements from (B) showing a partially occluded vessel adjacent to a vessel with mycelium, assumed to be V. dahliae. (D,E) Cross-sections from midribs of leaves showing (D) open xylem vessels in the leaves of healthy plants and (E) occluded vessels in leaves with advanced symptoms of Verticillium wilt. Photographs A,B,C courtesy of Dr. D.S. Mikkelsen; C,D after Misaghi et al (1978).

ture, nematode infections of roots, and plant nutrition needs clarification.

REFERENCES

Abd-El-Rehim MA, Ibrahim IA, Michail SH and Fadel FM (1971) Serological and immunoelectrophoretical studies on resistant and susceptible watermelon varieties to Fusarium semitectum Berk and Rav. Phytopath Z 71:49-75

Abeles FB (1973) Ethylene in plant biology. Academic Press, New York

Abu-Goukh AA and Labavitch JM (1983) The in vivo role of "Bartlett" pear fruit polygalacturonase inhibitors. Physiol Plant Pathol 23:123-135

Alba APC and DeVay JE (1985) Detection of cross-reactive antigens between Phytophthora infestans (Mont) de Bary and Solanum species by indirect enzyme-linked immunosorbent assay. Phytopath Z 112:97-104

Alba APC, Guzzo SD, Mahlow MFP and Moraes WBC (1983) Common antigens in extracts of Hemileia vastatrix Berk. et Br. urediniospores and Coffea arabica L. leaves and roots. Fitopathol Brasil 8:473-483

Andrews J (1975) Distribution of label from ^{3}H-glucose and ^{3}H-leucine in lettuce cotyledons during the early stages of infection with Bremia lactucae. Can J Bot 53:1103-1115

Beckman CH (1964) Host responses to vascular infection. Ann Rev Phytopathol 2:231-252

Beckman CH (1987) The nature of wilt diseases of plants. APS Press, St Paul, MN

Bennett AB and DellaPenna D (1987) Polygalacturonase: its importance and regulation in fruit ripening. In: Thomson WW, Nothnagel EA and Huffaker RC (eds) Plant senescence: its biochemistry and physiology. Amer Soc Plant Physiol, Rockville, MD,p 98

Bugbee WM and Sappenfield WP (1968) Varietal reaction of cotton after stem of root inoculation with Fusarium oxysporum f.sp. vasinfectum. Phytopathology 53:212-214

Chalutz E and DeVay JE (1969) Production of ethylene in vitro and in vivo by Ceratocystis fimbriata in relation to disease development. Phytopathology 59:750-755

Charudattan R and DeVay JE (1972) Common antigens among varieties of Gossypium hirsutum and isolates of Fusarium and Verticillium species. Phytopathology 62:230-234

Charudattan R and DeVay JE (1981) Purification and partial characterization of an antigen from Fusarium oxysporum f.sp. vasinfectum that cross-reacts with antiserum to cotton (Gossypium hirsutum) root antigens. Physiol Plant Pathol 18:289-295

Cooper RM and Wood RKS (1980) Cell wall degrading enzymes of vascular wilt fungi. III. Possible involvement of endopectin lyase in Verticillium wilt of tomato. Physiol Plant Pathol

16:285-300

Damian RT (1979) Molecular mimicry in biological adaptation. In: Nickol BB (ed) Host-parasite interfaces. Academic Press, New York, p 103

Dazzo FB (1980) Determinants of host specificity in the Rhizobium-clover symbiosis. In: Newton WE and Orme-Johnson WH (eds) Nitrogen fixation Vol II. Univ Park Press, Baltimore, MD, p 165

DeVay JE (1986) Half a century dynamics and control of cotton diseases: Fusarium and Verticillium wilt. In: Brown J (ed) Proc Beltwide Cotton Prod Res Conf, Natl Cotton Council Amer, Memphis TN, p 35

DeVay JE and Adler HE (1976) Antigens common to hosts and parasites. Ann Rev Microbiol 30:147-168

DeVay JE, Charudattan R and Wimalajeewa DLS (1972) Common antigenic determinants as a possible regulator of host-pathogen compatibility. Amer Nat 106:185-194

DeVay JE, Duniway JM, Temple SH and Forrester LL (1972) Temperature, humidity, water stress, and pathogen distribution in cotton leaves as factors in the development of Verticillium wilt. In: Brown J (ed) Proc Beltwide Cotton Prod Res Conf, Natl Cotton Council Amer, Memphis TN, p 20

DeVay JE and Pullman GS (1985) Epidemiology and ecology of diseases caused by Verticillium species, with emphasis on Verticillium wilt of cotton. Phytopath Mediter 23:95-108

DeVay JE, Schnathorst WC and Foda MS (1967) Common antigens and host-parasite interactions. In: Mirocha CJ and Uritani I (eds) The Dynamic Role of Molecular Constituents in Plant-Parasite Interaction. APS Press, St Paul, MN, p 313

DeVay JE, Wakeman RJ, Kavanagh JA and Charudattan R (1981) The tissue and cellular location of a major cross-reactive antigen shared by cotton and soil-borne fungal parasites. Physiol Plant Pathol 18:59-66

Dimond AE and Waggoner PE (1953a) The water economy of Fusarium wilted tomato plants. Phytopathology 43:619-623

Dimond AE and Waggoner PE (1953b) The cause of epinastic symptoms in Fusarium wilt of tomatoes. Phytopathology 43:663-669

Doubly JA, Flor HH and Clagett CO (1960) Relation of antigens of Melamspora lini and Linum usitatissimum to resistance and susceptibility. Science 131:229

Duniway JM (1971) Water relations of Fusarium wilt in tomato. Physiol Plant Pathol 1:537-546

Duniway JM (1973) Pathogen-induced changes in host water relations. Phytopathology 63:458-466

Fedotova TI (1935) The serological method of determining varietal resistance of cotton to infection. (In Russian). Zasch Rast (Moscow) 5:11-32

Garber RH (1973) Fungus penetration and development. In: Ranney CD (ed) Verticillium wilt of cotton. ARS-S-19, USDA-ARS Natl Cotton Path Lab, College Station TX, p 69

Gentile IA and Matta A (1975) Production of and some effects of ethylene in relation to Fusarium wilt of tomato. Physiol Plant Pathol 5:27-37

Gerik JS and Huisman OC (1985) Mode of colonization of roots by Verticillium and Fusarium. In: Parker CA, Rovira AD, Moore KG, Wong PTW and Kollmorgen JF (eds) Ecology and Management

of Soilborne Plant Pathogens, APS Press, St Paul MN, p 80
Goodwin PH (1987) Host-parasite specificity and physiology of Vitis vinifera L. infected with Pierce´s disease bacteria. PhD Diss, Univ Calif, Davis CA
Goodwin PH, DeVay JE and Meredith CP (1988a) Roles of water stress and phytotoxins in the development of Pierce´s disease of the grapevine. Physiol Mole Plant Pathol 32:1-15
Goodwin PH, DeVay JE and Meredith CP (1988b) Physiological responses of Vitis vinifera cv. "Chardonnay" to infection by the Pierce´s disease bacterium. Physiol Mole Plant Pathol 32:17-32
Gromova BB-O, Guseva NN and Leantas ES (1975) Plant responses to immunologically similar and different proteins of a pathogen (In Russian). Mikolo i Fitopat 9:494-500
Hall R and Busch LV (1971) Verticillium wilt of chrysanthemum: colonization of leaves in relation to symptom development. Can J Bot 49:181-185
Heide H and Smedegaard-Petersen V (1985) Common antigens between barley and barley powdery mildew and their relation to resistance and susceptibility. Can J Plant Path 7:341-346
Howell CR (1976) Use of enzyme-deficient mutants of Verticillium dahliae to assess the importance of pectolytic enzymes in symptom expression of Verticillium wilt of cotton. Physiol Plant Pathol 9:279-283
Jeffers DP, Garber RH and Roberts PA (1985) Factors affecting Fusarium wilt development in four cotton selections. Phytopathology 75:1347
Ludwig RA (1952) Studies on the physiology of hadromycotic wilting in the tomato plant. MadDonald College Bull 20:17-23
Mace ME, Bell AA and Beckman CH (eds) (1981) Fungal wilt diseases of plants. Academic Press, New York, NY
MacHardy WE, Busch LV and Hall R (1976) Verticillium wilt of chrysanthemum: quantitative relationship between increased stomatal resistance and local vascular disfunction preceding wilt. Can J Bot 54:1023-1034
McClure MA, Misaghi I and Nigh ELJr (1973) Shared antigens of parasitic nematodes and host plants. Nature 244:306-307
Misaghi IJ, DeVay JE and Duniway JM (1978) Relationships between occlusion of xylem elements and disease symptoms in leaves of cotton plants infected with Verticillium dahliae. Can J Bot 56:339-342
Moreau M, Catesson A, Péresse M and Czaninski Y (1978) Dynamique comparée des réactions cytologiques du xylème de l´oeillet en présence de parasites vasculaires. Phytopath Z 91:289-306
Mussell HW, Stilwell P and Peck S (1982) The possible origin of ethylene in Verticillium wilt of tomato. Phytopathology 72:968
Mussell HW and Strand LL (1977) Pectic enzymes: Involvement in pathogenesis and possible relevance to tolerance and specificity. In: Solheim B and Raa J (eds) Cell wall biochemistry related to specificity in host-plant pathogen interactions. Universitetsforlaget, Oslo,p 31
Oku H, Shiraishi T and Ouchi S (1987) Role of specific suppressors in Pathogensis of Mycosphaerella species. In: Nishimura S, Vance CP and Doke N (eds) Molecular determinants of plant diseases, Japan Sci Soc Press, Tokyo, Springer, Berlin Heidelberg New York, p 145

Pegg GF (1981) The involvement of growth regulators in the diseased plant. In: Ayres PG (ed) Effects of disease on the physiology of the growing plant, Cambridge Univ Press Cambridge, England, p 149

Pitt D and Golpin M (1973) Role of lysosomal enzymes in pathogenicity. In: Byrd RJW and Cutting CV (eds) Fungal Pathogenicity and the Plant Response, Academic Press, New York, p 449

Pitt D and Coombes C (1969) Release of hydrolytic enzymes from cytoplasmic particles of Solanum tuber tissues during infection by tuber-rotting fungi. J Gen Microbiol 56:321-329

Purkayastha RP and Hhosal A (1985) Analysis of cross-reactive antigens of Acrocylindrium oryzae and rice in relation to sheath rot disease. Physiol Plant Pathol 27:245-252

Ries SM and Strobel GA (1972) Biological properties and pathological role of a phytotoxic glycopeptide from Corynebacterium insidiosum. Physiol Plant Pathol 2:133-142

Robb J, Busch LV and Lu BC (1975) Ultrastructure of wilt syndrome caused by Verticillium dahliae. I. In: Chrysanthemum leaves, Can J Bot 53:901-913

Scheffer RP and Walker JC (1953) The physiology of Fusarium wilt of tomato. Phytopathology 43:116-125

Tzeng DD and DeVay JE (1985) Physiological responses of Gossypium hirsutum L. to infection by defoliating and nondefoliating pathotypes of Verticillium dahliae. Physiol Plant Pathol 26:57-72

Tzeng DD, Wakeman RJ and DeVay JE (1985) Relationships among Verticillium wilt development, leaf water potential, phenology, and lint yield in cotton. Physiol Plant Pathol 26:73-81

Van Alfen NK and Turner NC (1975) Changes in alfalfa stem conductance induced by Corynebacterium insidiosum toxin. Plant Physiol 55:559-561

VanderMolen GE, Beckman CH and Rodehurst E (1977) Vascular gelation: A general response phenomenon following infection. Physiol Plant Pathol 11:95-100

VanderMolen GE, Labavitch JM and DeVay JE (1986) Fusarium-induced vascular gels from banana roots - a partial chemical characterization. Physiol Plant Pathol 66:298-302

VanderMolen GE, Labavitch JM, Strand LL and DeVay JE (1983) Pathogen-induced vascular gels: Ethylene as a host intermediate. Physiol Plant 59:573-580

Wiese MV and DeVay JE (1970) Growth regulator changes in cotton associated with defoliation caused by Verticillium albo-atrum. Plant Physiol 45:304-309

EXOPOLYSACCHARIDES PRODUCED BY *PSEUDOMONAS SOLANACEARUM*

D. Trigalet-Démery
Laboratoire de Biologie Moléculaire
INRA-CNRS
BP 27
31326 Castanet-Tolosan Cedex
France

INTRODUCTION

Bacterial wilt caused by *Pseudomonas solanacearum* is one of the most important diseases of plants since it attacks many plants of economic value in most of the tropical and subtropical countries. The main characteristic of the disease is a rapid wilting of infected plants as the result of a heavy colonization and multiplication of the bacteria within the xylem vessels. Despite numerous publications since the beginning of the century, there remains a lack of truly effective controls and understanding of the detailed mechanism of patnogenesis. Among the bacterial molecules which could be involved in virulence, the exopolysaccharides appear as some of the most likely candidates.

CORRELATION BETWEEN VIRULENCE AND EXOPOLYSACCHARIDES

P. solanacearum is known to synthesize a polysaccharide which lies outside the cell wall and is secreted into the environment (culture media, soil, plant tissues). This polysaccharide is usually referred to as slime, mucus or exopolysaccharide (EPS). It can be readily isolated from the culture liquid medium by simple centrifugation followed by ethanol precipitation. The relationship between virulence and the presence of EPS in the xylem vessels of infected plants has long been known (Husain and Kelman 1985a). These authors held the EPS to be the primary wilt-inducing factor

NATO ASI Series, Vol. H28
Vascular Wilt Diseases of Plants
Edited by E. C. Tjamos and C. Beckman

by non specific reduction or event total blocking of water flow. In addition, P. solanacearum synthesizes extracellular enzymes: polygalacturonase and cellulase (Husain and Kelman 1985b; Schell 1987). The role of EPS and/or bacterial enzymatic degradation of the plant cell wall in vivo remains controversial (Wallis and Truter 1978).

Additional roles have been attributed to the slime of P. solanacearum. The slime could be used as a source of carbon under starvation on the one hand, and could help the bacterium to survive under dessication, on the other hand (Kulkarni and Patil 1982). Moreover, it has been demonstrated that soluble EPS from P. solanacearum could prevent bacteria from binding to plant cell walls and thus allow bacteria to multiply quickly and become systemic by xylem transport (for a review see Sequeira 1982). It has been shown that the EPS of P. solanacearum binds quantitatively to a potato cell wall lectin at relatively low ionic strengths in vitro. Whether that phenomenon could occur in planta is still questionable (Durick and Sequeira 1984).

CORRELATION BETWEEN VIRULENCE AND COLONY MORPHOLOGY

A correlation between virulence and colony morphology has long been demonstrated and rough avirulent mutants of P. solanacearum can be readily distinguished from the smooth wild type on agar medium that contains 2,3,5 triphenyl tetrazolium chloride TZC (Kelman 1954). The shift from smooth to rough type in culture varies with storage conditions and a procedure has been described that allows isolation of stable, spontaneous rough mutants (Kelman and Hruschka 1973). The virulent wild type forms irregularly-round fluidal white colonies with pink centers whereas avirulent mutants form small butyrous colonies with dark red centers. The fact that both rough and EPS deficient mutants (EPS^-) may form colonies of similar appearance on TZC medium (Hendrick and Sequeira 1984) is a source of confusion in the literature. Nevertheless one should bear in mind that defects in the O-antigen or core-region of the lipopolysaccharide (LPS) will induce rough mutants whereas defects in the production of extracellular soluble polysaccharide will induce EPS⁻ mutants; and that some mutants with rough LPS can cause wilting if they produce sufficient EPS (Hendrick and Sequeira 1984). All EPS⁻ mutants are strictly avirulent, i.e. unable to cause any wilting.

BIOCHEMICAL CHARACTERIZATION OF WILD TYPE EXOPOLYSACCHARIDE

The composition and structure of the LPS of P. solanacearum is rather well documented (Whatley et al 1980; Hendrick and Sequeira 1984; Akiyama et al 1984; Drigués et al 1985). In contrast information about the EPS fraction, mostly limited to composition, is rather confusing. Galactosamine was identified as the main sugar component by several groups (Dudman 1959; Whatley et al 1980; Drigués et al 1985; Akiyama et al 1986a), and more precisely it has been pointed out by the two latter groups that the galactosamine residues are acetylated. Glucose (Husain and Kelman 1958a), or glucose and mannose (Godwa and Rai 1980) were identified as main components in the absence of N-acetylgalactosamine. Additional minor sugar components have been reported, namely rhamnose (Dudman 1959; Drigués et al 1985; Akiyama et al 1986b), xylose and ribose (Kulkarni and Patil 1982), galactose (Balasubramanian and Nilakatan 1978; Kulkarni and Patil 1982), gluconic acid (Godwa and Rai 1980; Kulkarni and Patil 1982), fructose and xylose (Dudman 1959). Moreover, several of these authors pointed out the presence of some unidentified sugars. In no case was any structure established.

Several reasons could account for those discordant results, namely differences in the bacterial strain or culture conditions or both. It is striking to notice that EPS extraction from late stationary growth phase (8 to 20 days) never led to the identification of galactosamine. In addition, while some groups focused on a neutral polysaccharide fraction (Dudman 1959; Drigués et al 1985; Akiyama et al 1986a), some others pointed out the weak acidic character (Godwa and Rai 1980; Kulkarni and Patil 1982; Duvick and Sequeira 1984) without being actually able to define it.

Another reason for discordance lies in the heterogeneity of the EPS of P. solanacearum as demonstrated by permeation chromatography that revealed two high molecular weight fractions (Godwa and Rai 1980) or fractions of high and low molecular weight (Driguès et al 1985). The last result could be predicted from the extraction procedure with different concentrations of ethanol and acetone, as described by Dudman (1959).

A last but not an unimportant discordance remains: the qualitative and quantitative variation in amino acids (Dudman 1959; Godwa and Rai 1980; Kulkarni and Patil 1982; Driguès et al 1985; Akiyama et al 1986a). This suggests that some fractions of the EPS could be true glycopeptides.

BIOCHEMICAL CHARACTERIZATION OF EXOPOLYSACCHARIDES OF HIGHLY PLEIOTROPIC AVIRULENT MUTANTS

Numerous morphological, biological and biochemical changes characterize the spontaneous rough mutants beyond their slimeless aspect on TZC agar medium, namely loss of virulence (Kelman 1954), acquired ability to induce a hypersensitive reaction (HR) in tobacco leaf (Lozano and Sequeira 1970), increased motility (Kelman and Hruschka 1973), increased indoleacetic acid production (Sequeira and Williams 1963), increased piliation (Stemmer and Sequeira 1985), decreased cellulase production (Kelman and Cowling 1965), defective LPS (Whatley et al 1980; Hendrick and Sequeira 1984; Driguès et al 1985). These results indicate that this spontaneous mutation is pleiotropic. In addition, these rough mutants produce little or no EPS which explains the lack of data about biochemical characterization of the slime (Driguès et al 1985; Morales et al 1985). These mutants are now referred to as non fluidal mutants (Nfl) (Boucher et al 1988).

Another type of avirulent mutant has been generated by the selection of acridine orange resistance i.e. Acr^r mutants (Message et al 1978). These mutants have lost virulence and the ability to induce a HR. This mutation is due to the presence of a large deletion in the megaplasmid (Boucher et al 1986). Moreover this mutation is pleiotropic: these avirulent mutants in addition to their innocuity to their host plant, are auxotropic for methionine, release a brown diffusible pigment in the medium and have altered LPS (Drigués et al 1985). In addition, these Acr^r mutants produce noticeable amounts of EPS, whose sugar composition is about the same as that of the wild type. Nevertheless, permeation chromatography, NMR spectra and relative viscosity measurements revealed marked differences between Acr^r mutants and wild type EPS. However, due to the pleiotropic nature of the mutation no precise correlation between EPS and virulence could be drawn.

BIOCHEMICAL CHARACTERIZATION OF EXOPOLYSACCHARIDE OF Tn5-INDUCED AVIRULENT MUTANTS

Random mutagenesis of Nf1 mutant with transposon Tn5 led to identification and characterization of a virulence locus involved in EPS production in P. solanacearum (Staskawicz et al 1983). The same procedure in the genome of P. solanacearum K60 wild type allowed the isolation of avirulent prototrophic mutants (Xu et al 1988). Most of these mutants were identical to the wild type in that they produce slime on TZC agar. Two additional avirulent mutants were non fluidal. In addition, two other mutants which retained their virulence to their susceptible hosts were slimeless on plate culture and did not produce EPS. Thus, the role of EPS in pathogenicity was questioned by the latter authors. In no case was any biochemical characterization of EPS done. Random mutagenesis of a wild type strain with transposon Tn5 led to the isolation of 13 mutants which had lost their pathogenicity towards susceptible host plants (Boucher et al 1985), although most of them retained, to a certain extent, their ability to colonize and to multiply within the plant (Trigalet and Demery 1986). All of the Tn5-induced avirulent mutants exhibited the wild type colony morphology on TZC agar medium and released comparable amounts of EPS into the liquid culture medium as did the wild type.

We now report preliminary results of a comparative study on the sugar composition of EPSs from the wild type and from some of the above mentioned Tn5 mutants.

The extraction procedure and analysis methods have been described elsewhere (Drigués et al 1985). Ethanol precipitation and subsequent ultrafiltration yielded two types of fractions named HMW (high molecular weight) and LMW (low molecular weight) that correspond to 80% and 20% in weight respectively. Each of these fractions was successively chromatographed through cation and anion-exchange columns. Retained and unretained materials were further analyzed by permeation chromatography, gas chromatography of their trimethylsilyl and alditol acetate derivatives and by infra-red, NMR and mass spectroscopy. Different hydrolysis and reduction conditions for crude extract and subfractions had been previously assayed.

First characterization of the HMW and LMW fractions after chloroform delipidation and extensive dialysis revealed less than 3% nucleic acids and about 2% proteins as residual contaminants. Nevertheless further analysis of both fractions through permeation and ion exchange chromatography and spectroscopy (infra-red, NMR and mass) strongly suggested that sugar components were in close association with other non carbohydrate material. Further analysis showed 36% and 66% amino acids in the HMW and LMW fractions, respectively. Whereas cation exchange chromatography lowered this percentage to about 7% for the HMW fraction, the same procedure was ineffective for the LMW one (down to about 45%). Proteinase treatment was more efficient for this latter fraction (down to 23%). Further separation could not be achieved by any procedure tried. This strongly suggests that the LMW fraction is a glycopeptide.

Materials unretained by the cation exchange column were further identified by gas chromatography (GC) to contain galactosamine, mannose, glucose, rhamnose and galactose in the HMW fraction, but only mannose, glucose and galactose in the LMW one, with additional, non-identified and somewhat variable peaks in both fractions. These latter peaks correspond to acid-sensitive components since they have mostly been detected after mild hydrolysis and methanolysis. In addition, the acidic character of both fractions was established by the noticeable amount of material which was retained in anion exchange columns, and confirmed by infra-red spectroscopy ($\lambda max.cm^{-1}$ 1730) of crude HMW and LMW fractions. That the EPSs of P. solanacearum wild type exhibit an acidic character was already revealed by examination in transmission electron microscope after glutaraldehyde-ruthenium red-uranyl acetate staining of bacterial slime exuded from wilted tomato (J. Vasse personal communication).

CONCLUSION

The extensiveness of our present results could appear somewhat unsatisfactory in that more than 50% of sugar components in the wild type EPS are still unknown. Nevertheless all possibilites are far from being tested. Extremely mild methanolysis conditions appear to ensure sufficient protection of the acid sensitive glycosidic linkages and stable GC detection of the previously variable peaks. At this stage, coupled GC-mass spectrometry may lead to complete identification of unknown peaks and give a clue to structural features. Although usual spectrophotometric analysis (i.e. carbazole/sulfuric acid and m/phenylphenol/sulfuric acid) were unsuccessful in detecting uronic acids, as were the usual procedures for reduction, of potential uronic acids to their corresponding hexoses before any subsequent hydrolysis, the presence of uncommon uronic sugars (i.e. amino uronic) should not be definitively ruled out, nor the presence of non carbohydrate acidic removable substituents.

Preliminary comparative studies of Tn5 mutants EPSs indicate that sugar composition of the HMW fractions are roughly comparable to that of the wild type. However, differences could be noticed in the relative composition of sugar components after passage in the cation exchange column. An extensive comparative study of EPSs of wild type strain and Tn5 mutants, without a previous identification of all carbohydrate and non carbohydrate components of the wild type EPS would be meaningless. Therefore, our present work focuses on the biochemical characterization of these unknown components. As long as the biochemical composition and structure of EPS of P. solanacearum are not fully elucidated it appears premature to argue about its possible biological role.

REFERENCES

Akiyama Y, Eda S, Kato K and Tanaka H (1984) Structural studies of the O-antigen of the lipopolysaccharide from an avirulent strain (M_4S) of Pseudomonas solanacearum. Carbohydr Res 133: 289-296

Akiyama Y, Eda S, Nishikawaji S, Tanaka H, Fujimori T, Kato K and Ohnishi A (1986a) Extracellular polysaccharide produced by a virulent strain (U-7) of Pseudomonas solanacearum. Agric Biol Chem 50:747-751

Akiyama Y, Eda S, Nishikawaji S, Tanaka H and Ohnishi A (1986b) Comparison of extracellular polysaccharide produced by 17 virulent strains of Pseudomonas solanacearum. Ann Phytopath Soc Japan 52:741-744

Balasubramanian A and Nilakantan G (1978) Effect of a pesticide on the extracellular slime production and pathogenicity of a non-target phytopathogen. J Nucl Agric Biol (India) 7:147-149

Boucher C, Barberis P, Trigalet A and Demery D (1985) Transposon mutagenesis of Pseudomonas solanacearum: isolation of Tn5-induced avirulent mutants. J Gen Microbiol 131:2449-2457

Boucher C, Martinel A, Barberis P, Alloing G and Zischek C (1986) Virulence genes are carried by a megaplasmid of the plant pathogen Pseudomonas solanacearum. Mol Gen Genet 205:270-275

Boucher C, Arlat M, Zischek C and Boistard P (1988) Genetic organisation of pathogenicity determinants of ps. Physiology and Biochemistry of Plant-Microbial Interactions ASPP (in Press)

Drigués P, Demery-Lafforgue D, Trigalet A, Dupin P, Samain D and Asselineau J (1985) Comparative studies of lipopolysaccharide and exopolysaccharide from a virulent strain of Pseudomonas solanacearum and from three avirulent mutants. J Bact 162: 504-509

Dudman W (1959) Comparison of slime from tomato and banana strains of Pseudomonas solanacearum. Nature 184:1969-1970

Duvick J and Sequeira L (1984) Interaction of Pseudomonas solanacearum lipopolysaccharide and extracellular polysaccharide with agglutinin from potato tubers. Appl Environ Microbiol 48: 192-198

Godwa SS and Rai VP (1980) Phytotoxic glycopeptides produced by Pseudomonas solanacearum. Phytopath Z 98:68-75

Graham T and Sequeira L (1977) Interaction between plant lectins and cell wall components of Pseudomonas solanacearum role in pathogenicity and induced disease resistance. In Solheim B and Raa J (eds) Cell wall biochemistry related to specificity in host-plant interactions. Universitetsforlaget, Oslo, Norway

Hendrick C and Sequeira L (1984) Lipopolysaccharide-defective mutants of the wilt pathogen Pseudomonas solanacearum. Appl Environ Microbiol 48:94-101

Husain A and Kelman A (1958a) Relation of slime production to mechanism of wilting and pathogenicity of Pseudomonas solanacearum. Phytopathology 48:155-164
Husain A and Kelman A (1958b) The role of pectic and cellulolytic enzymes in pathogenesis by Pseudomonas solanacearum. Phytopathology 48:377-386
Kelman A (1954) The relationship of pathogenicity in Pseudomonas solanacearum to colony appearance on a tetrazolium chloride medium. Phytopathology 44:693-695
Kelman A and Cowling EB (1965) Cellulase of Pseudomonas solanacearum in relation to pathogenesis. Phytopathology 55:148-155
Kelman A and Hruschka J (1973) The role of motility and aerotaxis in the selective increase of avirulent bacteria in still broth cultures of Pseudomonas solanacearum. J Gen Microbiol 76:177-188
Kulkarni JH and Patil RB (1982) Production and utilization of extracellular slime by Pseudomonas solanacearum and its role on survival at different relative humidities. Acta Microbiol Pol 31:159-166
Lozano CJ and Sequeira L (1970) Differentiation of races of Pseudomonas solanacearum by a leaf infiltration technique. Phytopathology 60:833-838
Message B, Boistard P, Pitrat M, Schmit J and Boucher C (1978) A new class of fluidal avirulent mutants of Pseudomonas solanacearum unable to induce a hypersensitive reaction. In : Proceed 4th Int Conf Plant Path Bact. Station de Pathologie Végétale INRA Beaucouze, France, p 823-833
Morales VM, Stemmer WPC and Sequeira L (1985) Genetics of avirulence in Pseudomonas solanacearum. In : Sussex I (ed) Current communications in molecular biology: plant cell/cell interactions. Cold Spring Harbor Laboratory, Cold Spring Harbor NY, p 89-96
Patil R and Kulkarni J (1982) Production and utilization of extracellular slime by Pseudomonas solanacearum and its role on survival at different relative humidities. Acta Microbiol Pol 31 :159-165
Sequeira L and Williams PH (1963) Synthesis of IAA by wild and mutant strains of Pseudomonas solanacearum Plant Physiology 28:27 (Abstract)
Sequeira L and Graham T (1977) Agglutination of avirulent strains of Pseudomonas solanacearum by potato lectin. Physiol Plant Pathol 11:43-54
Sequeira L (1982) Determinants of plant response to bacterial infection.In : Wood RKS (ed) NATO Advanced Study Institute on Active Defense Mechanisms in Plants 1980. Plenum Press, New-York
Schell M (1987) Purification and characterization of an endoglucanase from Pseudomonas solanacearum. Appl Environ Microbiol 53:2237-2241
Staskavicz B, Dalbeck D, Miller J and Damm D (1983) Molecular analysis of virulence genes in Pseudomonas solanacearum. In : Molecular Genetic of Bacteria-Plant Interactions, Pühler A (ed) Springer Verlag Berlin Heidelberg:345-351
Stemmer WPC and Sequeira L (1985) Possible role of fimbriae in attachment of plant-associated bacteria to plant cell walls. In Proceed 2nd Int Symp Mol Gen Bacteria-Plant Interaction

Szalay AA and Legocky RP (eds): 199. Media Services Cornell University Ithaca NY

Trigalet A and Demery D (1986) Invasiveness in tomato plants of Tn5-induced avirulent mutants of Pseudomonas solanacearum. Physiol Mol Plant Pathol 28:423-430

Wallis FM and Truter J (1978) Histopathology of tomato plants infected with Pseudomonas solanacearum with emphasis on ultrastructure. Physiol Plant Pathol 13:307-317

Whatley MH, Hunter N, Cantrell MA, Hendrick L, Keegstra K and Sequeira L (1980) Lipopolysaccharide composition of the wild pathogen Pseudomonas solanacearum. Plant Physiology 65:557-559

Xu P, Leong S and Sequeira L (1988) Molecular cloning of genes that specify virulence in Pseudomonas solanacearum. J Bact 170:617-622

CROSS PROTECTION AMONG STRAINS OF VERTICILLIUM DAHLIAE ON SUNFLOWER

Doris Price and W.E. Sackston
Plant Science Department
MacDonald College of McGill University
21111 Lakeshore Road
Ste. Anne de Bellevue, Que., H9X 1C0
Canada

INTRODUCTION

Verticillium wilt (Verticillium dahliae Kleb.) is a major disease of sunflowers (Helianthus annuus L.), one of the world's most important sources of edible vegetable oil (Putt 1978; Sackston 1981; Zimmer and Hoes 1978). New strains of the pathogen able to attack previously resistant cultivars have appeared (Bertero and Vazquez 1982), making it necessary to utilize new sources of resistance, and advisable to investigate other methods of control.

Protection of susceptible host plants against virulent strains of a pathogen by avirulent strains has been reported in various fungi, including Fusarium (Gessler and Kuč 1982; Rouxel et al 1979) and Verticillium (Melouk and Horner 1975; Schnathorst and Mathre 1966; Tjamos 1979). The development of such protection may be influenced by the method of inoculation (Davis 1967, 1968) and the time interval between inducing and challenge inoculations (Kuc 1982; Kuc et al 1975; Melouk and Horner 1975; Schnathorst 1966; Wymore and Baker 1982). We present in this paper the results of some experiments with simultaneous and sequential inoculations of sunflower with V. dahliae. A preliminary report has been published (Price and Sackston 1983).

MATERIALS AND METHODS

The sunflower cultivar used was the inbred line CM 162 susceptible to *V. dahliae*, provided by Dr. W. Dedio, Agriculture Canada Research Station, Morden, Manitoba. The isolates of *V. dahliae* used were: P6, avirulent, from cotton in Australia, obtained from the Cotton Pathology Research Laboratory, College Station, Texas, U.S.A.; and VM, virulent, provided from Morden. Sunflowers were grown in a commercial peat: vermiculite: perlite mixture (Promix) in 10 cm diameter pots on growth benches maintained at 23/18 C day/night temperature, daylength 14 hours, and light intensity about 175 $\mu Em^{-2}s^{-1}$ from VHO cool white fluorescent tubes supplemented with incandescent bulbs. Five seeds were sown per pot, thinned to one plant per plot prior to inoculation at the two-to-four leaf stage, about 2 weeks after sowing. Fertilizer was added at 2-week intervals as a 0.1% aqueous solution of N:P:K, 20:20:20. Inoculum was prepared by washing spores from the surface of 2-week-old cultures of *V. dahliae* on potato dextrose agar with sterile distilled water (SDW), and straining through gauze to remove mycelial fragments. Spore concentrations were adjusted to approximately 5 x 10^5 per ml. Plants were inoculated by injection (HI) of a drop of spore suspension into each plant with a hypodermic syringae with a 26 1/2 gauge needle (Moser and Sackston 1973). Inducing inoculations by HI were made at the cotyledonary node, and challenge inoculations about 1.5 cm above it. Inducing inoculations were followed by challenge inoculations immediately in some experiments, or 2, 4, or 7 days later in others. Height of all plants was recorded each week in some experiments, and 3 to 5 weeks after the last inoculation in others. Percentage of leaves showing symptoms and symptom severity were recorded 3 to 5 weeks after inoculation in various experiments, on a scale of 0 to 4: 0, no symptoms; 1, less than 10% of total leaf area affected; 2, 11 to 50% of leaf area affected; 3, more than 50% leaf area affected; 4, plant dead (Sedun and Sackston 1982). Reisolation was attempted from the stem apex, the cotyledonary node, and the petiole of the lowest green leaf of plants without symptoms, and from stems at the level of leaves with symptoms. All treatments were replicated four or five times in a randomized complete block design, and all experiments were repeated. All data were analyzed statistically, including Duncan's multiple range analysis. As severity ratings are non-parametric on an ordinal scale, they were subjected to rank analysis.

RESULTS

Simultaneous Inoculations

Inoculation with isolate P6 increased the height of the plants, the percentage of leaves showing symptoms, and severity ranking above the controls, but none of the differences were statistically significant (Table 1). Inoculation with VM significanlty

Table 1. Effects of simultaneous inducing and challenge inoculations by HI with *Verticillium dahliae* isolates P6 and VM on sunflower line CM 162

Isolate		Mean height[1] (cm)	Mean % leaves affected	Mean severity rank[2]
Inducing	Challenge			
Control (SDW)[3]		28 abc[4]	0 h	10 e
SDW	P6	32 ab	19 efgh	20 e
SDW	VM	19 e	82 a	58 a
P6	VM	19 e	69 ab	47 ab

1. Means of 5 replicates; results tended to be similar when experiment was repeated.
2. Severity ratings from 0 to 4.
3. SDW = sterile distilled water.
4. Values with a letter in common do not differ significantly at P = 0.05, Duncan's multiple range test.

reduced plant height and increased percentage of leaves affected and severity ranking (Table 1). An inducing inoculation with P6 tended to lessen some of the effects of a challenge inoculation with VM, but the differences were not statistically significant (Table 1). Verticillium was reisolated from all plants inoculated with VM, but only from about 50% of plants inoculated with P6.

Sequential Inoculations

Inoculation with P6 in these experiments had no effect on plant height. It increased the percentage of leaves with symptoms and severity rating above those of the controls, but the differences were not statistically significant (Table 2).

Table 2. Effects of sequential inducing and challenge inoculations by HI with *Verticillium dahliae* isolates P6 and VM on sunflower line CM 162

Isolate/ interval (days)		Mean height[1] (cm)	Mean % leaves affected	Mean severity rank[2]
Inducing	Challenge/ interval			
Control (SDW)[3]		82 ab[3]	0 e	8 d
SDW	P6/0	81 ab	18 cde	23 bcd
SDW	VM/2	48 cd	77 a	50 a
SDW	VM/7	48 cd	53 abc	46 a
P6/0	VM/2	67 abc	27 cde	23 bcd
P6/0	VM/7	62 bcd	45 abcd	42 a

1. Means of 5 plants, results tended to be similar when experiment was repeated.
2. Severity ratings from 0 to 4.
3. SDW = sterile distilled water.
4. Values in a column having a letter in common do not differ significantly at P = 0.05, Duncan's multiple range test.

Inoculation with VM 2 or 7 days after an "inducing inoculation" with SDW significantly reduced plant height and increased percentage of affected leaves and severity ranking compared to plants injected with SDW or P6. An inducing inoculation with P6 reduced the adverse effects of a challenge inoculation with VM 2 or 7 days later. The protective effect on plant height was not statistically significant. The effect on percentage of leaves affected and severity ranking was statistically significant when the challenge inoculation was made 2 days, but not 7 days after the inducing inoculation (Table 2). Verticillium was consistently reisolated from plants with symptoms and occasionally from symptomless plants inoculated with VM. Verticillium was not isolated from plants inoculated only with P6, nor from the SDW controls.

DISCUSSION

Simultaneous inoculations with inducing and challenge organisms usually involve mixtures of the avirulent and virulent organisms (Schnathorst and Mathre 1966; Tigchelaar and Dick 1975), where possible antagonism between them must be considered. We did not mix our inocula before injecting plants, and did not observe any antagonism between the isolates in culture (Price and Sackston 1983). Resistance may be induced in simultaneous inoculations if the inducing organism grows more rapidly then the virulent pathogen (Bell and Mace 1981). P6 grew faster in culture than VM (Price and Sackston 1983). It is possible that the difference in growth rate may have been a factor in the limited protection observed in our experiments with simultaneous inoculations. The interval between inducing and challenge inoculations may be critical for the development of resistance. The critical period may vary from a few hours (Matta and Garibaldi 1977) to 2 days (Melouk and Horner 1975; Wymore and Baker 1982). The effect tends to reach its maximum in 7 to 9 days (Melouk and Horner 1975; Schnathorst and Mathre 1966), or the period normally required for symptom expression (kuč 1982). Since Verticillium wilt symptoms can appear on sunflowers within 5 days of inoculation by HI (Moser and Sackston 1973), induction of resistance could be expected to occur at about this time. In our experiments protection appeared to be greater when the interval between induction and challenge inoculations was 2 rather than 7 days, indicating that the protection may be of short duration. The concentration of inducing and challenge inocula may affect the development of cross protection in some cases (Schnathorst and Mathre 1966; Tigchelaar and Dick 1975; Wymore and Baker 1982). We used the same concentration for both inocula in all but one experiment, in which only two concentrations were tried. The effect of relative concentrations should be investigated in this system. We did not attempt to determine the mechanisms involved in the apparent protection observed. Cross protection in other systems has been attributed to the production of phytoalexins, and to formation of vascular occlusions (Bell 1969; Melouk and Horner 1975; Schnathorst and Mathre 1966). We have done preliminary work with other sunflower cultivars, other isolates of V. dahliae, and other methods of inoculation. More needs to be done to determine if cross protection in sunflowers against Verticillium wilt is a consistent phenomenon, and the parameters within which it occurs, before more complicated studies are undertaken on responsible mechanisms.

ACKNOWLEDGEMENTS

We are grateful to the Natural Sciences and Engineering Research Council of Canada for a scholarship to D.P. and for an operating grant to W.E.S. We are also grateful to Professor M. Fanous for his advice on statistical design of the experiments and on analysis of the data.

REFERENCES

Bell AA (1969) Phytoalexin production and Verticillium wilt resistance in cotton. Phytopathology 59:1119-1127

Bell AA and Mace ME (1981) Biochemistry and physiology of resistance. In: Mace ME, Bell AA and Beckman CH (eds) Fungal wilt diseases of plants. Academic Press, New York, pp 431-486

Bertero AB and Vazquez AN (1982) A new race of Verticillium

dahliae Kleb. In: 10th Int Sunflower Conf Surfer's Paradise Australia
Davis D (1967) Cross-protection in Fusarium wilt diseases. Phytopathology 57:311-314
Davis D (1968) Partial control of Fusarium wilt in tomato by formae of Fusarium oxysporum. Phytopathology 58:121-122
Gessler C and Kuč J (1982) Induction of resistance to Fusarium wilt in cucumber by root and foliar pathogens. Phytopathology 72:1439-1441
Kuč J (1982) Plant immunization-mechanisms and practical implications. In: Wood RKS (ed) Active defence mechanisms in plants Plenum U.S.A.,pp 157-158
Kuč J, Shockley G and Kearney K (1975) Protection of cucumber against Colletotrichum lagenarium by Colletotrichum lagenarium. Physiol Plant Pathol 7:195-199
Matta A and Garibaldi A (1977) Control of Verticillium wilt of tomato by preinoculation with avirulent fungi. Neth J Plant Pathol 83 (Suppl 1):457-462
Melouk HA and Horner CE (1975) Cross protection in mints by Verticillium nigrescens against V. dahliae. Phytopathology 65: 767-769
Moser PE and Sackston WE (1973) Effect of concentration of inoculum and method of inoculation on development of Verticillium wilt of sunflowers. Phytopathology 63:1521-1523
Price D and Sackston WE (1983) Cross protection against Verticillium in sunflower. Phytopathology 73:372-373 (Abstr)
Putt ED (1978) History and present world status. In: Carter JF (ed) Sunflower science and technology. Agronomy 19 Amer Soc Agron, Madison,pp 1-29
Rouxel F, Alabouvette C and Louvet J (1979) Recherches sur la résistance des sols aux maladies. IV Mise en évidence du role dès Fusarium autochtones dans la resistance d'un sol à la fusariose vasculaire du melon. Ann Phytopathol 11:199-207
Sackston WE (1981) The sunflower crop and disease: progress, problems, and prospects. Plant Disease 65:643-648
Schnathorst WC (1966) Cross protection in cotton with two strains of Verticillium albo-atrum. Phytopathology 56:151 (Abstr)
Schnathorst WC and Mathre DE (1966) Cross-protection in cotton with strains of Verticillium albo-atrum. Phytopathology 56: 1204-1209
Sedun FS and Sackston WE (1982) Effect of daylength, host line, and pathogen isolate on development of Verticillium wilt of sunflower. Can J Plant Pathol 4:109-114
Tigchelaar EC and Dick JB (1975) Induced resistance from simultaneous inoculation of tomato with Fusarium oxysporum Sacc. and Verticillium albo-atrum Reinke and Berth. Hort Sci 10: 623-624
Tjamos EC (1979) Induction of resistance to Verticillium wilt in cucumber (Cucumis sativus) Physiol Plant Pathol 15:223-227
Wymore LA and Baker R (1982) Factors affecting cross-protection in control of Fusarium wilt of tomato. Plant Disease 66:908-910
Zimmer DE and Hoes JA (1978) Diseases In Carter JF (ed) Sunflower science and technology. Agronomy 19 Amer Soc Agron, Madison, pp 225-262

ACCUMULATION OF PHYTOALEXINS IN SUSCEPTIBLE AND RESISTANT NEAR-ISOGENIC LINES OF TOMATO INOCULATED WITH *VERTICILLIUM ALBO-ATRUM* OR *FUSARIUM OXYSPORUM* F. SP. *LYCOPERSICI*

D.M. Elgersma and J.I. Liem
Willie Commelin Scholten Phytopathological Laboratory
Javalaan 20
3742 CP Baarn
The Netherlands

INTRODUCTION

The mechanism of resistance of tomato plants to Verticillium and Fusarium wilt has been subject of many studies in recent years (Beckman 1987). Results obtained by various research groups, however, are sometimes contradictory, especially those concerning the role of phytoalexins (Tjamos and Smith 1974; McCance and Drysdale 1975; Elgersma 1980; Hutson and Smith 1980; Harrison and Beckman 1987). These contradictory results are difficult to explain and more research is urgently needed.

In this paper we describe a study of the accumulation of the two main phytoalexins, the terpenoid rishitin and the polyacetylene cistetradeca-6-ene-1,3-diyne-5,8-diol, in susceptible and resistant near-isogenic lines of tomato plants after infection with *Verticillium albo-atrum* or *Fusarium oxysporum* f. sp. *lycopersici*.

MATERIALS AND METHODS

Near-isogenic tomato lines susceptible or resistant to Verticillium or Fusarium wilt were used: cv. Gem S and cv. Craigella (susceptible to both pathogens), cv. Gem R (monogeneically resistant to Verticillium wilt, containing the Ve gene) and cv.

Craigella FR (containing the homozygous I-1 gene, conferring resistance to Fusarium wilt for race 1).

Plants were grown in a glasshouse at 22 to 25 C. Six-week-old plants at the 7-8 leaf stage were inoculated and kept at 20 to 23 C during the experiment with V. albo-atrum or at 22 to 25 C when inoculated with F. oxysporum f. sp. lycopersici. Both fungal isolates, V. albo-atrum (isolate WCS800) and F. oxysporum f. sp. lycopersici race 1 (isolate WCS801) were pathogenic to tomato and maintained on potato dextrose agar.

In preparing of inoculum both fungi were cultured in Tchernoff's medium (Tchernoff 1965) on a reciprocal shaker for 5 days at 23 C. Mycelial fragments were removed by filtering through sterile glass wool. The filtered conidial suspensions were washed twice in sterile tap water and adjusted to the desired concentration by means of a haemocytometer. Stem inoculations of tomato plants were performed as described previously (Elgersma 1980).

For vacuum infiltration of stem parts, stem parts of 4 cm length were cut from the lower part of the plant just above the first leaf. After surface sterilization with 70% ethanol the parts were vacuum-infiltrated with 1 ml suspension of 10^7 conidia / ml. The inoculated stem parts were incubated at 23 C and a 16 hours photoperiod (fluorescent light, 1200 Lx).

The extraction and estimation of the two main phytoalexins of tomato, rishitin and cis-tetradeca-6-ene-1,3-diyne-5,8-diol were performed as previously described (Elgersma 1980; Elgersma et al 1984). Ten stem pieces of 4 cm length were taken from 0.5 to 4.5 cm above the site of inoculation and pooled. The cortex was removed and vascular tissue excised, frozen in liquid nitrogen and stored at -20 C till used for extraction.

Vascular tissue sap was collected by means of a 1-inch French Pressure Cell with hydraulic press. Sap and expressed tissue were immediately stored at -20 C.

RESULTS AND DISCUSSION

Fig. 1 shows a time course study of phytoalexin accumulation in the Verticillium wilt susceptible cv. Gem S and the resistant cv. Gem R after inoculation with a conidial suspension of 10^7 conidia/ml. At each sampling date 5 samples of 10 stem pieces were collected from both susceptible and resistant plants.

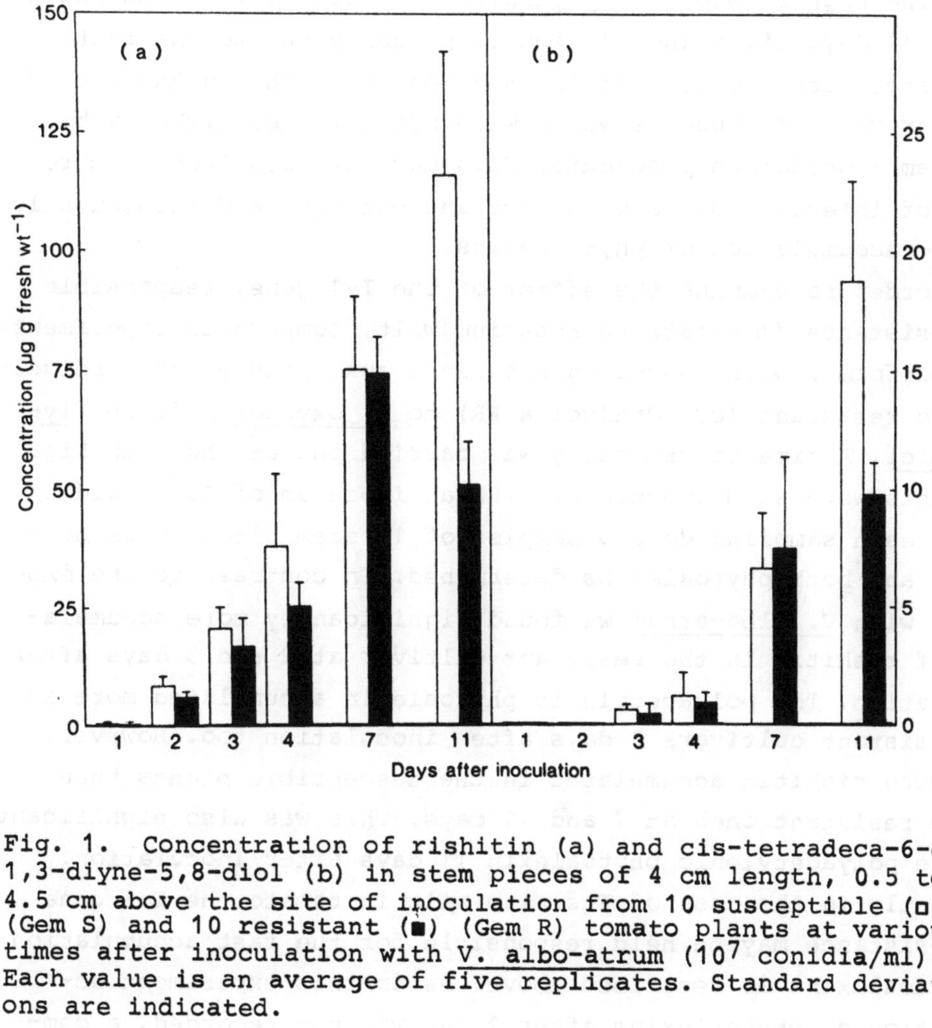

Fig. 1. Concentration of rishitin (a) and cis-tetradeca-6-ene-1,3-diyne-5,8-diol (b) in stem pieces of 4 cm length, 0.5 to 4.5 cm above the site of inoculation from 10 susceptible (□) (Gem S) and 10 resistant (■) (Gem R) tomato plants at various times after inoculation with V. albo-atrum (10^7 conidia/ml). Each value is an average of five replicates. Standard deviations are indicated.

Resistant plants did not accumulate more phytoalexins than susceptible ones, even 11 days after inoculation significantly ($\alpha < 0.05$) more accumulation of phytoalexins was observed in susceptible than in resistant plants. Accumulation of the polyacetylenic phytoalexin could hardly be detected at 1 and 2 days after inoculation. This result confirms our earlier data (Elgersma 1980) which were, however, obtained by using hybrid tomato plants. No convincing evidence was found that the Ve gene is responsible for a hypersensitive reaction, resulting in a faster accumulation of phytoalexins compared to plants lacking this gene. The higher amount of phytoalexins isolated from susceptible plants 11 days after inoculation is probably due to the fact that after such a period of infestation the pathogen had spread to other vascular bundles which were not infected directly by the stem inoculation procedure. This may have resulted in more sites of interactions between host and pathogen and subsequently in more accumulation of phytoalexins.

In order to examine the effect of the I-1 gene, responsible for resistance in tomato to Fusarium wilt, comparable experiments were performed with near-isogenic lines susceptible (cv. Craigella) and resistant (cv. Craigella FR) to F. oxysporum f. sp. lycopersici. A time course study was carried out as shown in Fig. 2. Plants were stem-inoculated with an inoculum of 10^7 conidia/ml. At each sampling date 5 samples of 10 stem pieces were harvested and both phytoalexins determined. In contrast to the experiment with V. albo-atrum we found significantly more accumulation of rishitin in the resistant cultivar at 2 and 3 days after inoculation. The polyacetylenic phytoalexin accumulated more in the resistant cultivars 3 days after inoculation too. However, much more rishitin accumulated in the susceptible plants than in the resistant ones at 7 and 11 days. This was also significant for the polyacetylenic phytoalexin 11 days after inoculation. Apparently in the case of Fusarium wilt in tomato the I-1 gene for resistance may be held responsible for the fast accumulation of phytoalexins as described above. As in this experiment accumulation of phytoalexins after 1 day was not recorded, a comparable experiment was performed in which 5 samples of 10 stem pieces of both cultivars were collected 1 day after inoculation with either 10^7 or 10^8 conidia/ml (Fig. 3). Only rishitin accumulation is shown as accumulation of the polyacetylenic phytoalexin was too low to be estimated accurately. In this experiment we could not record a differential accumulation of rishitin in both cultivars. Apparently first initiation of synthesis of ri-

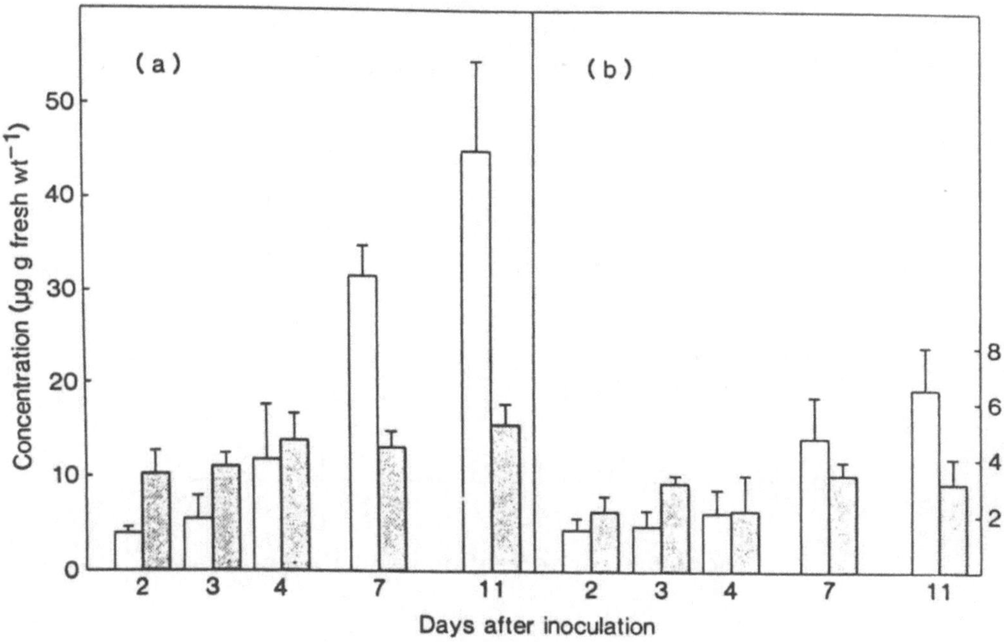

Fig. 2. Concentration of rishitin (a) and cis-tetradeca-6-ene-1,3-diyne-5,8-diol (b) in stem pieces of 4 cm length, 0.5 to 4.5 cm above the site of inoculation from 10 susceptible (□) (Craigella) and 10 resistant (■) (Craigella FR) tomato plants at various times after inoculation with F. oxysporum f. sp. lycopersici (10^7 conidia/ml). Each value is an average of five replicates. Standard deviations are indicated.

shitin (and probably also of the polyacetylene) is comparable in both, but in resistant plants accumulation was faster after 2 and 3 days than in susceptible plants, and at 4 days no differences could be detected anymore. The I-1 gene seems to be involved with the speed of phytoalexin synthesis shortly after the first interaction between host and pathogen.

Some authors (Tjamos and Smith 1974; McCance and Drysdale 1975) described experiments in which they vacuum-infiltrated stem segments of tomato plants with a conidial suspension of Verticillium albo-atrum or Fusarium oxysporum f. sp. lycopersici and subsequently incubated the stem parts for various periods and determined the accumulated phytoalexins. In order to compare the results obtained by this and our method of inoculation and

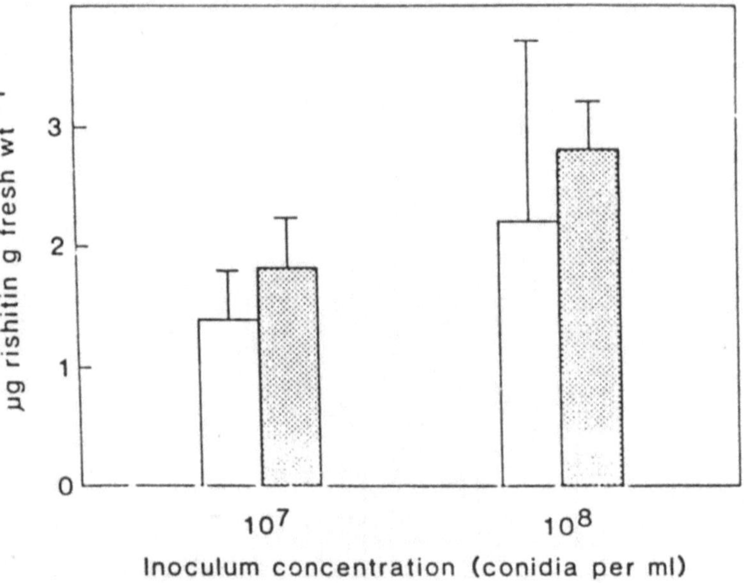

Fig. 3. Concentration of rishitin in stem pieces of 4 cm length, 0.5 to 4.5 cm above the site of inoculation from 10 susceptible (□) (Craigella) and 10 resistant (■) (Craigella FR) tomato plants at 1 day after inoculation with F. oxysporum f. sp. lycopersici (10^7 and 10^8 conidia/ml). Each value is an average of five replicates. Standard deviations are indicated.

incubation, we also performed an experiment with stem parts vacuum-infiltrated with a suspension of 10^7 conidia/ml of F. oxysporum f. sp. lycopersici using cv. Craigella and cv. Craigella FR. This experiment was performed in triplicate and as control stem parts were infiltrated with tap water only (Fig. 4). At none of the sampling dates was a significantly higher amount of phytoalexins determined in stem parts of the resistant cultivar compared to the susceptible one. This discrepancy in results between these two methods may be due to the rather unnatural way of inoculation by means of vacuum infiltration and incubation of excised stem parts.

It is supposed that the phytoalexins are produced in the parenchyma cells close to the infected xylem vessels and diffuse into the xylem fluid, subsequently inhibiting fungal growth. In this respect it will be of paramount importance to know which quantities of the phytoalexins are diffused into the xylem fluid.

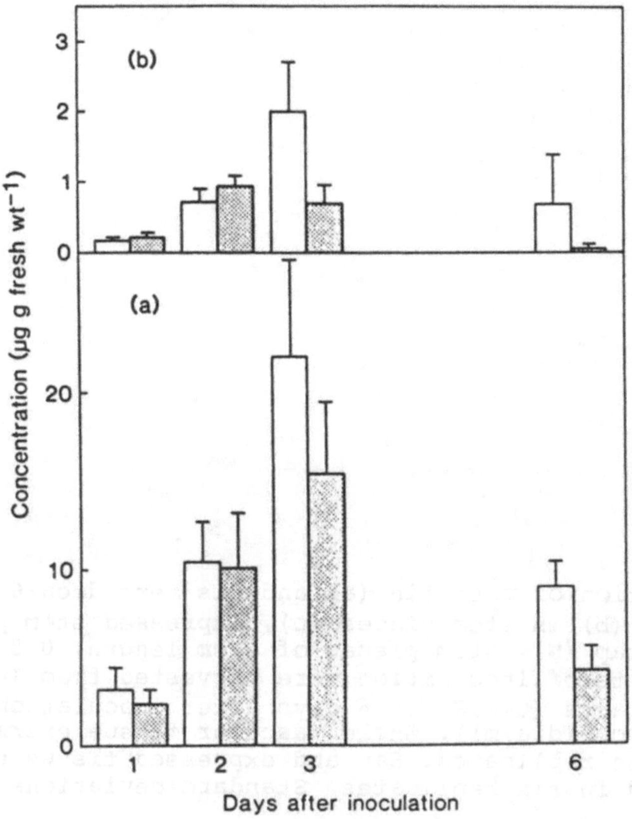

Fig. 4. Concentration of rishitin (a) and cis-tetradeca-6-ene-1,3-diyne-5,8-diol (b) in stem pieces of 4 cm length harvested just above the first leaf from 10 susceptible (□) (Craigella) and 10 resistant (■) (Craigella FR) tomato plants, vacuum-infiltrated with 1 ml of a conidial suspension of F. oxysporum f. sp. lycopersici and incubated for various times. Each value is an average of three replicates. Standard deviations are indicated.

As it is rather difficult to obtain xylem fluid from infected vascular tissue, we determined the amount of phytoalexins in sap expressed of infected vascular tissue. By comparing the amount of phytoalexins in sap with the amount left in the expressed vascular tissue we may obtain insight in the possible concentration of these substances in the xylem fluid. As Fig. 5 shows only a relatively small amount could be collected from the sap (not more than about 5%). Therefore we expect also a relatively low concentration in the xylem fluid, which might not have a

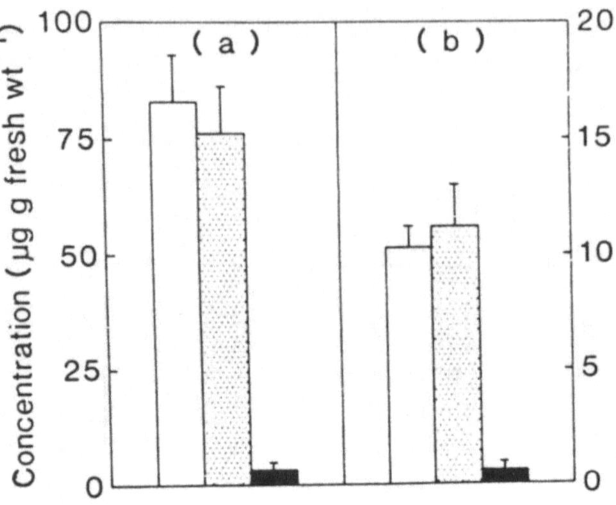

Fig. 5. Concentration of rishitin (a) and cis-tetradeca-6-ene-1,3-diyne-5,8-diol (b) in stem pieces (□), expressed stem pieces (▦) and expressed sap (■). Stem pieces of 4 cm length, 0.5 to 4.5 cm above the site of inoculation were harvested from 10 susceptible tomato plants (Gem S) at 6 days after inoculation with V. albo-atrum (10^7 conidia/ml). Whole vascular tissue extraction was performed in ten replicates. Sap and expressed tissue extraction were performed in six replicates. Standard deviations are indicated.

great impact on growth of the fungus in the first day after inoculation. Indeed, comparable growth of vascular pathogens in susceptible and resistant plants has been demonstrated during the first days after inoculation (Beckman 1968; Elgersma et al 1972; Elgersma 1980, 1982). It is therefore difficult to believe that the differences in accumulation of phytoalexins observed between susceptible and resistant cultivars can be held responsible for the localization of the pathogen in resistant plants. They may, however, play a role in protecting the parenchyma cells from being invaded, as generally during the early stages of disease development the pathogen is confined to the vessel elements. It may even be possible that the observed faster accumulation of phytoalexins in Fusarium wilt resistant plants is only a part, but not of crucial importance, of a series of defence mechanisms to localize the infection such as gummosis, tylosis, callose deposition, lignification and suberization of cell walls, synthesis of chitinase or β-1,3-glucanase etc. The speed and magnitude of these mechanisms may be regulated by genes for resistance and explain the differences observed in Verticillium wilt resistance (Ve gene) and Fusarium wilt resistance (I-1 gene). In our opinion most crucial in resistance to both wilt diseases is the effectiveness of the sealing-off process to localize the infection (Beckman et al 1972; Beckman 1987; Harrison and Beckman 1987). Infusion of the infested area by phytoalexins is important to prevent escape of the fungus from that area, but a faster accumulation of these stress metabolites in resistant plants is not necessary for the plant to successfully defend itself against the pathogen.

ACKNOWLEDGEMENT

We thank Mr. H.J. Roeymans and Drs A.C.M. Weijman and G.W. van Eijk of the Centraalbureau voor Schimmelcultures, Baarn, for the GLC-MS analyses.

REFERENCES

Beckman CH (1968) An evaluation of possible resistance mechanisms in brocolli, cotton, and tomato to vascular infection by Fusarium oxysporum. Phytopathology 58:429-433

Beckman CH (1987) The nature of wilt diseases of plants, APS Press, St Paul, Minnesota

Beckman CH, Elgersma DM and MacHardy WE (1972) The localization of fusarial infections in the vascular tissue of single-dominant-gene resistant tomatoes. Phytopathology 62:1256-1260

Elgersma DM (1980) Accumulation of rishitin in susceptible and resistant tomato plants after inoculation with Verticillium albo-atrum. Physiol Plant Pathol 16:149-153

Elgersma DM (1982) Susceptibility and possible mechanisms of resistance to Dutch elm disease. In: Heijbroek HM, Stephan BR and von Weissenberg K (eds) Resistance to diseases and pests in forest trees. Pudoc, Wageningen, 143-153

Elgersma DM, MacHardy WE and Beckman CH (1972) Growth and distribution of Fusarium oxysporum f. sp. lycopersici in near- isogenic lines of tomato resistant or susceptible to wilt. Phytopathology 62:1232-1237

Elgersma DM, Weijman ACM, Roeymans HJ and van Eijk GW (1984)

Occurrence of falcarinol and falcarindiol in tomato plants after infection with Verticillium albo-atrum and characterization of four phytoalexins by capillary gas chromatography-mass spectrometry. Phytopath Z 109:237-240

Harrison NA and Beckman CH (1987) Growth inhibitors associated with fusarium wilt of tomato. Physiol and Mol Plant Pathol 30:401-420

Hutson RA and Smith IM (1980) Phytoalexins and tyloses in tomato cultivars infected with Fusarium oxysporum f. sp. lycopersici or Verticillium albo-atrum. Physiol Plant Pathol 17:245-257

McCance DJ and Drysdale RB (1975) Production of tomatine and rishitin in tomato plants inoculated with Fusarium oxysporum f. sp. lycopersici. Physiol Plant Pathol 7:221-230

Tchernoff V (1965) Methods for screening and for the rapid selection of elms for resistance to Dutch elm disease. Acta Bot Neerl 14:409-452

Tjamos EC and Smith IM (1974) The role of phytoalexins in resistance of tomato to Verticillium wilt. Physiol Plant Pathol 4:249-259

FUSARIUM WILT OF OIL PALM: TRANSMISSION, ISOLATE VARIATION, RESISTANCE

R.M. Cooper, J. Flood and R. Mepsted
School of Biological Sciences
University of Bath
Bath BA2 7AY
Avon
U.K.

INTRODUCTION

The oil palm (Elaeis guineensis Jacq.) is grown throughout the equatorial regions of the world including West Africa, South America and South East Asia and provides approximately 15% of the world's vegetable oil. Up to 6 tonnes of oil per hectare per year can be obtained which far exceeds the yield of annual oil crops such as soya, but even within the best progenies, yields can still be improved (Jones 1983). Palm oil production from South East Asia, notably Malaysia, has continued to increase and the introduction of high yielding clonal material suggests even higher production in the future. In contrast, production in much of West Africa has fallen,due in part to diseases, of which the most serious is considered to be vascular wilt caused by Fusarium oxysporum f. sp. elaeidis (Foe). Annual losses in some areas of ca 10% of palms have been attributed to the disease (Renard et al 1972).

Vascular wilt was probably first observed in the 1940's in West Africa, where in Nigeria, it was described as 'lemon frond' but no pathogen was described and the disease was considered to be a physiological disorder (Turner 1981). Later Wardlaw described a new vascular disease in Zaire (Belgian Congo) and Fusarium was isolated from diseased plants (Wardlaw 1948). The disease was confirmed in Nigeria in 1948 and has subsequently been reported in many West African countries including Cameroon,

Ivory Coast and Benin (Renard et al 1972). Vascular wilt has also been reported from South America; however reports from Surinam and Columbia were unconfirmed, until more recently F. oxysporum was identified in Brazil (see below). The disease has never been reported in Malaysia (Colhoun 1981).

The serious losses caused by F. oxysporum f. sp. elaedis in West Africa can only be prevented by introduction of resistant varieties. However, breeding programmes are constrained by the lack of a rapid reproducible screen for resistance. Also lack of understanding of mechanisms and genetics of resistance and pathogenicity, of isolate variability and disease spread restricts logical deployment of resistance genes. Some of these factors are under investigation here which is facilitated by availability of clonal oil palm material (Jones 1983).

SCREENING FOR WILT RESISTANCE

Rate of disease progression

Fusarium wilt of oil palm is slow to develop and often takes 4-6 months to seriously affect even young plants. Symptoms include leaf chlorosis and necrosis, stunting and internal vascular necrosis. Hyphal colonization of vessels is usually sparse until the host is moribund (unpublished results). In mature palms fronds may flag, desiccate and eventually break to remain hanging around the stem as a cluster (Turner 1981). In the field the disease is sometimes reported to occur over several years in a chronic form and less frequently in a rapid acute form.

In seedlings and clonal plantlets we have slightly reduced the time for disease expression either by root wounding (although infection of intact roots does occur) or by increasing the commercially used inoculum from 10^7 conidia/ml to 1.5×10^9; in the latter case symptoms were obvious by 15-17 weeks and resistance of one progeny was still operative (as judged by symptoms and by growth assessment). Nevertheless the disease is still slow to develop when compared with Fusarium wilt of dicot hosts such as pea and tomato, which exhibit severe symptoms by 2-3 weeks. The apparently slow progress of infection may be a reflection of the unsuitability for colonization of the vascular system of monocots in general and of palms more specifically.

Palm vascular tissue is entirely primary and must remain functional throughout the life of the plant. It is therefore provided with considerable redundancy to provide alternative pathways in case of injury or infection. Also, in monocots, vessels are scattered (Zimmermann and Tomlinson 1972) rather than in adjacent bundles which facilitates lateral movement of vascular pathogens in dicots. This feature may also explain the relative paucity of vascular diseases of monocots; Fusarium wilt of banana, oil and date palms are notable exceptions.

Screening of seedlings and clonal plantlets for resistance

Axiomatically, to be of value resistance obtained following inoculation of young plants must also be expressed by mature palms under plantation conditions. The validity of screening with young plants is suggested by (i) the occurrence of Fusarium wilt in nurseries as well as in plantations and (ii) correlations between resistance in seedlings and mature palms in plantations (de Franqueville 1984; Renard 1976); Risbeth (1957) reported a similar correlation with Fusarium wilt of banana. Suitable isolates of <u>Foe</u> must be used but the existence of races is not yet clear (see below).

Clonal material with potential wilt resistance is now being developed and requires similar screening procedures before commercial release. It was considered necessary first to investigate the development of root anatomy with age to ensure that potential structural barriers to invasion were fully formed before inoculation. Light and scanning electron microscopy revealed that a lignified hypodermis formed rapidly in all clones; also the outer cortex, endodermis and pith soon lignified. After 2 months in perlite: compost medium the root system was anatomically mature and very similar to that of seedlings (Flood 1986). Variation with age and progeny between vessel lengths and structure of pneumathodes is also under study; vessel length dictates rapid transport of propagules and pneumathodes may be the main points of entry for Foe (Locke and Colhoun 1977).

Screening for disease resistance or tolerance in vitro

Identification of toxin(s) from Foe as determinants of pathogenicity and virulence offers the possibility of in vitro screening for resistance and tolerance with tissue-cultured oil palm cells. The feasibility of this approach has been successfully demonstrated for several diseases including Helminthosporium/ maize (Brettell et al 1980), F. oxysporum/lucerne (Hartman et al 1984) and F. oxysporum/potato (Behnke 1980).

Fungal toxins are known from F. oxysporum but most evidence, albeit contradictory, centers on fusaric acid (5-n-butyl picolinic acid) (Pegg 1981). This toxin is produced by Foe in vitro, and has been detected from infected (6 month) palms by extraction and t.l.c. (Davis 1969) and identified from Foe - infected callus by mass spectroscopy (unpublished data). Action of a diffusible toxin was suggested by the death of callus cells up to 10 mm in advance of hyphal invasion; this observation coincides with ultrastructural evidence for the necrotrophic habit of other F. oxysporum formae speciales in host roots (Bishop and Cooper 1983).

Oil palm callus from several clones was incubated on sub-lethal levels of crude culture fluids or on pure fusaric acid. Sub-lethal concentrations were established by growth experiments (Fig. 1), staining with vital dyes and subsequent culture of treated callus on media lacking toxin. Fast growing sectors were removed from the sub-lethal medium and cultured on higher concentrations of toxin several times. Stability of toxin-resistant callus was confirmed by growth on control medium for several months and then again on toxin medium. Currently attempts are being made to regenerate plants which may have some level of resistance or tolerance to Foe.

ISOLATE VARIATION

Pathogenicities of a range of Foe isolates from West Africa, two from Malaysia and one from Brazil were investigated. This was to (i) ascertain any variation in pathogenicity and/or virulence to susceptible and resistant clones or seedling progeny (ii) choose representative isolates suitable for screening for disease resistance (iii) critically determine pathogenicity of F. oxysporum from Malaysia, where the disease is claimed not to occur and from Brazil, where evidence for Fusarium wilt is tenuous (iv) test for possible differential virulence using a resistant clone. Unlike previous studies a qualitative and quantitive determination of colonization was performed.

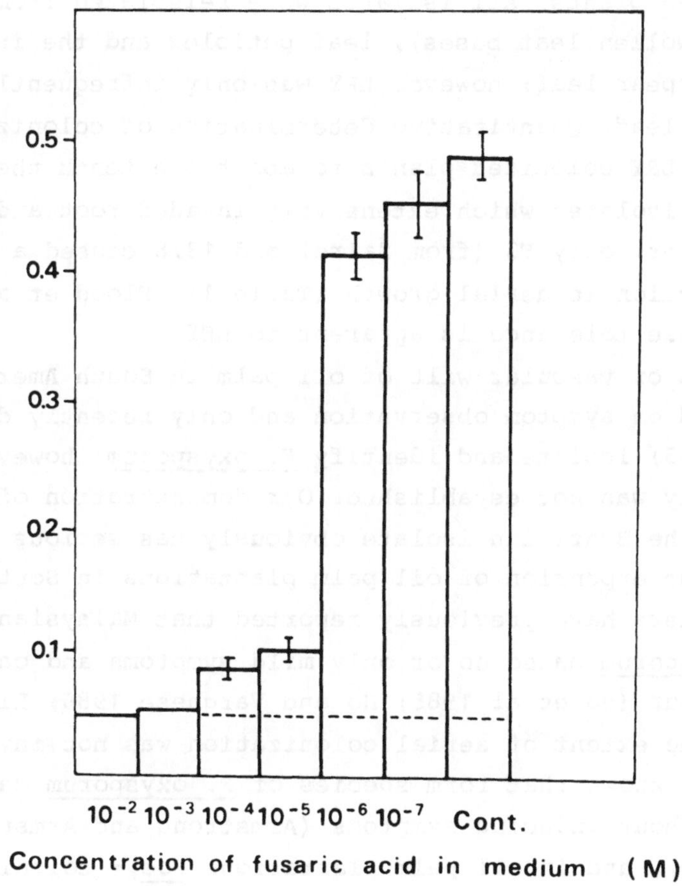

Fig. 1. Growth of oil palm callus on medium containing fusaric acid (Callus was grown for 10 weeks at 28 C).

All six isolates from West Africa were highly pathogenic to a susceptible clone and could not be distinguished in their level of virulence. West African isolates have been previously shown to be virulent to seedlings by Ho et al (1985) and Locke and Colhoun (1974). The Brazilian isolate (1378) caused a similar rate and degree of leaf chlorosis, stunting and extensive vascular necrosis to the West African isolates. However, one Malaysian isolate (LRY from roots of a symptomless palm) caused no symptoms and the other (LEY from plantation soil) caused mild chlorosis and stunting and some vascular necrosis in about 30% of infected plants. All isolates were reisolated from roots, "bulb" (swollen leaf bases), leaf petioles and the innermost, youngest spear leaf; however LRY was only infrequently found in the spear leaf. Quantitative determination of colonization revealed that LRY colonized plants to about one tenth the level of the other isolates which extensively invaded root and bulb tissue. However, only F3 (from Zaire) and 1378 caused a significant reduction in aerial growth (Table 1) (Flood et al 1988). Thus disease tolerance is apparent to LEY.

Reports of vascular wilt of oil palm in South America have been based on symptom observation and only recently did Van de Lande (1983) isolate and identify F. oxysporum; however its pathogenicity was not established. Our demonstration of the virulence of the Brazilian isolate obviously has serious implications for the expansion of oil palm plantations in South America. Other workers have previously reported that Malaysian isolates of F. oxysporum cause no or only mild symptoms and can reduce plant vigour (Ho et al 1985; Ho and Varghese 1986; Lim 1971), however the extent of aerial colonization was not investigated. It is well known that form species of F. oxysporum can enter plants without inducing symptoms (Armstrong and Armstrong 1948; Katan 1971), and in oil palm plantations Foe - colonized weeds may act as symptomless carriers (Oritsejafor 1986). In view of the long term importation of oil palm material into Malaysia it is perhaps remarkable that Fusarium wilt has not been introduced. It is conceivable that avirulent but parasitic isolates of F. oxysporum indigenous to Malaysia occupy an ecological niche (including the vascular system of the oil palm) which prevents establishment and colonization by virulent forms of Foe. We are presently investigating this possibility by preinoculating palms with Malaysian isolates before inoculation with virulent isolates. Renard (1976) has previously demonstrated that preinoculation with avirulent isolates reduces symptom development. An alternative interpretation was offered by Colhoun (1981) who suggested that the lack of a prolonged dry season in Malaysia could suppress disease expression; however, Ho et al (1985a) found that Malaysian isolates were non-pathogenic even under conditions of water stress.

Table 1. Colonization and effect on growth of a susceptible oil palm clone by F. oxysporum isolates from West Africa, Brazil and Malaysia

	Fungal Colonization (cfu)		Host growth (dry weight, g)	
Isolate	Root	Bulb*	Roots	Aerial tissue
LRY	118[a]	452[a]	3.1[a]	10.31[a]
LEY	790[ab]	1,495[b]	2.0[b]	10.2[a]
F3	1,200[b]	2,567[b]	1.1[b]	2.5[b]
1378	1,293[b]	3,174[b]	1.2[b]	2.3[b]
Control	-	-	3.3[a]	10.4[a]

Values represent means of 12 replicate plants 6 months after inoculation. Within each column, figures with the same letter are nsd at 1% level. Isolates: LRY (roots of symptomless palm - Malaysia), LEY (plantation soil - Malaysia), F3 (infected palm - Zaire), 1378 (infected palm - Brazil). *Bulb = swollen leaf bases.

Inoculation of a putative resistant clone (UF28) with two virulent isolates (F3 and Y1) from Zaire has suggested the possibility of differential interactions, better known with major genes for resistance to F. oxysporum of temperate crops such as pea and tomato. F3 showed only very limited colonization of UF28 and no symptoms or reduction in growth of palms. However, colonization by Y1 was not significantly different from that in the susceptible clone UF4, although no symptoms or effect on growth occurred. Clone UF28 appears to exhibit resistance to one virulent isolate and tolerance to another. Hardon (1976) had previously found that isolates from different regions did not appear to interact differentially with a range of progenies.

The increasing availability of resistant clonal material will allow a more extensive and critical study of the possible existence of races of Foe. The genetic control of wilt resistance is insufficiently understood to predict the development of virulent races following introduction of genes for resistance but the suggested polygenic control (Hardon 1976) lends some hope for stability of resistance in the field. The models described in other chapters to account for disease resistance must be adapted to explain the frequently reported phenomenon of tolerance.

TRANSMISSION AND SPREAD

Oil palm pollen and seed are routinely transported between palm oil producing countries and this movement of plant material might allow the introduction of the pathogen into new areas. For this reason Malaysia imposes strict quarantine regulations on oil palm products. Foe has previously been found on the outside of oil palm seeds (Locke and Colhoun 1973) but internal contamination of seeds or infestation of pollen has not been reported.

Thirty samples of pollen were collected from different parts of a plantation (Binga, Zaire) and commercially processed (desiccation and freeze-drying). Dilution plating onto Fusarium-selective medium Papavizas (1967) revealed contamination with a range of Fusarium spp. of which the most common were F. oxysporum and F. solani. F. oxysporum was present in 50% of the samples at levels of 3 to 3,600 cfu/10 mg pollen. The numerous isolates are being tested for pathogenicity but one has already been verified as virulent on oil palm.

Four consignments of seeds were also tested and included seeds from wilt resistant and wilt susceptible crosses. Seeds had been treated as for commercial use with thiram and DDT. No

Fusarium spp. were found on seed surfaces; presumably this is to be expected in view of the use of thiram as an eradicant. However *F. solani* was frequently (30%) found within seeds on the surface of kernels and *F. oxysporum* was detected within 3% of seeds from one cross. It has not yet been established if these are virulent isolates of *Foe*.

Contamination of breeding material requires that strict quarantine is observed and that every attempt is made to eradicate potentially pathogenic fungi fom pollen and seed. Work along these lines is currently in progress here.

The sources of contamination (systemic or external contamination) were investigated in the field by reisolation from the air and from male and female inflorescences at different stages of development on symptomless and wilted plants. Profuse white sporulating mycelium often occurred on mature male inflorescences (from which most pollen had been shed (but see Turner 1981)) on symptomless palms and on those with typical wilt symptoms. Both *F. oxysporum* and *F. solani* were identified but only the latter was detected within male spikelets and anthers. Every pollen sample collected was contaminated with both fungi. In contrast *F. oxysporum* was absent from surfaces and tissues of female spikelets, ovaries and seed. Also it was not detected in the peduncles of inflorescences; thus systemic infection by *Foe* does not appear to be the source of the pathogen in pollen or seed; although systemic infection of seed by *F. oxysporum* has been reported for *Mathiola incana* by Baker (1945). Stimulation of fungal growth *in situ* by factors from pollen has been shown for *Fusarium graminearum* on wheat anthers (Strange and Smith 1978).

Contamination of seed probably occurs after harvest when fruit bunches may be left on the ground for several days then retted in water to remove the pericarp. These conditions allow rapid saprophytic growth and sporulation of *Foe* (unpublished results).

Although-generally considered as a soil-borne organism *F. oxysporum* was found as the most frequent of four *Fusarium* spp. detected at high levels in the atmosphere from areas with or without wilt-infected palms; nevertheless numbers of propagules were greater in air around trees with symptoms. Spore numbers were markedly reduced following rain (Table 2). Air-borne spore dispersal of F. oxysporum has been previously linked with recolonization of steamed soil in tomato glasshouses (Rowe et al 1972) and with the spread of date palm wilt caused by f. sp. albedinis (Salvaraj 1978).

Table 2. Number of colony forming units (cfu) of Fusarium oxysporum and F. solani in plantation air samples

	cfu/m^3 air	
	Wilt area	Non-wilt area
F. oxysporum	96 (22)	36 (13)
F. solani	75 (26)	51 (23)

Samples were obtained from a portable Biotest Centrifugal Air Sampler which samples at 40 l/min onto strips of Fusarium - selective agar. Samples were taken before and after rain (parentheses).

Prima facie rapid, long distance aerial dispersal of Foe can occur; also inoculum level in soil around trees could be enhanced considerably by rain washing off spores from saprophytic mycelium on male inflorescences. This may explain previous observations of increased concentrations of F. oxysporum in the soil and of more diseased roots near the base of palms (Locke 1972; Wardlaw 1948). A further possible vector is the pollinating weevil (Elaeidobius sp.); all individual insects tested were externally contaminated with Fusarium spp.

These results have obvious implications for the epidemiology of Fusarium wilt and for the rapid spread of any possible new races which might arise following the widespread introduction of genes for wilt resistance.

REFERENCES

Armstrong GM and Armstrong JK (1948) Non susceptible hosts as carriers of wilt Fusaria. Phytopathology 38:808-826

Baker KF (1948) Fusarium wilt of garden stock (Mathiola incarna). Phytopathology 38:399-403
Behnke M (1980) Selection of dihaploid potato callus for resistance to the culture filtrate of Fusarium oxysporum. Z Pflanzenzuchtg 85:254-258
Bishop CD and Cooper RM (1983) An ultrastructural study of root invasion in three vascular diseases. Phys Plant Pathol 22:15-27
Brettell RIS and Ingram DS (1979) Tissue culture in the production of novel disease resistant crop plants. Biological Reviews 54: 329-345
Colhoun J (1981) Vascular wilt disease of oil palms. In:Nelson PE, Toussoun TA and Cook RJ (eds) Fusarium:Diseases, Biology and Taxonomy. Pennsylvania State University Press, p 21
Davis D (1969) Fusaric acid in selective pathogenicity of Fusarium oxysporum. Phytopathology 59:1391-1395
Flood J (1986) Resistance of oil palm (Elaeis guineensis) to Fusarium wilt. In:Proceedings of 2nd International Conference on Plant Protection in the Tropics. Kuala Lumpur (Abstr)
Flood J, Cooper RM and Lees P (1988) An investigation of pathogenicity of four isolates of Fusarium oxysporum from South America, Africa and Malaysia to clonal oil palm. Phytopath Z (in Press)
de Franqueville H (1984) Vascular wilt of the oil palm: relationship between nursery and field resistance. Oleagineux 39:516-518
Hardon JJ (1976) Oil palm breeding - Introduction. In: Corley RHV, Hardon JJ and Wood BJ (eds) Oil Palm Research. Elsevier, Amsterdam, p 89
Hartman CL, McCoy TJ and Knous TR (1984) Selection of alfalfa (Medicago sativa) cell lines and regeneration of plants resistant to the toxin(s) produced by Fusarium oxysporum f. sp. medicaginis. Plant Science Letters 34:183-194
Ho YW and Varghese G (1986) Pathogenic potential of soil fusaria from Malaysian oil palm habitats. Phytopath Z 115:325-331
Ho YW, Varghese G and Taylor GS (1985) Pathogenicity of Fusarium oxysporum isolates from Malaysia and Fusarium oxysporum f. sp. elaeidis from Africa to seedlings of oil palm. Phytopath Z 114: 193-202
Jones LH (1983) The oil palm and its clonal propagation by tissue culture. Biologist 30:181-188
Katan J (1971) Symptomless carriers of the tomato Fusarium wilt pathogen. Phytopathology 61:1213-1217
Lim WH (1971) Fusarium species from West Malaysia soil: A study of their distribution, biology and pathogenicity. MSc Thesis. University of Malaysia, Malaysia
Locke T (1972) A study of vascular wilt disease of oil palm seedlings Ph.D. thesis, University of Manchester
Locke T and Colhoun J (1973) Fusarium oxysporum f. sp. elaeidis as a seed-borne pathogen. Trans Br Mycol Soc 60:594-595
Locke T and Colhoun J (1974) Contributions to a method of testing oil palm seedlings for resistance to Fusarium oxysporum f. sp. elaeidis. Phytopath Z 79:77-92
Locke T and Colhoun J (1977) A process of infection of oil palm seedlings by Fusarium oxysporum f. sp. elaeidis . Phytopath Z 88 :18-28
Oritsejafor JJ (1986) Weed hosts of F. oxysporum f. sp. elaeidis.

Oleagineaux 41:1-5
Papavizas G (1967) Evaluation of various media and antimicrobial agents for isolation of Fusaria from soil. Phytopathology 57: 848-852
Pegg GF (1981) Biochemistry and physiology of pathogenesis. In: Mace ME, Bell AA and Beckman CH (eds) Fungal Wilt Diseases of Plants Academic Press, New York p 193
Renard JL (1976) Disease in Africa and South America. In: Corley, RHV, Hardon JJ and Wood BJ (eds) Oil Palm Research Elsevier, Amsterdam, p 447
Renard JL, Gascon JP and Bachy A (1972) Research on vascular wilt of the oil palm. Oleagineaux 27:581-591
Rishbeth J (1957) Fusarium wilt of bananas in Jamaica. II Some aspects of host-parasite relationships. Ann Bot 82:215-245
Rowe RC, Forley JD and Coplin DL (1977) Airborne spore dispersal and recolonization of steamed soil by Fusarium oxysporum in tomato greenhouses. Phytopathology 67:1513-1517
Selvaraj JC (1978) Systemic fungicides in the control of bayoud disease of date palms. World Crops 30:116-120
Strange RN and Smith H (1978) Specificity of choline and betaine as stimulants of Fusarium graminearum. Trans Br Mycol Soc 70: 187-192
Turner PD (1981) Oil Palm Disease and Disorders. Oxford University Press Oxford
Van de Lande HL (1983) Vascular wilt disease of oil palm (Elaeis guineensis Jacq) in Para, Brazil. Oil Palm News 27:6-10
Wardlaw CW (1948) Vascular wilt disease of the oil palm in Nigeria. Nature (London) 162:850-851
Zimmermann MHP and Tomlinson PB (1972) The vascular system of monocotyledonous stems. Botanical Gazette 133:141-155

IMPLICATIONS OF GENETIC/MOLECULAR EVIDENCE WITH RESPECT TO VIRULENCE/AVIRULENCE OF FUNGAL WILT PATHOGENS

J.B. Heale
Plant Cell and Molecular Sciences Group
Biosphere Sciences
King's College London, Kensington Campus
Campden Hill
London W8 7AH
United Kingdom

INTRODUCTION

This review deals with the genetics of virulence/avirulence in Verticillium albo-atrum, V. dahliae and Fusarium oxysporum f. sp. lycopersici. Any such study must take account of the complex interactions occurring between the vascular wilt fungus and its host; these involve different sequential events, host cells of distinct types and morphological/physiological changes in the pathogen, all within a single infection cycle. It is also necessary to draw on our rapidly developing understanding of the molecular mechanisms underlying race-specific and 'generalised' resistance in host/parasite relationships generally and to see how far we can 'fit' these in to the available information for the vascular wilt fungi with their unique and complex 'life-style'. The term pathogenicity is used here to denote the general capacity of a population of individual isolates of a pathogen (comprising a species, forma specialis or 'strain'), to cause disease symptoms in a particular host. Virulence is used to indicate quantitatively the ability of a particular isolate or entity of a pathogen, to cause disease in a susceptible host under standard conditions.

GENE-FOR-GENE CONCEPT

The gene-for-gene concept (Flor 1955, 1971) is a powerful model in probing the nature of host-specific interactions exhibited between plants and their biotrophic microbial pathogens. As usually interpreted, this involves a hypersensitive response determined by a single host 'resistance' gene following recognition of the product of a corresponding avirulent gene in the pathogen. According to this model, virulent isolates cause disease in the host because recognition fails to occur in the absence of the particular avirulent gene corresponding to the specific resistance gene. Such specific interactions in e.g. rust or powdery mildew diseases, result in the complete failure of avirulent races to colonise their hosts due to active defence mechanisms induced subsequent to such recognition events.

SELECTION FOR VIRULENCE DURING THE BIOTROPHIC PHACE OF VASCULAR WILT PATHOGENS

How far do such mechanisms operate in determining virulence/ avirulence in vascular wilt diseases? Let us briefly consider the pathogen's life cycle in Verticillium and Fusarial wilts. During the initial colonisation of the host, these fungi traverse the root cortex growing inter- or intracellularly through parenchyma cells without causing cell death. After endodermal penetration, there is extensive upward colonisation of the xylem vessels, but only when systemic colonisation is advanced and after full wilt symptoms are observed, does tissue necrosis occur as the pathogen leaves the vascular system. Thus, these pathogens behave essentially as biotrophs throughout the crucial phases of interaction with their hosts and evolutionary selection pressures have resulted in the effective limitation of their potentially destructive effects to the final stages of infection.

SINGLE GENES CONTROL RECOGNITION EVENTS IN VASCULAR WILTS

Examples of single dominant genes controlling recognition events in vascular wilt diseases include those of the Ve and I genes in tomato, controlling resistance to V. dahliae and F. oxysporum f. sp. lycopersici (F.o.lycopersici) respectively. Beckman (1987) has pointed out that for the vascular wilts, quantitative differences in the recognition mechanism itself, or in the rate of extent of host responses, determine the outcome of infection. These host responses will be dealt with comprehension elsewhere in this symposium; it is sufficient here to state that a multigenically-determined series of active defence responses which include wall appositions, coatings, phenolics, phytoalexins, gels/tyloses, production of chitinases etc, usually occur even in susceptible hosts, but the rate at which the particular response occurs, and its ultimate extent in individual cultivars depends upon the presence/absence of such single recognition genes, which therefore limit the quantitative expression of virulence by the pathogen. We must not forget also that preformed resistance is a highly complex, polygenically-controlled system in the host, consisting of traits affecting structural and preformed chemical defence which determine absolute resistance and susceptibility, but will also contribute to the selection pressures operating between moderately virulent isolates and partially susceptible hosts, i.e. where some degree of compatibility occurs.

THE PHENOMENON OF TOLERANCE IN VASCULAR WILTS

An intriguing aspect of vascular wilt fungi is their ability frequently to colonise hosts (including weed species) without inducing symptoms; thus, virulence is not expressed and the host is regarded as 'tolerant'. This occurs even in the presence of a recognition gene e.g. Race 1 strains of V. dahliae will infect Race 1-resistant tomato plants containing the Ve gene to a limited extent without causing symptoms (Grogan et al 1979). Here, the quantitative response of the host following recognition is sufficient to contain the pathogen at a low level and at the same time those fungal genes responsible for pathogenic effects e.g. production of enzymes, toxins etc are not induced, or are suppressed.

THE POSSIBLE RELEVANCE OF MOLECULAR BIOLOGY STUDIES IN OTHER HOST/PARASITE INTERACTIONS TO VASCULAR WILTS

Recognition genes modulate the expression of multiple disease-response genes in pea

Evidence that single genes responsible for recognition events in race-specific resistance are involved in the regulation of multiple disease-response genes associated with the general expression of resistance has come from very recent studies involving peas challenged with F. solani f. sp. phaseoli (non-host resistance), or virulent/avirulent races of Pseudomonas syringae pv. pisi (race-specific resistance). Kendra et al (1987) and Daniels et al (1987) reported that several cloned pea genes were activated in temporal correlation with the expression of disease resistance in both non-host and specific resistance. In the latter case, mRNAs for 5 pea genes that accumulated in a non-host resistance response also showed sustained accumulation in response to avirulent races of the bacterium. These mRNAs again accumulated within the first 10 hours after inoculation with virulent races, but this decreased later as the susceptible interaction progressed.

Mutation from avirulence to virulence is a single transposon insertion event in a bacterial disease

Very recently, Staskawicz and co-workers (Kearney et al 1988) have shown that in the bacterial spot disease of peppers caused by Xanthomonas campestris pv. vesicatoria, spontaneous mutation of the avrBs1 gene determining avirulence resulted in a failure of the recognition system and to virulence being expressed. Sequence analysis revealed that different mutants were characterised by the particular positions of 1.2 kb transposon inserts with-in the avrBs1 locus. These workers also showed that a plasmid containing the cloned avrBs1 gene converted the virulent mutants by transconjugation to avirulent strains. Interestingly, some of the *Xanthomonas* spontaneous mutants caused intermediate reactions suggesting partial recognition; one of these mutants resulted in a delay of 24 hours in the HR response and here the insertion of the transposed element was demonstrated to be in a regulatory region of the locus rather than a coding region, possibly causing lower levels of avirulent protein product. Similar phenomena might explain the modulation of generalised host responses in Ve- and I-gene determined systems, but because of the more complex nature of these fungal genomes and the interactions involved, we have no direct observations to comment upon as yet.

GENERALISED MODELS FOR GENE-FOR-GENE INTERACTIONS UNDERLYING BOTH BASIC COMPATIBILITY AND RACE-SPECIFICITY

The pathogenic ability of a fungal parasite towards a particular host species (i.e. its basic compatibility at the species level of specificity, in relation to that host) will be governed by its capacity to overcome or avoid general defences (both preformed and induced) (Heath 1981). Evolution forces further accommodation in the compatible association explaining why a given fungal pathogen rarely shows a marked change in host species range since random mutation is unlikely to produce simultaneously the multigenic changes required. Within the compatible association, selection pressures in the host lead to recognition of a specific fungal molecule and to the system of avirulent race/ resistant host or incompatible interaction (race-specific resistance operating at the cultivar level), usually a much more ephemeral resistance mechanism for the reasons outlined previously. There are numerous possible mechanisms based on gene-for-gene interactions between pathogen and host at both levels of specificity. Heath (1981) and Bushnell and Rowell (1981) have developed models incorporating the important role of supressors ('blockers') of defence reactions or detoxification of antifungal metabolites (pre-formed or induced). The latter workers propose that corresponding dominant genes in both pathogen and host determine basic compatibility. The model is based partly on the release of non-specific elicitors (from the walls of the fungal parasite) which become bound to receptors in the host, thus 'triggering' HR responses, phytoalexin accumulation, wall changes etc. At the same time, compatible pathogens secrete a species-specific suppressor which matches a series of corresponding sites on the receptor, preventing the action of the non-specific elicitors and 'inducing susceptibility'.

Robb et al (1987) have reported recently that the presence of the dominant allele at the Ve locus in a tomato cultivar interacted with an isolate of V. albo-atrum, conferred resistance to a pathogen-produced suppressor of the vascular coating response. The latter involved the phenyl propanoid pathway and synthesis of lignin-like materials detected histochemically in the petiolar xylem. Here, virulence is determined partially by suppressor production which, in cultivars lacking the Ve gene, essentially 'induces susceptibility'. Unfortunately no evidence was presented for suppressors in different strains of the pathogen.

VIRULENCE GENES AND THEIR MECHANISMS IN VASCULAR WILTS

Avirulence gene function and change to virulence

Let us now speculate on the nature of virulence genes in vascular wilt fungi. Firstly, the models described previously predict that virulent pathogens fail to incite recognition, or only do so partially, in hosts with the potential to react. Thus two near-isogenic lines of tomato:"Improved Pearson" (IP), and "Pearson VF-11" (VF) carrying the I gene, are susceptible and resistant respectively to Race 1 of F. o. lycopersici (Beckman 1987). We can assume that an avirulent gene product/'elicitor' from Race 1 binds with a specific receptor determined by the I recognition gene carried by the VF line (absent or 'impaired' in IP), and that this leads to activation of generalised resistance responses.

It is established that a process of vascular occlusion is an important aspect of the expression of this single dominant resistance gene, resulting in localization of the pathogen. Recently, Harrison and Beckman (1987) demonstrated that a host-produced inhibitor of tylose formation (tentatively identified as the phytoalexin rishitin) goes on accumulating in susceptible tomato cvs after 4 days, whereas it declines in resistant cvs, as also occurs after inoculation with *Verticillium* (Elgersma 1980). They interpret these events as being due to the rapid sealing-off of infected xylem vessels by gels in the *I* gene-containing cv, thus preventing an upward flow of stress metabolites which would otherwise (in susceptible plants) cause an inhibition of tylosis formation. Race 2 isolates of this pathogen can overcome the resistance of *I* gene cvs but not of the *I-2* gene cvs (Beckman 1987), presumably because of a genetic change at the corresponding avirulence gene locus carried by Race 1 isolates. It seems unlikely that this could be due to a single mutation or transposon insertion *alone*, as described earlier for the *Xanthomonas* system, since resistance to Race 1 provided by the *I* gene completely controlled wilt in tomato caused by *F. o. lycopersici* in Florida for 20 years before Race 2 became a serious problem in the 1960's; similarly the *I-2* gene has been effective for about 20 years, although currently Race 3 is spreading in Australia and Florida (Beckman 1987).

The distinction between (i) severe 'defoliating' and mild 'non-defoliating' strains of *V. dahliae* attacking cotton, on the one hand, and (ii) severe 'progressive' and mild ('fluctuating') strains of *V. albo-atrum* attacking hop on the other, can be construed as involving a change in the relatively avirulent pathogens at one or more virulence loci corresponding to recognition sites determined by 'major genes' in the host, thus allowing the now compatible strains to colonise the particular host without *rapid* generalised defence reactions, followed by evolution towards different degrees of basic compatibility (Fig.1).

1. INITIAL MUTATION AT AV LOCUS IN PATHOGEN

RACE 1-------------------------→'PROTO RACE 2'

AV LOCUS---------------------------→ av

AVIRULENT GENE PRODUCT/RECOGNITION	CHANGED GENE PRODUCT PARTIAL RECOGNITION?
	ALLOWS COLONISATION AND
RESISTANCE EXPRESSED	PARTIAL SUSCEPTIBILITY

2. FURTHER MUTATIONS AT AV LOCUS

av--------------------→ av^{NR}

--------→ NON-RECOGNITION:TOLERANCE EXTENSIVE COLONISATION

PARTIAL SUSCEPTIBILITY — TOLERANCE

3. MUTATIONS AT MANY LOCI DETERMINING NON-SPECIFIC INTERACTIONS; SELECTION FOR INCREASED VIRULENCE:-

(i) v^{a}----------------→ v^{a1}

(ii) v^{b}----------------→ v^{b1}

(iii) v^{c}----------------→ v^{c1}

EVENTUALLY?-------------→ HIGHLY VIRULENT RACE 2

COMPLICATING FACTORS, DELAYING SERIOUS OUTBREAK OF WILT DUE TO A NOVEL RACE

(a). MORE THAN A SINGLE GENE MUTATION REQUIRED FOR HIGH VIRULENCE TO BE EXPRESSED

(b). SELECTION PRESSURES FOR FITNESS

(c). SLOW BUILD UP (EPIDEMIOLOGICAL FACTORS)

STEPS 1 AND 2 (ABOVE) COULD OCCUR WITH EQUAL FREQUENCY AND ARE NOT NECESSARILY SEQUENTIAL; 3 FOLLOWS 1

THE RANDOM NATURE OF THE MUTATIONS INVOLVED AT NUMEROUS LOCI WILL RESULT IN A CONTINUUM OF VIRULENCE BEING EXPRESSED BY A LARGE POPULATION OF ISOLATES INDEPENDENTLY OBTAINED FROM DIFFERENT SITES/GEOGRAPHICAL AREAS

Fig. 1. Model for evolution of novel races Fusarium/Verticillium attacking tomato (Assumes Race 1 is non-pathogenic to cvs with I-1 of Ve single dominant recognition genes).

Durability of resistance and evolution of virulence

The phenomenon of prolonged durability of apparently single gene resistant host cvs to vascular wilt fungi appears to be a general one and deserves some explanation. Thus, the Ve gene in tomato conferring resistance to Race 1 of V. dahliae was effective in N. America for at least 10 years before Race 2 was reported there (Alexander 1962), and even later in Europe (Cirulli 1969). There are several aspects here, not necessarily mutually exclusive. The resistance of particular cvs carrying such demonstrable 'major gene' type of resistance will be enhanced by the presence of a modifying genetic background which complicates its effect. Beckman (1987) has discussed this in detail in relation to the tomato/F. o. oxysporum interaction. Thus we may infer that the genetic difference between highly virulent isolates of Race 1 and Race 2 is likely to involve several (numerous?) loci involving basic compatibility 'adjustments' as discussed earlier and not just a change at the 'recognition locus' determining avirulence to I gene cvs in the Race 1 isolates. A second factor may be that selection pressures towards increased adaptation in pathogenicity towards particular host cvs may result in decreased fitness e.g. in the necrotrophic or dormant soil phases in the life cycle. An example of reduced fitness is also exhibited by Race 2 isolates of V. dahliae in that they are less virulent than Race 1 isolates on tomato cvs lacking the Ve gene (Grogan et al 1979). Finally, it may be suggested that the variation in these pathogens is such that novel pathotypes will be generated fairly rapidly, but that the 'build up' of the new Race in the field is very slow given the epidemiological limitations that soil-borne, vascular wilt fungi are subject to.

Breeding programmes to produce cvs with increased resistance to vascular wilt fungi e.g. cotton/V. dahliae, alfalfa/V. albo-atrum hop/V. albo-atrum, have generally produced cvs with various degrees of 'tolerance' based on only partially characterised 'major genes', with evidence for additional multigenic resistance (Heale 1988). Again, there is little evidence for the rapid spread of 'novel' strains causing severe wilt and economic losses in such 'tolerant' cvs, although they have been detected in the field in some cases shortly after the introduction of the new cvs. Thus, novel pathotypes (PV2 and PV3) of the progressive strain of V. albo-atrum attacking hops were observed within a few years of the introduction of more tolerant cvs (Sewell and Wilson 1984). Here, initial and potentially serious outbreaks of progressive wilt on the cv 'Target' (popular with Growers because of its high bittering qualities as well as high tolerance to most isolates of the pathogen) by the PV3 pathotype, have not led to serious spread of wilt in this cv, partly at least through good field management of the disease, but altered fitness may play a part here also.

Virulence and basic compatibility

Some examples of interactions where selection could occur in the pathogen for increased virulence in an already at least partly compatible situation are: 1. Ability to detoxify or become resistant to preformed inhibitors. 2. Production of suppressors of the vascular coating (callose/suberin/lignin) response. 3. Inhibition of gels/tyloses. 4. Inhibition of biosynthetic pathways leading to phytoalexin/stress metabolite accumulation and/or ability to detoxify or become resistant to them. 5. Increased pectic enzyme activity/resistance to effects of host inhibitors of pectic enzymes. 6. Increased toxin production or activity. 7. Increased *in vivo* secondary spore spread. (This is not an exhaustive list!- see Beckman 1987). The main point here is that random genetic changes (whether by mutation or parasexual recombination as discussed later) at any of these or other relevant loci, which are then subject to selection pressures in individual compatible cvs, will produce over time a population of individual entities. If 'sampled' by random re-isolation, the resulting isolates are likely to exhibit a range of different overall pathogenicity when tested against standard cvs under the same environmental conditions. Thus we can predict that if carried out on a sufficiently large number of isolates, such procedures will detect a continuum of virulence as reported for *V. dahliae* in cotton by Bell (1973) and for *V. albo-atrum* in hop by Sewell and Wilson (1984). This probably explains the difficulties experienced by Ministry scientists in the UK when attempting to establish if new outbreaks of wilt in hop are due to either the progressive or mild strain. Here standardised pathogenicity tests often fail to distinguish between a small proportion of the isolates obtained. Some correlations have been established for differential pectic enzyme levels in mild and progressive hop isolates but exceptions occurred (Carder et al 1987). We initially attempted to distinguish between these strains by their in vitro toxin production tested against cells and protoplasts of hop cvs (Connell and Heale 1986), but more extensive investigation has failed to detect consistent differences that would reliably distinguish large numbers of isolates into such strains on these criteria (Heale, Legg and Brar, unpublished data). Bender and Shoemaker (1984) suggested that one determinant separating Race 1 and Race 2 pathotypes of V. dahliae attacking tomato was the type of wilt toxins produced. Recently, Nachmias et al (1987) have detected differences in the amino acid composition of phytotoxic peptides produced by these two races. If so, we can again interpret this as an example of further evolution of basic compatibility following the initial change from avirulence to virulence, although we should still expect variability in this trait in different isolates within a Race designation.

PARASEXUAL RECOMBINATION

Parasexual recombination has been reported in host plants between heterokaryon compatible isolates/Races of different virulence in F. oxysporum (e.g. in tomato, MacNeil et al 1980), in V. dahliae (e.g. in cotton, Kasyanenko and Portenko 1978); in V. albo-atrum (e.g. in antirrhinum, Clarkson and Heale, 1985a,b). A recombinant prototrophic haploid of V. albo-atrum recovered by the latter authors from antirrhinum (as a universal suscept) which had been inoculated with the two auxotrophic mutants: Mild strain (isolate M18) nic-4, cob-26, and Progressive strain (isolate PV3) arg-8, pyr-2, from hop, showed an intermediate virulence between the parent strains when tested against hop cvs. The strong selection pressure to prototrophy here would favour recombinants with virulence genes linked in coupling to wild-type growth loci and select against those linked similarly to auxotrophic alleles. Similar parasexual crosses in hop strains which led to recombinants with reduced virulence have been interpreted as being due to the disruption of specially-adapted gene complexes controlling virulence (Clarkson and Heale 1985c; Heale 1988) which is in accord with the ideas presented in this review. Rather suprisingly therefore McGeary and Hastie (1982) reported that of 21 independent prototrophic recombinant haploids derived from a cross between lucerne and tomato diauxotrophs on agar media, 13 of these displayed attributes of both parents causing severe or moderate symptoms in both hosts (the parent tomato isolate was completely avirulent in lucerne although it achieved partial colonisation of the roots). Nevertheless, the extent of the contributions made by parasexual recombination in the evolution of virulence/pathogenicity as compared with mutation, remains a matter of conjecture and future research.

HET-C/VEGETATIVE COMPATIBILITY GROUPINGS AND GENE EXCHANGE AFFECTING VIRULENCE

Heterokaryon incompatibility in Verticillium spp. has been reviewed recently (Heale 1988) and will be dealt with only briefly here. Most intraspecific paired combinations of diauxotrophs of either V. albo-atrum or V. dahliae, even of generally incompatible pairings, produced at least rare heterozygous diploids (Typas and Heale 1978). A different method of assessing incompatibility in V. dahliae was developed by Puhalla and Hummel (1983) viz. the ability to produce a line of black microsclerotia between paired microsclerotial pigment-deficient mutants; the results gave some correlations with pathogenicity groupings in that all defoliating isolates from hop were placed in the same het-c group. Typas (1983) used protoplast fusion and microsurgery to circumvent hyphal wall-determined compatibility factors in V. dahliae and V. albo-atrum. From these and other studies (Heale 1988), it appears that under natural conditions gene exchange via the parasexual cycle in Verticillium is probably very restricted, and particularly so between isolates adapted to different hosts, but although probably occurring as an extremely rare event, it could have considerable significance in the evolution of virulence and the survival of the pathogen, particularly in the temporary absence of the favoured host. It has been suggested (Heale 1988) that evolution of this pathogen may occur in broad-leaved weeds which often harbour the fungus and presumably mixtures of different strains. On the other hand, Puhalla (1985), Correll et al (1986), Bosland and Williams (1987), and Katan and Katan (1988) have shown agreement in demonstrating that different strains of F. oxysporum fall into apparently 'clear-cut' pathogenicity groups on the basis of ability to produce heterokaryons ('vegetative compatibility') by pairing complementary nitrate non-utilising mutants ('nit mutants'). Thus, gene flow between form species of F. oxysporum may be even more restricted than is the case for Verticillium, facilitating evolution of highly adapted virulences to the specific host at the expense of the ability to infect a different host.

MOLECULAR BIOLOGY APPROACHES IN FUSARIUM OXYSPORUM

Progress has recently been reported from several laboratories in the application of the developing techniques of molecular biology and DNA analysis in form species of F. oxysporum. Coddington et al (1987) demonstrated that restriction enzyme digest patterns of total DNA from different races of F. oxysporum f. sp. pisi were closely related to race pathotype. Manicom et al (1987) using a similar approach, and also random DNA probes generated from total DNA of one isolate, achieved differentiation at species, form species and isolate levels, RFLP analyses of mitochondrial DNA in strains of F. oxysporum from crucifers by Kistler et al (1987) gave patterns consistent with form species, whereas ribosomal DNA RFLP patterns were similar regardless of pathotype. Kistler and Leong (1986) identified a linear plasmid or mycovirus consisting of double-stranded 1.9 kb DNA molecules in strains of F. o. conglutinans, including one fragment that was apparently autonomously replicating; its presence/absence was closely correlated to host specificity and they concluded that since these plasmid-like DNAs were host-specific, they may carry genes that determine host specialisation. This appears to be a very promising line of research and possibly opens a new avenue to our understanding of virulence in vascular wilt fungi along with a recent report of a transformation system being developed for F. oxysporum by Kistler and Benny (1988). Earlier, Yoder et al (1986) reported regarding techniques and cloning strategies for the transformation of wild types using selectable markers, and the cloning of genes determining virulence in the maize plant-pathogenic fungus Cochliobolus heterostrophus (anamorph: Bipolaris maydis=Helminthosporium maydis). No similar reports are apparently available for Verticillium except that of Barbara et al (1987) who found that presence/absence of double-stranded, virus-like molecules in V. albo-atrum isolates from hop was not correlated with either progressive or mild pathotype.

ACKNOWLEDGEMENTS

Research by the author is supported in part by the Leverhulme Trust, Allied Breweries Ltd (UK) and the Agricultural and Food Research Council, UK.

REFERENCES

Alexander LJ (1962) Susceptibility of certain Verticillium-resistant tomato varieties to an Ohio isolate of the pathogen. Phytopathology 52:998-1000

Barbara DJ, Ashby SC and Morton A (1987) Double-stranded RNA in Verticillium albo-atrum . Plant Pathology 36:229-231

Beckman CH (1987) The nature of wilt diseases in plants. The American Phytopathology Society, pp 1-175

Bell AA (1973) Nature of disease resistance. In Verticillium wilt of cotton. Proc Work Conf Nat Cotton Path Res Lab Texas 1971, ARS-S-19:47-62

Bender CG and Shoemaker PB (1987) Prevalence and severity of Verticillium wilt of tomato and virulence of Verticillium dahliae Kleb. isolates in Western North Carolina. Proc Am Phytopath Soc (Abstr) 4, p 152

Bosland PW and Williams PH (1987) An evaluation of Fusarium oxysporum from crucifers based on pathogenicity, isozyme polymorphism, vegetative compatibility and geographic origin. Can J Bot 65:2067-2073

Bushnell WR and Rowell JB (1981) Suppressors of defense reactions: a model for roles in specificity. Phytopathology 71: 1012-1014

Carder JH, Hignett RC and Swinburne TR (1987) Relationship between the virulence of hop isolates of Verticillium albo-atrum and their in vitro secretion of cell-wall degrading enzymes. Physiol Mole Plant Pathol 31:441-452

Cirulli M (1969) Un isolato di Verticillium dahliae Kleb. virulento verso varieta resistenti di pomorodoro. Phytopatol Mediterr 8:132-136

Clarkson JM and Heale JB (1985a) Pathogenicity and colonisation studies on wild-type and auxotrophic isolates of Verticillium albo-atrum from hop. Plant Pathology 34:119-128

Clarkson JM and Heale JB (1985b) Heterokaryon compatibility and genetic recombination within a host plant between hop wilt isolates of Verticillium albo-atrum. Plant Pathology 34:129-138

Clarkson JM and Heale JB (1985c) A preliminary investigation of the genetics of pathogenicity in hop wilt isolates of Verticillium albo-atrum. Trans Br mycol Soc 85:345-350

Coddington A, Mathews PM, Cullis C and Smith KH (1987) Restriction digest patterns of total DNA from different races of Fusarium oxysporum f. sp. pisi - an improved method for race classification. J Phytopathol 118:9-20

Connell SA and Heale JB (1986) Development of an in vitro selection system for novel sources of resistance to Verticillium wilt in hops. In Withers L and Alderson (eds), Tissue Culture and Agriculture, Butterworths UK, pp 451-459

Correll JC, Puhalla JE and Schneider RW (1986) Vegetative compatibility groups among non-pathogenic root-colonizing strains of Fusarium oxysporum. Can J Bot 64:2358-2361

Daniels CH, Fristensky B, Wagoner W and Hadwiger LA (1987) Pea genes associated with non-host disease resistance to Fusarium are also active in race-specific disease resistance to Pseudomonas. Plant Mole Biol 8:309-316

Elgersma DM (1980) Accumulation or rishitin in susceptible and resistant tomato plants after inoculation with Verticillium albo-atrum. Physiol Plant Pathol 16:149-153

Flor HH (1955) Host-parasite interaction in flax rust-its genetics and other implications. Phytopathology 45:680-685

Flor HH (1971) Current status of the gene-for-gene concept. Annu Rev Phytopathol 9:275-296

Grogan RG, Ioannou N, Schneider RW, Sall MA and Kimble KA (1979) Verticillium wilt on resistant tomato cultivars in California: Virulence of isolates from plants and soil and relationship of inoculum density to disease incidence. Phytopathology 69: 1176-1180

Harrison NA and Beckman CH (1987) Growth inhibitors associated with Fusarium wilt of tomato. Physiol Mole Plant Pathol 30: 401-420

Heale JB (1988) Verticillium spp., the cause of vascular wilts in many species. In Ingram DS and Williams PH (series eds), Advances in Plant Pathology, Vol 6, Sidhu GS (ed) Genetics of Pathogenic Fungi. Academic Press, London, pp 291-312

Heath MC (1981) A generalized concept of host-parasite specificity. Phytopathology 71:1121-1123

Kasyanenko AG and Portenko LG (1978) Somatic recombination as a source of new races of cotton Verticillium wilt causal agent. In Maksumov A (ed) Genetic variability of the wilt causing agent and prospects for increasing cotton wilt resistance. Donish Pub House, Dushanbe, USSR, pp 107-109 (in Russian, English summary)

Katan T and Katan J (1988) Vegetative compatibility groupings of Fusarium oxysporum f. sp. vasinfectum from tissue and the rhizosphere of cotton plants. Phytopathology (in Press)

Kearney B, Ronald PC, Dahlbeck D and Staskawicz BJ (1988) Molecular basis for evasion of plant host defence in bacterial spot disease of pepper. Nature 332:541-543

Kendra DF, Fristensky B, Daniels CH and Hadwiger LA (1987) Disease resistance response genes in plants: expression and proposed mechanisms of induction. In Arntzen CJ and Ryan C (eds) Molecular strategies for crop protection. Alan R Liss, Inc.New York, p 13

Kistler HC and Leong SA (1986) Linear plasmid-like DNA in the plant pathogenic fungus Fusarium oxysporum f. sp. conglutinans. J Bacteriol 167:587-593

Kistler HC, Bosland PW, Benny UK, Leong SA and Williams PH (1987) Relatedness of strains of Fusarium oxysporum from crucifers measured by examination of mitochondrial and ribosomal DNA. Phytopathology 77:1289-1293

Kistler HC and Benny UK (1988) Genetic transformation of the fungal plant wilt pathogen Fusarium oxysporum. Curr Genet 12 (in Press)

MacNeil BH, Romero-Munoz F and Chilton S (1980) Genetic recombination in Fusarium oxysporum f. sp. lycopersici in vivo and in vitro. Phytopathology 70:465

Manicom BQ, Bar-Joseph M, Rosner A, Vigodsky-Haas H and Kotze JM (1987) Potential applications of random DNA probes and restriction length polymorphisms in the taxonomy of the Fusaria. Phytopathology 77:669-672

McGeary FM and Hastie AC (1982) Hybridisation of Verticillium albo-atrum strains from tomato and lucerne. Physiol Plant Pathol 21:437-444

Nachmias A, Buchner V, Tsror I, Burstein Y and Keen N (1987) Differential phytotoxicity of peptides from culture fluids of Verticillium dahliae Races 1 and 2 and their relationship to pathogenicity of the fungi on tomato. Phytopathology 77:506-510

Puhalla JE and Hummel M (1983) Vegetative compatibility groups within Verticillium dahliae. Phytopathology 73:1305-1308

Puhalla JE (1985) Classification of strains of Fusarium oxysporum on the basis of vegetative compatibility. Can J Bot 63:179-183

Robb J, Powell DA and Street PFS (1987) Time course of wall-coating secretion in Verticillium-infected tomatoes. Physiol Mole Plant Pathol 31:217-226

Sewell GWF and Wilson JE (1984) The nature and distribution of Verticillium albo-atrum strains highly pathogenic to hop. Plant Pathology 33:39-52

Typas MA and Heale JB (1978) Heterozygous diploid analyses in the parasexual cycle and a cytoplasmic pattern of inheritance in *Verticillium* spp. Genet Res Camb 31:131-144

Typas MA (1983) Heterokaryon incompatibility and interspecific hybridisation between *Verticillium albo-atrum* and *V. dahliae* following protoplast fusion and microinjection. J gen Microbiol 129:3043-3056

Yoder OC, Weltring K, Turgeon BG, Garber RC and VanEtten HD (1986) Technology for molecular cloning of fungal virulence genes. In Bailey JA (ed) Biology and Molecular Biology of Plant-Pathogen Interactions Springer, pp 371-384

MOLECULAR BASES FOR VIRULENCE AND AVIRULENCE OF FUNGAL WILT PATHOGENS

N.K. Van Alfen
Department of Biology
Utah State University
Logan, UT 84322-5305
USA

INTRODUCTION

Virulence expression by fungal wilt pathogens is an interesting phenomenon. Compared with many other pathogens, fungal wilt pathogens are, at best, weakly virulent. Wilt pathogens do not vigorously invade healthy host tissue. Once within the xylem of susceptible plants, probably after entering through a wound, their colonization is limited to the water-conducting vessels of the plant. Only when the host succumbs to pathogen-induced water stress are the wilt pathogens able to colonize other host tissues. Thus, although fungal wilt pathogens cause some of the most devastating plant diseases, they have a limited ability to directly damage plant cells. They damage the plant by colonizing the water-conducting vessels of the plant, and in the process, affect the plant's water status.

We must be careful in distinguishing between wilt pathogens and the pathogens that invade living tissues of plants. The microorganisms that inhabit the vascular system are specialized parasites able to enter, move, and grow within plant vessels or tracheids. These characteristics alone, however, do not define a pathogen. Endophytes, microorganisms which reside within the vessels of plants without stressing the plant, are being discovered (Siegel et al 1985). So, how do parasites, which cause wilting symptoms, differ from endophytes, which do not? Can endophytes be considered avirulent pathogens? Answers to these questions will define the differences between organisms that cause serious damage to plants and those that apparently do little damage, and these are the questions that will be addressed in this paper. There are excellent literature reviews of all aspects of fungal wilt physiology (Mace et al 1981; Beckman 1987), so this paper emphasizes my interpretation of the literature.

UNIQUE ASPECTS OF FUNGAL WILT DISEASES

Organisms live on the surfaces of plants because the plant releases reduced carbon and nitrogen (Alexander 1977; Windels and Lindow 1985). Organisms on plant surface have evolved to compete in this environment and can survive because they successfully compete with other microorganisms. Plants may also affect the nature of the surface microbial community by the release of specific metabolites or macromolecules. The effects of different root surface saprophytes on the plant are just now being unraveled (Schippers et al 1987). Some appear to be beneficial,and these mutually beneficial interactions between plants and surface microorganisms have probably limited the evolution of mechanisms to exclude microorganisms from root surfaces. Defense mechanisms in non-suberized root epidermal cells and root hairs, important in water transport, prevent extensive damage by root saprophytes, but the signals that initiate the defense response of these cells probably differ from those of mesophyll cells. Leaf mesophyll cells probably do not normally coexist with microorganisms, but epidermal cells near root tips do. Although this hypothesis has not been confirmed, I believe that different plant tissues must have evolved different levels of signal recognition in their coevolution with microorganisms. Microorganisms can cause considerable damage in leaf mesophyll tissue if allowed to grow there, but some that grow on the surfaces of roots may be beneficial (Schippers et al 1987).

How would I classify xylem in the defense responses of plant tissues? Researchers would have once agreed that, because microorganisms are interlopers in the xylem and are likely to damage the plant, the plant probably evolved mechanisms similar to those in the leaf mesophyll to prevent colonization. However, it now appears that mechanisms to limit vessel colonization are not as effective as those that have evolved to limit mesophyll or xylem parenchyma colonization. Xylem parenchyma effectively prevent colonization of rays in secondary wood, but are not nearly as effective in preventing the colonization of vessels (Shigo 1984). Perhaps plants tolerate the microbial colonization of vessels, just as they seem to tolerate microbial contact with root hairs more than contact with mesophyll cells.

This general discussion of host defenses is related to pathogen virulence mechanisms because a successful pathogen must first evade host defenses. The tissue specialization of vascular wilt fungi is probably the result of their inability to evade the defenses of root cells, xylem parenchyma, and leaf mesophyll cells. Wilt pathogens colonizing the xylem are certainly positioned to invade each of the afore-mentioned tissues, yet they do not.

What genes and gene products of wilt pathogens affect colonization of the vessels? I define those genes that allow colonization of a living host as the genes for pathogenesis. The genes encoding products that cause enough damage to result in symptom expression are the virulence genes of vascular wilt pathogens. These definitions of genes for pathogenesis and virulence are commonly used. For the wilt diseases, however, virulence genes may not be necessary for the successful colonization and growth of the organism within the host. This is unlike most known pathogens of plant tissue since in most cases living cells are either invaded by or killed in advance of the growing pathogen. Microorganisms can grow, however, in the plant's vessels and tracheids without causing damage to living plant tissues (Siegel et al 1987).

FUNCTIONS DEFINING A PATHOGEN IN WILT DISEASES

The gene functions associated with the colonization of the vessels or tracheids of a plant are those which enable the microorganism to avoid host defense mechanisms, to move, and to extract nutrients from the vessels. How vascular wilt fungi gain entry into vessels is not clearly understood. Most studies have indicated that fungi are seldom able to directly penetrate the root stele (Beckman 1987). Most vascular wilt fungi probably enter the xylem through wounds. It would be unusual if the fungal wilt pathogen directly penetrated the root stele to enter the xylem. If this occurred, it would mean that the fungus directly penetrates root cortical tissue and the various cells which constitute the vascular stele of roots, but could not colonize these tissues after it entered the vascular system. It is far more likely that the fungi enter the stele through wounds.

The ability of a pathogen to evade host defenses appears to be the result of a lack of signal recognition by the host (Ebel 1986). The recent studies of the genes governing avirulence in bacteria, which have a gene-for-gene relationship with their hosts, indicate that avirulence is the result of a gene product, probably a signal polypeptide which triggers host defense responses (Staskawicz et al 1984). Evidence for gene-for-gene relationships in wilt is limited. There appears to be such a gene-for-gene system present in Fusarium and Verticillium wilts of tomato, where races of the pathogens corresponding to resistant genes of the host have been identified (Beckman 1987). The limited numbers of races of the pathogens and genes for resistance in the host make it difficult to arrive at definitive conclusions concerning gene-for-gene relationships in wilt diseases, however. Gene-for-gene relationships may be the genetic manifestations of the ability of the host and pathogen to exchange recognizable signals. Microorganisms in the vascular system can probably inhabit the vessels of a plant because adjacent xylem parenchyma cells either do not respond to signals of the pathogen, or do so slowly enough to allow the pathogen to survive. These organisms are probably unable to invade other tissues because these tissues have defensive responses to the signals emitted by the pathogen. The timing, properties, and mechanisms of signal recogniction are not known for wilt disease pathogens and are only poorly understood in systems that have been intensively studied (Ebel 1986). Obviously, much remains to be learned about the nature and extent of molecular signaling between wilt pathogens and their hosts.

Several hypotheses have been advanced concerning the nutrition of wilt pathogens living within host xylem. There are probably three sources of nutrients for microorganisms within the vessels of plants. The first is the xylem fluid. Although the xylem functions primarily to transport water and mineral nutrients from the roots, the fluid also contains low concentrations of sugars and amino acids (Elgersma 1969; Singh and Smalley 1963). It has been suggested that the wilt pathogens growing within the xylem can utilize these dilute nutrients (Pegg 1981), presumably in a manner similar to microorganisms growing in chemostats. This hypothesis does not resolve whether or not vessels containing microorganisms remain functional (Hall and MacHardy 1981). Microorganisms enter vessels by passing through the vessel walls. Zimmermann (1983) speculated that the resulting wounds would allow embolisms to form, and the vessel would be disfunctional. It is generally agreed that wilt pathogens induce water stress by interfering with water conduction, so the vessels that contain pathogens must quickly become non-functional, even if embolisms do not immediately occur when pathogens enter into the vessel. I believe that normal xylem fluids provide little, if any, of the pathogen's nutritional requirements.

Plant cell wall components may be readily available to the pathogen. These complex carbohydrates and proteins may provide some of the nutrients needed by the pathogen. If so, substantial wall degradation should be clearly apparent. However, most studies have indicated that vessel walls remain intact. Micrographs showing vessels with intact walls filled with fungal hyphae suggest that there is no substantial transfer of biomass from the plant cell walls into fungal tissues (Mace et al 1981; Beckman 1987). This issue will be resolved by careful studies of pathogens that do not produce specific cell wall degrading enzymes.

Recent studies with plant pathogenic bacteria suggest an alternative source of nutrients for vascular inhabiting pathogens. Atkinson and Baker (1987a) used a series of mutants of Pseudomonas syringae pv. syringae with different effects on plant membrane K^+/H^+ exchange, to show that the ability to stimulate such an exchange is directly correlated with the ability of the bacterium to multiply within host tissue. Mutants unable to stimulate K^+ efflux/H^+ influx were unable to grow in leaves of Phaseolus vulgaris. In another study Atkinson and Baker (1987b) showed that wild-type bacterial cells caused an efflux of sucrose from bean leaf discs, but that a Tn5 mutant unable to affect membrane K^+/H^+ exchange did not cause the sucrose efflux. They concluded that the bacterial induction of K^+/H^+ exchange and the resultant efflux of sucrose from bean cells provided nutrients for the bacteria to multiply in leaf intercellular spaces. If so, it may be that pathogens growing within vessels could affect the direction of transport of sucrose and/or amino acids in adjacent xylem parenchyma cells. Such non-toxic effects on host cells could be a rich source of nutrients in the regions where pathogen is growing. If the pathogen stimulates such nutrient leakage into infected vessels, it would be advantageous for the pathogen to be in a non-functional vessel; in a functional vessel, most of the nutrients would be swept away in the rapidly moving xylem fluid.

VIRULENCE MECHANISMS IN WILT DISEASES

The virulence of wilt disease pathogens was defined above as the ability to induce the water stress symptomatic of these diseases. This is a rather specific definition, but it is consistent with the generally accepted definition of virulence as a quantitative measure of the relative severity of disease. Wilt pathogens that are able to grow within plants, without causing water stress, would be considered avirulent. If such pathogens are mutants of microorganisms that normally cause disease symptoms, the mutated gene(s) is considered to be a gene for virulence. If we are to understand the mechanisms of virulence of pathogens, we must identify and understand the function of the products of virulence genes.

Traditionally, attempts to understand how wilt pathogens cause symptom expression have focused on the use of bioassays. Bioassays were designed to reproduce specific symptoms or effects of the pathogen on host tissue. The testing of molecules for their ability to wilt plant cuttings is one example of a bioassay that has been used (and frequently misused) to identify pathogen molecules with wilt-inducing properties (Van Alfen and McMillan 1982). However, demonstration of bioactivity of a pathogen molecule does not mecessarily mean that the molecule has similar functions in the host-pathogen relations. While it is not particularly difficult to demonstrate the presence of pathogen molecules whose bioactivity mimicks some symptom of disease, it has been very difficult to demonstrate that the molecules function in the host-pathogen interaction as postulated. It has been particularly difficult to sort out the molecular bases of virulence expression in fungal wilt diseases. The fungal wilt literature is replete with references regarding the possible role of various wilt toxins, extracellular enzymes, and plant hormones as virulence factors. In spite of considerable research concerning how fungal wilt pathogens cause water stress, we still do not fully understand how a pathogen molecule functions as a virulence factor. I will not review this literature since some excellent current reviews are available (Mace et al 1981; Beckman 1987).

MUTATIONAL AND MOLECULAR APPROACHES TO THE STUDY OF VIRULENCE MECHANISMS

Recent advances in fungal genetics and molecular biology, however, have provided new methods of studying virulence of fungal wilt diseases from a different perspective. The first requirement in any such study is the development of mutants of the fungus. Mutational analysis is a very powerful tool in the study of any microbial process, and, unfortunately, it has been underutilized by plant pathologists seeking to understand how fungi cause disease. Virulence genes have not yet been unambiguously identified for any fungal wilt disease (Puhalla and Bell 1981). The roles of specific gene products in virulence expression have been studied. Puhalla and Howell (1975) selected mutants of <u>V. dahliae</u> that did not produce a constitutive endopolygalacturonase, but were virulent. They concluded that the fungus does not need this enzyme for virulence. The careful studies of Cooper and Durrands reported in this volume are exemplary of the type that should be conducted with other putative virulence factors. While these studies may not be conclusive, they illustrate the care required in the mutational analysis of enzymes that occur as isozymes. In their studies of the role of pectinases as virulence factors of V. dahliae, Copper and Durrands conclude that pectinases may be virulence factors but are not determinants of pathogenicity.

Mutational analysis of putative virulence factors, such as described for pectinases, is one way to study virulence. Another approach is to seek virulence factors by assaying random mutants for virulence expression. Most studies of host-pathogen interactions start with some preconceived idea as to what class of pathogen products may affect the outcome of the interaction. Seeking random mutants unable to express virulence, based upon a direct virulence assay, is an approach that does not prejudice the outcome of the study, but one that requires a good virulence assay. Unfortunately, the fungal wilt diseases are not as easy to assay for virulence as some other pathogens. Mutations to avirulence in wilt pathogens, however, appear to be fairly common. Spontaneous virulence mutants are commonly observed during culturing of Fusarium and Verticillium wilt pathogens (Puhalla and Bell 1981). These mutants have not been genetically defined, but appear to be pleiotropic mutants. Such mutants may be useful for virulence studies since they could represent mutants of regulatory genes involved in virulence expression.

Analysis of pleiotropic mutants is more complicated than an analysis of mutants with simple phenotypes. Pleiotropic mutants may be the result of major chromosomal rearrangements, multiple mutations, mutations of regulatory genes, or mutation of a gene with multiple functions. Genetic analysis of the mutant is necessary to determine the number of mutated loci. When mutants of known gene products are selected, such as described for the pectinase mutant of V. dahliae (Puhalla and Howell 1975), it is relatively easy to demonstrate that the known gene product was mutated. When the gene product is not known, as when seeking mutations leading to avirulence, identification of the gene product is much more difficult. The most direct approach to identification of the gene product is to isolate the gene by complementation, clone the gene into an expression vector, make antibody to the expressed protein, and then seek information concerning the gene product using the antibody. I do not know of a case where this approach has been used to identify a fungal virulence gene, but variations of this procedure are used to isolate genes identified by mutations (Vollmer and Yanofsky 1986) and to characterize the products of genes (Shatzman and Rosenberg 1987). Obviously, before such procedures are attempted, care should be taken to eliminate mutants such as auxotrophs, which could affect virulence, but are not virulence genes.

My laboratory has been using another approach that is amenable to the analysis of mutants with pleiotrophic phenotypes. We have been attempting to identify genes involved in virulence expression of the fungus <u>Cryphonectria</u> (<u>Endothia</u>) <u>parasitica</u>, the causal agent of chestnut blight. <u>C. parasitica</u> is a pathogen similar in many respects to wilt pathogens. It is unable to directly penetrate its hosts, requiring a wound for entry. It colonizes wounds and then forms mycelial fans which penetrate host periderm tissues, which have formed in response to the wound (Hebard et al 1984). Although a number of mechanisms of virulence expression have been proposed for this pathogen, as for the wilt pathogens, we do not yet understand how the pathogen is able to invade and kill host cells (Van Alfen 1988). <u>C. parasitica</u> is not host specific and can invade a variety of oaks as well as chestnuts and chinquapins. As with many wilt pathogens, there is no evidence of gene-for-gene relations with its hosts (Roane et al 1986). <u>C. parasitica</u> can be infected with dsRNA viruses which cause the fungus to lose virulence. We use these dsRNAs in a manner similar to the use of pleitropic mutants in mutational analysis to investigate how the dsRNAs affect virulence expression.

We have constructed strains of the fungus that are isogeneic except for the presence or absence of various dsRNA viruses. The strains containing a virus are hypovirulent and sporulate poorly, while those lacking a virus are virulent and sporulate normally.

A variety of approaches are available to compare how strains, which are identical except for the character of interest, differ in gene expression. In this case we wanted to identify gene products that are expressed by virulent fungi, but not expressed by hypovirulent ones. We have used two approaches; one was to identify differentially expressed mRNAs and the other was to identify differentially expressed proteins.

To identify differentially expressed mRNAs we modified a differential hybridization method described by Timberlake (1986). A genomic library of the fungus was cloned into the lambda cloning vector EMBL3. Poly(A) RNA was then isolated from a virulent strain of the fungus and from a hypovirulent strain. cDNA was prepared from the poly(A) RNA of each strain. Radioisotope labeled cDNA from the virulent strain was mixed with a 10-fold excess of poly(A) RNA from the hypovirulent strain, and vice versa, and the mixtures were then hybridized to identical membranes containing DNA transferred from plaques of the EMBL3 genomic library of the fungus. Clones containing genes expressed by virulent, but not hypovirulent, strains of the fungus were identified by the hybridization of the unique sequences of cDNA (from poly(A) RNA) from virulent strains to the plaque DNA containing the genomic sequence. We used this method to identify two poly(A) RNAs that are highly expressed in virulent strains but not in hypovirulent ones (Powell and Van Alfen 1987a). We are currently attempting to isolate the proteins translated from these mRNAs and to identify the function of the proteins.

The second approach involved the comparison of two dimensional gels of total proteins extracted from virulent and hypovirulent strains of the fungus. From these gels we identified about 10 major proteins that are expressed by virulent, but not hypovirulent, strains of the fungus (Powell and Van Alfen 1987b). We have isolated one of these proteins, prepared antibody to it, and used the antibody to help clarify its function. The protein we isolated is a hydrophobic protein of about 22 kD. Its properties are similar in many respects to those of cerato-ulmin, a small hydrophobic protein, reported to be a phytotoxin, involved in virulence expression by the Dutch elm disease pathogen Ophiostroma ulmi (Takai 1974). We are convinced that the 22 kD protein, which we have named cryparin, is not a phytotoxin, although it can cause plant cuttings to wilt. Cryparin accumulates on the surface of sporulation competent tissues (as does cerato-ulmin (Takai et al 1980)), but it is not found in mycelial fans, the structures involved in penetration of host tissue by the fungus (Hebard et al 1984). Cryparin has lectin-like activities. It is able to agglutinate certain types of red blood cells and this agglutination is inhibited by cell walls of the fungus. The results of our studies lead us to believe that this protein is involved in sporulation, but not virulence.

The general approach that we used to study expression of virulence of C. parasitica is amenable to other fungi. In the cases of Fusarium and Verticillium, a variety of avirulent mutants could be compared to determine if specific proteins or mRNAs are expressed only in virulent strains. This approach makes it possible to study virulence without preconceived notions as to what types of gene products are virulence factors. Most approaches used to date have relied upon bioassays of molecules thought to have a role in virulence expression. This traditional approach facilitates prejudice. Analysis of differentially expressed proteins and mRNAs also has pitfalls. Identification of proteins and their functions from mRNA sequences is not easy. And some differentially expressed mRNAs may be spurious. It is encouraging, however, to have new tools available to study old problems.

REFERENCES

Alexander M (1977) Soil microbiology, 2nd edn. John Wiley & Sons, New York

Atkinson MM and Baker CJ (1987a) Association of host plasma membrane K^+/H^+ exchange with multiplication of Pseudomonas syringae pv. syringae in Phaseolus vulgaris. Phytopathology 77: 1273-1279

Atkinson MM and Baker CJ (1987b) Alteration of plasmalemma sucrose transport in Phaseolus vulgaris by Pseudomonas syringae pv. syringae and its association with K^+/H^+ exchange. Phytopathology 77:1573-1578

Beckman CH (1987) The nature of wilt diseases of plants. APS Press, St Paul

Ebel J (1986) Phytoalexin synthesis: the biochemical analysis of the induction process. Ann Rev Phytopathol 24:235-264
Elgersma DM (1969) Resistance mechanisms of elms to Ceratocystis ulmi. Meded Phytopathol Lab Willie Commelin Scholten, Baarn No 77, pp 84
Hall R and MacHardy WE (1981) Water relations. In: Mace ME, Bell AA and Beckman CH (eds) Fungal wilt diseases of plants. Academic Press, New York
Hebard FV, Griffin GJ and Elkins JR (1984) Developmental histopathology of cankers incited by hypovirulent and virulent isolates of Endothia parasitica on susceptible and resistant chestnut trees. Phytopathology 74:140-149
Mace ME, Bell AA and Beckman CH (1981) Fungal wilt diseases of plants. Academic Press, New York
Pegg GF (1981) Biochemistry and physiology of pathogenesis. In: Mace ME, Bell AA and Beckman CH (eds) Fungal wilt diseases of plants. Academic Press, New York
Powell WA and Van Alfen NK (1987a) Differential accumulation of poly(A) RNA between virulent and double-stranded RNA-induced hypovirulent strains of Cryphonectria (Endothia) parasitica. Mol Cell Biol 7:3688-3693
Powell WA and Van Alfen NK (1987b) Two nonhomologus viruses of Cryphonectria (Endothia) parasitica reduce accumulation of specific virulence associated polypeptides. J Bact 169:5324-5326
Puhalla JE and Bell AA (1981) Genetics and biochemistry of wilt pathogens. In: Mace ME, Bell AA and Beckman CH (eds) Fungal wilt diseases of plants. Academic Press, New York
Puhalla JE and Howell CR (1975) Significance of endopolygalacturonase activity to symptom expression of Verticillium wilt in cotton, assessed by the use of mutants of Verticillium dahliae. Physiol Plant Pathol 7:147-152
Roane MK, Griffin GJ and Elkins JR (1986) Chestnut blight, other Endothia diseases, and the genus Endothia. APS Press, St Paul
Schippers B, Bakker AW and Bakker PAHM (1987) Interactions of deleterious and beneficial rhizosphere microorganisms and the effect of cropping practices. Ann Rev Phytopathol 25:339-358
Shigo AL (1984) Compartmentalization: a conceptual framework for understanding how trees grow and defend themselves. Ann Rev Phytopathol 22:189-214
Shatzman AR and Rosenberg M (1987) Expression, identification, and characterization of recombinant gene products in Escherichia coli. In: Berger SL and Kimmel AR (eds) Guide to molecular cloning techniques, Academic Press, San Diego
Siegel MR, Latch GCM and Johnson MC (1987) Fungal endophytes of grasses. Ann Rev Phytopathol 25:293-315
Singh D and Smalley EB (1969) Changes in amino acid and sugar constituents of the xylem sap of American elm following inoculation with Ceratocystis ulmi. Phytopathology 59:891-896
Staskawicz BJ, Dahlbeck D and Keen NT (1984) Cloned avirulence gene of Pseudomonas syringae pv. glycinea determines race-specific incompatibility on Glycine max (L.) Merr.Proc Nat Acad Sci USA 81:6024-6028
Takai S (1974) Pathogenicity and cerato-ulmin production in Ceratocystis ulmi. Nature 252:124-126
Takai S, Hiratsuka Y, Krywienczyk J, Richards WC and Davies YP

(1980) Evidence for the presence of the toxin cerato-ulmin in the synnema head fluid of Ceratocystis ulmi. Can J Bot 58: 669-675

Timberlake WE (1986) Isolation of stage- and cell-specific genes from fungi. In: Bailey J (ed) Biology and molecular biology of plant-pathogen interactions, NATO ASI series vol HI. Springer-Verlag KG, Berlin

Van Alfen NK (1988) Viruses of Endothia parasitica. In: Koltin Y and Leibowitz MJ (eds) Viruses of fungi and simple eukaryotes, Marcel Dekker, New York Basel

Van Alfen NK and McMillan BD (1982) Macromolecular plant-wilting toxins: artifacts of the bioassay method? Phytopathology 72: 132-135

Vollmer SJ and Yanofsky C (1986) Efficient cloning of genes of Neurospora crassa. Proc Natl Acad Sci, USA 83:4869-4873

Windels CE and Lindow SE (1985) Biological control on the phylloplane. The American Phytopathological Society, St Paul

Zimmermann MH (1983) Xylem structure and the ascent of sap, Springer-Verlag, Berlin Heidelberg New York Tokyo

VIRULENCE AND AVIRULENCE OF BACTERIAL PATHOGENS AND DESIGNATION OF RACES

A. Trigalet
Laboratoire de Biologie Moléculaire
CNRS - INRA
BP 27
31326 Castanet-Tolosan Cedex
France

INTRODUCTION

There is a general consensus among plant pathologists that distinguishes two types of pathogen specificity, namely the plant specificity which determines the host species range and the cultivar specificity which characterizes the cultivar range within a given host species. There is a tendency to consider that both specificities are intimately related and that the latter one is superimposed on the first one. Although the genetic background of these interactions has long been established there has been a lack of suitable tools to isolate genes and define their functions in disease and/or resistance in plants. Rapid progress is now being made with the use of recombinant DNA techniques for the genetic analysis of bacterial pathogenicity. Thus, some bacterial genes have been cloned and subsequently analysed. The molecular basis of host-pathogen specificity is becoming better understood and it appears pertinent to review our knowledge about virulence and avirulence and designation of races of bacterial pathogens in the light of recent studies.

DEFINITION OF BASIC TERMS

A wealth of literature has long been devoted to the definition of basic terms and phenomena which describe and characterize the

NATO ASI Series, Vol. H28
Vascular Wilt Diseases of Plants

plant-pathogen interaction. Therefore we shall limit ourselves to a brief repeat of the essential terms and references.

A bacterial pathogen which can cause disease to a plant, i.e. a phytopathogenic bacterium, is called virulent if the symptoms of disease are well developed on the host plant; this latter plant being called susceptible. After an initial contact between bacterium and plant, if the symptoms of disease fail to develop, the bacterium is called avirulent and the plant resistant. These terms are quantitative and characterize all to none interactions between bacterial pathogens and plants. The term aggressiveness is often used to account for qualitative variations of virulence, i.e. to qualify differences in the severity of disease symptoms induced by various virulent strains on the same host plant. Thus, although they are all virulent, these strains (isolates) may differ between themselves by their respective aggressiveness towards a given host plant.

The interaction between the host and the bacterial pathogen may be very short because both organisms are not suited for each other. This inability of a plant and bacteria to coexist results in incompatibility and absence of disease. Plants exhibit a non-host resistance towards such non adapted pathogens. This resistance is generally non specific for a given pathogen and may be due to the presence of preformed physical and/or chemical factors and/or active mechanisms as the result of an elicitation from the bacterial pathogen (for a review see Halverson and Stacey 1986). This non-host resistance involves a hypersensitive reaction (HR) which is a non specific pleiotropic response of the host plant involving numerous physiological and morphological changes (for reviews see Kiraly 1980; Sequeira 1984).

In some cases bacteria may adapt and overcome the non-host resistance and establish a compatible reaction; both partners are then able to coexist at least during the early stage of the interaction, i.e. disease of symbiosis.

Some bacteria are pathogenic to a broad host range (Pseudomonas solanacearum, Agrobacterium tumefaciens) whereas others are limited to certain cultivars of a host species (Pseudomonas sp., Xanthomonas sp.). There could exist variations in the virulence among the different isolates of a bacterial pathogen. These isolates are then classified into separate groups (pathovars or physiological races) (for a review see Young et al 1978). These pathovars or races have established compatibility with their host plants at the species level of specificity. This compatibility is often referred as to basic compatibility (Ellingboe 1976; Heath 1981). Another compatibility can be developed at the race-cultivar level which is attributed to a single gene for gene relationship (Flor 1956) and has been demonstrated for bacterial pathogens (for a review see Panopoulos and Peet 1985).

These two types of pathogen specificity will be further detailed below in the section Pseudomonas solanacearum (basic compatibility and physiological races) and in Pseudomonas syringae pathovar (pv) glycinea and Xanthomonas campestris pv. malvacearum (cultivar specific races). Genetics of pathogenesis of Erwinia and Agrobacterium have been reviewed elsewhere (Mills and Gonzales 1982; Collmer and Keen 1986; Kotoujanski 1987).

DESIGNATION OF RACES

Pseudomonas solanacearum is a heterogeneous group whose differentiation at the sub-specific level is extremely confusing. On the basis of biochemical and serological properties, and phage and bacteriocin reactions, P. solanacearum isolates from Japan could be divided into 40 groups (Okabe and Goto 1961, 1963). These groups do not correlate with the 13 pathovars previously described. Later, three races have been designated: race 1 that attacks Solanaceous and numerous other plants; race 2 restricted to Musaceae (triploid Banana); and race 3 for isolates adapted to relatively low temperature (16 C) and pathogenic to potato (Buddenhagen et al 1962). On the ability to utilize three hexose alcohols (mannitol, sorbitol and dulcitol) and/or to produce acid from three disaccharides (cellobiose, maltose and lactose) four biovars were later established (Hayward 1964). Most workers then referred either to the three races and/or to the four biovars typing. However, it soon appeared that within each of these races or biovars they were numerous subtypes (for a review see

Buddenhagen and Kelman 1964; Hayward 1976) that could be associated with particular geographical locations and/or new host plants (Digat 1968; Zehr 1970; Keshwal and Joshi 1976; He et al 1983; Prior and Steva 1988). Temptation to define new races is great (He et al 1983).

At the present time it appears that race 1 is a storage group that includes every strain that does not belong either to race 2 or to race 3. The biovar classification does not correlate with the race typing, except for race 3 and biovar II. This latter biovar II contains all isolates from potato that have a narrow host range (potato, tomato), are poorly aggressive on tobacco and possess lower temperature optima than other isolates. Increasing the number of phenotypical tests renders the distinction between biovars less clear. Biovar I could be divided into nine sub-types (Harris 1972; Hayward 1976). Thus, the biovar classification is unable to account for the complex situation within P. solanacearum. However, numerous reports have pointed out that the different biovars behave under natural conditions in the fields as separate entities, i.e. biovar II is never found in the same location as biovar III and biovar IV, and never occured with biovar I. Distribution of these biovars seems to be determined by environmental and biological factors that are still ill-defined. Whatever the official race designation may be in P. solanacearum, it remains that precise identification of the pathovars or races is of great importance in providing sufficient information and predictive value on the ecology and etiology of infection and movement between territories.

Pseudomonas syringae pv. glycinea has long been subdivided into races as determined by inoculation onto a collection of differential cultivars of soybean (Gross et al 1966). Although up to nine races have been reported (Long et al 1984) only five can now be clearly defined (Staskawicz et al 1985).

No relationship between race specificity and phenotypic characters such as cell surface charge, hydrophobicity (Fett 1985), or LPS composition and structure has been yet demonstrated, with the remarkable exception of race 4 whose O-antigen fraction differs significantly from those of race 1,5 and 6 which appear similar to each other (Barton-Willis et al 1984). However, LPS of

P. syringae pv. glycinea are not directly involved in race glyceollin elicitation of soybean (Barton-Willis 1987).

A GENETIC UNDERSTANDING OF VIRULENCE vs AVIRULENCE OF PHYTOPATHOGENIC BACTERIA

Recent advances in the genetics of phytopathogenic bacteria have provided evidence for different classes of genes involved in pathogenicity.

The first group of genes are required for both pathogenicity on homologous host plants (compatible reaction) and induction of hypersensitive necrosis on heterologous host plants (incompatible reaction) and on resistant cultivars of the homologous host (incompatible reaction). These genes have been first identified in Pseudomonas syringae pv. phaseolicola and since referred to as hrp, i.e. Hypersensitive Reaction and Pathogenicity (Lindgren et al 1986). Such genes have been later identified in Erwinia amylovora (Bauer and Beer 1987) and in P. solanacearum (Boucher et al 1987).

The second group of genes are responsible for the control of disease development on host plants (compatible reaction) without affecting the hypersensitive reaction. These can be referred as to dsp for disease specific (Boucher et al 1987). Such genes exist in Erwinia stewartii (Coplin et al 1986, 1988), in Xanthomonas campestris pv. campestris (Daniels et al 1984, 1987) and in P. solanacearum (Boucher et al 1987; Arlat and Boucher 1988).

The third group of genes correspond to avirulence genes (avr) and are involved in the control of race-cultivar specific interaction. These have no concern with the induction of hypersensitive reaction on heterologous plant. These have been described in Pseudomonas syringae pv. glycinea (Stakawicz et al 1984) and in Xanthomonas campestris pv. malvacearum (Gabriel et al 1986). Identification of these genes demonstrated that the gene for gene pattern of interaction exists in bacterial pathogen-plant interactions.

The genetic organization of pathogenicity determinants of P.

solanacearum has received increasing interest in the past ten years. The structure of the bacterial genome has become elucidated: a megaplasmid is detected in most of the strains regardless of race or geographic origin (Rosenberg et al 1982). It has been further demonstrated that some virulence genes are carried by this megaplasmid (Boucher et al 1986) and that hrp genes controlling both pathogenicity on tomato and hypersensitivity on tobacco (Boucher et al 1985) are clustered in a 21 kb of DNA on the megaplasmid (Boucher et al 1987, 1988). A 3 kb region adjacent to the right end of this hrp cluster consist of dsp genes specifically involved in aggressiveness towards tomato. Additional dsp genes located outside the previous described cluster and adjacent region are present on the megaplasmid (Boucher et al 1988). Sequence homologies to this hrp cluster of P. solanacearum have been found in all the pathovars of Xanthomonas campestris tested (Boucher et al 1987). Presently neither information about the products of the above mentioned, plasmid-borne genes, nor about their regulation is available. A chromosomal genetic map of P. solanacearum has been elaborated (Zischek et al, manuscript in preparation). No pathogenicity determinant has been identified on the chromosome so far. By hybridization of genomic DNA of numerous strains of P. solanacearum with several probes, it has been established that homologous sequences are present in all strains but that Restriction Fragment Length Polymorphism (RFLP) exists that defines 7 specific patterns; thus, virulent strains of P. solanacearum differ from each other by changes in the structure of their genome (Boucher et al 1988). These 7 RFLP groups do not correspond to the race or the biovar typing previously described in P. solanacearum.

In addition to the previous studies transposon Tn5 mutagenesis has been widely used in P. solanacearum to generate point insertion avirulent mutants. The identification of a virulence locus involved in the exopolysaccharide (EPS) production has been reported (Staskawicz et al 1983). A similar procedure was used which allowed subsequent cloning of eight distinct loci that seem to specify virulence; six of these Tn5-induced avirulent mutants exihibit colony morphology identical to that of the wild type whereas the last two were altered in EPS production (Xu et al 1988). In addition, these authors isolated two virulent, EPS defective mutants. Their results suggest that the role of EPS in pathogenicity is questionable. However, as long as the EPS of P. solanacearum has not been fully characterized, arguing about its possible role in pathogenicity is premature (Trigalet-Démery 1988). A chromosomal gene encoding the major endoglucanase (egl) excreted by many P. solanacearum strains (Schell 1987) has been cloned and evidence has been presented that the egl gene although not essential for pathogenicity, is involved in the aggressiveness of the bacterium (Roberts et al 1988).

Genetics of race-specific avirulence genes has been abundantly documented in fungul pathogens since Flor's theory appeared (1956), but genetic evidence for the existence of the gene for gene relation is recent in bacterial pathogens. Five phenotypes (races) of Pseudomonas syringae pv. glycinea (Psg) have been strictly separated from each other on the basis of their phenotypic reaction on soybean leaf, i.e. timing of appearance of HR, intensity of necrosis in leaf, measure of bacterial growth in vivo and accumulation of glyceollin (Long et al 1984). For instance, Psg4 (race 4) is virulent on all cultivars of soybean, whereas Psg0 (race 0) is virulent only on Centennial cv. A cosmid clone pPg6L3 from race 6, when introduced into other races, namely PsG1, Psg4 and Psg5, changes the specificity of these races from virulence to avirulence on the appropriate soybean cvs (Staskawicz et al 1984). The gene cloned from Psg6 is named avirulence gene A (avrA) and has been tentatively identified (Napoli and Staskawicz 1987). Unfortunately, the resistance locus in soybean which corresponds to avrA gene of Psg6 has not been genetically mapped. The only genetic locus (Rpg1) that confers resistance to race 1 has been identified in cvs Norchief and Harosoy.

The fact that Psg0 (race 0) is avirulent to all soybean cvs but Centennial was exploited to construct a new wide host range cosmid. Thus, two avirulence genes (avrB0 and avrC) from Psg0 and one avirulence gene (avrB1) from Psg1 have been identified. When introduced into Psg4 (race 4) which is normally virulent on all cvs of soybean, these latter three genes specify incompatibility and control qualitatively the phenotype of the HR (Staskawicz et al 1987). Another important study conducted on Xanthomonas campestris pv. malvacearum demonstrated that five avirulence genes of the pathogen specifically interacted with individual resistant genes in the congenic cotton lines and that some A/R gene interactions appeared qualitatively different from others (Gabriel et al 1986). In the latter example, it clearly appeared that different alleles of the same locus could be responsible for different incompatibility phenotypes. These observations support the hypothesis, previously put forward in fungal pathogens that resistance response controlled by the gene for gene relation is not unique (for a review see Tepper and Anderson 1984).

GENETIC MECHANISMS FOR RACE AND VIRULENCE VARIATIONS

The possible genetic mechanisms involved in race and virulence variation have been reviewed (Mills and Gonzales 1982; Panopoulos and Peet 1985) and it is beyond the scope of this article to describe them again. However, recent studies illustrate perfectly well the involvement of genetic mechanisms in this race variation and in the shift from virulence to avirulence.

It has long been known that the pepper race 2 strains of Xanthomonas campestris pv. vesicatoria can spontaneously mutate from avirulence to virulence at a high frequency ($5x10^{-4}$) on pepper carrying the Bs_1 dominant resistance locus (Dahlbeck and Stall 1979). It has just been demonstrated that a self transmissible plasmid of 200 kb coding for copper resistance is responsible for this race variation, because it carries an active transposable 1,2 kb element which can insert at high frequency in the avirulence gene avr Bs_1 specifying the resistance of pepper lines carrying Bs_1. Thus, copper sprays to control Xanthomonas campestris pv. vesicatoria in the field facilitate the spread of a highly mutagenic element responsible for the shift from resistance to susceptibility towards the pathogen of interest.

These results reinforce the evidence that plasmid-borne traits can be disseminated at a high frequency by transferable plasmids and that this mechanism can be of economic significance.

Studies describing the presence of genomic rearrangements in

P. solanacearum are being published (Boucher et al 1988). The comparison of XbaI restriction fragments of the entire genomes of a collection of independant non fluidal mutants and of the wild type by field inversion gel electrophoresis has showed that modifications of the banding patterns occur, and that these rearrangements are localized in the megaplasmid. Thus genomic rearrangements are related to change in virulence.

CONCLUDING REMARKS

Our knowledge and identification of the bacterial genes which are involved in pathogenicity is rapidly increasing. In phytopathogenic bacteria like Pseudomonas and Xanthomonas spp. it appears that virulence vs avirulence is a complex mechanism. Bacterial genes involved control either the compatible (dsp genes) or the incompatible (avr genes) host relations. Moreover some genes (brp genes) have been shown to simultaneously control both the host compatible or incompatible and the non-host incompatible relation.

These hrp genes are clustered in Xanthomonas and Pseudomonas spp. and some nucleic acid sequence homologies have been detected between the hrp cluster of P. solanacearum and the hrp cluster of all pathovars of X. campestris so far tested. In addition the hrp cluster of P. syringae appears common to all pathovars within the species. However the hrp cluster of P. solanacearum and of X. campestris group show no homology with the one of the P. syringae group. Since all these different types of hrp clusters control the expression of the resistance phenotype within the broad host range of these phytopathogenic bacteria, they are probably not involved in specific relations with the host plant. Moreover it has been recently demonstrated that pea genes controlling the non-host resistance to a fungal pathogen can also been implicated in the phenotypic expression of resistance in race specific reaction against different races of a bacterial pathogen (Daniels et al 1987). Thus a set of common plant genes is involved in the whole spectrum of specific and non specific resistance against fungal and bacterial pathogens. However some studies tend to indicate that differences in the activation and expression of plant genes could exist depending on the type of pathogen plant interaction. The notion that a few common plant pathogen genetic systems could account for the phenotypic resistance in compatible and incompatible combinations is emerging.

REFERENCES

Arlat M and Boucher C (1988) Identification of a cluster of dsp genes controlling agressivity in Pseudomonas solanacearum. In: Proceed 4th Int Symp Mol Gen Plant Microbe Interactions Acapulco Mexico May 15-20 1988 (in Press)

Barton-Willis PA, Wang MC, Holliday MJ, Long MR and Keen NT (1984) Purification and composition of lipopolysaccharides from Pseudomonas solanacearum syringae pv. glycinea. Physiol Plant Pathol 25:387-398

Barton-Willis PA, Wang MC, Staskawicz B and Keen NT (1987) Structural studies on the O-chain polysaccharides of lipopolysaccharides from P. syringae pv. glycinea. Physiol Mol Plant Pathol 30:187-197

Boucher C, Barberis P, Trigalet A and Demery D (1985) Transposon mutagenesis of Pseudomonas solanacearum: isolation of Tn5-induced avirulent mutants. J Gen Microbiol 131:2449-2457

Boucher C, Martinel A, Barberis P, Alloing G and Zischek C (1986) Virulence genes are carried by a megaplasmid of the plant pathogen Pseudomonas solanacearum. Mol Gen Genet 205:270-275

Boucher C, Van Gijsegem F, Barberis P, Arlat M and Zischek C (1987) Pseudomonas solanacearum genes controlling both pathogenicity on tomato and hypersensitivity on tobacco are clustered. J Bact 169:5626-5632

Boucher C, Arlat M, Zischek C and Boistard P (1988) Genetic organization of pathogenicity determinants of Pseudomonas solanacearum. In: Physiology and Biochemistry of Plant Microbial Interactions. ASPP (in Press)

Buddenhagen I, Sequeira L and Kelman A (1962) Designation of races in Pseudomonas solanacearum. Phytopathology 52:726 (Abstract)

Buddenhagen I and Kelman A (1964) Biological and physiological aspects of bacterial wilt caused by Pseudomonas solanacearum. Ann Rev Phytopath 2:203-230

Ciampi L and Sequeira L (1980) Influence of temperature on virulence of race 3 strains of Pseudomonas solanacearum. Amer Potato J 57:307-317

Collmer A and Keen NT (1986) The role of pectic enzymes in plant pathogenesis. Ann Rev Phytopath 24:383-409

Coplin DL, Frederick RD, Majerczak DR and Hass ES (1986) Molecular cloning of virulence genes from Erwinia stewartii. J Bact 168:

619-623
Cross JE, Kennedy BW, Lambert JW and Cooper RL (1966) Pathogenic races of the bacterial blight pathogen of soybeans, Pseudomonas glycinea. Plant Disease Reporter 50:557-560
Daniels MJ, Barber CE, Turner PC, Sawczyc MK, Byrde RJW and Fielding AH (1984) Cloning of genes involved in pathogenicity of Xanthomonas campestris using the broad host range cosmid pLAFR1. Embo J 3:3323-3328
Daniels M, Collinge D, Maxwell Dow J, Osborn A and Roberts I (1987) Molecular biology of the interaction of Xanthomonas campestris with plant. Plant Physiol Biochem 25:353-359
Digat B (1968) Why and how to distinguish the Pseudomonas solanacearum strains, causal agent of the bacterial wilt of Solanaceous and musaceous crop in the Caribbean zone. In Proceed Caribbean Food Crops Society 6th Annual Meeting St-Augustine Trinidad July 7-13 1968, VI:86-91
Dolph PJ, Majerczak DR and Coplin DL (1968) Characterization of a gene cluster for exopolysaccharide biosynthesis and virulence in Erwinia stewartii. J Bact 170:865-871
Ellingboe AH (1976) Genetics of host-parasite interactions. In: Heitefuss R and Williams PH (eds). Encyclopedia of plant Physiology 4 Physiological Plant Pathology Springer Verlag Berlin, Heidelberg New-York:761-778
Fett WF (1985) Relationship of bacterial cell surface hydrophobicity and charge to pathogenicity, physiologic race, and immobilization in attached soybean leaves. Phytopathology 75: 1414-1418
Flor HH (1959) The complementary genic systems in flax and flax rust. Adv Gene 8:29-54
Halverson LJ and Stacey G (1986) Signal exchange in plant-microbe interactions. Microbiol Rev 50, 193-225
Hayward AC (1976) Systematics and relationships of Pseudomonas solanacearum. In: Sequeira L and Kelman A (eds) 1st Int Planning conf Workshop on the Ecology and Control of Bacterial Wilt Raleigh July 18-24. North Carolina State Univ:6-21
He LY, Sequeira L and Kelman A (1983) Characteristics of strains of Pseudomonas solanacearum from China. Plant Disease 67:1357-1361
Heath MC (1981) A generalized concept of host-parasite specificity. Phytopathology 71:1121-1123
Kearney B, Ronald PC, Dahlbeck D and Staskawicz J (1988) Molecular basis for evasion of plant host defence in bacterial spot disease of pepper. Nature 332:541-543
Keshwal RL and Joshi LK (1976) Variation in isolates of Pseudomonas solanacearum EFS. Ind J Microbiol 16:94-96
Kiraly Z (1980) Defenses triggered by the invader: hypersensibility: In Horsfall JG and Cowling EB (eds) Plant Disease an advanced treatise. V How plants defend themselves. Academic Press Inc New-York London Toronto Sidney San Francisco:201-224
Kotoujansky A (1987) Molecular genetics of pathogenesis by soft-rot Erwinias. Ann Rev Phytopath 25:405-430
Lindgren PB, Peet RC and Panopoulos NJ (1986) Gene cluster of Pseudomonas syringae pv. phaseolicola controls pathogenicity of bean plants and hypersensitivity on nonhost plants. J Bact 168:512-522

Long M, Barton-Willis P, Staskawicz B, Dahlbeck D and Keen NT (1984) Further studies on the relationship between glyceollin accumulation and the resistance of soybean leaves to P. syringae pv. glycinea. Phytopathology 75:235-239

Mills D and Gonzales CF (1982) The evolution of pathogenesis and race specificity. In: Mount MS and Lacy GH (eds). Phytopathogenic prokaryotes 1:77-119. Academic Press

Napoli C and Staskawicz B (1987) Molecular characterization and nucleic acid sequence of an avirulence gene from race 6 of P. syringae pv. glycinea.J Bact 169:572-578

Okabe N and Goto M (1961) Studies on Pseudomonas solanacearum XI Pathotypes in Japan. Shizvoka Univ Fac Agr Rept 11:25-42

Okabe N and Goto M (1963) Bacteriophages of plant pathogens. Ann Rev Phytopath 1:397-418

Panopoulos NJ and Peet RC (1985) The molecular genetics of plant pathogenic bacteria and their plasmids. Ann Rev Phytopath 23: 381-419

Prior P and Steva H (1988) Characteristics of isolates of Pseudomonas solanacearum from the French West Indies (Martinique and Guadeloupe). Plant Disease (in Press)

Sequeira L (1980) Defenses triggered by the invader: recognition and compatibility phenomena: In Horsfall JG and Cowling EB (eds) Plant Disease an advanced treatise. V How plants defend themselves. Academic Press Inc New York London Toronto Sidney San Francisco:179-200

Sequeira L (1984) Plant-Bacterial Interactions. In: Pirson A and Zimmermann MH (eds) Encyclopedia of Plant Physiology New Series 17:187-211

Staskawicz B, Dahlbeck D, Miller J and Damm D (1983) Molecular analysis of virulence genes in Pseudomonas solanacearum. In: Pühler A (ed) Molecular Genetic of Bacteria Plant Interactions. Springer Berlin Heidelberg New-York:345-352

Staskawicz BJ, Dahlbeck D and Napoli C (1985) Molecular genetics of race-specific avirulence genes in P. syringae pv. glycinea. In: Sussex E (ed) Current communications in molecular biology: plant cell-cell interactions. Cold Spring Harbor Laboratory Cold Spring Harbor New-York:109-114

Staskawicz B, Dahlbeck D, Keen NT and Napoli C (1987) Molecular characterization of cloned avirulence genes from race 0 and Race 1 of P. syringae pv. glycinea. J Bact 169:5789-5794

Tepper CS and Anderson AJ (1984) The genetic basis on plant-pathogen interaction. Phytopathology 74:1143-1145

Young JM, Dye DW, Bradbury JF, Panagopoulos CG and Robbs CF (1978) A proposed nomenclature and classification for plant pathogenic bacteria. N Z J Agric Res 21:153-177

Zehr FI (1970) Cultural, physiological and biochemical properties of isolates of Philippine Pseudomonas solanacearum. Philippine Phytopathology 6:29-43

ORGANIZATION AND EXPRESSION OF THE hrp GENE CLUSTER IN PSEUDOMONAS SYRINGAE PV. PHASEOLICOLA

L. Rahme, M. Mindrinos, C. Grimm, R. Frederick, P. Lindgren and N. Panopoulos
Department of Plant Pathology
University of California
Berkeley, CA 94720
USA

INTRODUCTION

Infection of plants by pathogens generally leads to one or the other of two mutually exclusive outcomes, either the development of disease symptoms characteristic of the pathogen-host-environment combination (compatible interactions) or the triggering of the host resistance response (incompatible interaction). The resistance response is often typified by necrosis of host cells directly contacted by the pathogen or cells in their immediate vicinity at early stages of infection. Restriction of pathogen growth and development may result as a direct consequence of substances that are inhibitory to the pathogen, termed phytoalexins, as an indirect consequence of cell death per se, in the case of obligate pathogens, or via other less well-understood biochemical or cyto-developmental events that are characteristic of this response at the biochemical, cellular, or histological level. The triggering of hypersensitive death may be effected by pathogen- or plant-cell-derived necrosis-eliciting molecules whose nature, role and mechanisms of action are poorly understood at present.

From a genetic perspective, the expression of disease resistance or susceptibility is jointly controlled by host and pathogen genes. Until recently, Mendelian crosses between (fungal) pathogen "races" and host cultivars were the only means by which the presence of these genes could be demonstrated. In recent years, molecular genetics and other modern analytical tools have been applied in the study of genes and gene products involved in pathogenesis and host resistance reactions, chiefly with bacterial pathogens. This paper summarizes the present state of molecular genetic analysis of pathogenicity in the bean halo blight pathogen Pseudomonas syringae pv. phaseolicola.

CONSERVED PATHOGENICITY GENES IN P. SYRINGAE PATHOVARS ALSO CONTROL THE ELICITATION OF THE HYPERSENSITIVE REACTION

Previous studies in our laboratory (Panopoulos et al 1985; Lindgren and Panopoulos 1986; Lindgren et al 1986) led us to identify and clone a group of genes, designated hrp, which controls the ability of the bean halo blight pathogen, Pseudomonas syringae pv. phaseolicola, to elicit the hypersensitive reaction (HR) on tobacco and other non-host plants and its ability to produce the primary disease symptoms (watersoaked lesions) on bean, its normal host. The existence of such genes in other strains of this bacterium was independently reported by other groups (Daesy et al 1987; Somlyai et al 1986). Subsequent studies (Lindgren et al 1988; Grimm and Panopoulos, unpublished data) showed that hrp DNA sequences are conserved among several pathovars of Pseudomonas syringae but not in other taxonomic groups of bacterial plant pathogens, symbionts or saprophytes. Functional analogues of hrp genes exist in closely related members of the P.syringae group and have been cloned from other pathovars of the P. syringae group (Huang et al 1988; Malik et al 1987; Niepold et al 1984; Staskawicz et al, pers. commun.; Smith and Fulbright, pers. commun.).

Genes that exert a similar (dual) control on host- and non-host-pathogen interactions apparently also exist in other phytopathogenic bacteria that are taxonomically unrelated and pathologically distinct from P. syringae pathovars. Thus, Boucher et al (1987) and Morales et al (1987) described similar genes in the vascular wilt pathogen, Pseudomonas solanacearum. Many of these genes are also organized as a cluster which is located on a megaplasmid (Boucher et al 1987), whereas the hrp genes of P. s. phaseolicola are chromosomal (Rahme et al, this laboratory, unpubl.). The P. solanacearum hrp cluster does not have detectable homology to hrp genes of P. syringae group members (Boucher et al 1987) and failed to complement Hrp⁻ mutants of P.s. phaseolicola (Rahme, this laboratory, unpublished). Accordingly, these genes provisionally define a separate hrp homology group which may be shared by some Xanthomonas pathogens. Functionally similar genes have also been identified in Erwinia amylovora (Bauer and Beer 1987; Steinberger and Beer 1984). These genes may constitute a third hrp homology group among the phytopathogens.

GENETIC ORGANIZATION OF THE hrp CLUSTER OF P. S. PHASEOLICOLA

In P.s. pv. phaseolicola NPS3121 the hrp genes are organized in two linkage groups. One of these apparently comprises a single locus, mutations in which lead to a mucoid morphology. The other group, which comprises the majority of known hrp genes, defines a region of approximately 20 kb which is referred to as the "hrp cluster". This region has been analyzed further by insertional mutagenesis with newly constructed "reporter" transposons (see below), marker exchange and complementation analysis. The 20 kb cluster encodes hrp functions nearly across its entire length except for a short (ca. 0.5-1.0 kb) segment that appears not to encode such functions and subdivides the region into two approximately equal halves ("left" and "right"). At least five complementation groups have been defined in this cluster. Complex complementation patterns have been obtained in some of these groups, suggesting either a need for further functional subdivision or a transcriptionally complex organization.

A NOVEL ASSAY FOR MEASURING BACTERIAL GENE EXPRESSION IN VITRO AND IN PLANTA

We have developed a novel and highly sensitive method for

the quantitative analysis of gene expression in bacteria (Lindgren et al 1988). Because this method may be more broadly applicable in the study of pathogen gene expression at the infection court, a brief description is given below. Unlike conventional methods which involve the measurement of enzymatically active "reporter proteins", our method is based on a physicochemical assay, namely the freezing of water droplets in the presence of the ice nucleation protein as a catalyst. The method requires the construction of genetic fusions between the gene of interest and a promoterless ice nucleation gene, inaZ (Wolber et al 1986). This can be accomplished either in vivo, by insertional mutagenesis with our newly constructed ice nucleation gene transposons (Lindgren et al 1988; Mindrinos, unpublished data, this laboratory) or in vitro by cloning promoter carrying DNA fragments in vector plasmids carrying the promoterless inaZ gene (Staskawicz and Dahlbeck, unpublished). The fused genes are then transferred to appropriate bacterial hosts which are then used for inoculation or are exposed to various environmental, nutritional, or other conditions which may affect the expression of the target genes. The ice nucleation activity of the cultures or infected tissue is determined after grinding or serial dilution by dispensing 10-µl droplets onto sheets of aluminum foil precoated with paraffin and floated on an alcohol cooling bath that has been adjusted at an appropriate subzero temperature (e.g. -9 C). The number of droplets that freeze is used to calculate the number of bacteria carrying ice nucleation sites active at that temperature (Lindgren et al 1988).

TRANSCRIPTIONAL ORGANIZATION AND INDUCIBLE EXPRESSION OF hrp GENES

We have successfully applied the above method to the analysis of hrp gene transcription in P. s. phaseolicola. A recombinant plasmid carrying ca. 18 kb insert from the hrp cluster was mutagenized with an ice nucleation transposon, Tn3-Spice (Lindgren et al 1988). Several of the mutations were marker exchanged into the genome of P. s. phaseolicola, the resulting mutants were examined for their hrp phenotype and the ice nucleation activities of both haploid and merodiploid strains were determined during growth in vitro as well as in planta at various times post-inoculation. We have thus determined that an extended segment of the hrp region of P. s. phaseolicola (ca. 9 kb) carries genes that are expressed at a much higher level in planta and in minimal media, compared to nutritionally complex medium (Fig. 1). Induction occurs rapidly after the cultures are transferred from complex (King's B) medium to mineral salts media as well as upon host (bean) or non-host (tobacco) leaves. Similar induction has been described for an avirulence gene, $avrB_0$ of P. s. glycinea (Huynh et al in preparation). Our complementation, induction and subcloning data further indicate that this segment is organized into at least three transcriptional units that are transcribed in the same direction from separate promoters (Fig. 1; Mindrinos, Rahme and Panopoulos in preparation).

IDENTIFICATION OF hrps AS A REGULATORY LOCUS

In a previous study (Grimm et al 1988; Grimm and Panopoulos 1988), nucleotide sequence analysis of a 5.6 kb hrp segment was determined. Computer-assisted analysis of this data revealed extensive homology between the predicted protein product of an open reading frame present in this region and a conserved domain present in several transcriptional regulatory proteins of enteric and plant symbiotic bacteria (Table 1). Among these proteins are NtrC, DctD, NifA and TyrR, which control the expression of genes or operons involved in nitrogen assimilation, dicarboxylic acid transport, nitrogen fixation, and tyrosine biosynthesis, respectively (Drummond et al 1986; Henikoff and Wallace 1988; Nixon et al 1986; Ronson et al 1986). We have designated this protein HrpS and have obtained direct evidence that the encoding locus indeed controls the expression of several other hrp genes. For example, marker exchange mutants of a hrpS::Tn5 strain carrying hrp::Tn3-Spice fusions in the 9 kb segment comprising the

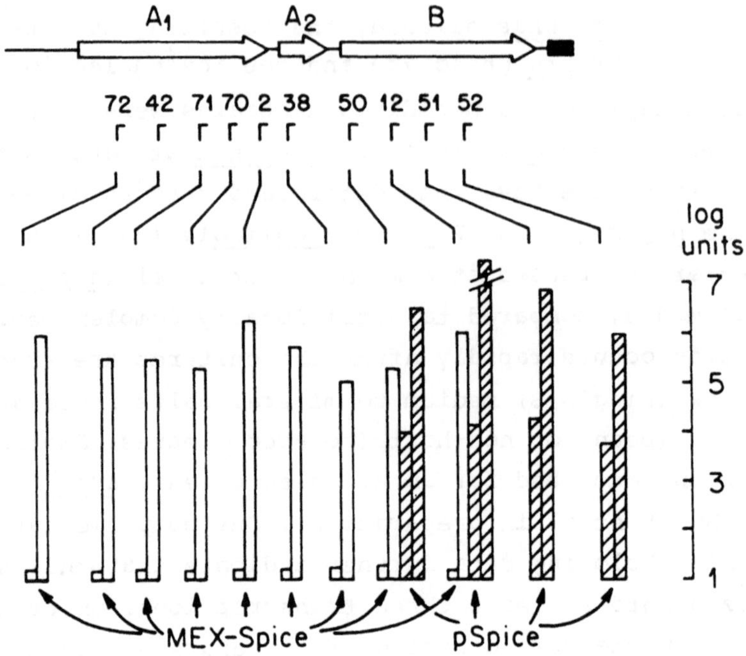

Fig. 1. Transcriptional organization and inducible expression of genes located in the ca. 915 kb segment ("left half") of the 20 kb hrp region (Lindgren et al 1986). A_1, A_2, B correspond to separate complementation groups (Lindgren et al 1988; Rahme et al in preparation). Strains carrying inaZ reporter insertions either in the chromosome (MEX-Spice, open bars) or in plasmid pPL11 (pSpice, cross-hatched bars) were transferred from KB medium to Mg salts medium supplemented with 0.2% glycerol or were infiltrated under vacuum into "Red Kidney" bean leaves as described in arbitary units on a logarithmic scale. The left bar in each pair is the "constitutive" level of activity and the adjacent bar on the right the "induced" level. Two reporter insertions (#50 and 12) were studied both in marker exchange mutants (single-copy level constructs) and NPS3121 (pPL11-Tn3-Spice) merodiploids (high copy number constructs).

Table 1. Amino acid sequence homology between HrpS and five different regulatory proteins

Proteins compared	Amino acid position	% Identity
HrpS / KpNtrC	22-235 / 152-368	44.4
HrpS / KpNifA	6-234 / 213-440	40.8
HrpS / RmNifA	20-233 / 200-426	44.1
HrpS / EcTyrR	20-235 / 206-425	43.8
HrpS / R1DctD	32-236 / 158-373	47.8

"left" half of the hrp cluster fail to induce during nutritional downshifts. Thus, approximately half of the hrp cluster encodes genes that require the function of another hrp gene in order to be expressed.

BIOCHEMICAL INSIGHTS INTO hrp GENE AND PROTEIN FUNCTIONS

Our present knowledge of the biochemical functions of the hrp genes and their protein products is limited. One clue as to the possible function of hrp genes comes from the finding that Tn5-induced hrp^- mutants of P. syringae pv. syringae lose the ability to induce plasma membrane K^+/H^+ exchange in tobacco (Baker et al 1987). Furthermore, the preliminary results presented allow us to investigate in more detail the biochemical mechanisms of hrp gene regulation. For example, NtrC and many of its sequence homologues require two additional proteins as co-activators, a distinct sigma factor (σ^{54}, the subunit of RNA polymerase that determines promoter specificity) and another protein that receives and transmits an environmental signal or intracellular signals generated in response to it (Drummond et al 1986; Nixon et al 1986; Ronson et al 1986; Winans et al 1986). In the case of NtrC, the external signal (limited nitrogen availability) leads to phosphorylation of the protein by another protein, NtrB, which is a specific NtrC phosphorylase-dephosphorylase (Keener and Kustu 1988; Ninfa and Magasanik 1986). The phosphorylated form of NtrC interacts with either the RNA polymerase holoenzyme or the sigma factor associated with it and directs the transcriptional complex to promoters under its control, ultimately producing a response appropriate to nitrogen limitation. It would be interesting to determine what the hrp gene inducer molecules are, whether particular promoter structures are characteristic of Hrps-regulated genes, and whether (an)other regulatory protein(s), protein phosphorylation event(s) and distinct sigma factor(s) are involved in the activation of hrp genes. Furthermore, the identification of transcriptional units in the inducible segment of the hrp region will greatly assist in the construction of clones that overproduce the inducible proteins so that they may be identified and utilized in antibody production, cellular localization, interaction with plant cells and a variety of other studies.

PERSPECTIVE AND PROSPECTS

The finding that a large number of genes are required for the production of primary disease symptoms by the bean halo blight pathogen, would not in itself be surprising, except for the fact that the same genes also determine the pathogen's ability to elicit the hypersensitive response. Molecular genetic analysis in the closely related soybean blight pathogen, P. syringae pv. glycinea (Staskawicz et al 1984, 1987) had shown that single avirulence genes (avr) determined whether the pathogen would elicit the hypersensitive reaction (incompatible interaction) on soybean differential cultivars. At least one of these cultivars carried a single Mendelian gene for resistance. Thus, the outcome of plant-bacterium interaction in the soybean blight system is exactly as predicted from the gene-for-gene hypothesis. However, as reported in our previous studies (Lindgren et al 1986; Lindgren et al 1988) P. s. glycinea carries a structurally and functionally equivalent hrp cluster and hrp gene function is required in addition to avr genes for HR to be elicited on the resistant cultivars of its host. This is not only true for the single hrp locus initially studied by us but applies to all hrp genes presently known in this pathogen (Huynh et al in preparation). The interaction and interdependence between hrp and avr genes and/or their products suggested by these findings are only beginning to be understood. For instance, two or three hrp loci are required for the transcriptional activation of an avr locus ($avrB_0$) in P. s. glycinea (Huynh et al in preparation). The remaining hrp loci or their encoded products must presumably interact with $avrB_0$ or its protein product at (an)other level(s). Although regulatory control of other avr genes in this or other pathogens has not yet been established, the epistatic effect of hrp mutations on avr-dependent hypersensitive necrosis is not restricted to $avrB_0$ but applies also to other avr genes. Thus, the expression of disease resistance and susceptibility to bacterial infection must be ultimately understood in the context of hrp and avr gene and protein functions, their mutual interactions and the nature of the plant's response.

Our findings further suggest that hrpS is an evolutionary relative of other prokaryotic regulatory loci that control the expression of diverse sets of genes and operons. These loci presumably evolved from a common progenitor sequence during the course of prokaryote speciation and evolution. It is interesting to note that the hrpS protein is the shortest member of its evolutionary relatives found in other bacteria, consisting of only one domain which is highly conserved in all members of this protein family (Drummond et al 1986; Grimm and Panopoulos 1988). We speculate that the lack of significant amino terminal expressions in this protein suggests a less highly adapted state and different, possibly less complex regulatory control requirements for pathogenic functions in comparison with nitrogen fixation or other prokaryotic functions that are regulated by proteins related in sequence to HrpS. Another hrp gene (hrpR) encoding a protein homologous to HrpS, is also found in P. s. phaseolicola (Grimm et al in preparation) and the closely related pathovars P. s. glycinea (Staskawicz and Dahlbeck, pers. commun.) and P.s. savastanoi (our unpublished data). The nucleotide and predicted amino acid sequences between the HrpS and HrpR proteins in each pathovar are less similar to each other than each is to its homologue in the other pathovars, suggesting that the evolutionary origin of hrpS and hrpR apparently predates the divergence between the pathovars themselves. Similar comparisons between hrpS hrpR homologues between more distantly related P. syringae pathovars as well as bacterial pathogens of other taxonomic groups that may carry such homologues might provide similar clues about the evolutionary history of pathogenic functions in phytopathogenic prokaryotes.

ACKNOWLEDGMENTS

This work was supported in part by Grants from the National Science Foundation (PCM-8313052 and PCM-8409723), by post-doctoral Fellowships to C. Grimm from the European Economic Community and the Deutsche Forschugsgemeinschaft and by the Secretary of Research and Technology, Greece. We acknowledge the assistance of S. E. Lindow in analyzing ice nucleation data and B. J. Staskawicz, T. Huynh and D. Dahlbeck and other colleagues for sharing data with us prior to publication.

REFERENCES

Baker CJ, Atkinson MM and Colmer A (1987) Concurrent loss in Tn5 mutants of Pseudomonas syringae pv. syringae of the ability to induce the hypersensitive response and host plasma membrane K^+/H^+ exchange in tobacco. Phytopathology 77:1268-1272

Bauer DW and Beer SV (1987) Cloning of a gene from Erwinia amylovora involved in induction of hypersensitivity and pathogenicity. pp 425-429 In: Civerolo EL, Collmer A, Davis RE and Gillaspie AG (eds) Plant Pathogenic Bacteria. Proceedings of the Sixth International Conference of Plant Pathogenic Bacteria, Maryland, June 2-7, 1985. Martinus Nijhoff, Dordrecht, The Netherlands

Boucher CA, Van Gijsegem F, Barberis PA, Arlat M and Zischek C (1987) Pseudomonas solanacearum genes controlling both pathogenicity on tomato and hypersensitivity on tobacco are clustered. J Bacteriol 169:5626-5632

Deasy MC, Stapleton MJ and Mathysse AG (1987) Isolation and characterization of Tn5 mutants of Pseudomonas syringae pv. phaseolicola which fail to elicit a hypersensitive response. pp 444-447 In: Civerolo EL, Collmer A, Davis RE and Gillaspie AG (eds) Plant Pathogenic Bacteria. Proceedings of the Sixth International Conference on Plant Pathogenic Bacteria, Maryland, June 2-7, 1985. Martinus Nijhoff, Dordrecht, The Netherlands

Drummond M, Whity P and Wotton J (1986) Sequence and domain relationships of ntrC and nifA from Klebsiella pneumoniae: homologies to other regulatory proteins. EMBO J 5:441-447

Henikoff S and Wallace JC (1988) Detection of protein similarities using nucleotide sequence databases. Nucl Acids Res (in Press)

Grimm C and Panopoulos NJ (1988) The predicted protein product of a pathogenicity locus of Pseudomonas syringae pv. phaseolicola is homologous to several prokaryotic regulatory proteins (submitted)

Grimm C, Rahme L, Frederick R, Mindrinos M, Lindgren PB and Panopoulos NJ (1988) The common pathogenicity genes of Pseudomonas syringae pathovars. In: Staskawicz BJ, Yoder OC and Alquist P (eds) Molecular Biology of Plant-Pathogen Interactions, Vol. 101 UCLA Symposia on Molecular and Cellular Biology, Alan R. Liss, Inc., New York, NY (in Press)

Huang HC, Schuurink R, Denny TP, Atkinson MM, Baker CJ, Yucel I, Hutcheson SW and Collmer AC (1988) Molecular cloning of a Pseudomonas syringae pv. syringae gene cluster that enables Pseudomonas fluorescens to elicit the hypersensitive response in tobacco. J Bacteriol (in Press)

Keener J and Kustu S (1988) Protein kinase and phosphoprotein phosphatase activities of nitrogen regulatory proteins NTRB and NTRC of enteric bacteria: roles of the conserved amino-terminal domain of NTRC. Proc Natl Acad Sci USA 85:4976-4980

Lindgren PB, Frederick R, Panopoulos NJ, Govindarajan A, Staskawicz BJ and Lindow SE (1988) An ice nucleation reporter gene system: identification of inducible pathogenicity genes in Pseudomonas syringae pv. phaseolicola (submitted)

Lindgren PB and Panopoulos NJ (1986) Genetic control of hypersensitivity and pathogenicity in P. syringae pv. phaseolicola. pp 298-305, In: Bailey JA (ed) Biology and Molecular Biology of Plant-Pathogen Interactions. Springer-Verlag

Lindgren PB, Peet RC and Panopoulos NJ (1986) A gene cluster of Pseudomonas syringae pv. phaseolicola controls pathogenicity on bean and hypersensitivity on non-host plants. J Bacteriol 168:512-522

Lindgren PB, Panopoulos NJ, Staskawicz BJ and Dahlbeck D (1988) Genes required for pathogenicity and hypersensitivity are conserved and interchangeable among pathovars of Pseudomonas syringae. Molec Gen Genet 211:499-506

Malik AN, Vivian A and Taylor JD (1987) Isolation and partial characterization of three classes of mutants in Pseudomonas syringae pv. pisi with altered behavior toward their host, Pisum sativum. J Gen Microbiol 133:2393-2399

Morales VM, Stemmer WPC and Sequeira L (1985) Genetics of avirulence in Pseudomonas solanacearum. pp 89-96 In: Malmberg M, Crouch M, Sussex I and Ellingboe AH (eds) Plant Cell/Cell Interactions. Current Communications in Molecular Biology, Cold Spring Harbor Laboratory, Cold Spring Harbor, NY

Niepold F, Anderson D and Mills D (1984) Cloning determinants of pathogenesis from Pseudomonas syringae. Proc Natl Acad Sci USA 82:406-410

Ninfa AJ and Magasanik B (1986) Covalent modification of the glnG product, NR_I, by the glnL product, NR_{II}, regulates transcription of the glnALG operon in Escherichia coli. Proc Natl Acad Sci USA 83:5909-5913

Nixon TB, Ronson CW and Ausubel FM (1986) Two-component regulatory systems responsive to environmental stimuli share strongly conserved domains with the nitrogen assimilation regulatory genes ntrB and ntrC. Proc Natl Acad Sci USA 83:7850-7854

Panopoulos NJ, Lindgren PB, Willis DK and Peet RC (1985) Clustering and conservation of genes controlling the interactions of Pseudomonas syringae pathovars with plants. pp 69-77 In: Malmberg R, Crouch M, Sussex I and Ellingboe AH (eds) Plant Cell/Cell Interactions. Current Communications in Molecular Biology, Cold Spring Harbor Laboratory, Cold Spring Harbor, NY

Ronson CW, Astwood PM, Nixon BT and Ausubel FM (1987) Deduced products of C4-dicarboxylate transport regulatory genes of Rhizobium leguminosarum are homologous to nitrogen regulatory gene products. Nucl Acids Res 15:7921-7934
Somlyai GM, Hevesi M, Banfalvi A, Klement Z and Kondorosi A (1986) Isolation and characterization of non-pathogenic and reduced virulence mutants of Pseudomonas syringae pv. phaseolicola. Physiol Mol Plant Pathol 29:369-380
Staskawicz BJ, Dahlbeck D, Keen N and Napoli C (1987) Molecular characterization of cloned avirulence genes from race 0 and race 1 of Pseudomonas syringae pv. glycinea. J Bacteriol 169: 5789-5794
Staskawicz BJ, Dahlbeck D and Keen N (1984) Cloned avirulence gene of Pseudomonas syringae pv. glycinea determines race-specific incompatibility on Glycine max. Proc Natl Acad Sci USA 81:6024-6028
Steinberger EM and Beer SV (1984) Isolation and mapping of Tn5 mutations in pathogenicity genes of Erwinia amylovora. Phytopathology 74:797-798
Winans SC, Ebert PR, Stachel SE, Gordon MP and Nester EW (1986) A gene essential for Agrobacterium virulence is homologous to a family of positive regulatory loci. Proc Natl Acad Sci USA 83:8278-8282
Wolber PK, Deininger CA, Southworth MW, Vandekerckhove J, Van Montague M and Warren GJ (1986) Identification and purification of a bacterial ice nucleation protein. Proc Natl Acad Sci USA 83:7256-7260

THE ANALYSIS OF VERTICILLIUM STRAIN RELATIONSHIP

A.C. Hastie
Department of Biological Sciences
University of Dundee
Dundee
DD1 4HN
Scotland

INTRODUCTION

The genetic versatility of Verticillium species is now firmly established through laboratory studies but the extent to which that versatility applies to field conditions is uncertain. The thesis of this paper is that research should now be directed at natural variability through the use of spontaneous molecular genetic markers. The primary objective of that research being the investigation of Verticillium population structure, we may anticipate results relevant to host specificity and epidemiology. However the techniques to be advocated can also be harnessed to research into the mechanism of infection, wilt pathogenesis and the etiology of disease. Advances in DNA technology have so far had no real impact on Verticillium research. Research initiatives in that direction should be rewarding.

THE PARASEXUAL CYCLE

The parasexual cycle of Verticillium has most recently been reviewed by Hastie and Heale (1984). Intraspecific parasexual gene recombination has been demonstrated in V. albo-atrum (Hastie 1962), V. dahliae (Puhalla and Mayfield 1974) and V. tricorpus (Molchanova et al 1978). V. albo-atrum is the species which has been most intensively studied and all the standard features of the parasexual cycle in it have been identified and

NATO ASI Series, Vol. H28
Vascular Wilt Diseases of Plants

analysed in some detail (Hastie and Heale 1984). Heterozygous diploid nuclei are formed in unstable heterokaryons synthesised from haploid strains and containing few heterokaryotic cells. Once formed, these heterozygous diploids tend to outgrow heterokaryons in appropriate selective environments and the heterozygous diploids can therefore be isolated for the purposes of genetic analysis.

New haploid and diploid genotypes are formed during the growth of heterozygous diploids and these provide the information about gene segregation and recombination which can be used for genetic analyses. Such analyses have been described by Hastie (1964, 1981) and also by Typas and Heale (1978). Five gene linkage groups have been detected. These may correspond with the five chromosomes which are believed to comprise the haploid chromosome set of V. albo-atrum (Heale et al 1968), but that remains to be confirmed. Recombination frequencies have been used to map the relative positions of gene loci within linkage groups and the loci of some centromeres with respect to linked genes have been mapped by mitotic recombination (Hastie 1967; Typas and Heale 1978).

Interspecific parasexual gene recombination has also been detected in the laboratory between V. albo-atrum and V. dahliae (Hastie 1973) and also between V. tricorpus and each of these species (Hastie 1978). Haploid genotypes were formed by these interspecific diploids, but both the variety and frequency of recombinant haploid genotypes was restricted. These observations are consistent with the existence of chromosomal non-homology between the parental genotypes and therefore justify their distinctive species status.

All these observations of intraspecific and interspecific genetic recombination rely on the use of laboratory induced auxotrophic mutants and associated selective devices. They reveal the potential genetic versatility of Verticillium species but their relevance to the field situation is questionable and uncertain (Hastie and Heale 1984). If parasexual recombination is of regular occurrence in the field then we may expect wild diploid strains to be found in natural populations. They can be distinguishable from haploids by their conidial dimensions. Conidia from haploids are generally relatively short (<8μ) while those containing diploid nuclei are longer (>8μ) (Hastie and Heale 1984). In fact few authors have reported the isolation of diploid strains from natural environments (Table 1).

The likelihood of detecting wild diploids clearly depends on numerous factors which will include the frequency with which they are formed, their persistence as diploids, their pathogenicity and their capacity to conidiate. The first two factors will dictate the equilibrium frequency of diploids present in any population. The latter two factors are significant because it is surely true that most wild Verticillium strains examined are from diseased plants and conidial dimensions are the most routinely applied distinguishing criterion of diploid strains. It is uncertain whether the naturally occurring wild diploid strains listed in Table 1 were homozygous or heterozygous. Consequently they are not necessarily evidence of somatic nuclear fusions and therefore of the natural occurrence of the parasexual cycle. These rare spontaneous diploids could be homozygotes formed as restitution nuclei in aberrant haploid mitosis. Tolmsoff (1972) believes that homozygous diploid nuclei are a regular feature in the life cycle of several pathogenic Verticillium species. Investigations of natural genetic variation will provide the only reliable criteria for distinguishing between wild homozygotes and heterozygotes. Isozyme variation will almost certainly provide relevant genetic markers because these variations are generally inherited without dominance. Diploid strains, heterozygous for allelic isozyme markers, can therefore be expected to express both alleles and will consequently be distinguishable from both homozygous diploids and haploids.

HETEROKARYON INCOMPATIBILITY

Pathogenic Verticillium species can only engage in genetic recombination by asexual mechanisms. Puhalla and Hummel (1983) have described how we can regard V. dahliae either as a single

Table 1. Verticillium diploids isolated from natural populations

Species	Reference	Host	Country
V. dahliae	Stark 1961	Horseradish	Germany
V. tricorpus	Tolmsoff 1972		U.S.A.
V. albo-atrum	Typas et al 1976	Tomato	U.K.
V. dahliae	Puhalla et al 1983	Rape	Sweden
V. dahliae	Puhalla et al 1983	Beet	Sweden

interbreeding population, within which new genotypes are freely formed, or a collection of genetically isolated sub-populations each of which is characterised by "specific virulence capabilities". The results obtained by Puhalla and Hummel (1983) certainly favour the latter view, and indicate that the vegetative incompatibility groups they identified in V. dahliae may represent genetically isolated sub-populations of that species. Each vegetative incompatibility (v - c) group is comprised of strains between which hyphal fusion are readily formed, and between which genetic recombination is therefore possible. Conversely, strains belonging to different v - c groups do not make heterokaryons and parasexual gene recombination between them is thereby generally excluded. It is still to be estimated whether each v - c group (or sub-population) is a clone within which new genotypes arise only by mutation. Alternatively, and at the other extreme, a v - c group may be a more or less freely parasexually interbreeding sub-population. The scarcity of reports of diploid strains in wild populations may indicate that the parasexual cycle is rare in nature. Puhalla and Hummel (1983) recognised that more extensive pathogenicity testing is required to determine the degree of genetic variability within each v - c group of strains. Additional wildtype genetic markers, other than host range and pathogenicity, merit assessment in that context.

The detection of v - c groups within the species V. dahliae was made by the use of complementary induced mutants which specifically affect the colour of microsclerotia (Puhalla 1979).

Complementary and vegetatively compatible microsclerotial colour mutants from black microsclerotia where the mutant strains come in contact on an unselective medium, incompatible mutants do not. Thus the formation of wildtype black microsclerotia in such paired cultures indicates that at least some heterokaryotic cells have been formed and therefore that the paired strains engage in making hyphal fusions. No comparable unselective method has been applied to the detection of v - c groups in other pathogenic Verticillium species. All relevant experiments with V. albo-atrum for the detection of heterokaryons have relied on the use of paired complementary auxotrophic mutants, grown on selective minimal media (Hastie 1962, Typas and Heale 1976b). In these circumstances the intense selective pressure in favour of establishing heterokaryosis may over-ride the stability of v - c genes which may be present. A mutation of a v - c gene in either of two paired incompatible homokaryons could result in rare heterokaryons being formed and thereby obscure the incompatibility of the paired homokaryons (Hastie and Heale 1984). It is therefore still uncertain whether V. albo-atrum is subdivided into v - c groups comparable to those which have been found in V. dahliae, and this is also true of other pathogenic Verticillium species. It is certainly true however that strains of V. albo-atrum have been found which apparently do not make hyphal fusions even with themselves. Hastie (unpublished) has found at least one such strain isolated from wilted tomato in the U.K. That strain is analogous to the six strains of V. dahliae recorded by Puhalla and Hummel (1983) which did not make any heterokaryons.

PROSPECTS FOR THE ANALYSIS OF VERTICILLIUM POPULATIONS

There is little evidence relating to the genetic structure of wild populations of Verticillium species apart from that reviewed earlier and briefly in this paper. There is no compelling evidence of the natural occurrence of parasexual recombination. Even if it does occur in the field we still need to know whether it contributes significantly to the origin of new pathotypes. Is V. dahliae comprised of numerous sub-populations (v - c groups) which are effectively clones, and between which there is no gene flow? Is V. albo-atrum one heterogenous and potentially interbreeding population? Is genetic material exchanged between Verticillium species in the field? All these questions are relevant to the epidemiology of Verticillium wilt diseases which in total afflict a large proportion of all dicotyledonous crop species. The definitive analysis of Verticillium population structure will utilise naturally occurring genetic variation. In the past we have relied upon variation in pathogenicity and host range, resistance to fungicides and other toxic chemicals, and also variation in assorted morphological traits as indicators of natural genetic variation. Kerr (1987) insists "plant pathologists are going molecular". The prospects for the analysis of Verticillium populations require that we follow that lead.

The choise of appropriate genetic molecular markers for Verticillium would seem to be between isozyme differences detectable by gel electrophoresis and DNA sequence differences detectable by restriction fragment length analysis (Michelmore and Hulbert 1987). Analytical systems using either type of molecular marker could provide answers to the questions posed above. The detection of genetic polymorphisms within the v - c groups of V. dahliae could enable us to decide whether asexual hybridisation and gene recombination occur in nature.

The elegant analysis of isozyme variability in Phytophthora cinnamoni by Old et al (1984) serves as a model of what isozyme analysis of populations can achieve. They were able to demonstrate that effective sexual reproduction of that fungus does not regularly occur in the field. Similarly, Burdon and Roelfs (1985) used isozyme markers to compare the genetic structure of sexually and asexually reproducing populations of Puccinia graminis f. sp. tritici. They found that the structure of an asexual population deviated significantly from that expected of a panmictic regime, but that of a sexual population did not. The asexual population was effectively a clone.

Hall (1969) has used electrophoresis of crude protein extracts from both *V. albo-atrum* and *V. dahliae* to show that these species produce characteristic patterns of protein bands. Pelletier and Hall (1971) extended these studies to show that the conidia of *V. albo-atrum* produce at least three proteins not detected in mycelium extracts of the same species. Refinements of their techniques to visualise substrate specific enzymes is essentially what is required for the detection and characterisation of any isozyme polymorphism in *Verticillium* and provide valuable information about population structure. Systems for detecting over fifty isozymes are available (Tanksley and Orton 1983), but only a few would be required for these purposes.

Restriction fragment length polymorphisms (RFLP) result from differences in DNA nucleotide sequences in otherwise homologous genetic material (Beckman and Soller 1983). These differences are known to exist in very many eukaryotes, perhaps in all that have been examined. They are found in both translated and non-translated genetic materials and their existence in *Verticillium* is virtually guaranteed. However, the application of RFLP analyses to *Verticillium* would involve a greater input of resources than the use of isozymes. The requirements for the detection of RFLP´s are these:

1. Development of cloned radioactive DNA probes each containing a *Verticillium* DNA fragment,
2. Preparation of *Verticillium* DNA digested (fragmented) by appropriate restriction endonucleases,
3. Separation of these DNA fragments by agarose gel electrophoresis and ultimately their immobilisation on a suitable substrate,
4. Hybridisation of the separated DNA fragments to radioactive DNA probes,
5. Autoradiography to detect DNA fragments homologous to the probe.

DNA extraction procedures are generally more difficult to perfect than the protein extractions required for isozyme characterisation by electrophoresis. Consequently, a large investment of resources should be anticipated even before one could being to apply that protocol routinely to wildtype strains of *Verticillium*. The development of suitable probes is itself likely to be a progress limiting step. Once produced however they are routinely usable. Experience indicates that the most efficient probes for detecting and analysing RFLP's are about 20 kilobases long and devoid of repeated DNA sequences (Michelmore and Helbert 1987). The technology for producing these probes is now well developed in other organisms and continues to advance.

The choice between analysing Verticillium genetic diversity and strain relationships by either isozyme systems or RFLP's will depend upon our objectives in relation to the required commitment of resources. Given the limited objective of deciding whether Verticillium v - c groups are either clones or interbreeding sub-populations then the isozyme approach should be the first choice. That technology is less demanding of time, labour and expertise. However, if one wishes to extend our objectives to include the molecular analysis of virulence, aggressiveness and other agriculturally relevant traits then the approach should be at the DNA level. The input is then greater, but so are the rewards.

REFERENCES

Beckman JS and Soller M (1983) Restriction fragment length polymorphisms in genetic improvement: Methodologies, mapping and costs. Theor Appl Genet 67:35-43

Burdon JJ and Roelfs AP (1985) The effect of sexual and asexual reproduction on the isozyme structure of populations of Puccinia graminis. Phytopathology 75:1068-1073

Hall R (1969) Verticillium albo-atrum and V. dahliae distinguished by acrylamide gel electrophoresis of proteins. Can J Bot 47:2100-2111

Hastie A (1962) Genetic recombination in the hop-wilt fungus Verticillium albo-atrum. J gen Microbiol 27:373-382

Hastie A (1964) The parasexual cycle in Verticillium albo-atrum. Genet Res 5:305-315

Hastie A (1967) Mitotic recombination in conidiophores of Verticillium albo-atrum. Nature 214:249-252

Hastie A (1973) Hybridisation of Verticillium albo-atrum and V. dahliae. Trans Br mycol Soc 60:511-523

Hastie A (1978) Genetic analysis in Verticillium. In Sadasivan T, Subramanian C, Kalyanasundaram R and Saraswathi-Devi L (eds) Pathological wilting of plants. The University of Madras, Madras, p 44

Hastie A (1981) The genetics of conidial fungi. In Cole G and

Kendrick B (eds) Biology of conidial fungi, Vol 2. Academic Press, New York, p 511

Hastie A and Heale J (1984) Genetics of Verticillium. Phytopath Medit 23:130-162

Heale J, Gafoor A and Rajasingham K (1968) Nuclear division in conidia and hyphae of Verticillium albo-atrum. Can J Genet Cytol 10:321-340

Kerr A (1987) The impact of molecular genetics on plant pathology. Ann Rev Phytopathol 25:87-110

Michelmore R and Hulbert S (1987) Molecular markers for genetic analysis of phytopathogenic fungi. Ann Rev Phytopathol 25:383-404

Molchanova R, Kasyanenko A and Azizova Z (1978) Polyploidisation and haploidisation as one of the possible mechamisms of genetic variability of Verticillium tricorpus. In Maksumov (ed) Genetic variability of the wilt causing agent and prospects for increasing cotton wilt resistance. Donish Publishing House, Dushanbe, p 130

Old K, Moran G and Bell J (1984) Isozyme variability among isolates of Phytophthora cinnamoni. Can J Bot 62:2016-2122

Pelletier G and Hall R (1971) Relationships among species of Verticillium: Protein composition of spores and mycelium. Can J Bot 49:1293-1297

Puhalla J (1979) Classification of isolates of Verticillium dahliae based on heterokaryon incompatibility. Phytopathology 69:1186-1189

Puhalla J and Hummel M (1983) Vegetative compatibility groups within Verticillium dahliae. Phytopathology 73:1305-1308

Stark C (1961) Das auftreten der Verticillium-Tracheomykosen in Hamburger Cartenbaukulturen. Gartenbauwissenschaft 26:493-528

Tanksley S and Orton T (1983) Isozymes in plant genetics and breeding. Elsevier, Amsterdam Oxford New York

Tolmsoff W (1972) Diploidisation and heritable gene repression-derepression as a major source of variability in morphology, metabolism and pathogenicity of Verticillium species. Phytopathology 62:407-413

Typas M and Heale J (1976a) Acriflavine-induced hyaline variants of Verticillium albo-atrum and V. dahliae. Trans Br Mycol Soc 66:15-25

Typas M and Heale J (1976b) Heterokaryosis and the role of cytoplasmic inheritance in dark resting structure formation in Verticillium spp. Molec gen Genet 146:17-26

Typas M and Heale J (1978) Heterozygous diploid analyses in the parasexual cycle and a cytoplasmic pattern of inheritance in Verticillium spp. Genet Res Camb 31:131-144

SELECTION, CHARACTERIZATION, PATHOGENICITY AND VIRULENCE OF PECTINASE - DEFICIENT MUTANTS OF *VERTICILLIUM ALBO-ATRUM*

R.M. Cooper and R.K. Durrands
School of Biological Sciences
University of Bath
Bath
Avon
U.K.

INTRODUCTION

Evidence for a role in pathogenesis of endo-polygalacturonan hydrolases and lyases of vascular parasites is plentiful, but circumstantial (Cooper and Wood 1980; Durrands and Cooper 1988b; Pegg 1981).

Thus (i) Polygalacturonase (PG) and pectin lyase (PL) are produced by most fungal vascular parasites.

(ii) Conditions in xylem vessels are conducive to enzyme synthesis in terms of availability of inducers and absence of catabolite repression.

(iii) Gels in vessels can arise from the polygalacturonide-rich middle lamella-primary wall complex of vessel-vessel paired pits and can precede or coincide with wilting or reduced vascular flow.

(iv) Enzymes (mainly PL) have been detected in vascular tissue before disease symptoms.

(v) Disease symptoms have been reproduced with pure enzymes and may result from (a) vascular occlusion, (b) release of physiologically active wall-bound components such as peroxidase (unpublished results). Contradictory results have been reported from different laboratories on (iii)-(v) and the role of pectinases was questioned by Puhalla and Howell (1975) and Howell (1976) based on studies with enzyme-deficient mutants of *Verticillium dahliae*.

NATO ASI Series, Vol. H28
Vascular Wilt Diseases of Plants

The most pragmatic approach in resolving the controversy is through isolating pectinase mutants and testing their relative pathogenicity and virulence. Such mutants would also be adjuncts to molecular cloning techniques as these are developed for fungal vascular parasites(Heale this volume).

Pectinase mutants of wilt pathogens have previously been isolated following conventional mutagenesis but with only limited success, partly because of insufficient awareness of the complexity of pectinase types and production (e.g. Mann 1962; Puhalla and Howell 1975). Also, much improved techniques are now available for detecting and characterizing pectinase isozymes (Cooper 1987).

This paper will report on the selection and characterization of three pectinase mutants of V. albo-atrum in vitro, then will describe their relative pathogenicities.

SELECTION MEDIA AND POTENTIAL RANGE OF MUTANTS

We found that detection media used previously to isolate PG -or PL- deficient mutants were relatively non-specific for PG or PL because of pH fluctuation; also amendments could effect catabolite repression (e.g. Hankin and Anagnostakis 1975; Howell 1976). Therefore two media were devised for separate detection of PG and PL production by V. albo-atrum colonies, by exploiting the contrasting conditions for their production and activities; i.e. pH, substrate methylation and Ca^{2+} requirement (Durrands and Cooper 1988a). The selectivity of these media was confirmed and other aspects of enzyme regulation were revealed by analysis of the cleared zones around colonies for specific enzyme activities, pH, concentrations of substrate degradation products and any sugars added to the medium.

As well as screening for enzyme-deficient mutants it is feasible to detect a range of regulatory mutants by judicious use of inducers and repressors and detection overlays as described by Cooper (1987). In practice however, "self catabolite-repression" and end product inhibition may be caused by accumulation around colonies of high levels of galacturonides; thus further addition of glucose to media for the selection of derepressed mutants has little effect. Nevertheless, any hyperproductive mutant detected on the conventional media could be insensitive to catabolite repression on producing pectinases resistant to end product inhibition (Durrands and Cooper 1988a).

PECTIN LYASE MUTANTS

Three PL - deficient mutants of V. albo-atrum were isolated from about 30,000 survivors (treated with EMS to give 10% survival). No PG mutants or hyperproductive mutants were detected (Durrands and Cooper 1988b).

In vitro characterization of the putative mutants

The relative sizes of cleared zones on the PL - detection medium gave a guide to the reduced PL production which was confirmed in liquid culture (Fig. 1, Table 1). Mutants 24d, C23 and 34i produced PL at rates of ca. 43, 9 and 3% of the wild type respectively when grown on liquid pectin medium.

Growth, conidial production, and morphology were similar for all isolates and wild type on a range of media with the exception of 34i which grew very slowly on galacturonides (Fig. 1).

Mutant 34i was apparently unable to utilize saturated or unsaturated galacturonides. This was evident from the high levels of galacturonides (e.g. 600 µg/ml) which accumulated and remained in culture fluids under a range of pH conditions, whereas levels in cultures of other isolates were negligible or considerably less. Presumably basal PG (under acidic conditions) and basal PL (under alkaline conditions) of 34i resulted in sufficient substrate breakdown to cause the build up. The apparent inability to use galacturonides was confirmed by inoculating cultures with galacturonic acid as sole carbon source in which growth of 34i was negligible, in contrast to that of other isolates which grew at normal rates.

All isolates were also grown for 7 days on tomato cell walls extracellular fluids until visible autolysis of mycelium).

Table 1. Relative production of extracellular enzymes by V.albo-atrum wild type and mutants on host cell walls

Enzyme Isolate	PL	PG	Cellulase	β-galactosidase	β-glucosidase	L-leucine arylamidase
Wild type	100	100	100	100	100	100
C23	3	9	3	30	31	13
24d	43	95	110	101	94	100
34i	9	7	164	34	30	17

Intracellular PL constituted 5.5 - 6% of total (intracellular and extracellular) PL in wild type and 24D, but in comparable extracts from C23 levels were more than fourfold greater.

Pectin lyase and polygalacturonase isozyme profiles of mutants and wild type

Individual isozymes or specific combinations of them may be of greater importance in pathogenesis than overall activity; therefore PG and PL isozyme profiles were determined by isoelectric focussing for each isolate.

Wild type V. albo-atrum produced about six PL isozymes (pIs 10.6, 9.0, 8.0, 6.6, 5.4, 4.5). C23 and 34i showed an overall reduction of all forms of PL but in particular both mutants produced relatively low levels of the two main alkaline isozymes. 24d produced wild type levels of the major PL form (pI 10.6) but all other isozymes were present at much reduced levels (Fig. 2).

At least 25 apparent PG isozymes (pIs 3 - 9.5) are produced by wild type in inducing (glacturonide) media, whereas a single isozyme is released basally (on glucose or cellulose media). All isozymes were produced by mutants C23 and 24d, but 34i produced only the single basal form. This result further confirms the inability of 34i to utilise galacturonides which function as specific inducers of pectic enzyme synthesis (Cooper 1983).

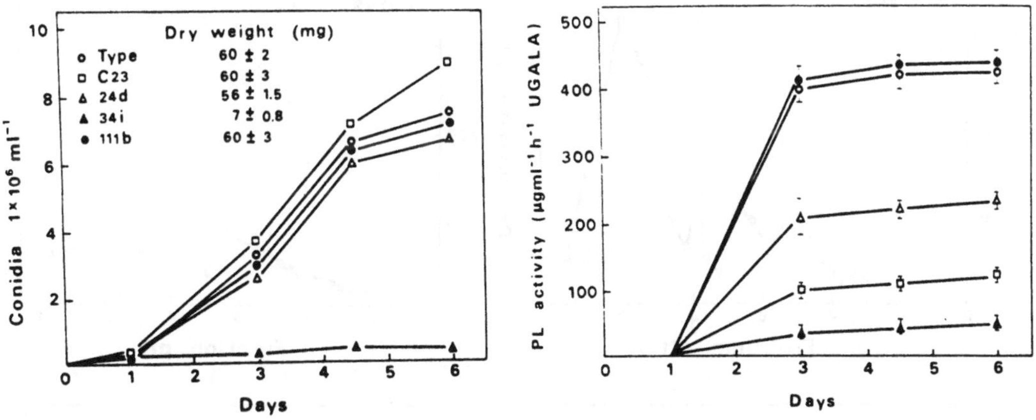

Fig. 1. Growth (as conidia and mycelium dry weight) and pectin lyase production by V. albo-atrum wild type and mutants in pectin-salts medium (mean results of 4 cultures).

as composite carbon source to induce a range of enzyme activities (Table 1). Relative production of PL and PG was similar to the differences attained on pectin media. Thus 24d was partly deficient in PL but produced wild type levels of PG, cellulase, β-galactosidase, β-glucosidase and L-leucine arylamidase. In contrast C23 produced low levels of all enzymes which suggested that it might be a secretory mutant. Low production by 34i probably reflected its initially slow growth on cell walls, although it is noteworthy that the cellulase level was about 1.6-fold greater than that of the wild type.

Determination of intracellular PL in the putative secretory mutant

Reduced production of a range of extracellular enzymes by mutant C23 is suggestive of a mutation affecting a common secretory mechanism. This defect is the most common cause of enzyme-deficient phenotypes (Cooper 1987) in which the active enzyme often accumulates intracellularly. Therefore PL was determined within mycelium and conidia of each isolate. Phosphatases served as indicators of intracellular activity (phosphatases are not found in

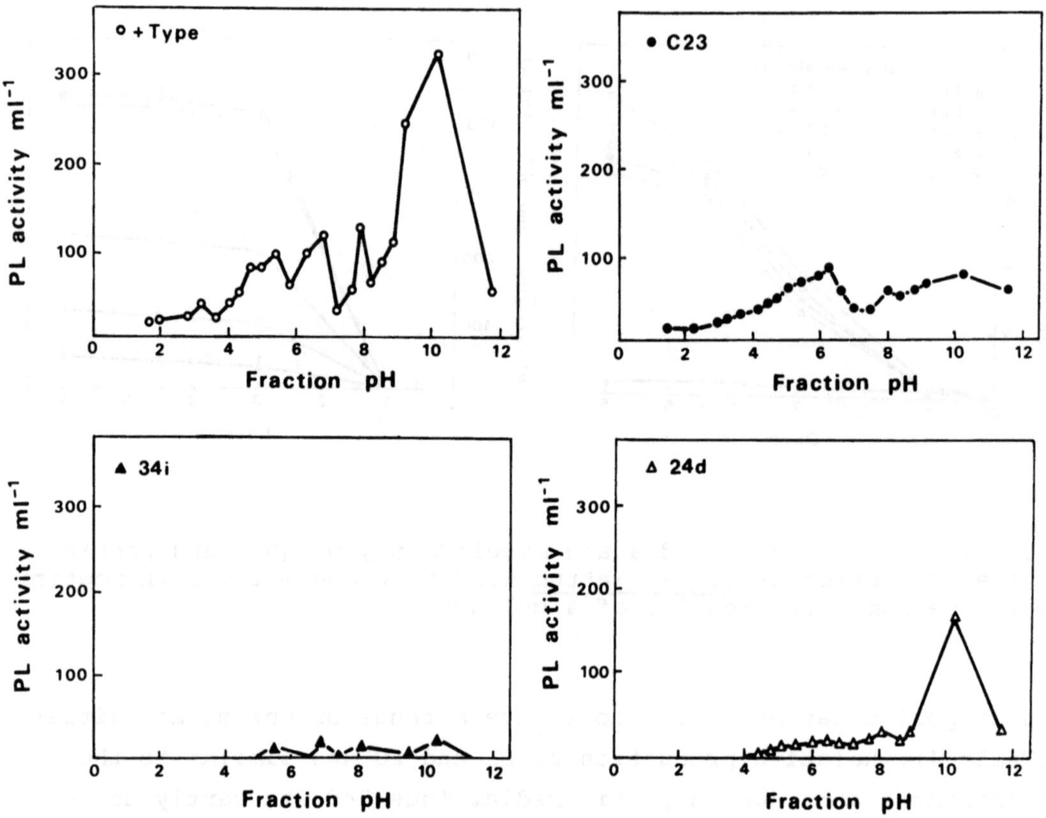

Fig. 2. Pectin lyase isozymes resolved by column isoelectric focussing (pH 3-10). Culture fluids were from 7 day pectin-salts medium concentrated with ammonium sulphate (95% w/v).

The remarkable multiplicity of PG (although some bands may represent artifacts of isoelectric focussing or post-translational modifications) reveals why no PG - deficient mutants were obtained.

Structural gene mutants are required to give the critical information needed to determine the role of enzymes in pathogenesis, but they are notoriously difficult to detect phenotypically because of masking by the numerous other unaffected isozymes. To the authors' knowledge no structural gene mutants for cell wall- degrading enzymes have been shown; they have only been clearly obtained by site-directed mutagenesis of *Erwinia chrysanthemi* pectate lyase (pel) genes (e.g. Roeder and Collmer 1985). It follows that *V. dahliae* which has a much simpler pattern of 3-5 PG isozymes would appear more suitable for genetic analysis (Durrands et al 1988).

It is not surprising therefore that from numerous studies on virulence factors of bacterial pathogens of plants and animals, secretory (Out^-) mutants has been the phenotype most often detected (Cooper 1987). C23 falls into this category and the intracellular accumulation of PL resembles that of pectate lyase in the periplasm of *E. chrysanthemi* Out^- mutants (Andro et al 1984).

Similarly, if several isozymes are under joint control by regulatory genes there should be an increased chance of obtaining hypoproductive mutants. The only specific PL mutant, 24d, had a reduced production of five out of six PL isozymes, which might suggest a defect in a common control mechanism. These three mutants of *V. albo-atrum* should be useful tools in studying enzyme regulation and secretion, and uronide uptake or metabolism. However, our primary aim was to establish the significance of PL in vascular parasitism by comparing the virulence to and colonization of tomato plants by the mutants with that of a wild type isolate. Plants were inoculated *via* intact roots in order to judge the effect of enzyme deficiency at all stages of pathogenesis. Howell (1976) circumvented the normal means of entry by direct injection of stems with *V. dahliae* mutants, which unfortunately reduces the value of this work.

PATHOGENICITY AND VIRULENCE OF PL-DEFICIENT MUTANTS

Symptoms of epinasty, chlorosis and wilting were absent, reduced or delayed in plants inoculated with the mutants (Durrands and Cooper 1988c). Plants inoculated with C23 remained healthy in appearance apart from mild chlorosis on lower leaves and slight wilting in 50% of plants.

All symptoms were slower to develop following inoculation with 24d and 34i but were no less extensive after 16 days (data were analyzed by a two way analysis of variance based on 10 replicate plants). Each mutant was then reisolated from stem segments which showed they had not reverted; this was evident from maintained reduced capacity to produce PL. Host colonization was determined microscopically after 19 days. In internodes 1,3 and 5, 30-40% of vessels had been colonized to similar extents by wild type and mutants 24d and 34i. In contrast C23 was present in only < 2% of vessels. The corresponding figures in a comparable experiment were 55% and 10% (Fig. 3).

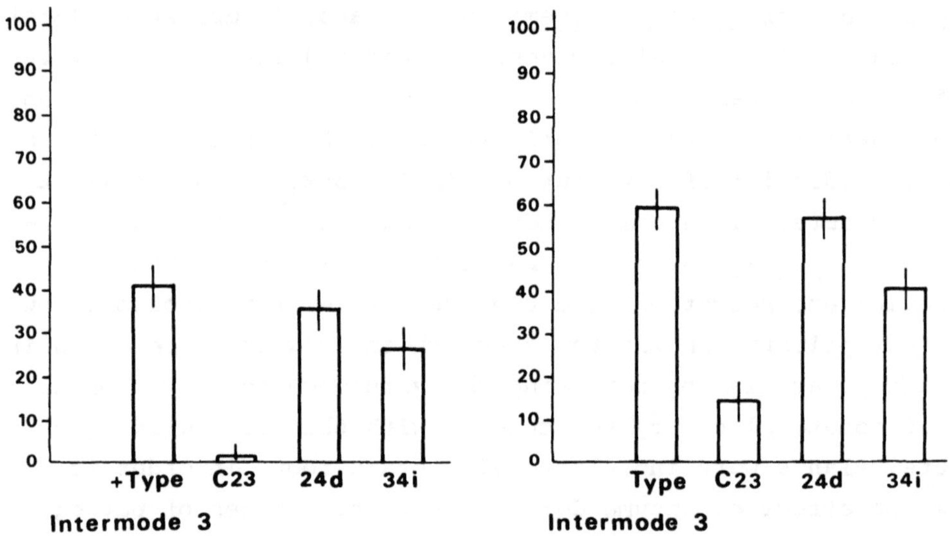

Fig. 3. Colonization of tomato plants by wild type and PL mutants of V. albo-atrum. Vascular infection was assessed as the mean percentage of vessels containing hyphae in sections taken from internodes 1,3,5 and 7 from 5 replicate plants, 19 days after inoculation. Only results from internode 3 are shown from two comparable experiments.

CONCLUSIONS

Prima facie at induced levels PL functions as a determinant of virulence but possibly not of pathogenicity because host colonization can proceed in its (relative) absence and symptoms are reduced or delayed. However, it should be stressed that none of the mutants completely lacked pectinases.

34i produced only basal PG and PL yet must have succeeded in frequent cell wall penetration in roots and of xylem paired pits to match the level of invasion achieved by wild type. Ultrastructural studies of root invasion by V. albo-atrum show enzymic wall degradation is highly localized around penetration hyphae and accompanies assentially biotrophic invasion (Bishop and Cooper 1983a). Thus high induced synthesis may not be required, at least at this stage. Analogously, pathogenicity of an uninducible E. chrysanthemi mutant (deficient in the intracellular inducer-generating enzyme oligogalacturonide lyase) was unaffected (Collmer et al 1982); however, basal synthesis by this soft rot pathogen is proportionately much higher than that of V. albo-atrum (Cooper 1987).

C23 produced higher levels of PG and PL than 34i but was much less virulent. Secretory mutants of several pathogens, including E. chrysanthemi, E. carotovora and Xanthomonas campestris has all been found to be non-pathogenic (Cooper 1987). Avirulence of C23 did not result from inability to degrade host walls because growth in vitro on cell walls was equivalent to that of wild type. However, many other enzymes were also much reduced, including PG and C_4, C_8 and C_{14} esterases. The latter could play a crucial role in suberin degradation in endodermal walls; the importance to overall colonization of the extravascular phase of invasion has been stressed by Talboys (1958) and others (Beckman this volume).

The lack of symptoms induced by C23 probably relates to the sparse level of vascular colonization. Not only pectinases but other potential virulence factors such as high molecular weight polymers and toxins would also be secreted in proportionately reduced amounts.

Virulence of 24d was also reduced even though it had a full complement of PG isozymes and secreted over 40% of potential PL levels. However, the PL derived mainly from one alkaline isozyme. Acidic isozymes are likely to be more mobile and therefore more destructive in vessels because of adsorption of positively charged alkaline forms onto host cell walls (Cooper 1983).

Additionally, results with 34i provide critical information on nutrition during vascular colonization by V. albo-atrum. Xylem fluids are very low in organic nutrients and it has often been suggested that enzymically released galacturonides may provide a key source of carbon (Pegg 1981). However, 34i was unable to metabolize saturated or unsaturated galacturonides, yet succeeded in extensively colonizing the host xylem. Nutrition may therefore be mainly through utilization of amino acids as indicated by Dixon and Pegg (1972); alternatively breakdown of secondary wall polymers may contribute. V. albo-atrum produces appropriate enzymes such as xylanase and cellulase which are capable of degrading host secondary walls (Cooper et al 1978) and ultrastructural evidence reveals some degree of erosion of xylem vessel secondary walls (Bishop and Cooper 1983b). Also,despite poor growth 34i produced enhanced levels of cellulase in comparison to other isolates when grown on host cell walls; presumably this observation reflects enforced cellulose breakdown in the absence of its pectolytic capacity.

Further progress will depend on isolation and cloning of genes for specific isozymes of PL and PG, but a vascular parasite with reduced enzyme multiplicity is required to make the task practicable.

This work was supported by an SERC studentship to PKD.

REFERENCES

Andro T, Chambost J-P, Kotoujansky A, Cattaneo J, Bertheau Y, Barras F, Van Gijisegem F and Coleno A (1984) Mutants of Erwinia chrysanthemi defective in secretion of pectinases and cellulase. Journal of Bacteriology 160:1199-1203

Bishop CD and Cooper RM (1983a) An ultrastructural study of root invasion in three vascular wilt diseases. Phys Plant Pathol 22:15-27

Bishop CD and Cooper RM (1983b) An ultrastructural study of vascular colonisation in three vascular wilt diseases. I. Colonisation of susceptible cultivars. Phys Plant Pathol 23:323-343

Cooper RM (1983) The mechanisms and significance of enzymic degradation of host cell walls. In: Callow JA (ed) Biochemical Plant Pathology. John Wiley & Sons, p 101

Cooper RM (1987) The use of mutants in exploring deploymerases as determinants of pathogenicity. In: Day PR and Jellis GJ (eds) Genetics and plant pathogenesis. Balckwell Scientific, Oxford, pp 253

Cooper RM, Rankin B and Wood RKS (1978) Cell wall degrading enzymes of vascular wilt fungi. II. Properties and modes of action of polysaccharidase of *Verticillium albo-atrum* and *Fusarium oxysporum* f. sp. *lycopersici*. Phys Plant Pathol 13:101-134

Cooper RM and Wood RKS (1980) Cell wall-degrading enzymes of vascular wilt fungi. III. Possible involvement of endopectin lyase in Verticillium wilt of tomato. Phys Plant Pathol 16:285-300

Durrands PK and Cooper RM (1988a) Development and analysis of pectic screening media for use in detection of pectinase mutants. Applied Microbiology and Biotechnology 28:463-467

Durrands PK and Cooper RM (1988b) Selection and characterisation of pectinase-deficient mutants of the vascular wilt pathogen *Verticillium albo-atrum*. Phys and Mol Plant Pathol 32:343-362

Durrands PK and Cooper RM (1988c) The role of pectinases in vascular wilt disease as determined by defined mutants of *Verticillium albo-atrum*. Phys and Mol Plant Pathol 32:363-371

Durrands PK, Keen RA, Cooper RM and Clarkson JM (1987) Polygalacturonase isozyme profiles of *Verticillium dahliae* isolates race 1 and 2 from different geographical origins. Trans Br Mycol Soc (in Press)

Hankin L and Anagnostakis SL (1975) The use of solid media for detection of enzyme production by fungi. Mycologia 67:597-607

Howell CR (1976) Use of enzyme deficient mutants of *Verticillium dahliae* to assess the importance of pectolytic enzymes in symptom exrpession of Verticillium wilt of cotton. Phys Plant Pathol 9:279-283

Mann B (1962) Role of pectic enzymes in the Fusarium wilt syndrome of tomato. Trans Br Mycol Soc 45:169-178

Pegg GF (1981) Biochemistry and physioloy of pathogenesis. In: Mace ME, Bell AA and Beckman CH (eds) Fungal Wilt Disease of Plants. Academic Press, New York, p 193

Puhalla JE and Howell CR (1975) Significance of endopolygalacturonase activity to symptom expression of Verticillium wilt in cotton, assessed by the use of mutants of *Verticillium dahliae* Kleb. Phys Plant Pathol 7:147-152

Roeder DL and Collmer A (1985) Marker-exchange mutagenesis of a pectate lyase isozyme gene in *Erwinia chrysanthemi*. Journal of Bacteriology 164:51-56

NUTRITIONAL FACTORS AFFECTING FUSARIUM WILT INCIDENCE AND SEVERITY

A.W. Engelhard, J.P. Jones and S.S. Woltz
University of Florida
Gulf Coast Research & Education Center
5007 60th Street East
Bradenton, FL 34203
USA

INTRODUCTION

Fusarium wilt (Fusarium oxysporum Schlecht. f.sp. lycopersici (Sacc.) Snyder and Hansen) of tomato (Lycopersicon esculentum Mill.) was first reported in the United States in 1899 by E.F. Smith who said that the disease had put an end to the growing of tomatoes for the northern markets in certain areas of Florida. By 1920, the disease had become widespread and destructive throughout the southeast, midwest, and middle Atlantic states (Walker 1971).

The first research efforts on Fusarium wilt of tomato initiated *circa* 1910 by Norton (1912, 1914) and Essary (1912, 1920) concentrated on the development of tolerant varieties. Nearly concurrently, Wollenweber (1913) and Sherbakoff (1915) started their studies on the morphology and physiology of the genus Fusarium. A third phase of research was begun around 1915 by Edgerton and Moreland (1913, 1920) who investigated the effect of nutrition on disease development.

Fusarium wilt of chrysanthemum was recognized as a disease in 1963 by Armstrong and Armstrong (1964) when they demonstrated that F. oxysporum Schlecht. f. sp. tracheiphilum Armstrong and Armstrong race 1 caused wilt of the chrysanthemum cultivar Encore and certain cultivars of cowpea and soybean. Armstrong et al (1970) described F. oxysporum f. sp. chrysanthemi (Litt.,

Armstrong and Armstrong) as the cause of wilt in the cultivars Encore and Yellow Delaware. The latter forma specialis commonly is found in chrysanthemum.

Research on the effects of lime (pH), nitrogen source and systemic fungicides on the control of Fusarium wilt of chrysanthemum was first published by Engelhard and Woltz (1972, 1973) and Woltz and Engelhard (1973).

This review places special emphasis on investigations and concepts which led to the Florida system of Fusarium wilt control on vegetable crops and the lime (pH)-nitrate-chemotherapy system on ornamental crops. Research was conducted by an interdisciplinary group of scientists located mainly at the University of Florida,Gulf Coast Research and Education Center, Bradenton. Crop emphasis in Florida was on tomato and chrysanthemum because of the importance of these two crops, but research on watermelon, cucumber, muskmelon, King Aster and carnation are included also in the review.

LIME, CALCIUM, pH

Edgerton and Moreland reported in 1913 and 1920, respectively, in their work on tomatoes that "lime added to the soil in large amounts, such as ten tons to the acre, hinders the development of the wilt". Their finding was confirmed by numerous workers including Scott (1924), Sherwood (1923) and Fisher (1935). Through the years the association of higher pH and/or calcium with less wilt was shown by many other investigators on numerous crops, including the following:

Crop	Investigators
Tomato	Foster and Walker, 1947
	Edgington and Walker, 1958
	Gorden, 1965
	Jones and Overman, 1971
	Jones and Woltz, 1967,1968,1969, 1970,1972,1981,1983

	Standaert et al, 1978
	Woltz and Jones, 1973,1984
	Sarhan and Kiraly, 1981
Chrysanthemum	Engelhard and Woltz, 1972,1973,1978
	Woltz and Engelhard, 1973
	Raju, 1983
	Locke et al, 1985
Carnation	Engelhard, 1979
	Pergola et al, 1979
Cotton	Albert, 1946
Cucumber	Jones and Woltz, 1975
Flax	Keane and Sackston, 1970
Gladiolus	Woltz and Magie, 1975
	Engelhard and Woltz, 1978
King Aster	Engelhard, 1975
	Engelhard and Woltz, 1978
Muskmelon	Stoddard, 1947
Radish	Sun and Huang, 1985
	Huang et al, 1986
Strawberry	Yoshino and Hashimoto, 1978
Watermelon	Everett and Blasquez, 1967
	Jones and Woltz, 1975
	Sun and Huang, 1985

NITROGEN

Clayton (1923) started investigations to determine the effect of nitrate-nitrogen (NO_3-N) on the severity of Fusarium wilt of tomato. This work was followed by that of Ahmet (1933) and of Cook (1937). These workers, in general, compared NO_3-N to no nitrogen treatments and obtained inconclusive results. Fisher (1935) probably was the first to demonstrate that an increase in the ratio of ammonia-nitrogen (NH_4-N) to NO_3-N increased disease severity. However, he attributed the increased severity to an increase in total N.

Much of the research with nitrogen source was combined with pH studies. Albert (1946) reported that cotton plants grown in $Ca(NO_3)_2$ at high pH had less wilt injury than those grown in any other nitrogen containing solution. Jones and Wolts (1967, 1968, 1969, 1970, 1972, 1975) and Woltz and Jones (1972, 1973 a,b,c) greatly expanded the early work with nitrogen source and lime amendments on tomato. They consistently demonstrated that liming sandy soils to a high pH (6.5-7.5) greatly decreased the incidence and severity of Fusarium wilt of tomato as Scott and other workers had reported. They also demonstrated that NO_3-N decreased disease development compared to NH_4-N and that the beneficial effects of high soil pH could be overcome by the use of NH_4-N (Woltz and Jones 1973 b,c). They further demonstrated that effects of high soil pH and NO_3-N were additive so that the combination of the two almost invariably resulted in even better wilt control. Methods were developed by Jones, Woltz and Engelhard in Florida which resulted in consistent control in the field and greenhouse of Fusarium wilt of tomato (Jones and Woltz 1967, 1968, 1969, 1970, 1972, 1975; Woltz and Jones 1973 a,b,c) and other vegetables (Jones and Woltz 1975), and of chrysanthemum (Chrysanthemum morifolium (Ramat.) Hensl.) (Engelhard and Woltz 1972, 1973, 1978; Woltz and Engelhard 1973) and other ornamentals (Engelhard 1975, 1979; Engelhard and Woltz 1978). Jones and Woltz (1975) also showed that gross yields were increased and wilt development decreased when NO_3-N, as compared to NH_4-N was combined with high pH on watermelons and cucumbers. Also,

Stoddard (1947) demonstrated that less wilt occurred (Fusarium oxysporum f. sp. niveum Wr.) on muskmelon (Cucumis melo L.) with a NO_3-N source at pH 6.0. Schneider (1985) found more wilt with NH_4-N than with NO_3-N in celery wilt (F. oxysporum f. sp. apii). Trillas-Gay et al (1986) found significantly reduced disease levels at the highest concentration of calcium nitrate compared with an ammonium source on radish. Engelhard (1975, 1979), Engelhard and Woltz (1972, 1973, 1978) and Woltz and Engelhard (1973) consistently found less wilt in chrysanthemums, King Aster and carnation with NO_3-N compared to NH_4-N. Locke et al (1985) and Raju (1983) reported the same on chrysanthemum. Dick and Tisdale (1938) found that nitrogen and potash alone and especially in certain combinations reduced wilt and increased yield of cotton.

Hopkins and Elmstrom (1976), however, found no significant difference in wilt (F. oxysporum f. sp. niveum) control of watermelons between high pH-all NO_3-N treatments and the low pH NH_4-N treatments. Also, Pergola et al (1979) on carnation found the severity of Fusarium wilt decreased when NH_4-N was compared with NO_3-N but a high soil pH of 7.5 resulted in less wilt.

PHYSIOLOGY, MACRO-AND MICROELEMENTS, INTERACTIONS

Fusarium wilt severity decreased with an increase in calcium (Ca) concentration from 5 to 500 ppm with tomato in a nutrient solution experiment of Edgington and Walker (1958). Corden (1965) and Standaert et al (1975) also found that a Ca deficiency encouraged tomato wilt development and Corden (1965) suggested that the increased pathogenesis resulted because there was little Ca to inhibit the activity of the polygalacturonase produced by Fusarium.

Jones and Woltz (1968, 1969) found in greenhouse and field experiments that amending infested soil with calcium sulfate ($CaSO_4$-gypsum) did not increase the soil pH and did not reduce the occurrence of tomato Fusarium wilt, even though the Ca content of tissue of plants grown in gypsum-amended soil was as great as that of plants grown in soil amended with hydrated lime ($Ca(OH)_2$) which did control the disease. They concluded that high tissue Ca was not the major factor controlling Fusarium wilt.

Edgington and Walker (1958) in their work with calcium found that the influence of boron (B) on disease was dependent upon the Ca supply so that with 100 ppm Ca the disease decreased with increasing B, but at 500 ppm Ca, disease index increased with increases in B. Working with flax (Linum usitatissimum L.), Keane and Sackston (1970) found that a B deficiency exacerbated development of Fusarium wilt.

Walker and Foster (1946) conducted nutritional experiments on Fusarium wilt of tomato using solution culture techniques. They found that Fusarium wilt severity declined with an increase in salt concentration. Stoddard and Diamond (1948), MacKay (1959) and Standaert et al (1978) also reported development of tomato wilt to be inhibited with an increasing nutrient concentration.

Walker and Foster (1946) further reported that disease development in the highly susceptible cv. 'Bonny Best' tomato variety in nutrient cultures using all NO_3-N increased in severity in the following order : low N, high potassium (K), normal N + K, high N, and low K. This indicated that high N or low K favoured disease development, whereas low N or high K retarded it. In contrast, Foster and Walker (1947) found that tomato plants preconditioned with low N or high K prior to inoculation were more susceptible than those preconditioned with high N or low K. They also showed that a high pH inhibited wilt development.

Bloom and Walter (1955) applied urea-N to tomato foliage weekly for 30 days before inoculation and found a slight increase in disease severity with an increase in concentration.

High levels of N, regardless of source, were found by Woltz and Magie (1975) to encourage Fusarium corm rot of gladiolus (Gladiolus hortulanus Bailey) caused by F. oxysporum f. sp. gladioli (Massey) Snyder and Hansen. However, NH_4-N resulted in a greater incidence of Fusarium yellows of the mother plants than 50:50 NH_4:NO_3-N.

Woltz and Jones (1973 a,b,c; 1981) reported that a high level of phosphorus (P) increased the severity of Fusarium wilt of tomato in pot and field experiments and that the combination of high lime plus low P greatly curtailed disease development. Similarly, Sagdullaev and Berezhnova (1974) reported that P amendments increased the severity of Fusarium wilt of muskmelon (Cucumis melo L.).

Schneider (1985) reported inhibition of Fusarium yellows of celery (Apium graveolens L. var. dulce (Miller) Pers.) with a specific concentration of potassium chloride (KCl) and NO_3-N. He further attributed this control to the ratio of K and Cl in the plant tissue. Ramasamy and Prasad (1975) reported that increasing rates of KCl applied to muskmelon in pots led to a reduction in Fusarium wilt incidence. However, Spiegel and Netzer (1984), and Stoddard (1967) working with muskmelon, found that K level had very little effect on Fusarium wilt regardless of N source. Tharp and Wadleigh (1939) demonstrated in a sand-nutrient culture experiment that an increase in potassium was accompanied by a significant reduction in severity of wilt in cotton.

Fusarium oxysporum f. sp. lycopersici was shown by Jones and Woltz (1970, 1981) and Woltz and Jones (1968, 1973, 1981) to have a relatively high requirement for micronutrients. Deficiencies of copper, iron, manganese, molybdenum, and zinc reduced growth and sporulation of the fungus. Moreover, Fusarium grown in liquid cultures devoid of molybdenum or zinc was not as virulent as Fusarium grown on optimal levels of these two micronutrients. The response of Fusarium to manganese, iron, and zinc was very pronounced. Increasing amounts of these micronutrients (above those usually found in soil solutions for tomato culture) are increasingly beneficial to growth and sporulation of the pathogen, whereas concentrations below average inhibited growth and spore production.

Liming soil infested with F. oxysporum f. sp. lycopersici in the field or glasshouse to pH 7.0-7.5 greatly limits the availability of micronutrients and consistently decreases wilt in naturally low pH soils. However, when soil with a high pH was further amended with lignosulfonate metal complexes of zinc and manganese or iron and manganese (the metal ions in these complexes are available for plant growth at high soil pH values), the beneficial effect of pH elevation was reversed according to Jones and Woltz (1970). Consequently, the beneficial effects of liming apparently are not caused by an increased soil or tissue calcium content, but rather by the unavailability of micronutrients created by the high soil pH, which in turn limits the growth, sporulation, and virulence of the pathogen. Increases in soil pH limit not only the availability of micronutrients, but also of other elements essential for F. oxysporum, especially phosphorus and magnesium. Supplemental applications of superphosphate (in field plots) above the amount required for growth of tomato greatly increase occurrence of wilt in soils of pH 6.0. At pH 7.0 or 7.5 supplemental applications do not increase wilt occurrence because at these pH values phosphorus becomes sparingly available. Increasing amounts of available magnesium supplied above sufficiency levels increase stepwise the growth, sporulation, and virulence of F. oxysporum (Woltz and Jones 1973 a,b; 1981).

Iron in adequate supply has been found to be an important factor in the disease process, according to Scher and Baker (1982), Simeoni et al (1987), and Yuen and Schroth (1982). It has been suggested by Scher and Baker that the management of iron availability through iron competition can reduce wilt severity. Siderophores produced by bacteria complex Fe+++ iron and reduce its availability to Fusarium. Adding $FeCl_3$ to the soil overcame Fusarium suppression. A similar mechanism of iron starvation for the Fusarium pathogen was shown by Jones and Woltz (1969, 1970, 1972) and Woltz and Jones (1968) for the Fusarium wilt pathogen.

Schneider (1985), in his work on Fusarium yellows of celery, found that NH_4-N was more conducive to disease development than NO_3-N and that chloride salts of calcium or magnesium resulted in even more severe disease when added in combination with KNO_3 than when KNO_3 was added alone. Calcium nitrate added to K resulted in "almost complete disease control". Schneider attributed disease suppression to a specific tissue concentration ratio of K:Cl. Disease was least when the K:Cl ratio was about 3.5, with more severe disease developing at ratios above or below 3.5. Factors affecting the ratio were N source, the availability of competing cations (Ca and NH_4) and anions (SO_4 and NO_3) and the concentration of available K and Cl in the soil. NH_4 repressed uptake of NO_3 and K and stimulated Cl uptake, thereby favouring disease development.

The work of Schneider was preceded by that of Davet et al (1966) who reported that a cool weather temperature strain of Fusarium oxysporum f. sp. lycopersici existed in Marocco that produced severe wilt symptoms especially if affected plants were irrigated with sodium chloride or magnesium chloride. Calcium nitrate and calcium sulfate protected against the pathogen, they hypothesized, because the calcium ions replaced the sodium and magnesium ions of the polygalacturonic chains of the cell membrane.

DISCUSSION AND CONCLUSIONS

The mechanism(s) of control of Fusarium wilt with lime, NO_3-N, and low phosphorus is not known with certainty. However, there is little doubt that the regime is highly effective. Although conflicting observations on plant nutrition effects are encountered in the literature certain consistencies are apparent. First, liming the soil to a pH of 6.0-7.5, in the absence of organic matter or NH_4-N, routinely decreases the incidence and severity of Fusarium wilt of many crops. In fact, Jones and Overman (1971) found that raising the pH of EauGallie fine sand to 7.0 controlled tomato wilt as well as fumigating soil adjusted to pH 6.6 with a broadspectrum soil fumigant. Second, N source almost invariably affects disease development. The level of wilt decreases when NO_3-N is used in the fertilizer and increases when NH_4-N is used. Finally, high phosphorus encourages wilt development. Engelhard (1979) and Engelhard and Woltz (1972, 1973) further modified the system for ornamentals by including the fungicide benomyl which alone gives partial symptom control. They combined the additive effects of NO_3-N, high pH, and benomyl drenches to obtain complete control of Fusarium wilt of chrysanthemum and aster with the integrated lime-nitrate-chemotherapy system. Thus, systems were developed at the Gulf Coast Research and Education Center,

Bradenton, FL, that have proven to be eminently successful in obtaining a high level of control of Fusarium wilt of tomato (Jones and Woltz 1981; Woltz and Jones 1981), cucumber, watermelon (Citrullus vulgaris Schrad.) (Jones and Woltz 1975), gladiolus (Engelhard and Woltz 1978; Woltz and Magie 1975), chrysanthemum (Engelhard and Woltz 1972, 1973, 1978; Woltz and Engelhard 1973), aster (Callistephus chinensis (L.) Nees.)(Engelhard 1975; Engelhard and Woltz 1978), and carnation (Dianthus caryophyllus L.) (Engelhard 1979). Raju (1983) duplicated the results of Engelhard and Woltz under commercial field conditions with several chrysanthemum cultivars in Florida. They reported that a calcium nitrate based fertilizer combined with a soil pH of 7.0, and benomyl or thiophanate M drenches provided excellent wilt control. Locke et al (1985) demonstrated the same results in Maryland with greenhouse chrysanthemums.

Sarhan and Kiraly (1981) reported that NO_3-N drenched onto potted tomatoes grown on a high lime and benomyl regime provided complete control of Fusarium wilt in Hungary. They said each factor contributed to partial disease control, but by using an integrated approach near perfect control resulted, confirming the work of Engelhard and Woltz (1972, 1973), Locke et al (1985) and Raju (1983).

The work of Schneider (1985) and of Huang and Sum (1982) paralleled the NO_3:NH_4-N research of Jones, Woltz and Engelhard. They obtained excellent control of Fusarium yellows of celery in California and of Fusarium wilt of watermelons, respectively. Jones and Woltz (1983) found that the regime resulted in control of Fusarium wilt of tomato caused by race 3 in Florida. Morgan and Timmer (1984) reported that Fusarium wilt severity of Mexican limes (Citrus aurantifolia (Christm.) Swingle) and the numbers of propagules were less (and the pH higher) in media fertilized with NO_3-N. Spiegel and Netzer (1984) in Israel demonstrated that NH_4-N fertilized plants showed 30% more wilt compared to NO_3-N applications on Fusarium wilt of muskmelon. And again Keim and Humphrey (1984) reported excellent control of Fusarium wilt of Hebe (Hebe spp.) in California by using NO_3-N compared to NH_4-N. They also observed that where benomyl was used with NO_3-N, all plants were healthy, whereas when benomyl was used with NH_4-N, only 31% were healthy. Ebben (1979) in England showed that urea increased Fusarium wilt of carnation compared to calcium nitrate and that even less wilt occurred when a high soil pH was combined with calcium nitrate.

The effect of N source may be fourfold: first, NO_3-N increases and NH_4-N decreases soil pH; second, F. oxysporum f. sp. lycopersici grown on NH_4-N is far more virulent than Fusarium grown on NO_3-N; third, tomato seedlings grown on NO_3-N are preconditioned and are more resistant to Fusarium wilt than seedlings grown on NH_4-N; and fourth, NH_4-N represses K uptake and stimulates Cl uptake, thereby favouring disease development. The predominant factor seems to be the effect of nitrogen source on soil pH. However, NH_4-N applied to soil weekly may reverse the usual disease preventative effect of high soil pH without a concomitant decrease in soil pH.

Liming inhibits development of Fusarium wilt probably not only because of its effect on the availability of micronutrients, but also because of its effect on soil microflora. Actinomycete and bacterial populations are favoured by high soil pH values (Waksman 1927). Certain of these microorganisms are antagonistic to F. oxysporum f. sp. lycopersici, preventing spore germination and vegetative growth by means of toxic compounds. Often zones of Fusarium inhibition extend beyond the boundaries of the inhibiting organism. Bacteria and actinomycetes also compete with Fusarium for organic and inorganic nutrients in the soil solution.

Other factors have been reported to have an effect on Fusarium wilt. Miyake and Takahashi (1983) reported that silicon amendments to the soil remarkebly lowered the incidence of Fusarium wilt of cucumber (Cucumis sativus L.). Loffler et al (1986 a,b) showed that nitrite inhibited chlamydospore formation while urea or ammonium chloride reduced chlamydospore formation and lysis. They concluded that nitrite but not ammonia was involved.

Florida Agricultural Experimental Station Journal Series No 9118

REFERENCES

Ahmet K (1933) Untersuchgen uber tracheomykosen. Phytopath Zeitschr 6:49-101

Albert WB (1946) The effects of certain nutrient treatments upon the resistance of cotton to Fusarium vasinfectum. Phytopathology 36:703-716

Armstrong GM and Armstrong JK (1964) Wilt of chrysanthemum caused by race 1 of the cowpea Fusarium. Phytopathology 54:886 (Abstr)

Armstrong GM, Armstrong JK and Littrell RH (1970) Wilt of chrysanthemum caused by Fusarium oxysporum f. sp. chrysanthemi, forma specialis nov. Phytopathology 60 :496-498

Bloom JR and Walter JC (1955) Effect of nutrient sprays on Fusarium wilt of tomato. Phytopathology 45:443-444

Clayton EE (1923) The relation of soil moisture to the Fusarium wilt of the tomato. Amer J Bot 10:133-147

Cook WS (1937) Relation of nutrition of tomato to disposition to infectivity and virulence of Fusarium lycopersici. Bot Gaz 98: 647-669

Corden CE (1965) Influence of calcium nutrition on Fusarium wilt of tomato and polygalacturonase activity. Phytopathology 55: 222-224

Davet P, Messiaen CM and Rieuf P (1966) Interpretation of winter manifestation of Fusarium wilt of tomato in North Africa, favoured by irrigation water salts. (Trans. title in French). Proc First Cong Mediterr Phytopath Union:407-416

Dick JB and Tisdale HB (1938) Fertilizers in relation to incidence of wilt as affecting a resistant and a susceptible variety. Phytopathology 28:666-667 (Abstr)

Ebben MH (1979) Carnation wilt caused by Fusarium oxysporum f. sp. dianthi. Ann Rep Glasshouse Crops Res Inst, p 207

Edgerton CW and Moreland CC (1913) Diseases of the tomato in Louisiana. La Agr Exp Sta Bull 142

Edgerton CW and Moreland CC (1920) Tomato wilt. La Agr Exp Sta Bull 174

Edgington LV and Walker JC (1958) Influence of calcium and boron nutrition on development of Fusarium wilt of tomato. Phytopathology 48:324-326

Engelhard AW (1975) Aster Fusarium wilt: Complete symptom control with an integrated fungicide-NO_3-pH control system. Proc Am Phytopath Soc 2:62 (Abstr)

Engelhard AW (1979) Control of Fusarium wilt of carnation with an integrated nitrate-nitrogen and systemic fungicide control program. Phytopathology 69:1027 (Abstr)

Engelhard AW and Woltz SS (1972) Complete control of Fusarium wilt of chrysanthemum with chemotherapeutants combined with a high lime and nitrate-nitrogen culture regime. Phytopathology 62:756

Engelhard AW and Woltz SS (1973) Fusarium wilt of chrysanthemum: complete control of symptoms with an integrated fungicide-lime-nitrogen regime. Phytopathology 63:1256-1259

Engelhard AW and Woltz SS (1978) Effect of temperature, nitrogen source, lime and benomyl treatments on Fusarium wilt of chrysanthemum, aster, and gladiolus. Third Int Congress of Plant Pathology, p 375 (Abstr)

Essary SH (1912) Notes on tomato diseases with results of selection for resistance. Tenn Agr Exp Sta Bull 95
Essary SH (1920) Report of the botanist. Tenn Agr Exp Sta Ann Rept 1919-20:15-16
Everett PH and Blasquez CH (1967) Influence of lime on the development of Fusarium wilt of watermelon. Proc Fla State Hort Soc 80:143-148
Fisher PL (1935) Physiological studies on the pathogenicity of Fusarium lycopersici Sacc. for the tomato plant. Md Agr Exp Sta Bull 374
Fisher PL (1935) Responses of tomato in solution cultures with deficiencies and excesses of certain essential elements. Md Agr Exp Sta Bull 375
Foster RE and Walker JC (1947) Predisposition of tomato to Fusarium wilt. J Agri Res 74:165-185
Hopkins DL and Elmstrom GW (1976) Effect of soil pH and nitrogen source on Fusarium wilt of watermelon on land previously cropped to watermelons. Proc Fla State Hort Soc 89:141-143
Huang JW and Sun SK (1982) The effects of nitrogenous fertilizers on disease development of watermelon fusarial wilt. Plant Prot Bull (Taiwan ROC) 24:101-110
Huang JW, Sun SK and Juang CF (1986) Studies on the integrated control of radish yellows, caused by Fusarium oxysporum f. sp. raphani. Plant Prot Bull (Taiwan ROC) 28:81-90
Jones JP and Overman AJ (1971) Control of Fusarium wilt of tomato with lime and soil fumigants. Phytopathology 61:1415-1417
Jones JP and Woltz SS (1967) Fusarium wilt (race 2) of tomato: effect of lime and micronutrient soil amendments on disease development. Plant Disease Reporter 51:645-648
Jones JP and Woltz SS (1968) Field control of Fusarium wilt (race 2) of tomato by liming and stake disinfestation. Proc Fla State Hort Soc 81:187-191
Jones JP and Woltz SS (1969) Fusarium wilt (race 2) of tomato: calcium, pH, and micronutrient effects on disease development. Plant Disease Reporter 53:276-279
Jones JP and Woltz SS (1970) Fusarium wilt of tomato: interaction of soil liming and micronutrients on disease development. Phytopathology 60:812-813
Jones JP and Woltz SS (1972) Effect of soil pH and micronutrient amendments on Verticillium and Fusarium wilt of tomato. Plant Disease Reporter 56:151-153
Jones JP and Woltz SS (1975) Effect of liming and nitrogen source on Fusarium wilt of cucumber and watermelon. Proc Fla State Hort Soc 88:200-203
Jones JP and Woltz SS (1981) Fusarium-incited diseases of tomato and potato and their control. In: Nelson PE, Toussoun TA and Cook RJ (eds) Fusarium: Diseases, Biology and Taxonomy. Penn State Univ Press, University Park, pp 157-168
Jones JP and Woltz SS (1983) Cultural control of Fusarium wilt race 3 of tomato. Proc Fla State Hort Soc 96:82-83
Keane EM and Sackston WE (1970) Effects of boron and calcium nutrition of flax on Fusarium wilt. Can J Plant Sci 50:415-422
Keim R and Humphrey WA (1984) Fertilizer helps control Fusarium wilt of Hebe. Calif Agr 38:12-14
Locke JC, Marois JJ and Papavizas GC (1985) Biological control of Fusarium wilt of greenhouse-grown chrysanthemums. Plant

Disease 69:167-169
Loffler HJM, Cohen EB, Oolbekkink GT and Schippers B (1986) Nitrite as a factor in the decline of Fusarium oxysporum f. sp. dianthi in soil supplemented with urea or ammonium chloride. Neth J Plant Pathol 92:153-162
Loffler HJM, Koelman A, Nielander HB and Schippers B (1986) Reduced chlamydospore formation and enhanced lysis of chlamydospores of Fusarium oxysporum in soil with added urea or ammonium chloride. Biol Fert Soils 2:1-6
Mackay JHE (1952) Fusarium wilt of tomato - the effects of level of nutrition on disease development. J Austral Inst Agr Sci 17:207-211
Miyake Y and Takahashi E (1983) Effect of silicon on the growth of cucumber plants in soil culture. Soil Sci Plant Nutr 29: 463-471
Morgan KT and Timmer LW (1984) Effect of inoculum density, nitrogen source and saprophytic fungi on Fusarium wilt of Mexican lime. Plant and Soil 79:203-210
Norton JBS (1912) Differences in varieties of fruit and truck crops in reference to disease. Rep Maryland State Hort Soc 15:62-67
Norton JBS (1914) Tomato diseases. Md Agr Exp Sta Bull 180:102-114
Pergola G, Guda CD and Garibaldi A (1979) Fusarium wilt of carnation: effect of soil pH and nitrogen source on disease development. Med Fac Landbouww Rijksuniv Gent 44(1):414-421
Raju BC (1983) Fusarium wilt of mums. Acta Horticulturae 152:65-76
Ramasamy K and Prasad NN (1975) Effect of potassium nutrition on phenol metabolism of melon wilt. Madras Agri J 62:313-317
Sagdullaev MM and Berezhnova W (1974) The effects of phosphorus fertilizers on the physiological properties, yield, and resistance of melons to fusarium wilt. AgroKhimiya 4:36-40
Sarhan ART and Kiraly Z (1981) Control of Fusarium wilt of tomato with an integrated nitrate-lime-fungicide regime. Acta Phytopath Acad Scientiarum Hung 16:9-14
Scher FM and Baker R (1982) Effect of Pseudomonas putida and a synthetic iron chelator on induction of soil suppressiveness to Fusarium wilt pathogen. Phytopathology 72:1567-1573
Schneider RW (1985) Suppression of Fusarium yellows of celery with potassium, chloride, and nitrate. Phytopathology 75:40-48
Scott IT (1924) The influence of hydrogen ion concentration on the growth of Fusarium lycopersici and on tomato wilt. Mo Agri Exp Sta Res Bull 64, p 32
Sherbakoff CD (1915) Fusaria of potatoes. Cornell Univ Agr Exp Sta Memoir 6
Sherwood EC (1923) Hydrogen-ion concentration as related to the Fusarium wilt of tomato seedlings. Amer J Bot 10:537-553
Simeoni LA, Lindsay WL and Baker R (1987) Critical iron level associated with biological control of Fusarium wilt. Phytopathology 77:1057-1061
Smith EF (1899) Wilt diseases of cotton, watermelon, and cowpea. US Dept Agr Bur Plant Ind Bull 17
Spiegel Y and Netzer D (1984) Effect of nitrogen form at various levels of potassium on the Meloidogyne-Fusarium wilt complex in muskmelon. Plant and Soil 81:85-92
Standaert JY, Myttenaere C and Meyer JA (1973) Influence of sodium/calcium ratios and ionic strength of the nutrient solution

on Fusarium wilt of tomato. Plant Sci Letters 1:413-420

Standaert JY, Maraite H, Myttenaere C and Meyer JA (1978) A study of the effect of salt concentration and sodium/calcium ratio in the nutrient medium on the susceptibility of tomatoes to Fusarium wilt. Plant and Soil 50:269-286

Stoddard DL (1947) Nitrogen, potassium, and calcium in relation to Fusarium wilt of muskmelon. Phytopathology 37:875-884

Stoddard EM and Diamond AE (1948) Influence of nutritional level on the susceptibility of tomatoes to Fusarium wilt. Phytopathology 38:670-671

Sun SK and Huang JW (1985) Formulated soil amendment for controlling Fusarium wilt and other soilborne diseases. Plant Disease 69:917-920

Tharp WH and Wadleigh CH (1939) The effects of nitrogen, phosphorus, and potassium nutrition on the Fusarium wilt of cotton. Phytopathology 29:756

Trillas-Gay MI, Hoitink HAJ and Madden LV (1986) Nature of suppression of Fusarium wilt of radish in a container medium amended with composted hardwood bark. Plant Disease 70:1023-1027

Waksman SA (1927) Principles of soil microbiology. The Williams and Wilkins Co Baltimore MD, p 897

Walker JC (1971) Fusarium wilt of tomato. The American Phytopath Soc Monograph No 6, p 56 Am Phytopath Soc St Paul MN

Walker JC and Foster RE (1946) Plant nutrition in relation to disease development. III Fusarium wilt of tomato. Amer J Bot 33:259-264

Wollenweber HW (1913) Studies on the Fusarium problem. Phytopathology 3:24-50

Woltz SS and Engelhard AW (1973) Fusarium wilt of chrysanthemum: Effect of nitrogen source and lime on disease development. Phytopathology 63:155-157

Woltz SS and Jones JP (1968) Micronutrient effects on the *in vitro* growth and pathogenicity of *Fusarium oxysporum* f. sp. *lycopersici*. Phytopathology 58:336-338

Woltz SS and Jones JP (1972) Control of Fusarium wilt of tomato by varying the nutrient regimes in the soil. Phytopathology 62:799 (Abstr)

Woltz SS and Jones JP (1973) Interactions in source of nitrogen fertilizer and liming procedure in the control of Fusarium wilt of tomato. Hort Science 8:137-138

Woltz SS and Jones JP (1973) Tomato Fusarium wilt control by adjustments in soil fertility. Proc Fla State Hort Soc 86: 157-159

Woltz SS and Jones JP (1973) Tomato Fusarium wilt control by adjustments in soil fertility: A systematic approach to pathogen starvation. Agr Res Ed Center Bradenton Res Rept GC1973-7, p 4

Woltz SS and Jones JP (1981) Nutritional requirements of *Fusarium oxysporum*: Basis for a disease control system. In: Nelson PE, Toussoun TA and Cook RJ (eds) Fusarium: Diseases, Biology, and Taxonomy. Penn State Univ Press, University Park, pp 340-349

Woltz SS and Jones JP (1984) Effects of aluminum, lime, and phosphate combinations on Fusarium wilt (race 3) of tomato. Phytopathology 74:629 (Abstr)

Woltz SS and Magie RO (1975) Gladiolus Fusarium disease reduction

by soil fertility adjustments. Proc Fla State Hort Soc 88:559-562
Yoshino M and Hashimoto K (1978) Studies on the ecology of strawberry yellows and its control. Bull Saitama Hort Exp Sta No 7, pp 13-34
Yuen GY and Schroth MN (1986) Inhibition of Fusarium oxysporum f.sp. dianthi by iron competition with an Alcaligenes sp. Phytopathology 76:171-176

SOIL TEMPERATURE INTERACTIONS WITH THE BIOTIC COMPONENTS OF VASCULAR WILT DISEASES

J. Katan
Department of Plant Pathology and Microbiology
The Hebrew University of Jerusalem
Faculty of Agriculture
Rehovot 76100
Israel

INTRODUCTION

Temperature is an environmental factor which strongly affects wilt diseases since it has an influence on each of the three biotic components involved in the production of the syndrome of wilt diseases, i.e., the pathogen, the host plant and soil microorganisms. Surface temperatures vary with latitude and altitude and change cyclically each day and each year. Soil temperature is affected by the presence of vegetation, water content and depth (Griffin 1972) and therefore, may vary under natural conditions, and in extreme cases from far below freezing point to even above 60 C. Thus, temperature determines not only the relative prevalence of soil-borne diseases at different times of the year, and from one year to another, but also their geographical distribution (Garrett 1960). A better understanding of the effect of temperature on the three biotic components of wilt diseases may enable us to better predict disease development and also to develop means for disease control, such as by thermal eradication of the pathogen, plant thermotherapy, disease escape, and through thermal manipulation of biocontrol agents in the soil.

In the earlier days of plant pathology, emphasis was given to the study of pathogen-temperature relationships, usually by determining the growth of the pathogen in culture at various temperatures. The interpretation was that a severe disease is expected at a temperature similar to that required for optimal

NATO ASI Series, Vol. H28
Vascular Wilt Di...

growth of the pathogen. However, this is an oversimplified approach for studying disease-temperature interactions, since the effect of temperature on the two other biotic components might be different from that on the pathogen. Indeed, cases where the optima for pathogen growth and for disease incidence are different, were recorded and will be discussed below. Certain generalization can be made only if the exceptions also are considered. For example, the optimal temperature for the growth of fungi belonging to various formae speciales of Fusarium oxysporum is approximately 27-28 C (Walker 1968), for Verticillium dahliae it is about 21-24 C, and for V. albo-atrum is approximately 21 C (Schnathorst 1981). Frequently, but not always, the diseases incited by the above three pathogens are prevalent in warm climates, cooler climates and cool climate regions, respectively. Pseudomonas solanacearum does not persist in cool regions and its distribution is primarily between 45 N and 45 S latitude (Bruehl 1987).

As early as 1905 Ball studied in Egypt the interactions among temperature, Rhizoctonia solani and the cotton plant (quoted in Bruehl 1987). He concluded that the disease is most severe at temperatures which are lower than the optimal for the growth of the pathogen and the host. This was verified in the following decades by e.g. Hayman (1969). A very comprehensive study of the relationships between soil temperatures and soil-borne seases were initiated by L.R. Jones and his associates at the University of Wisconsin at about 1916 (Walker 1969). These workers used a thermostatically controlled water bath, an apparatus known as the Wisconsin soil-temperature tank. These studies were supplemented by a research on pathogen growth at various temperatures in an attempt to distinguish between the effect of temperature on the pathogen, and on the host's resistance. An extensive study of temperature-pathogen-host damping-off incidence interactions was carried out by Leach (1947). Unfortunately, only a few studies have taken into account also the effect of temperature on the antagonists.

THE RANGE OF TEMPERATURES

A soil-borne pathogen and its surrounding antagonists might be exposed in the soil, under field or greenhouse conditions, to a wide range of temperatures depending on the climatic conditions and crop management. These temperatures might range from far below zero during the winter in temperate zones, to as high as 60 C during summer in arid zones. In addition, the organisms might be exposed to higher temperatures when thermal control measures are employed, e.g. composting (Hoitink 1980), solarization (Katan et al 1976), heat treatment of seeds and other propagation material, and soil steaming (Baker and Cook 1974). An illustration of such temperature relationships is given in Fig. 1 in which a Fusarium pathogen was taken as a model. Nor-

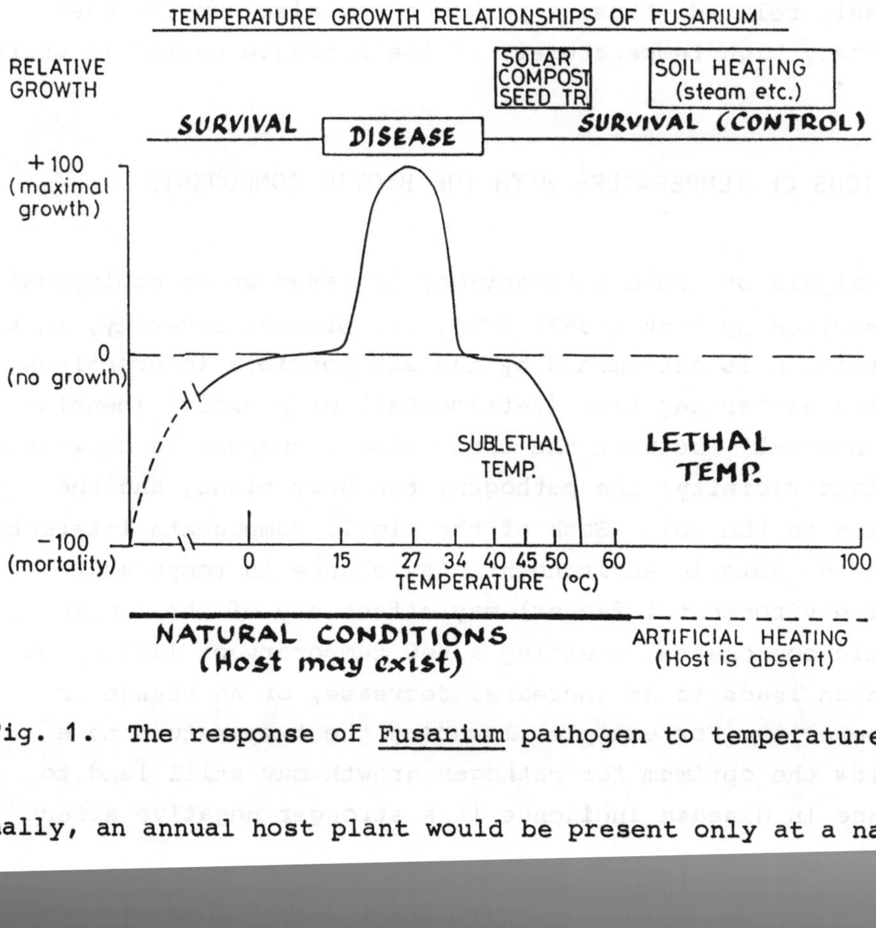

Fig. 1. The response of Fusarium pathogen to temperature.

mally, an annual host plant would be present only at a narrow range of temperature (approximately 5-40 C), although perennial crops might be exposed to lower temperatures during the winter. The range of temperatures in which disease occurs is even narrower, approximately 15-35 C. In the absence of the host, the pathogen is in a passive stage and temperatures, especially at the extreme range, affect its survival and consequently the capacity of the surviving inoculum to incite a disease in the following crop. In the presence of the host, temperatures affect the disease in the course of its development.

The following should be considered when analyzing the effect of temperatures on living organisms:

1. The effect of temperature is mediated by other factors e.g., inoculum density, soil moisture, host genotype, and antagonistic activity. For example, under favourable conditions, the pathogen may grow also at marginal temperatures.
2. The thermal effect (especially at the extreme range) depends on temperature level and on the length of exposure, which are inversely related. Thus, a pathogen may also survive even relatively high temperatures, if the exposure period is short.

INTERACTIONS OF TEMPERATURE WITH THE BIOTIC COMPONENTS

The analysis of these interactions is based on an ecological model described by Park (1963) (Fig. 2). Disease severity, in a given situation is determined by the six possible interactions each being either negative (detrimental) or positive (beneficial) or neutral between the three biotic components coexisting in the plant vicinity: the pathogen, the host plant, and the antagonists in the soil. Each of the biotic components interacts also with the abiotic environment. Any change in temperature (or other environmental factor) may affect any of the biotic or the abiotic components, creating a new temporary or lasting balance, which leads to an increase, decrease, or no change in disease severity. For example, lowering the temperature to a level below the optimum for pathogen growth may still lead to an increase in disease incidence if a stronger negative effect is exerted on the two other components, host resistance and antagonistic activity. These interactions may occur at all sites where any of the involved organisms exist: rhizosphere, rhizoplane, plant residues, root tissues, non-rhizosphere soil, and above ground plant tissues and organs.

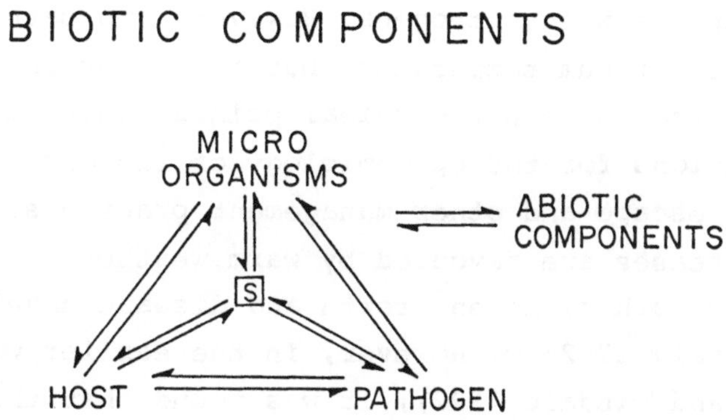

Fig. 2. The interactions of biotic and abiotic components involved in wilt diseases.
S = An external factor interacting with the above components.

a. Effect of temperature on the pathogen

Temperature may affect any activity at any stage of the life cycle of the pathogen, e.g. germination (fungistasis), growth in the soil, ectotrophic growth on the root surface, tissue penetration, growth and reproduction inside plant tissues, formation of toxins and enzymes, formation of resting structures and survival (longevity). Temperature may affect various activities of the same pathogen in opposite directions. Moreover, it may affect growth of the pathogen (or any other activity) in different ways when tested in different media, e.g. culture <u>versus</u> soil. Thus, measuring linear growth of the fungus on an agar medium at various temperatures does not necessarily reveal the total effect of temperature on pathogen. Growth of a pathogen starts at a minimum temperature, increases to an optimum, and then declines and stops at a maximum temperature. These are the cardinal temperatures. The optimum temperature is usually only a little below the maximum for growth. Deverall (1965) reviewed this concept and gave examples of fungi where the optimum temperatures for growth may vary depending on the period of incubation.

The optimum temperature serves as a convenient parameter for comparing between pathogens. However, two pathogens can have the same optimum temperature but may nevertheless differ in regard to the two other cardinal points. This may have important implications for the epidemiology of the disease, and its control by escape and other management practices. Various Fusarial wilt diseases are favoured by warm weather, the optimum temperature for both pathogen growth and disease development being approximately 27-28 C. However, in the earlier works in Wisconsin (Jones and Tisdale 1922), it was found that the minimum temperatures for the development of disease in flax, cabbage and tomato are about 14, 16-17 and 19 C, respectively. This finding is also in agreement with data on the geographical distribution of the diseases in the USA. Since the critical points for the three pathogens are similar, these investigations stimulated further studies of the possible effects of temperature on the host and on soil microorganisms.

b. Effect of temperature on host resistance

Host tissues are not merely a growth medium for the pathogen. Since temperature (similar to all environmental factors) affects the growth and metabolism of the host plant, it may consequently affect its resistance or susceptibility to the pathogen. The main difficulty in determining the effect of temperature on the host stems from the fact that exposing an inoculated plant to a certain temperature regime may affect not only the pathogen and the host, but also the soil microorganisms if the test is carried out in nonsterile soil. This possibility was not always taken into account in studies of temperature-disease relationships.

Various approaches were developed in order to elucidate the

effect of temperature on host resistance. For example, if the effect of temperature on pathogen growth under axenic conditions (providing that the system represents the natural conditions) is different from that on disease development, then an effect on host resistance may be assumed. Further evidence can be obtained if the effect of temperature on disease is determined with various hosts inoculated with the same pathogen. Such a classical study was done by Dickson (1923) with seedling blight of cereals caused by Gibberella zeae. Seedling blight developed to the greatest extent in corn at low temperatures (12 C or below) and in wheat seedlings inoculated with the same pathogen at relatively high temperatures (16-28 C). The optimum for fungal growth on agar is at 24-28 C. Thus, enhanced disease development occurs at temperatures far from the optimum for root growth of each host. A correlation between reduced host resistance and adverse thermal conditions for plant growth cannot be regarded as a universal rule. The optimum temperature for the operation of the various defence mechanisms in the host plant does not necessarily coincide with that for plant growth. On the contrary, certain resistance mechanisms operate mainly under stress conditions. Leach (1947) carried out extensive studies of the effect of temperature on incidence of damping-off diseases of various hosts inoculated with pathogens e.g. Rhizoctonia and Pythium. He found that disease incidence was affected by the ratio between velocity of seedling emergence and growth rate of the pathogen. These studies explained why the incidence of damping-off in watermelon inoculated with Pythium was high at low temperatures in spite of the fact that growth of the pathogen was also adversely affected at low temperature (though to a lesser extent). This classical study clearly demonstrated that measuring the effect of temperature on pathogen growth is not sufficient for understanding the effect on disease incidence. However, enhanced plant growth at certain temperature does not necessarily indicate greater resistance. In the above study, the possible effect of temperature on antagonistic activity was not studied.

Another approach for elucidating the effect of temperature on the host is to determine the effect of temperature on certain chemical or structural mechanisms which play a role in resistance.

Dickson et al (1923) found that at the optimum temperature for growth of wheat and corn, lignification of the sheath is the most rapid. Resistance of banana plants to Fusarium wilt results from physical occlusion of the plant vascular system by induced gel and tyloses which prevent distribution of the pathogen and the spread of its spores (Beckman et al 1962). Thus, the invaded vessels are rapidly sealed off in advance of the invading fungus. The occlusion was rapid and continuous in the resistant Lactan cultivar at all tested temperatures (21, 27 and 34 C), whereas in the susceptible Gros Michel cultivar, disease resistance and the occlusion mechanism were temperature-dependent. They were both the least pronounced at 27 C, intermediate at 21 C and high at 34 C. Temperature may also affect the level of root exudates, which in turn affects pathogen and soil microorganism growth. Hayman (1969) found that at low temperatures both damping-off (caused by R. solani) and seed exudation were enhanced, as compared with higher temperatures.

Temperature may affect predisposition of the plant to diseases. Tomato plants were grown at various temperatures and subsequently inoculated with Fusarium and maintained at 28 C. The plants were most favourably predisposed to Fusarium wilt at 28 C, less so at 24 and 20 C, and the least at 16 C (Foster and Walker 1947). The temperature for predisposition to a certain disease is not necessarily identical to that which is favourable for disease development after inoculation. Predisposition may occur at extreme low or high temperatures. Water stress, which is enhanced at high temperatures, may also predispose plants to diseases. Hot water treatments (e.g. at 48-49 C for 30 seconds) of the roots and aerial parts of tomato plants, prior to inoculation, induced resistance to Fusarium wilt (Anchise et al 1985). This protection probably results from physiological changes in the host and resembles induced immunity.

c. Effect of temperature on antagonists

There is no doubt that temperature affects the activity and reproduction of antagonists in the soil, as of any other soil organism. However, in only a few studies dealing with the effect of temperature on the incidence of soil-borne diseases, was the direct effect of temperature on antagonists investigated. In a study of this question with root rot of wheat caused by Gaeumanomyces graminis Henry (1932) showed that maximum disease occured at 14-18 C in non-sterile soil and decreased as the temperature was increased, whereas in sterile soil disease was uniformly severe at all temperatures above the range 14-27 C. The conclusion was that disease was suppressed at high temperatures due to increased antagonism from soil microflora, in spite of the fact that the pathogen is able to grow well at these temperatures. Another approach for elucidating the effect of temperature on antagonists is to measure the various parameters of antagonisms, e.g., substrate colonization, competition, lysis, production of antibiotics. Temperature may affect the number of antagonists or their antagonistic capacity or both. This may result in an effect on inoculum density or inoculum potential. The use of soils artificially infested with pathogen grown on a rich medium (e.g. cornmeal) may give misleading results.

TEMPERATURE - GENOTYPE INTERACTIONS

Resistance of plants belonging to various genotypes can be affected by temperature in various ways, not necessarily following the same trend observed with susceptible genotypes. Generally, resistance of plants with monogenic or a high level of resistance is much less affected by temperature than resistance of plants with polygenic or partial resistance. Thus, resistance to Fusarium wilt of the resistant banana cultivar Lacatan (Beckman et al 1962) and of the monogenic resistant tomato cultivar Red Current (Foster and Walker 1947) was very high or complete at all temperatures tested. Temperature had a similar effect on wilt severity with both the susceptible tomato cultivar Bonny Best and the intermediate resistant Marglobe, though to a lesser extent with the latter cultivar. Similarly, resistance to Fusarium wilt in cabbage is more affected by temperature in the moderately level resistant plants (type B) than in the monogenic resistant ones (type A) (Walker and Smith 1930). At 24 C, all susceptible and multigenic resistant plants became diseased, whereas the monogenic resistant ones survived. At higher temperatures even monogenic resistant plants become partially diseased, and at low temperatures multigenic resistant ones may survive. At lower temperatures, incidence of Fusarium wilt in tomatoes decreased in the susceptible cultivar Marmande but increased in the resistant cultivars Homestead and Roma VF (II)and even more so with the heterozygous (Ii) hybrids (Alon et al 1974). This was regarded as a reduced penetrance of resistance at low temperatures. Fusarium wilt of chrysanthemum increased with temperature and was greatest at 35 C, which is rather unusual (Gardiner et al 1987). However, disease severity in the highly resistant cultivar was much less affected by temperature. Verticillium wilt development in susceptible mint (Mentha piperita) was slower at 10 than at 25 C, but the opposite was true with the highly resistant M. crispa (Berry and Thomas 1961). The defence mechanisms operating in plants of various genotypes might be different, and thus they may differ also in their response to temperature in the presence of pathogens.

INTERACTION OF TEMPERATURE WITH OTHER FACTORS

The effect of temperature on the biotic components of wilt diseases is not absolute since it is affected by other factors e.g. plant genotype, as discussed earlier. Inoculum density is such an important factor especially when dealing with susceptible cultivars. Fusarium wilt of tomato increased as temperatures rose from 15 to 27 C when inoculum concentration was medium, but reached nearly 100% at all temperatures tested, when inoculum concentration was high (Alon et al 1974). The effect of temperature on the incidence of Fusarium wilt of muskmelon depends also on soil moisture (Miller 1945). Thus, in moist soil disease level was the highest at low temperatures, while in dry soil it was high at both low and high temperatures. Such interactions may explain the different or even contradictory results obtained by various researchers. Irrigation affects both soil moisture and temperature. Apparently, there is some compensation between factors affecting disease, i.e., an excess of one factor may compensate for a deficiency in another. According to Cook and Baker (1983), plant-pathogenic fungi have the ability to grow at unusually high temperatures when the water potential is low.

An additional interaction exists between soil and air temperatures. The latter affects soil temperature, the temperature of the foliage, the water economy of the plant and its response to water stress. Soil temperature affected internal stem temperature of tomato plants and consequently Verticillium wilt incidence (Edgington and Walker 1957). Therefore, the relation between air temperature and incidence of wilt disease is very complicated. Moreover, the large fluctuations in daily temperatures complicate the evaluation of the relationship between air temperature and wilt incidence. Pullman and DeVay (1982) studied the relation between inoculum density of V. dahliae and mid-September incidence of wilt symptoms in cotton. Instead of using Julian days for relating the time course to disease progress they used physiological time-degree days determined as an accumulation of daily temperature above 54.5 F (11.9 C), the cotton developmental threshold.

Using computerized model systems to simulate soil heating under various climatic conditions and different agricultural practices is now common (Mahrer 1979). Such models might also be adapted to study the effect of temperature on soil-borne pathogens.

EFFECT OF TEMPERATURE ON PATHOGEN SURVIVAL AND THE USE OF HIGH TEMPERATURES TO CONTROL WILT DISEASES

Soil-borne pathogens survive in the absence of the host as resting structures either free in the soil or embedded in plant residues or in seeds or vegetative propagation material. These propagules may be naturally exposed to high temperatures for long or short periods. The upper layers of soil in hot climate regions may be solar heated to even 60 C (Brock 1978). The heat sensitivity of the propagules is much greater under moist conditions than dry ones (Katan 1981).

Increasing temperatures in order to control pathogens in the soil or in plant materials, by solar heating or by burning, was attempted centuries ago in India, Egypt and by the Romans, as well as in modern times (Cook and Baker 1983; Katan 1981). There are several approaches to pathogen control by heating: (a) heating seeds or vegetative propagation material at temperatures of 40-60 C; (b) composting organic residues; (c) heating the soil for short periods, usually by steam at 70-100 C; and (d) soil solarization, i.e., covering the soil with transparent polyethylene for approximately 30-40 days during the hot season, thereby heating the soil and killing the pathogens (Katan et al 1976; Katan 1981). Changing the sowing date to a time when the temperature is less favourable for the pathogen, may enable us to reduce disease incidence. Manipulating temperature may also enable us to increase host resistance.

Sublethal heating might be detrimental to the pathogens in the long run, this since the pathogen might be weakened at such temperatures and rendered vulnerable to the attack of antagonists in the soil (Katan 1981; Lifshitz et al 1983).

It is essential to analyze the effect of temperature on each of the biotic and abiotic factors involved in the formation of disease syndrome. Such information is of great value for developing nonchemical means to control wilt diseases.

REFERENCES

Alon A, Katan J and Kedar N (1974) Factors affecting penetrance of resistance to Fusarium oxysporum f. sp. lycopersici in tomatoes. Phytopathology 64:455-461

Anchisi M, Gennasi M and Matta A (1985) Retardation of Fusarium wilt symptoms in tomato by pre- and post-inoculating treatments of the roots and aerial parts of the host in hot water. Physiol Plant Pathol 26:175-183

Baker KF and Cook RJ (1974) Biological control of plant pathogens. Freeman, San Francisco

Beckman CH, Halmos S and Mace ME (1962) The interaction of host, pathogen and soil temperature in relation to susceptibility to Fusarium wilt of bananas. Phytopathology 52:134-140

Berry SZ and Thomas CA (1961) Influence of soil temperatures, isolates, and method of inoculum on resistance of mint to Verticillium wilt. Phytopathology 51:169-174

Brock TD (1978) Thermophilic microorganisms and life at high temperatures. Springer Verlag, New York

Bruehl GW (1987) Soilborne plant pathogens. McMillan, London

Cook RJ and Baker KF (1983) The nature and practice of biological control of plant pathogens. APS, St Paul

Deverall BJ (1965) The physical environment of fungal growth. I. Temperature. In: Ainsworth GC and Sussman AS (eds) The fungi - an advanced treatise Vol 1, Academic Press, New York, p 543

Dickson JG (1923) Influence of soil temperature and moisture on the development of seedling-blight of wheat and corn caused by Gibberella saubinetti.J Agr Res 23:837-870

Dickson JG, Eckerson SH and Link KP (1923) The nature or resistance to seedling blight of cereals. Proc Nat Acad Sci Washington, 9:434-439

Edgington LV and Walker JC (1957) Influence of soil air temperature on Verticillium wilt on tomato. Phytopathology 47:594-598

Foster RE and Walker JC (1947) Predisposition of tomato to Fusarium wilt. J Agr Res 74:165-185

Gardiner DC, Horst RK and Nelson PE (1987) Symptom enhancement of Fusarium wilt in chrysanthemum by high temperatures. Plant Disease 71:1106-1109

Garrett SD (1960) Biology of root-infecting fungi. Cambridge University Press, Cambridge

Griffin DM (1972) Ecology of soil fungi. Chapman and Hall, London

Hayman DS (1969) The influence of temperature on the exudation of nutrients from cotton seeds and on preemergence damping-off by Rhizoctonia solani. Can J Bot 47:1663-1669

Henry AW (1932) Influence of soil temperature and soil sterilization on the reaction of wheat seedlings to Ophiobolus graminis. Sacc. Can J Res 7:198-203

Hoitink HAJ (1980) Composted bark, a lighweight growth medium with fungicidal properties. Plant Disease 64:142-147

Jones LR and Tisdale WB (1922) The influence of soil temperature upon the development of flax wilt. Phytopathology 12:409-413

Katan J (1981) Solar heating (solarization) of soils for control of soilborne pests. Ann Rev Phytopathol 19:211-236

Katan J, Greenberger A, Alon H and Grinstein A (1976) Solar heating by polyethylene mulching for the control of soilborne diseases caused by soilborne pathogens. Phytopathology 66:683-688

Leach LD (1947) Growth rates of host and pathogen as factors determining the severity of pre-emergence damping-off. J Agr Res 75:161-179

Lifshitz R, Tabachnik M, Katan J and Chet I (1983) The effect of sublethal heating of sclerotia of Sclerotium rolfsii. Can J Microbiol 29:1607-1610

Mahrer Y (1979) Prediction of soil temperature of a soil mulch with transparent polyethylene. J Appl Meteor 18:1263-1267

Miller JJ (1945) Studies on the Fusarium of muskmelon wilt. II. Infection studies concerning the host of the organism and the effect of environment on disease incidence. Can J Res 23: 166-187

Park D (1963) The ecology of soil-borne fungal disease. Ann Rev Phytopathol 1:241-258

Pullman GS and DeVay JE (1982) Epidemiology of Verticillium wilt of cotton: a relationship between inoculum density and disease progression. Phytopathology 72:549-554

Schnathorst WC (1981) Life cycle and epidemiology of Verticillium. In Mace ME, Bell AA and Beckman CH (eds) Fungal wilt diseases. Academic Press, New York, pp 81-111

Walker JC (1969) Plant Pathology. 3rd edit, McGraw Hill

Walker JC and Smith R (1930) Effect of environmental factors upon the resistance of cabbage to yellows. H Agr Res 41:1-15

MICROBIAL POPULATIONS AND MECHANISMS DETERMINING SOIL-SUPPRESSIVENESS TO FUSARIUM WILTS

J. Louvet
Institut National de la Recherche Agronomique
Station de Recherches sur la Flore Pathogène dans le Sol
17 rue Sully
F21034 Dijon Cédex
France

INTRODUCTION

Basically a disease results from the intimate interaction between a plant and a pathogen. However it is also determined by a lot of environmental factors which affect the plant and the pathogen.

In case of diseases due to soil-borne plant pathogens it is obvious that the soil environment plays an important role to enhance or reduce disease in a crop. The best example is the existence of disease suppressive soils in which the disease severity or incidence is obviously low in spite of the presence of inoculum and conditions favourable for disease development.

Different soils suppressive to the most important soil-borne diseases have been reported in numerous areas in the world (Schneider 1982). Two main examples have been extensively studied during the two last decades; they concern soils suppressive respectively to take-all (Cook 1985) or to Fusarium-wilts (Alabouvette 1986). Generally soils became suppressive to take-all after continuous monoculture of susceptible wheat, a phenomenon well known as take-all decline. On the contrary, soil-suppressiveness to Fusarium-wilts is not correlated with the culture of the host-plant. It has been described as a "constitutive suppressiveness" in contrast with the "acquired suppressiveness". In fact, it is difficult to assess that the Fusarium-wilt suppressiveness is independant of crop rotation (Sneh et al 1987), but its observed

NATO ASI Series, Vol. H28
Vascular Wilt Diseases of Plants

level in fields appears stable with time. Therefore, most of the early studies were focused on relationships between the physico-chemical properties of soils and their level of suppressiveness to Fusarium wilts (Stover 1962; Stotzky and Martin 1963).

Another characteristic of all Fusarium-wilt suppressive soils described in the world is that the suppressiveness is higly specific to Fusarium-wilts due to different formae speciales but ineffective against other non wilt Fusarium diseases or diseases caused by other root pathogens (Smith and Snyder 1971; Toussoun 1975; Alabouvette et al 1980).

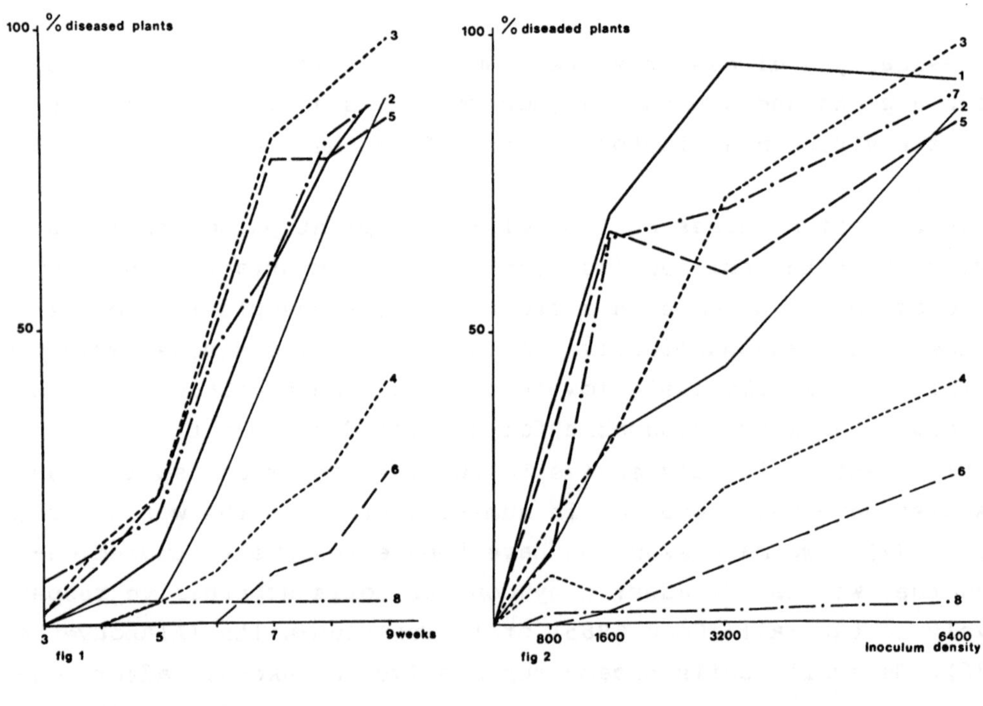

Methods for assessment of level of soil suppressiveness (soils 1-8).
Fig. 1. Evolution of percentage of wilted plants within time after infestation of the soils with only one inoculum concentration (6,400 CFU/g soil).
Fig. 2. Evolution of percentage of wilted plants after 9 weeks of culture with different infestations of each soil (800, 1,600, 3,200, 6,400 CFU/g soil) (adapted from Rouxel et al 1988).

Therefore it is possible to choose any plant susceptible to Fusarium-wilt to study soil suppressiveness. For example flax has been frequently retained as host-plant to assess accurately the level of soil suppressiveness under standardized conditions (Alabouvette et al 1982). Moreover the specificity of the suppressiveness indicates that all formae speciales of Fusarium oxysporum proceed to the same point during the different phases preceding invasion of the vascular tissues. The mechanisms of soil suppressiveness should apply during these phases, and studies of soil suppressiveness will help to better understand the processes of infection by vascular wilt Fusarium.

ASSESSMENT OF SOIL-SUPPRESSIVENESS TO FUSARIUM-WILTS

Since early studies of Smith and Snyder (1971), pathologist's attention has been focused on a few examples of soils with conspicuous Fusarium-wilt suppressing effect. However it is obvious that every soil has a potential for some microbiological disease suppression and this natural biological control can be considered as a continuum (Alabouvette et al 1982; Linderman et al 1983).

Introduction into various soils of the same inoculum under standardized conditions favourable to disease expression do not induce the same disease incidence on a population of susceptible plants. The relationship of disease incidence to time allows us to compare the behaviour of soils (Fig. 1). But the use of a unique inoculum concentration does not permit the differentiation of soils with close levels of suppressiveness. Another way is to infest soils with several concentrations of a given inoculum. Comparison of the percentage of diseased plants after a given time of culture did not improve the differentiation of close levels of suppressiveness (Fig. 2). But it is possible to correlate disease incidence with both time and inoculum concentrations by calculating the survival probability (Fig. 3). These survival probabilities can be compared by using a statistical model which permits a further calculation of one value which represents the level of suppressiveness of each soil and permits a ranking of soils in relation to their suppressive effect (Fig. 4) (Corman et al 1986; Rouxel et al 1988).

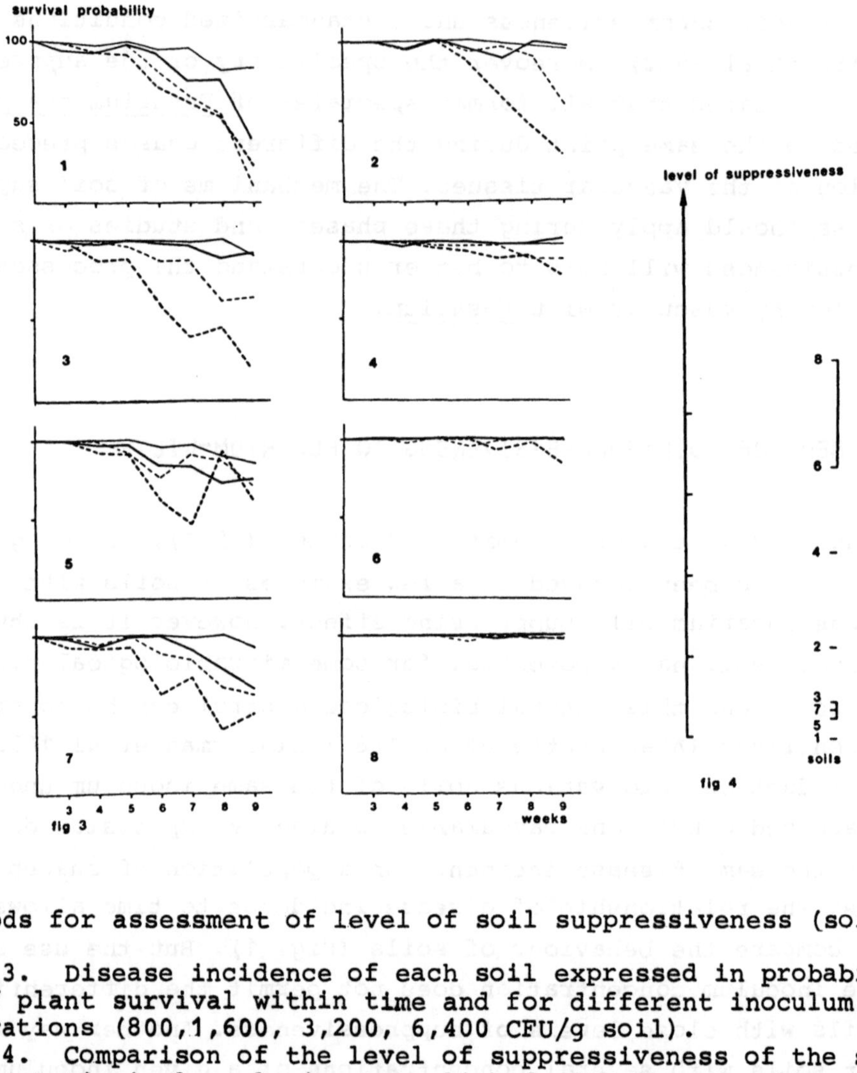

Methods for assessment of level of soil suppressiveness (soils 1-8).
Fig. 3. Disease incidence of each soil expressed in probability of plant survival within time and for different inoculum concentrations (800,1,600, 3,200, 6,400 CFU/g soil).
Fig. 4. Comparison of the level of suppressiveness of the soils after statistical analysis. Data linked with the same line are not significantly different at 0,95 level (adapted from Rouxel et al 1988).

Suppressiveness to Fusarium-wilts reflects the capacity of a soil to limit the saprophytic and/or the pathogenic abilities of the pathogen. It is a characteristic of the soil. Therefore, to assess the level of soil suppressiveness it is important to avoid any modification of the soil, for example by addition of a food base with the inoculum.

MICROBIOLOGICAL NATURE OF FUSARIUM-WILT SOIL-SUPPRESSIVENESS

Fusarium-wilt suppressiveness of soils studied so far is basically microbiological, it means it results from more or less complex microbial interactions between the pathogen and other saprophytic microbial populations. At first the role of the microflora was demonstrated indirectly by establishing that the suppressive effect disappears consequently to the destruction of living organisms induced by various biocidal treatments such as steam, methylbromide, γ-rays (Louvet et al 1976; Rouxel et al 1977; Alabouvette et al 1977; Scher and Baker 1980). Moreover suppressiveness can be transmitted by mixing a small quantity of disease suppressive soil into a previously heat-treated conducive soil (Komada and Ezuka 1970; Louvet et al 1976; Scher and Baker 1980; Schneider 1984). Only microorganisms are likely able to multiply and thus to confer a high level of suppressiveness to a conducive soil.

The fact that suppressiveness is microbiological in nature does not mean that soil physico-chemical properties are not involved in the mechanism. The current state of knowledge, however, does not allow evaluation of the role of these properties. One of the first correlations established between suppressiveness and physico-chemical factors concerned the presence of smectite-type clays in soils suppressive to Fusarium wilt of bananas in Central America (Reinking and Manns 1933; Stotzky and Martin 1963). However no progress has been made recently in this field, on the contrary several examples show that some sandy-loam soils are also suppressive to Fusarium wilts (Smith and Snyder 1971; Tamietti and Alabouvette 1986). Another correlation indicates that suppressiveness is associated with high pH (Louvet et al 1976; Scher and Baker 1980). PH does not play a direct role on Fusarium but influences the microbial interactions. Competition for nutrients depends on pH values because cation availability is greatly dependent on this factor. Moreover other characteristics of soils, such as clay and sand contents, aeration or carbon-nitrogen ratio, determine the activity of the total biomass which plays an important role in the "general suppression" (Cook and Baker 1983; Chaussod et al 1986).

INVOLVED MICROORGANISMS

Having indirectly demonstrated the role of the microflora in the Fusarium-wilt suppressiveness, workers try to identify microorganisms specifically involved. The aim is to establish correlations between induced modifications of some microbial populations and variations of the level of soil suppressiveness. Accurate methods to selectively modify the microbial balance in the soil are not available. Application of moderate heat-treatments, irradiations or antibiotics can be used for their selective effect on sensitive groups of microorganisms and addition of specific nutrients can be used to stimulate some groups of microorganisms. For example heat treatments at temperatures between 50 to 70 C were used to demonstrate the importance of non pathogenic Fusarium in the mechanisms of suppression in the Châteaurenard soils. Their involvement was confirmed, as in Koch's postulates, demonstrating their introduction into the steamed suppressive soil re-established suppressiveness (Rouxel et al 1979). But most of the microorganisms (Arthrobacter sp., Bacillus sp., Streptomyces sp., Pseudomonas sp.) proposed to be involved in mechanisms of suppression have not be retained after such an analytical experimentation. Usually workers showed that their massive introduction into a conducive soil increased the suppressive level (Fig. 5). This type of experiments did not demonstrate that these microorganisms are really responsible for suppressiveness in natural suppressive soils.

A literature review shows that only two types of microorganisms, non pathogenic Fusarium and fluorescent Pseudomonas (Scher and Baker 1980), have been extensively studied and their mechanism of action begins to be elucidated.

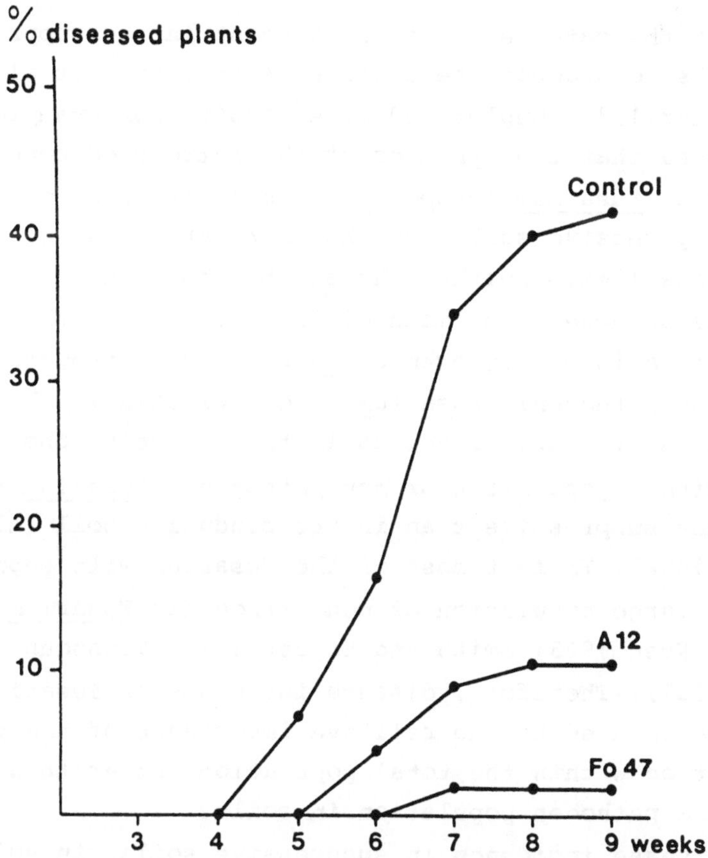

Fig. 5. Induction of soil-suppressiveness to Fusarium wilt of flax by addition of Pseudomonas putida A12 and non pathogenic Fusarium oxysporum F.o.47 (adapted from Lemanceau 1988a).

MECHANISMS OF SUPPRESSION

In their first book on biological control, Baker and Cook (1974) proposed three types of mechanisms to explain the disease suppressiveness of some soils: - the pathogen cannot establish- it establishes but fails to produce disease - it establishes and causes disease at first but diminishes with continuation of growing the same crop.

Results show that in most of the examples studied so far the mechanism of suppression to Fusarium wilts belongs to the second type in which the pathogen is still present in the suppressive soil but fails to produce the disease (Alabouvette et al 1984; Scher and Baker 1980; Dupler and Baker 1984). For example it was established that the dynamics of the introduced population of the pathogen Fusarium oxysporum was similar in a conducive and in the suppressive soil from Châteaurenard showing that the suppressiveness limits neither the establishment nor the persistance of the pathogen. The main difference between the two soils concerned the ratio of pathogenic Fusarium/non pathogenic Fusarium. When the pathogenic Fusarium is established in the suppressive soil at the same level as in the conducive one, it is confronted with a population of non pathogenic Fusarium ten times greater in the suppressive than in the conducive soil (Alabouvette et al 1984). In fact most of the Fusarium-wilt suppressive soils host a large population of non pathogenic Fusarium (Wensley and Mac Keen 1963; Smith and Snyder 1972; Toussoun 1975; Sneh et al 1987). Therefore, disease incidence of Fusarium wilts appears to be related to the relative importance of the pathogenic population within the total population rather than to the density of the pathogen population in soils.

The low disease incidence in suppressive soils, in spite of the presence of the pathogen, can be explained by the antagonism due to the soil microflora which limits its activity. In fact at the present time no example shows the role of either antibiosis or hyperparasitism as mechanisms of soil suppressiveness to Fusarium wilts. On the contrary there are clear evidences that competition for nutrients is involved. It was demonstrated that modifications of the level of nutrients available for the pathogen induce modifications of the level of soil suppressiveness. At present time two main nutrients, carbon and iron, have been proposed as limiting elements (Lemanceau, in this book). Addition of EDDHA which limits the concentration of iron available for Fusarium induces a decrease of the percentage of diseased plants. On the contrary addition of FeEDTA providing iron available for Fusarium induces an increase of disease incidence (Scher and Baker 1982; Lemanceau et al 1988b) (Fig. 6). Addition of glucose providing energy for Fusarium induces an increase of disease incidence in both conducive and suppressive soils but a higher concentration of glucose is needed in the suppressive soil indicating that the competition for carbon is more intense in suppressive than in conducive soils (Alabouvette et al 1985b; Lemanceau et al 1988b) (Fig. 7).

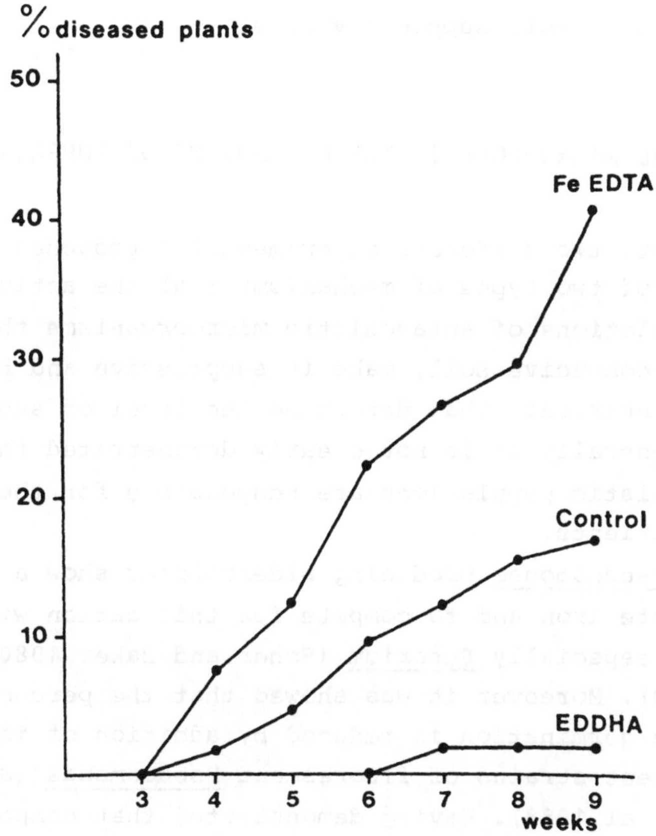

Fig. 6. Modification of suppressiveness to Fusarium wilt of flax by addition of FeDTA or EDDHA (adapted from Lemanceau 1988b).

In order to explain these induced modifications of disease incidence the behaviour of Fusarium in soils was followed by microscopic observations. Addition of available carbon and iron induces an increase in the percentage of chlamydospore germination in both conducive and suppressive soils. However, as the percentage of chlamydospore germination is lower in the suppressive than in the conducive soils it is necessary to add higher concentration of nutrients in the suppressive than in the conducive soils to induce the same level of chlamydospore germination and produce the same saprophytic growth of Fusarium (Sneh et al 1984; Alabouvette et al 1985a). These results showed that the competition for nutrients determines the level of activity of the pathogen in soils and consequently plays an important role in the mechanisms of soil suppressiveness.

ROLE OF THE TOTAL MICROFLORA IN THE MECHANISMS OF SUPPRESSION

As pointed out, two different experimental approaches led to the proposal of two types of mechanisms : a) the activity of specific populations of antagonistic microorganisms that, introduced in a conducive soil, make it suppressive and b) the competition for nutrients that determine the level of suppressiveness. But generally it is not clearly demonstrated that the specific antagonistic populations are responsible for the competition for nutrients.

Fluorescent Pseudomonas producing siderophores show a strong ability to chelate iron and to compete for this cation with other microorganisms, especially Fusarium (Scher and Baker 1980; Kloepper et al 1980). Moreover it was showed that the percentage of chlamydospore germination is reduced by addition of iron chelators or efficient strains of fluorescent Pseudomonas (Sneh et al 1984; Park et al 1988). Having demonstrated that competition for iron exists in suppressive soils the American workers attributed the suppressiveness of these soils to the activity of fluorescent Pseudomonas. However, there are no data concerning the density and the efficiency of the natural populations of Pseudomonas in the suppressive soils. In the case of the soil from Châteaurenard it has been recently established that even if competition for iron exists, populations of fluorescent Pseudomonas are neither more abundant nor more competitive in the suppressive soil than in a conducive control (Lemanceau et al 1988c).

On the contrary, analyses of the fungal populations showed that the densities of the Fusarium populations are generally

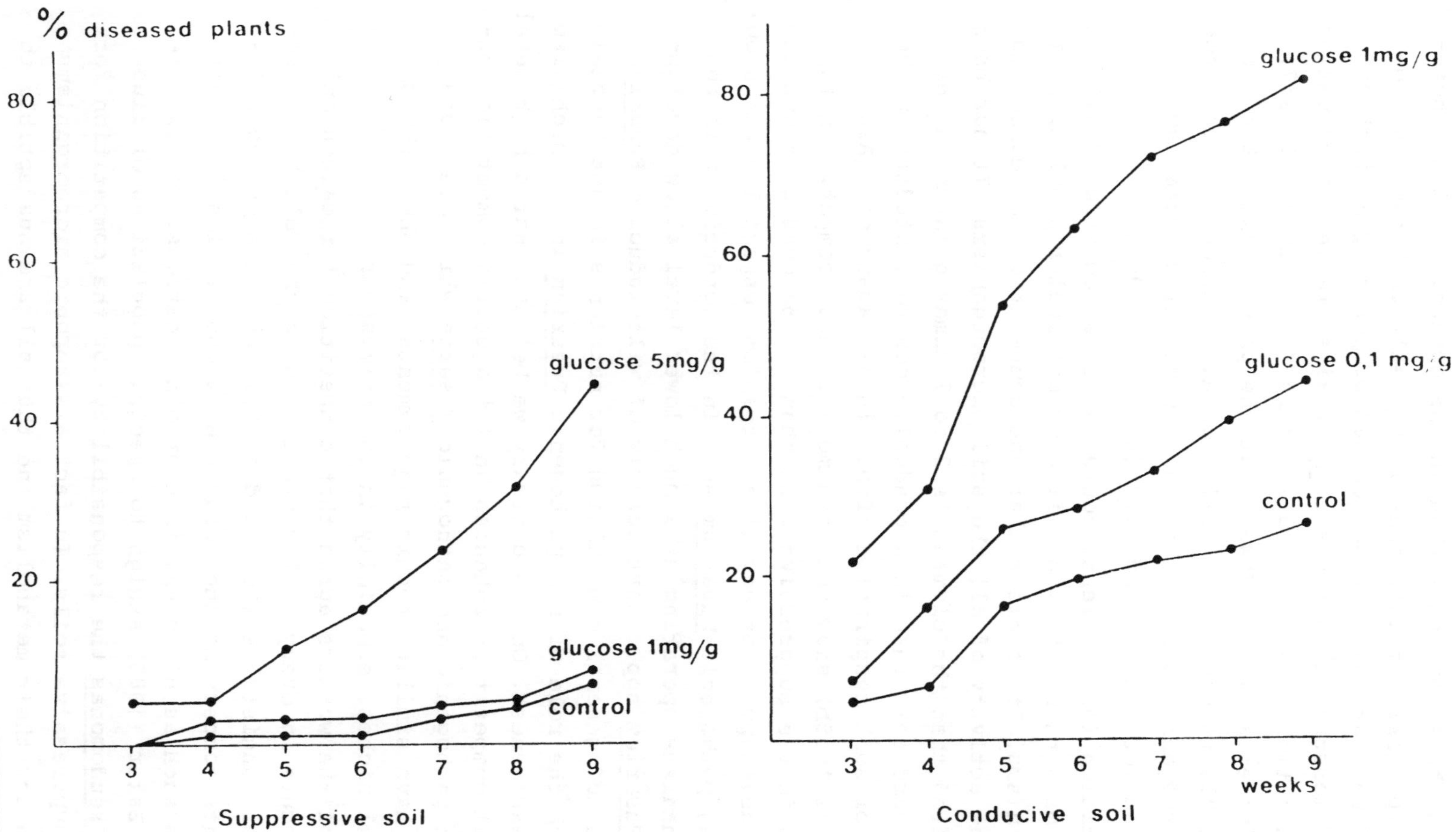

Fig. 7. Modification of soil-suppressiveness to Fusarium wilt of flax by increasing levels of glucose. Infestation of suppressive and conducive soils respectively with 1 x 10^4 and 0·5 x 10^4 CFU of Fusarium oxysporum f. sp. lini (adapted from Lemanceau 1988b).

greater in the suppressive than in the conducive soils. Therefore a hypothesis of intrageneric competition between pathogenic and non pathogenic Fusarium was proposed. Quantitative studies demonstrated that Fusarium strains are really competing for colonization of steam-treated soils (Zegerman et al 1987; Couteaudier et al 1988). However at the present time it has not been possible to establish a relationship between the abundance of non pathogenic Fusarium and the domonstrated mechanism of competition for glucose and iron (Lemanceau 1988a).

Competition for nutrients could be better correlated with total biomass than with a specific or particular population of microorganisms. It is clear that the competition for carbon is due to the activity of all the soil microorganisms. It has been demonstrated that the biomass is 3 to 7 times greater in the Châteaurenard soil than in a conducive control. Studies on the kinetics of soil respiration after glucose amendment (Alabouvette et al 1985b) show that the more intense competition for nutrients in the suppressive soil than in the conducive one limits the development of any given organism, especially the slow-responding pathogenic Fusarium spp. In the conducive soil the same mechanisms operating at a much lower level allow development of Fusarium spp., particularly of wilt-inducing Fusarium. This level of general competition for nutrients is not inconsistent with the role of non pathogenic Fusarium in the mechanisms of suppressiveness. On the contrary we believe that a high level of general competition enhances an intrageneric competition between non pathogenic and pathogenic Fusaria which are closely related, have similar trophic requirements, and share the same ecological niches, especially in the rhizosphere.

Most of the workers agree that competition for carbon and energy occurs in every soil where it causes the well known phenomenon of fungistasis (Lockwood 1981). But there are some disagreements concerning the occurence of competition for other nutrients such as nitrogen, iron or other cations. For example, Elad and Baker (1985) assign to specific populations of fluorescent Pseudomonas the responsibility for the competition for iron in suppressive soils. In fact all aerobic microorganisms need iron for their metabolism and thus all are susceptible to competition for this cation. As in the case of competition for carbon, competition for iron results from the activity of both the total microbial population and specific populations of Pseudomonas which produce siderophores with a high affinity for iron.

PROBLEMS AND PERSPECTIVES

Recent studies on Fusarium-wilt suppressive soils have proven the importance of general mechanisms of competition for nutrients in the establishment of the microbial balance in soils. Results show that the same type of interactions exist in all soils, but the intensity of these competitive interactions determine the level of soil-suppressiveness to Fusarium wilts.

The population of the pathogenic Fusarium represents only a tiny part of the total microbial populations of soils. Thus it is not clear how a general mechanism of competition concerning all the microflora can limit the activity of the pathogenic population in such a way that the disease is suppressed.

In fact, to better understand the relationships between the pathogen and the rest of the microflora it is necessary to study these interactions at close proximity to the root tips where colonization occurs (Huisman 1982; Gerik and Huisman 1985). It is clear that the nutrient status of soil is quite different in bulk soil and in the rhizosphere where the plant exudes nutrients available for microorganisms. Consequently, in the rhizosphere, chlamydospores of the pathogenic Fusarium are induced to germinate and the germ tubes have necessarily to grow saprophytically in soil, even for a very small distance, to reach the susceptible infection sites. According to Huisman (1982), to achieve a systemic infection a chlamydospore of Fusarium has only a few hours to germinate in response to root exudates and reach the passing root zone of elongation, where it can penetrate undifferentiated tissue. During this saprophytic phase of its life cycle the pathogenic Fusarium is very susceptible to the antagonistic influence of the microflora. Thus to realize a successful vascular infection the pathogenic Fusarium is limited both by restricted nutrient availability and by the very short time available to it to reach the moving infection court.

Most of the researches have concerned the studies on microbial activity in soil itself. It is clear that it is now necessary to increase the efforts to study the interactions between the pathogen and the saprophytic microflora in the rhizospheric soil and at the root surface. Until now most of the studies dealing with the behaviour of Fusarium strains have been conducted in raw soils amended with nutrients, iron chelators or antagonistic microorganisms. However Elad and Baker (1985), studying the effect of iron competition on the germination of Fusarium chlamydospores, showed that some iron chelators did not produce the same effect in raw soil and in rhizospheric soil. Moreover, strains of Fusarium oxysporum show different abilities to colonize the root cortex that is independent of their ability to colonize soil and to germinate in the rhizosphere (Nagao et al 1988). Then Schneider (1984) established that non pathogenic Fusarium could enhance or suppress root infection by the pathogen by means of "parasitic competition" in colonizing the root cortex in the same manner that it offers "saprophytic competition" for development in soil. These two mechanisms of competition do not exclude each other; on the contrary both could exist in the same soil and contribute to the soil suppressiveness.

Beside the mechanisms of competition acting outside or at the surface of the host-plant, cross protection involving non pathogenic Fusarium oxysporum colonizing the vascular tissues is another mechanism which controls Fusarium wilts (Ogawa and Komada 1984; Tamietti and Alabouvette 1986). However at the present time cross protection has never been proposed to explain Fusarium-wilt soil suppressiveness.

In conclusion, in his book "The Nature of the Wilt Diseases of Plants" (1987), C.H. Beckman proposed after Talboys (1957) that the progress of the Fusarium-wilt diseases could be divided in distinct phases including two determinative phases in the host-parasite interactions, respectively outside and inside vascular system. We propose to consider the preinfectious phase as another determinative one in which the success or the failure of the pathogen in germinating, competing in the rhizosphere and penetrating the root is determined by soil-parasite interactions. As pointed out very recently by Park et al (1988) in reference to Mitchell (1976) : "The impact of interaction of microbial populations in the root-soil interface on prepenetration activity, infection and resulting disease is a critical area that has received relatively little attention... (and) ...the interactions that occure are of major significance".

REFERENCES

Alabouvette C, Rouxel F and Louvet J (1977) Recherches sur la résistance des sols aux maladies. III. Effets du rayonnement sur la microflore d'un sol et sa résistance à la fusariose vasculaire du melon. Ann Phytopathol 9:467-471

Alabouvette C, Rouxel F and Louvet J (1980) Recherches sur la résistance des sols aux maladies. VI. Mise en évidence de la spécificité de la résistance d'un sol vis-à-vis des fusarioses vasculaires. Ann Phytopathol 12:11-19

Alabouvette C, Couteaudier Y and Louvet J (1982) Comparaison de la réceptivité de différents sols et substrats de culture aux fusarioses vasculaires. Agronomie 2:1-6

Alabouvette C, Couteaudier Y and Louvet J (1984) Recherches sur la résistance des sols aux maladies. IX. Dynamique des populations du Fusarium spp. et de Fusarium oxysporum f. sp. melonis dans un sol résistant et dans un sol sensible aux fusarioses vasculaires. Agronomie 4:729-733

Alabouvette C, Couteaudier Y and Louvet J (1985a) Recherches sur la résistance des sols aux maladies. XI. Etude comparative du comportement des Fusarium spp. dans un sol résistant et un sol sensible aux fusarioses vasculaires enrichis en glucose. Agronomie 5:63-68

Alabouvette C, Couteaudier Y and Louvet J (1985b) Recherches sur la résistance des sols aux maladies. XII. Activité respiratoire dans un sol résistant et un sol sensible aux fusarioses vasculaires enrichis en glucose. Agronomie 5:69-72

Alabouvette C (1986) Fusarium wilt suppressive soils from the Châteaurenard region : review of a 10 year study. Agronomie 6:273-284

Alabouvette C, Couteaudier Y and Lemanceau P (1986) Nature of intrageneric competition between pathogenic and non-pathogenic Fusarium in a wilt suppressive soil. In : Swinburne TR (ed) Iron, siderophores, and plant diseases NATO ASI Series Plenum Press 165-178

Baker R, Mauper CL and Maurer AA (1967) Ecology of plant pathogens in soil. VII. Mathematical models and inoculum density. Phytopathology 57:662-666

Baker KF and Cook RJ (1974) Biological control of plant pathogens. Ann Phytopathol Soc St Paul, MN, pp 433

Beckman CH (1987) The nature of wilt diseases of plants. Ann

Phytopathol Soc St Paul, MN, pp 175
Chaussod R, Nicolardot B, Catroux and Chretien J (1986) Relations entre les caractéristiques physico-chimiques et microbiologiques de quelques sols cultivés. Sci Sol 24:213-226
Cook RJ and Baker KF (1983) The nature and practice of biological control of plant pathogens. Ann Phytopathol Soc St Paul, MN, pp 539
Cook RJ (1985) Biological control of plants pathogens with special reference to the take-all fungus in suppressive soils. Plant Bot Bull Taiwan Roc 27:179-198
Corman A, Couteaudier Y, Zegerman M and Alabouvette C (1986) Réceptivité des sols aux fusarioses vasculaires: méthode statistique d'analyse des résultats Agronomie 6:751-757
Couteaudier Y, Zegerman M and Alabouvette C (1988) Evaluation de l'aptitude à la compétition des souches de Fusarium utilisables en lutte biologique. CR Agric Fr 73 7:129-136
Dupler M and Baker R (1984) Survival of Pseudomonas putida a biological control agent in soil. Phytopathology 74:195-200
Elad Y and Baker R (1985) Influence of trace amounts of cations and siderophore-producing pseudomonas on germination of Fusarium oxysporum chlamydospores. Phytopathology 75:1047-1052
Gerik JS and Huisman OC (1985) Mode of colonization of roots by Verticillium and Fusarium. In: Parker CA, Rovira AD, Moore KJ, Wong PTW and Kollmorgen JS (eds) Ecology and management of soil-borne plant pathogens. Ann Phytopathol Soc St Paul, MN 80-83
Huisman OC (1982) Interrelations of root growth dynamics to epidemiology of root-invading fungi. Ann Rev Phytopathol 20:303-327
Kloepper JW, Leong J, Teinzte M and Schroth MN (1980) Pseudomonas siderophores : a mechanism explaining disease suppressive soils. Curr Microbiol 4:317-320
Komada H and Ezuka A (1970) Ecological study of Fusarium diseases of vegetable crops. I. Survival of pathogenic fusaria in different soil types. Res Prog Rept Tokai-Kinki Natl Agric Exp Stn 6:1-6
Lemanceau P (1988a) Réceptivité des sols aux fusarioses vasculaires: étude cinétique des théories proposées. Thèse Université Lyon I, pp 95
Lemanceau P, Alabouvette C and Couteaudier Y (1988b) Recherches sur la résistance des sols aux maladies. XIV. Modifications du niveau de réceptivité d'un sol résistant et d'un sol sensible aux fusarioses vasculaires en réponse à des apports de fer ou de glucose. Agronomie 8:155-162
Lemanceau P, Samson R and Alabouvette C (1988c) Recherches sur la résistance des sols aux maladies. XV. Comparaison des populations de Pseudomonas fluorescents dans un sol résistant et un sol sensible aux fusarioses vasculaires. Agronomie 8:243-249
Linderman RG, Moore LW, Baker KF and Cooksey DA (1983) Strategies for detecting and characterising systems of biological control of soil-borne plant pathogens. Plant Disease 67:1058-1064
Lockwood JL (1981) Exploitation competition. In: Wicklow DT and Carroll G (eds) The fungal community. Its organization and role in the ecosystem M Dekker :319-350
Louvet J, Rouxel F and Alabouvette C (1976) Recherches sur la

résistance des sols aux maladies. I. Mise en évidence de la nature microbiologique de la résistance d'un sol au dévelopement de la fusariose vasculaire du melon. Ann Phytopathol 8: 425-436

Mitchell JE (1976) The effect of roots on the activity of soil-borne plant pathogens. In: Heitfuss RH and Williams PH (eds) Physiological Plant Pathology Springler-Verlag Berlin 104-128

Nagao H, Couteaudier Y and Alabouvette C (1988) Disparity in the ability of Fusarium strains to colonize steamed soil and flax roots (to be published)

Ogawa K and Komada H (1983) Biological control of Fusarium wilt of sweet potato by non-pathogenic Fusarium oxysporum. Ann Phytopath Soc Japan 50:1-9

Park CS, Paulitz TC and Baker R (1988) Biocontrol of Fusarium wilt of cucumber resulting from interactions between Pseudomonas putida and non pathogenic isolates of Fusarium oxysporum. Phytopathology 78:190-194

Reinking OA and Manns MM (1933) Parasitic and other fusaria counted in tropical soils. Z Parasitenkunde Inf 6:23-75

Rouxel F, Alabouvette C and Louvet J (1977) Recherches sur la résistance des sols aux maladies. II. Incidence de traitements thermiques sur la résistance microbiologique d'un sol à la fusariose vasculaire du melon. Ann Phytopathol 9:183-192

Rouxel F, Alabouvette C and Louvet J (1979) Recherches sur la résistance des sols aux maladies. IV. Mise en évidence du rôle des Fusarium autochtones dans la résistance d'un sol à la fusariose vasculaire du melon. Ann Phytopathol 11:199-207

Rouxel F, Alabouvette C and Masson JP (1988) Intérêt de la modélisation pour l'étude de la réceptivité des sols aux maladies d'origine tellrique : exemples d'applications à la hernie des crucifères et aux fusarioses vasculaires. In Symposium Modelisation et Protectio des Cultures. CR Acad Agric Fr 73: 137-149

Scher M and Baker R (1980) Mechanism of biological control in a Fusarium suppressive soil. Phytopathology 70:412-417

Scher M and Baker R (1982) Effect of Pseudomonas putida and a synthetic iron chelator on induction of soil suppressiveness to Fusarium-wilt pathogens. Phytopathology 72:1567-1573

Schneider RW (1982) (ed) Suppresive soils and plant disease. Ann Phytopathol Soc St Paul, MN, pp 88

Schneider RW (1984) Effects of non pathogenic strains of Fusarium oxysporum on celery root infection by F. oxysporum f. sp. apii and a novel use of the lineweaver-burk double reciprocal plot technique. Phytopathology 74:646-653

Smith SN and Snyder WC (1971) Relationship of inoculum density and soil types to severity of Fusarium wilt of sweet potato. Phytopathology 61:1049-1051

Sneh B, Dupler M, Elad Y and Baker R (1984) Chlamydospore germination of Fusarium oxysporum f. sp. cucumerinum as affected by fluorescent and lytic bacteria from Fusarium suppressive soil. Phytopathology 74:1115-1124

Sneh B, Pozniak D and Salomon D (1987) Soil suppressiveness to Fusarium wilt of melon, induced by repeated croppings of resistant varieties of melons. J Phytopathology 120:347-354

Stotzky G and Martin RT (1963) Soil mineralogy in relation to the spread of Fusarium wilt of banana in Central America. Plant Soil 18:317-338

Stover RH (1962) Fusarial wilt of bananas and other Musa species. Commonwealth Mycol Inst Phytopathol Paper 4:pp 117
Talboys PW (1957) The possible significance of toxic metabolites of Verticillium albo-atrum in the development of hop wilt symptoms. Trans Brit Mycol Soc 40:415-427
Tamietti G and Alabouvette C (1986) Résistance des sols aux maladies. XIII. Rôle des Fusarium oxysporum non pathogènes dans les mécanismes de résistance d'un sol de Noirmoutier aux fusarioses vasculaires. Agronomie 6:541-548
Toussoun TA (1975) Fusarium-suppressive soils. In: Bruehl (ed) Biology and control of soil-borne plant pathogens. Ann Phytopathol Soc St Paul, MN, pp 145-151
Wensley RN and Mac Keen CD (1963) Populations of Fusarium oxysporum f. sp. melonis and their relation to the wilt potential of two soils. Can J Microbiol 9:237-249
Zegerman M, Couteaudier Y, Alabouvette C and Pave A (1987) Modélisation du processus de compétition nutritive entre microorganismes: application aux Fusarium. Agronomie 7:797-806

ROLE OF COMPETITION FOR CARBON AND IRON IN MECHANISMS OF SOIL SUPPRESSIVENESS TO FUSARIUM WILTS

P. Lemanceau

INRA - Station de Recherches sur la Flore Pathogène dans le Sol
17 rue Sully - F21034 Dijon Cédex
France

INTRODUCTION

Many examples of soils suppressive to Fusarium wilts have been described. Two of them are extensively studied (Louvet in the same book): the suppressive soil from Salinas Valley (California) and the suppressive soil from Chateaurenard (France).

If all the workers agree on the fact that the suppressiveness is microbiological in nature and involves competition for nutrients between the pathogen and all or a part of the saprophytic microflora, the characteristics of this competition differ upon the model studied.

The suppressiveness of the Chateaurenard soil was attributed to competition for energy. The high level and activity of the microbial biomass of this soil act as a nutrient sink and then deprive the pathogenic _Fusarium_ of carbon (Alabouvette 1983). On the other hand, the suppressiveness of the Salinas Valley soil was attributed to competition for iron due to the activity of populations of fluorescent _Pseudomonas_ producing iron-acquiring chelators (siderophores) more efficient than those produced by _Fusarium_ (Scher and Baker 1982).

Considering that the chemical properties of both suppressive soils show some similarities: high pH and low available iron content, it seemed interesting to assess the validity of the hypothesis of competition for iron as a mechanism of suppressiveness in the Chateaurenard soil. It was unfortunately impossible to determine the role of competition for carbon in the suppressiveness of the soil from Salinas Valley. All the experiments were, therefore, only carried out with the suppressive soil from Chateaurenard in comparison to a conducive soil from Carquefou (France). This soil was chosen for its conduciveness to Fusarium wilts and for its physico-chemical properties highly different from those of the soil from Chateaurenard, especially low pH (pH = 5.5) and high content of available iron (ammonium oxalate extraction).

As described by Baker (1968), the effect of the nutrient competition on the level of suppressiveness was assessed by recording the mortality of plants grown in infested soil, after modification of the nutrient availability by amendements. As the level of suppressiveness is clearly higher in the soil from Chateaurenard than in the soil from Carquefou, the suppressive and the conducive soils were infested with respectively 10,000 and 5,000 colony forming units (CFU) of Fusarium oxysporum f. sp. lini/g of soil. Following the experimental methods used by Scher and Baker (1982), the concentration of iron available for the pathogen in soils was modified by addition of iron chelators (EDTA, EDDHA) and their chelates (FeEDTA, FeEDDHA). In the same way, the soils were amended with glucose to increase their carbon content (Alabouvette et al 1985). Then the soil treatments were placed in pots with six replicates per treatment. The pots were seeded with 10 sensitive flax seeds (Linum sativum "Hera"). The plants were cultivated during 9 weeks. Percentages of wilted plants were recorded every week.

INVOLVEMENT OF COMPETITION FOR IRON IN MECHANISMS OF SOIL SUPPRESSIVENESS

To demonstrate that competition for iron is involved in mechanisms of soil suppressiveness to Fusarium wilts, the level of iron available for the pathogen was increased by addition of FeEDTA or decreased by addition of EDDHA. According to Lindsay (1974), FeEDTA shows a low stability, thus Fusarium can use iron bound to EDTA. On the contrary, EDDHA shows a high affinity for iron and then decreases the concentration of iron available for the Fusarium.

a) Effect of increased concentrations of iron available for *Fusarium* on wilt incidence

Figure 1 shows the effect of addition of increasing amounts of FeEDTA into the suppressive (Fig. 1A) and conducive (Fig. 1B) soils. In both soils, the increased levels of iron available for microorganisms are correlated with an increase of wilted plants. The effectiveness of FeEDTA amendements to increase disease incidence is greater in the conducive than in the suppressive soil; it is mainly obvious for the lowest concentration of FeEDTA (100 µg/g).

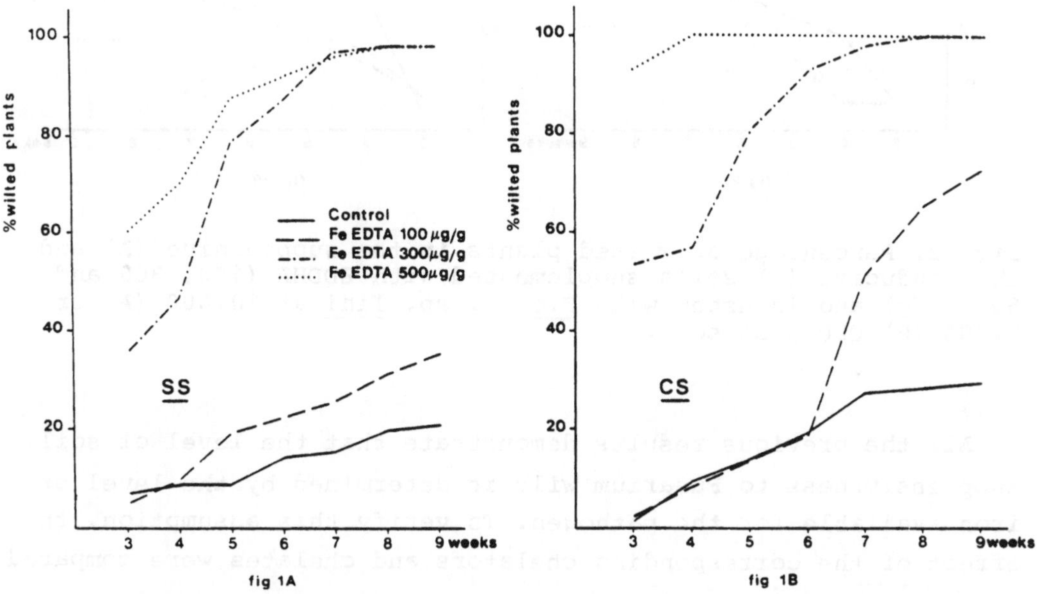

Fig. 1. Percentage of wilted plants in the suppressive (A) and the conducive (B) soils supplemented with FeEDTA (100, 300 and 500 µg/g) and infested with F.o. f. sp. *lini* at 10,000 (A) or 5,000 (B) CFU/g of soil.

b) Effect of decreased concentrations of iron available for *Fusarium* on wilt incidence

Figure 2 shows the effect of addition of increasing concentrations of EDDHA into the suppressive (Fig. 2A) and conducive (Fig. 2B) soils. In both soils, the decreased levels of iron available for microorganisms are correlated with decreased percentages of wilted plants.

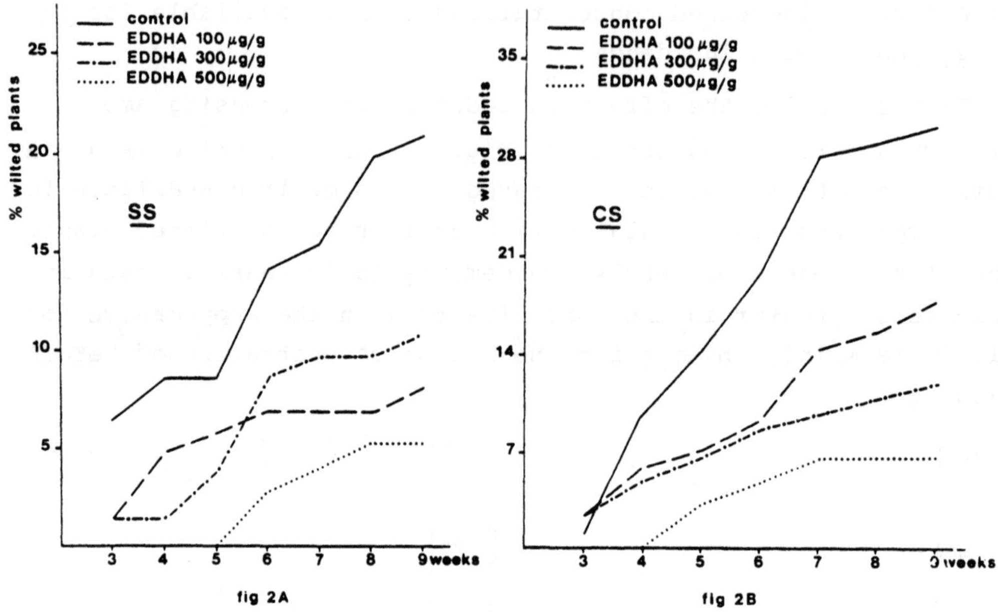

Fig. 2. Percentage of wilted plants in the suppressive (A) and the conducive (B) soils supplemented with EDDHA (100, 300 and 500 µg/g) and infested with F.o. f. sp. lini at 10,000 (A) or 5,000 (B) CFU/g of soil.

All the previous results demonstrate that the level of soil suppressiveness to Fusarium wilt is determined by the level of iron available for the pathogen. To verify this assumption, the effect of the corresponding chelators and chelates were compared in the suppressive soil.

Figure 3A shows that both EDTA and FeEDTA induce a similar increase in the percentage of wilted plants grown in infested suppressive soil. The cause of the plant mortality differs with the nature of the amendement. The increased mortality of plants is either due to Fusarium wilt in presence of FeEDTA or to phytotoxicity in presence of EDTA. As a matter of fact, EDTA induces mortality of plants even when grown in absence of pathogenic inoculum (Fig. 3B). The phytotoxic effect of EDTA was already pointed out by Van Driel (1964). On the other hand the plant mortality induced by FeEDTA is always greater in infested than in non infested soil. In fact, the comparison of the percentage of wilted plant in infested (Fig. 3A) and non infested

(Fig. 3B) soil leads to the conclusion that the mortality incidence in the infested suppressive soil is totaly (FeEDTA 100 µg/g) or partially (FeEDTA 300 µg/g) due to an increase of iron avaibility for the pathogenic Fusarium.

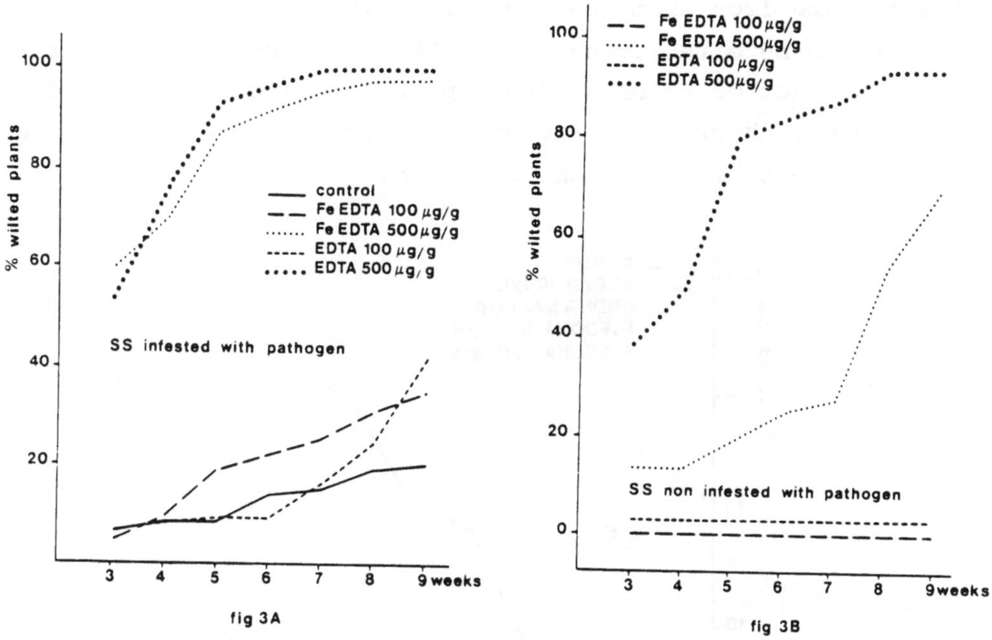

Fig. 3. Percentage of wilted plants in the suppressive soil supplemented with EDTA or FeEDTA and infested (A) or not (B) with F.o. f. sp. lini at 10,000 CFU/g of soil.

All these results demonstrate that it is possible to modify the level of soil suppressiveness by modification of the iron avaibility in soils. Thus the intensity of competition for iron, indeed, determines the level of soil suppressiveness to Fusarium wilts.

Figure 4 shows that addition of FeEDDHA induces a similar decreasing effect on disease incidence as EDDHA. Moreover, at the same concentration, effectiveness of FeEDDHA is greater than that of EDDHA. These results are in agreement with those of Scher and Baker (1982) but need to be carefully discussed. It has already been shown that FeEDDHA is efficient in reducing germination of chlamydospores in plant rhizosphere but inefficient in raw soil (Sneh et al 1984; Elad and Baker 1985). As both plant (Lindsay 1974) and fluorescent Pseudomonas (Scher et al 1984) are able to use iron chelated by EDDHA, these workers proposed the hypothesis that, once stripped of its iron, the chelator is able to bind iron from the iron pool. This hypothesis can explain the same effectiveness of EDDHA and FeEDDHA but not the greater effect of FeEDDHA. One hypothesis is that FeEDDHA favors the growth of the plant and thus contributes to a decrease in its susceptibility to the pathogen.

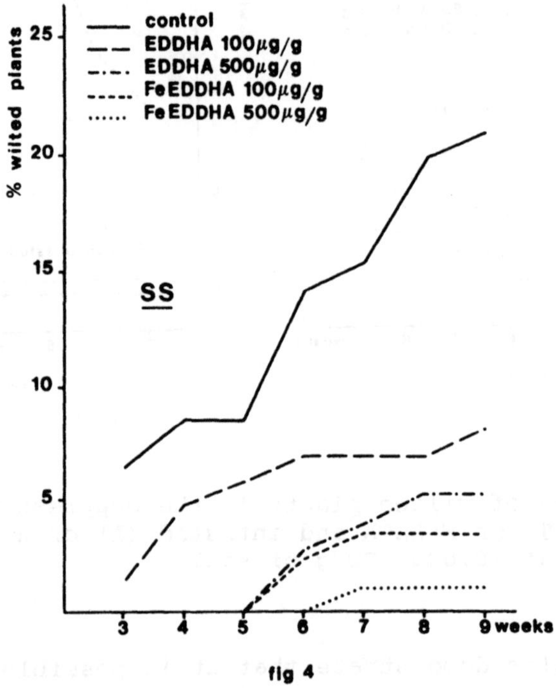

Fig. 4. Percentage of wilted plants in the suppressive soil supplemented with EDDHA or FeEDDHA and infested with F.o. f. sp. lini at 10,000 CFU/g of soil.

INVOLVEMENT OF COMPETITION FOR CARBON IN MECHANISMS OF SOIL SUPPRESSIVENESS

To show that competition for carbon is involved in the mechanisms of soil suppressiveness to Fusarium wilts, it would have been interesting to record the effect of increased and decreased levels of carbon available for the pathogen on disease incidence. Unfortunately, it is difficult to reduce this level experimentally. Therefore, the amount of available carbon was only increased by addition of increasing concentrations of glucose.

Figure 5 shows that addition of carbon induces an increase in the percentage of diseased plants in both suppressive (Fig. 5A) and conducive (Fig. 5B) soils. However, it is necessary to add a greater amount of glucose to reduce the level of suppressiveness in the suppressive than in the conducive soil. In fact, only the highest concentration of glucose (5 mg/g) makes the suppressive soil conducive. Meanwhile in the conducive soil, disease incidence is directly correlated with increasing levels of carbon added.

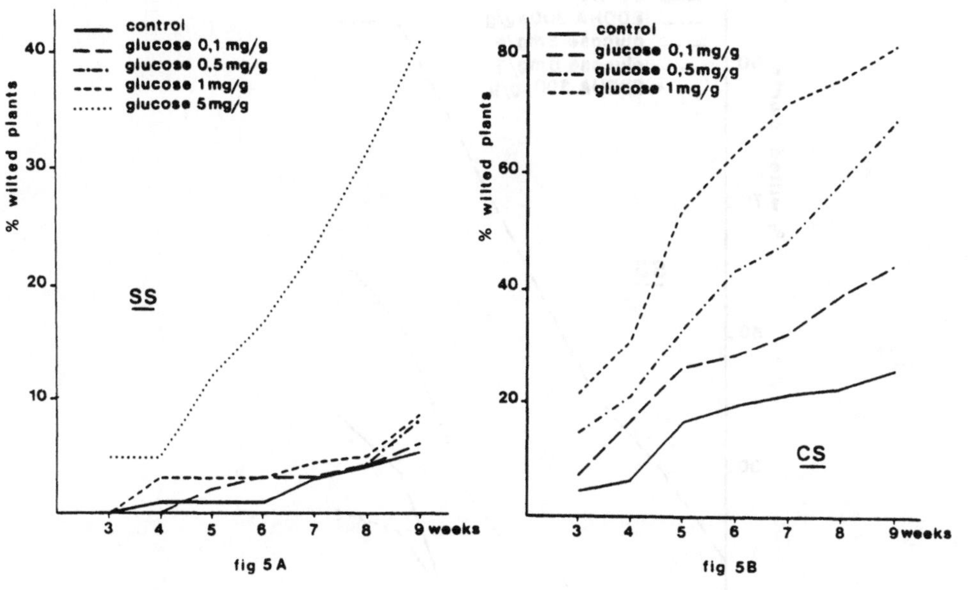

Fig. 5. Percentage of wilted plants in the suppressive (A) and conducive (B) soils supplemented with increasing levels of glucose and infested with F.o. f. sp. lini at 10,000 (A) or 5,000 (B) CFU/g of soil.

All these results demonstrate that the intensity of competition for carbon determines the level of soil suppressiveness to

Fusarium wilts.

COMPARATIVE INVOLVEMENT OF COMPETITION FOR CARBON AND IRON IN MECHANISMS OF SOIL SUPPRESSIVENESS

To assess the respective importance of competition for carbon and competition for iron in the suppressive soil, the concentration of these nutrients was modified by adding glucose and EDDHA together.

Figure 6 shows that, in accordance with previous results, addition of glucose induces an increase of disease incidence, whereas the addition of EDDHA induces a decrease of disease incidence. When glucose and EDDHA were added together, the percentage

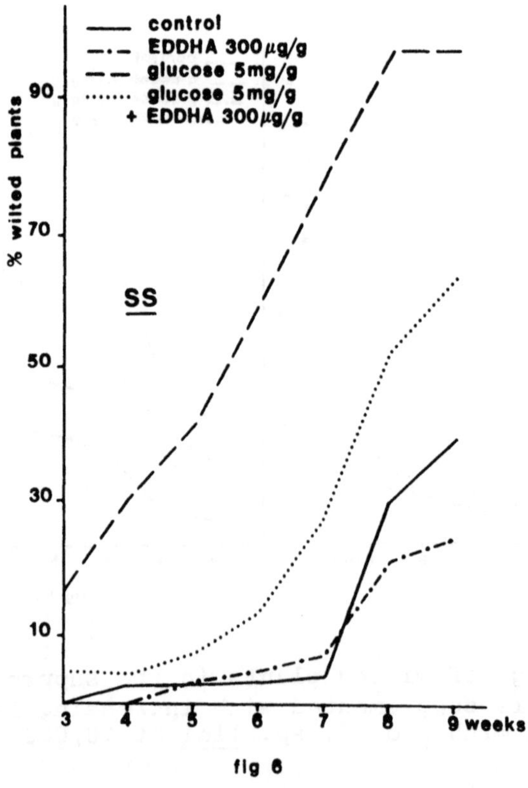

Fig. 6. Percentage of wilted plants in the suppressive soil supplemented with EDDHA and glucose and infested with F.o. f. sp. lini at 10,000 CFU/g of soil.

of diseased plants was intermediate between glucose treatment and EDDHA treatment. These results show that addition of available carbon induces an increase of disease incidence even when the level of available iron is reduced by addition of EDDHA. Thus, even if both competition for carbon and competition for iron play a role in the mechanisms of soil suppressiveness, competition for carbon seems to be the most important.

DISCUSSION

All these results presented above demonstrate that the level of soil suppressiveness to Fusarium wilts is correlated with both the levels of carbon and iron available for the pathogen. In fact, these results allow to extend the role of competition for iron already described in the suppressive soil from Salinas Valley to the suppressive soil from Chateaurenard (Scher and Baker 1982). Furthermore, the presented results show that this competition for iron takes place in both suppressive and conducive soils but with a greater intensity in the suppressive soil. This work confirms also the correlation between the intensity of competition for carbon and the level of soil suppressiveness already described by Alabouvette et al (1985). Thus, competition for carbon and competition for iron are expressed in the same soils and play a role in the mechanisms of soil suppressiveness. But it appears that the activity of the pathogen depends in the first place on the quantity of available carbon. As a matter of fact, addition of glucose always induces an increase of disease incidence. But, the concentration of iron available affects the efficiency of carbon metabolism of the pathogen. Indeed, disease incidence in soil amended with glucose is reduced by addition of EDDHA which decreases the availability of iron.

Suppressiveness and conduciveness are determined by the same microbial interactions which depend on the nutrient competition, though this competition takes place with a greater intensity in suppressive than in conducive soils (Alabouvette 1987). The intensity of the competition for a given nutrient depends on both the natural availability of this nutrient in soil and the density and activity of the microflora able to metabolize it. Consequently, levels of suppressiveness should be correlated with the nutrient content of the soil and with the microbial biomass. In fact, it has been demonstrated that the biomass is 3 to 7 times greater in the suppressive than in the conducive soils and furthermore was more responsive to added carbon although carbon contents of the two soils are similar (Alabouvette et al 1985; Lemanceau 1988).

It was also established that the concentration of iron available is 15 times greater in the conducive than in the suppressive soil (Lemanceau et al 1988a). This higher availability of iron for the pathogen has to be related to the lower pH as the total iron content is similar in these two soils. But, even in these conditions of low pH and high iron availability, it was possible to induce competition for iron at a level which can make the conducive soil suppressive. A major role was attributed to fluorescent _Pseudomonas_ in the iron competition (Kloepper et al 1980; Baker et al 1986). Indeed, these bacteria produce siderophores with a very high affinity for siderophores (Meyer and Abdallah 1978) and then contribute to decrease the availability of iron for the _Fusarium_. But, as all the aerobic microorganisms need iron for their metabolism, it seems obvious that the intensity of competition for iron is also determined by the density and the activity of the aerobic microflora (Lemanceau et al 1988b).

Up to now, the role of the plant in the mechanisms of soil suppressiveness has been neglected. As soil suppressiveness is assessed through disease incidence, it would be necessary to take into account the plant. Indeed the plant increases the availability of some nutrients by its exudates but reduces the availability of others by absorbing minerals and water. In fact the results obtained show that in reality, the addition of FeEDDHA does not only modify soil suppressiveness but also plant susceptibility.

REFERENCES

Alabouvette C (1983) La réceptivité des sols aux fusarioses vasculaires: rôle de la compétition nutritive entre microorganismes. Thèse Dr es Sciences Naturelles Université de Nancy, p 158

Alabouvette C (1986) Fusarium wilt suppressive soils from the Châteaurenard region: review of a 10 year study. Agronomie 6: 273-284

Alabouvette C, Couteaudier Y and Louvet J (1985) Recherches sur la résistance des sols aux maladies. XII. Activité respiratoire dans un sol résistant et un sol sensible aux fusarioses vasculaires enrichis en glucose. Agronomie 5:69-72

Baker R (1968) Mechanisms of biological control of soil-borne pathogens. Ann Rev Phytopath 6:263-294

Baker R, Elad Y and Sneh B (1986) Physical, Biological and host factors in iron competition in soils 77-84.In:Swinburne TR (ed) Iron,siderophores and plant diseases. NATO ASI Series A vol 117, p 351

Elad Y and Baker R (1985) Influence of trace amounts of cations and siderophore-producing pseudomonads on germination of *Fusarium oxysporum* chlamydospores. Phytopathology 75:1047-1052

Kloepper JW, Leong J, Teintze M and Schroth MN (1980) *Pseudomonas* siderophores: a mechanism explains disease suppressive soils. Curr Microbiol 4:317-320

Lemanceau P (1988) Réceptivité des sols aux fusarioses vasculaires: étude critique des théories proposées. Thèse Université Lyon I, pp 95

Lemanceau P, Alabouvette C and Couteaudier Y (1988a) Recherches sur la résistance des sols aux maladies. XIV. Modifications du niveau de réceptivité d'un sol résistant et d'un sol sensible aux fusarioses vasculaires en réponse à des apports de fer ou de glucose. Agronomie 8:155-162

Lemanceau P, Samson R and Alabouvette C (1988b) Recherches sur la résistance des sols aux maladies. XV. Comparaison des populations de *Pseudomonas* fluorescents dans us sol résistant et un sol sensible aux fusarioses vasculaires. Agronomie 8: 243-249

Lindsay WL (1974) Role of chalation in micro-nutrient availability,507-524.In:Carson EW (ed) The plant root and its environment. University Press of Virginia Charlottesville, pp 651

Meyer JM and Abdallah MA (1978) The fluorescent pigment of *Pseudomonas fluorescens*: biosynthesis, purification and physicochemical properties. J Gen Microbiol 107:319-328

Scher M and Baker R (1980) Effect of *Pseudomonas putida* and a synthetic iron chelator on induction of soil suppressiveness to Fusarium wilt pathogens. Phytopathology 72:1567-1573

Scher M, Dupler M and Baker R (1984) Effect of synthetic iron chelator on population densities of *Fusarium oxysporum* and the biological control agent *Pseudomonas putida* in soil. Can J Microbiol 30:1271-1275

Sneh B, Dupler M, Elad Y and Baker R (1984) Chlamydospore germination of *Fusarium oxysporum* f. sp. *cucumerinum* as affected by fluorescent and lytic bacteria from *Fusarium* suppressive soil. Phytopathology 74:1115-1124

Van Driel W (1964) The effect of iron EDTA on the growth and metabolism of tomato plants in water culture. Plant Soil 20:85-104

EPIDEMIOLOGY OF *FUSARIUM OXYSPORUM* F.SP. *RADICIS-LYCOPERSICI*

W.R. Jarvis
Agriculture Canada
Research Station
Harrow, Ontario NOR 1GO
Canada

INTRODUCTION

Fusarium crown and root rot (FCRR) of tomato caused by *Fusarium oxysporum* f.sp. *radicis-lycopersici* Jarvis & Shoemaker (FORL) is essentially a disease of fumigated or steam-sterilized soils (Jarvis 1977; Jarvis et al 1983; Rowe and Farley 1981; Sonoda 1976, 1979; Sonoda et al 1978). Before the disease was better understood, growers were advised to sterilize soil thoroughly and to carry out a complete hygiene programme by washing greenhouse superstructures, pots and equipment with formalin, but Jarvis (1977) noted that growers who did all this often had the worse affected crops. This proved to be a common experience (Rowe and Farley 1981), and suggested that FORL had a particularly efficient dispersal mechanism.

DISPERSAL

An important finding in that respect was that of airborne dispersal of microconidia (Rowe et al 1977). The fungus was readily recovered as microconidia from trash piles of tomato vines, soil and straw mulch, and these spores were assumed to be the source of microconidia trapped on agar plates in four naturally-infested and cropped greenhouses in Ohio. About 3% of isolates of *F. oxysporum* recovered from air in this way proved to be FORL. Following trapping, the crops were cleared, the soil steam-sterilized and the superstructure interior sprayed with

NATO ASI Series, Vol. H28
Vascular Wilt Diseases of Plants

formalin. Two weeks later, the fungus was not detected in air or soil, but it was trapped from the air at 7 weeks in numbers only about 10% of those at the end of the previous crop.

The long-range dispersal of the disease has long been the subject of conjecture: seedborne dispersal seems an obvious possibility (Krikun et al 1983) but more critical work needs to be done to be certain about such dispersal through commercial channels. Since it is easy to isolate the fungus from dust in and around greenhouses (Rowe et al 1977), there is also the possibility that chlamydospores may be carried on clothing and shoes, and on machinery, packing crates, etc., for shorter distances. Couteaudier et al (1985) reported isolating FORL from compost imported into France from the Netherlands.

Early work (Dirks and Jarvis 1975; Jarvis et al 1975) suggested that affected plants are randomly distributed in greenhouse crops and that spread tends to occur along the rows rapidly, as much as 1% per day. This suggests a plant-to-plant spread, perhaps aided by mechanical procedures or the sledge-mounted watering system of the 1970 decade in Ontario. However, such runs of infected plants can also be explained by the pattern of transplanting along rows from trays of infected plants. The within-crop pattern of disease spread needs further work for its elucidation as the basis of rational control design.

COMPETITIVE SAPROPHYTIC ABILITY

Rowe et al (1977) found that in sterilized soil artificially infected with microconidia, the population rose by 10^3 after 3 days and 10^4 after 7 days. Two weeks later, the population was steady at about 10^3 propagules/g soil, exclusively as chlamydospores. Used as inoculum for freshly sterilized soil, these did not result in a population increase.

These findings of Rowe et al (1977) explained why the usual soil-sterilizing and hygiene procedures fail to control the disease. Further, the fungus can survive as chlamydospores in soil groundbeds 50 cm below the surface (Jarvis 1977) and well below the level of effective soil sterilization by steam or fumigation (about 30-35 cm), as well as in thick roots and lumps of clay difficult to sterilize.

Marois and Mitchell (1981a, 1981b) and Marois et al (1981) investigated FORL populations and their infectivity in relation to other microorganisms in soil. In field soil, the numbers of FORL chlamydospores needed to cause infection of 50% of Bonny Best tomato plants were 300, 900 and 6,500/g, respectively, in fumigated soil, in soil not fumigated, and in soil infested with a mixture of conidia of Penicillium funiculosum, Trichoderma harzianum and Aspergillus ochraceus (Marois and Mitchell 1979). These fungi had been selected for their ability to multiply rapidly in fumigated soil, to become established in the rhizosphere of the tomato, and to reduce the pathogenic activity of FORL (Fig. 1).

Following fumigation of field soil, Marois and Mitchell (1981b) found that the saprophytic proliferation of FORL in the soil, sampled at 7-day intervals, and the incidence of tomato infection was correlated inversely with the logarithm of numbers of saprophytic fungal propagules. When conidia of T. harzianum, A. ochraceus and P. funiculosum were added to freshly-fumigated soil infested with chlamydospores of FORL, 25% of Bonny Best tomato seedlings were infected within 2 weeks, while in fumigated, unaugmented, FORL-infested soil, 90% were infected. When transplanted after 46 days, only 1% of plants were infected in both augmented and nonaugmented, FORL-infested soils. These results were explained by the building-up of an antagonistic, saprophytic microflora, competing with FORL in the soil ecology (Figs 2 and 3).

Under field conditions, Marois et al (1981) applied conidia of the same antagonistic fungi to the roots of tomato plants at trasnsplanting. In the presence of a nearby inoculum of chlamydospores of FORL, the incidence of crown and root rot increased with chlamydospore density in fumigated soil not augmented with the antagonists, but did not increase in fumigated, antagonist-augmented soil (Table 1). The pathogen pupulation decreased from 600 to 200 propagules/g in antagonist-augmented soils, but increased from 1000 to over 5 x 10^4 propagules/g in nonaugmented soils. Only 7% of plants were infected at harvest in augmented

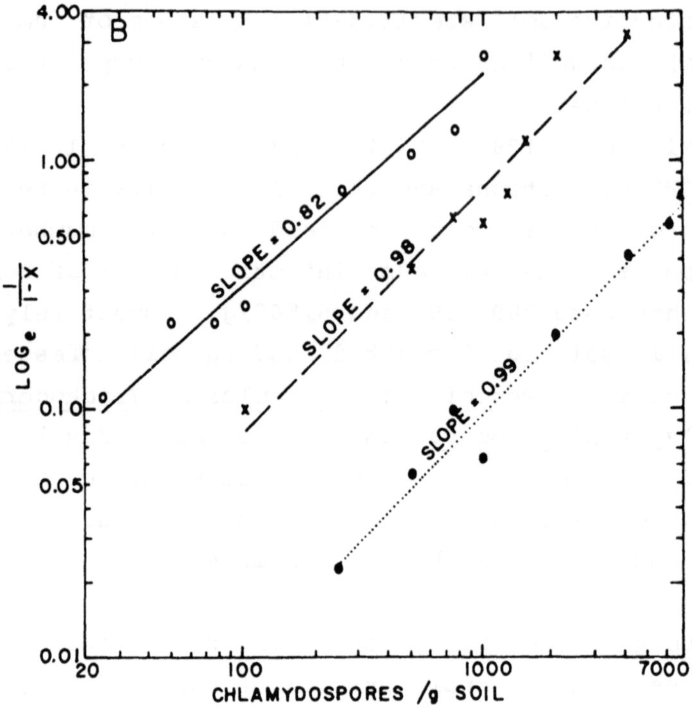

Fig. 1. The relationship of percentage of infection of tomato (cultivar Bonny Best) under growth-chamber conditions to densities of chlamydospores of Fusarium oxysporum f. sp. radicis-lycopersici in fumigated soil (———), nonfumigated soil (-----), and fumigated soil amended with Trichoderma harzianum, Aspergillus ochraceus, and Penicillium funiculosum (······): proportion of infection adjusted for multiple infections (logarithmic) and inoculum density (logarithmic). (From Marois and Mitchell 1981a).

soils as against 37% in nonaugmented soils. Evidently, biological control is mediated, in part at least, by microorganisms in the soil.

Similar conclusions were drawn by Sivan and Chet (1986) and Sivan et al (1987) using T. harzianum as a biological control.

In sterilized media with high populations of FORL, seedlings can be severely affected, with a collapse resembling Pythium

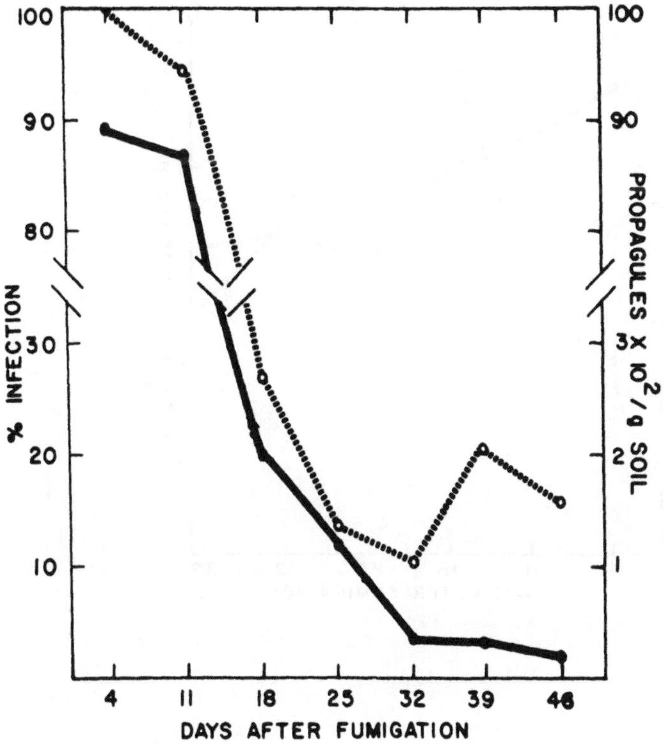

Fig. 2. The relationship of percent infection of roots of cultivar Bonny Best tomato plants (●———●) and inoculum density of <u>Fusarium</u> <u>oxysporum</u> f. sp. <u>radicis-lycopersici</u> (o-----o) to time after fumigation of soils which were allowed to be recolonized naturally; the pathogen was added at 1,000 chlamydospores per gram to 1,500 g samples of soil taken every 7 days and tomato plants were maintained in the infested soil for 14 days under growth chamber conditions before infection incidence and the inoculum density of the pathogen were determined. (From Marois and Mitchell 1981b).

damping-off. However, seedlings in media with low populations of <u>FORL</u> may become infected but the symptoms remain undetected. Tomato plants in a vegetative condition survive on a remarkably small root system and often on adventitious roots only, but when assimilates move to the fruit, the plant collapses, typically at the time of maximum fruit load, just before the first pick.

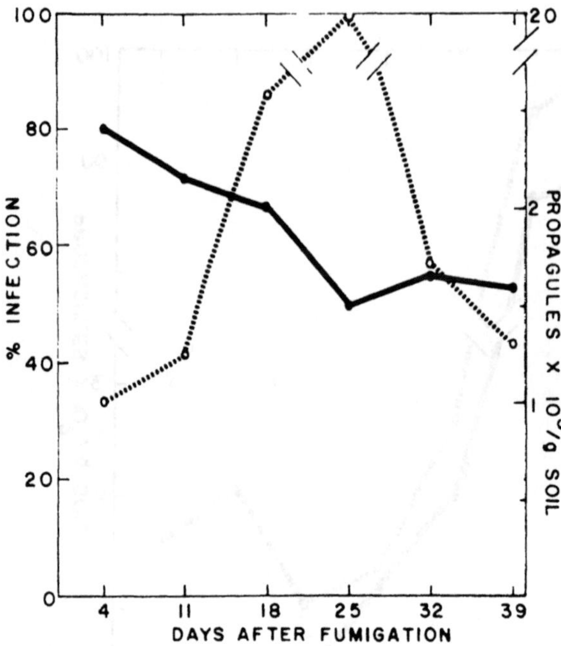

Fig. 3. The relationship of percent infection of roots of tomato cultivar Bonny Best tomato plant (●——●) and inoculum density of Fusarium oxysporum f. sp. radicis-lycopersici (o-----o) to time after fumigation of soils in which recolonization by other microorganisms was not inhibited; the pathogen was added at 1,000 chlamydospores per gram of soil 4 days after fumigation. (From Marois and Mitchell 1981b).

Control measures, therefore, must be directed to very early stages in seedling and transplant production; they are too late when applied to the wilting plant.

In contrast to Fusarium wilt caused by F. oxysporum f. sp. lycopersici, the optimum temperature for disease expression is about 15-18 C, or 10-20 C (Jarvis and Shoemaker 1978; Jarvis et al 1983; Rowe 1980; Sato and Araki 1974; Sonoda 1976, 1979; Yamamoto et al 1974; W.R. Jarvis and H.J. Thorpe, unpublished), whereas that for Fusarium wilt is about 27 C (Walker 1971). The disease is one of cool winter months in staked field tomatoes in California and Florida, and of very early spring greenhouse crops. Within continuous temperature regimes of 18 or 24 C, a 20-day chilling period at 15 C had no effect on symptom expression in tomato transplants (W.R. Jarvis and H.J. Thorpe, unpublished).

Table 1. Effect of initial inoculum density of *Fusarium oxysporum* f. sp. *radicis-lycopersici* and a composite of five antagonists on mean lesion length and percentage of plants with lesions under greenhouse conditions

Inoculum density (chlamydospores per pot)[x]	Antagonists	Mean lesion length (cm)	Plants with lesions (%)
0	-	0.00[y]	0[y]
500	-	0.987	50
5,000	-	1.73	69
50,000	-	2.22	70
mean		1.23a	47a
0	+[z]	0.14	4
500	+	0.73	33
5,000	+	0.68	33
50,000	+	0.96	44
mean		0.63b	28b

[x] Twenty milliliters of a chlamydospore suspension was injected into the soil 7 cm from the plant.
[y] Means in some column with different letters differ significantly ($P = 0.05$) as determined by t-test; percentage data analyzed after transformation to arcsine $\sqrt{x}$.
[z] + = conidia of each of five isolates (three isolates of *Trichoderma harzianum*, one isolate of *Penicillium funiculosum*, and one isolate of *Aspergillus ochraceus*) were added to the crown area of the transplant at 5×10^5 conidia of each isolate per pot. (From Marois et al 1981).

Again in contrast with Fusarium wilt, which is alleviated by lime with nitrate nitrogen (Jones and Woltz 1969; Walker and Foster 1946), nitrate N and ammonium N have similar effects of the incidence of Fusarium crown and root rot in the greenhouse (Jarvis and Thorpe 1980a), and yields are also comparable. In the field, however, Sonoda (1978) showed that ammonium N was associated with the higher disease severity but also with the highest yields in a series of fertilizer-fungicide treatments.

A further contrast is provided by the failure of the root nematode *Meloidogyne incognita* to predispose tomato to infection by FORL (Jarvis et al 1977), which it does in the case of *F. oxysporum* f. sp. *lycopersici* (Bergeson et al 1970; Goode and McGuire 1967; Jenkins and Coursen 1957). This is probably related to the mode of infection of roots directly through the epidermis (Brammall 1986; Charest 1983; Charest et al 1983, 1984) and at the sites of emergence of secondary roots (Jarvis and Shoemaker 1978).

CULTURAL PRACTICES

Cultural practices in both field and soil-borne greenhouse crops have profound effects on the course of the disease. Jarvis et al (1983) surveyed cultural practices in 119 greenhouses in the southwest of Ontario, where they were closely-spaced within an area of about 150 km^2, and comprising an area of 58.6 ha under glass and 22.1 ha under plastic. The disease first appeared in 23% of the houses in a small area in the district in 1974, and spread to outlying greenhouses, affecting 62% of them by 1978.

Pink-fruited cultivars were affected more than red-fruited cultivars mainly because the latter were planted later in warmer soils. Eighty percent of early spring crops (planted in late December and early January) were affected compared with 48% of late-spring crops (planted in March), and 14% of fall crops. The incidence of the disease within crops also decreased over the year; 5.2%, 0.6% and 0.1%, of plants respectively. These differences were attributed to rising soil temperatures. Similarly, of crops under glass, planted early, 63% were affected compared with 29% under plastic, which were planted in late spring, and again, the differences are probably related to soil temperature.

In the survey of Jarvis et al (1983), it was noted that growers who practised good general hygiene had crops with the disease as often and as severely as growers who did not. As Rowe et al (1977) showed, the pathogen is an effective colonizer of sterilized soils, and the close proximity of greenhouses in the survey area, often only 20-50 m apart, tended to override advantages of hygiene. The fungus was readily isolated from dirt on headerhouse floors and from trash piles inside and outside greenhouses and headerhouses, and the proximity, or not, of the surveyed greenhouses to field tomato crops made no significant differences to incidence or severity.

When the disease first appeared in Ontario and Ohio in 1974, a priori advice was for soil sterilization, followed by disinfection of the greenhouse superstructure with formalin applied by airblast sprayer. Surprisingly, experience proved the contrary: more conscientious growers tended to have more disease in their crops than their less sterilization-minded neighbours. Soil steaming through buried tile ducts was followed by the disease in 87% of crops; surface-steaming in 71% of crops; fumigation in 44%, fumigation followed by tile steaming in 57%; fumigation followed by surface steaming in 57%. Five fall crops following cucumber crops, and with no prior soil sterilization had no disease; these provided the clue to biological control. Cleansing the superstructure with formalin before and after soil sterilization proved unsuccessful; all crops were affected. Seventy four percent of crops following formalin treatment before sterilization were affected, 50% after steaming followed by formalin, and 50% with no formalin treatment.

The survey of Jarvis et al (1983) revealed no effective fungicide treatment, despite a wide range of practices tried by the Ontario growers; none was able to carry out the steaming-captafol treatment recommended in Ohio (Rowe and Farley 1978, 1981).

The source of irrigation water was important; more crops were affected, and more severely when irrigated by municipal water from Lake Erie than by water from creeks or wells. This was tentatively ascribed to its temperature, about 4 C when applied to the early spring crop, as against about 13 C for most ground water. Earlier, Jarvis and Thorpe (1976) had noted that overwatered potted plants tended to be more severely affected than plants not overwatered.

There was also an apparent temperature effect associated with straw mulch, whether it was sterilized or not. Straw places around the plants too early did not permit insolation of soil, and soil temperatures remained below or near the optimum for the disease during the critical pre-fruiting period of cropping.

Jarvis and Thorpe (1980a) found that the fungus occurred in a straw-amended (mulched) soil at populations of the order of 10^3 times those in unmulched soil, declining with the period of straw decomposition. FORL grew readily along straws in the soil, but poorly away from the straw into the soil.

Many growers in southwestern Ontario buy transplants from specialist growers; the survey could not detect differences among these specialists, nor whether the seedlings were raised under glass or plastic, nor among the several soil and soil-substitute media used for seedling and transplant production.

In Florida, Sonoda (1976) and Sonoda et al (1978) found the pathogen in field soils following a variety of cropping sequences, including peppers, and populations were highest in soils with recent histories of tomatoes.

Soil pH had no effect on the incidence of the disease in field plantings in Florida (Overman and Jones 1984), but Sonoda (1978) found that in the absence of fungicides, ammonium N was associated with high levels of disease severity, but high yields. No lesions were found in plants receiving no nitrogen fertilizer but yields were very poor. The lowest severity index was recorded in plants treated with a benomyl or captafol drench, lime and nitrate N.

SOILLESS CULTURE

The nutrient film technique (NFT) and rockwool substrates are two modern methods of greenhouse husbandry in which plants never come into contact with soil (Resh 1984). Nevertheless, Fusarium crown and root rot is common and serious in susceptible red-fruited tomato cultivars in hydroponic systems in Ontario (W.R. Jarvis and H.J. Thorpe, unpublished; S.K. Khosla, personal communication), but little work has been done on its epidemiology and control.

In recent years, Fusarium oxysporum, including f. sp. lycopersici, has been isolated, together with more than 35 other species of fungi from the roots of tomato in hydroponic systems (Evans 1979; Kegler et al 1982, 1983; Price 1980; Price and Fox 1984; Price and Maxwell 1980; Vanachter et al 1983). Although spores of F. oxysporum f. sp. lycopersici remained viable only to a limited extent in the nutrient solution, they have infected intact plants (Kegler et al 1982). The spread of Fusarium spores is very rapid in the circulatory system and soon the entire system becomes fully contaminated. In at least some cases reported in the literature, it might be assumed that the Fusarium oxysporum or Fusarium spp. isolated, but not completely identified, could have been F. oxysporum f. sp. radicis-lycopersici. In Quebec, Thériault et al (1987) found a higher incidence of reisolated FORL in plants grown in soil than in NFT or sawdust.

ALTERNATIVE HOSTS

Yamamoto et al (1974) found two isolates of FORL to incite an indistinct reaction on the roots of Solanum melongena L. and Capsicum frutescens L. but none on Citrullus vulgaris Schrad., Cucumis sativus L., Cucurbita spp., Brassica oleracea L., Raphanus sativus L., Nicotiana tabacum L., N. glutinosa L. or N. glauca R.C. Grah.

Rowe (1980) tested the pathogenicity of 42 isolates to 15 hosts in five families; cucurbits, crucifers and cereals were unaffected, but Glycine max (L.) Merr., Phaseolus vulgaris L., Pisum sativum L. and Arachis hypogea L. were variously infected by three isolates. Two of these isolates also infected Solanum melongena L. That some Leguminoseae are attacked and may act as alternative hosts raises the question of to what extent such plants may act as weed or crop reservoirs of inoculum, particularly in field tomato crops. There is no information on the role of these hosts, if naturally infected, in epidemiology.

CONCLUSION

F. oxysporum f. sp. radicis-lycopersici is unusual among Fusaria in that its airborne microconidia play a significant part in the epidemiology of Fusarium crown and root rot in tomatoes. FORL has poor competitive saprophytic ability but is a very effective colonizer of sterile soils. Fusarium crown and root rot is thus essentially a disease of sterilized soils and hydroponic systems, and is susceptible to biological control because of its poor competitive saprophytic ability.

REFERENCES

Bergeson GB, van Gundy SD and Thomason IJ (1970) Effect of Meloidogyne javanica on rhizosphere microflora and Fusarium wilt of tomato. Phytopathology 60:1245-1249

Brammall RA (1986) Host-parasite interactions in Fusarium crown and root rot disease of tomato. PhD Thesis, University of Toronto, pp 102

Charest P-M (1983) Études anatomiques et ultrastructurales de Fusarium oxysporum f. sp. radicis-lycopersici Jarvis and Shoemaker dans les tissus de son hôte (Lycopersicon esculentum Mill.) et en culture pure. Thèse PhD, Université Laval, Quebec, pp 167

Charest P-M, Ouellette GB and Pauzé FJ (1983) Fine structure of Fusarium oxysporum f. sp. radicis-lycopersici in root tissues of tomato. Can J Plant Path 5:202-203

Charest P-M, Ouellette GB and Pauzé FJ (1984) Cytological observations of early infection process by Fusarium oxysporum f. sp. radicis-lycopersici in tomato plants. Can J Bot 62:1232-1244

Couteaudier Y, Alabouvette C and Soulas M-L (1985) Nécrose du collet et pourriture des racines de tomate. Rev Hortic (Paris) 254:39-42

Dirks VA and Jarvis WR (1975) A basal stem and root rot of greenhouse tomatoes caused by Fusarium oxysporum and its spread in crops. Ann Proc Am Phytopathol Soc 2:36 (Abstr)

Evans SG (1979) Susceptibility of plants to fungal pathogens when grown by the nutrient film technique (NFT). Plant Pathology 28:45-48

Goode MJ and McGuire JM (1967) Relationship of root knot nematodes to pathogenic variability in Fusarium oxysporum f. sp. lycopersici. Phytopathology 57:812 (Abstr)

Jarvis WR (1977) Biological control of Fusarium. Can Agric 22: 28-30

Jarvis WR, Dirks VA, Johnson PW and Thorpe HJ (1977) No interaction between root-knot nematode and Fusarium foot and root rot of greenhouse tomato. Plant Disease Reporter 61:251-254

Jarvis WR and Shoemaker RA (1978) Taxonomic status of Fusarium oxysporum causing foot and root rot of tomato. Phytopathology 68:1679-1680

Jarvis WR and Thorpe HJ (1976) Susceptibility of Lycopersicon species and hybrids to the foot and root rot pathogen Fusarium oxysporum. Plant Disease Reporter 60:1027-1031

Jarvis WR and Thorpe HJ (1980a) The role of straw in the epidemiology of Fusarium foot and root rot of greenhouse tomatoes. Res Rep Can Hort Council, Ottawa, p 74

Jarvis WR and Thorpe HJ (1980b) Effects of nitrate and ammonium nitrogen on severity of Fusarium foot and root rot and yield of greenhouse tomatoes. Plant Disease Reporter 64:309-310

Jarvis WR, Thorpe HJ and MacNeill BH (1975) A foot and root rot disease of tomato caused by Fusarium oxysporum. Can Plant Dis Surv 55:25-26

Jarvis WR and Thorpe HJ (1983) Survey of greenhouse management practices in Essex County, Ontario, in relation to Fusarium foot and root rot of tomato. Plant Disease 67:38-40

Jarvis WR, Thorpe HJ and Meloche RB (1983) Survey of greenhouse management practices in Essex County, Ontario, in relation to Fusarium foot and root rot of tomato. Plant Disease 67:38-40

Jenkins WR, Coursen BW (1957) The effect of root-knot nematodes, Meloidogyne incognita acrita and M. hapla on Fusarium wilt of tomato. Plant Disease Reporter 41:182-186

Jones JR and Woltz SS (1969) Fusarium wilt (race 2) of tomato: calcium, pH, and micronutrient effects on disease development. Plant Disease Reporter 53:275-279

Kegler H, Griesbach E and Skadow K (1982) Ausbreitung von Pathogenen beim Tomatenanbau im NFT-Verfahren. Archiv Gartenbau 30: 325-337

Kegler H, Griesbach E, Skadow K, Fritzsche R and Weber I (1983) Ausbreitung von Krankheitserregern und Schädlingen der Tomate in NFT-Kultur und ihre Vorbeugung. Nachrichtenbl Dtsch Pflanzenschutzdienst (Berl) 37:28-29

Krikun J, Nachmias A, Offenbach R and Ucko O (1983) Establishment of the crown and root rot pathogen of tomato in the Negev region. Phytoparasitica 11:3-4

Marois JJ and Mitchell DJ (1979) The relationship of density of chlamydospores of Fusarium oxysporum to infection of tomatoes in fumigated and nonfumigated soils. Phytopathology 69:1037

Marois JJ and Mitchell DJ (1981a) Effects of fumigation and fungal antagonists on the relationships of inoculum density to infection incidence and disease severity in Fusarium crown rot of tomato. Phytopathology 71:167-170

Marois JJ and Mitchell DJ (1981b) Effects of fungal communities on the pathogenic and saprophytic activities of Fusarium oxysporum f. sp. radicis-lycopersici. Phytopathology 71:1251-1256
Marois JJ, Mitchell DJ and Sonoda RM (1981) Biological control of Fusarium crown rot of tomato under field conditions. Phytopathology 71:1257-1260
Overman AJ and Jones JP (1984) Soil fumigants for control of nematodes, Fusarium wilt and Fusarium crown rot of tomato. Proc Florida State Hort Soc 97:194-197
Price D (1980) Fungal flora of tomato roots in nutrient film culture. Acta Hortic (The Hague) 98:269-275
Price TV and Fox P (1984) Behaviour of fungicides in recirculating nutrient film hydroponic systems. Proc 6th Intern Congr Soilless Culture, p 511-523
Price TV and Maxwell MK (1980) Studies of disease problems and their control of hydroponics in Australia. Acta Hortic (The Hague) 98:307-316
Resh HM (1984) Hydroponic food production. 3rd ed Woodbridge Press, Sanda Barbara, pp 287
Rowe RC (1980) Comparative pathogenicity and host ranges of Fusarium oxysporum isolates causing crown and root rot of greenhouse and field grown tomatoes in North America and Japan. Phytopathology 70:1143-1148
Rowe RC and Farley JD (1978) Control of Fusarium crown and root rot of greenhouse tomatoes by inhibiting recolonization of steam disinfected soil with a captafol drench. Phytopathology 68:1221-1224
Rowe RC and Farley JD (1981) Strategies for controlling Fusarium crown and root rot in greenhouse tomatoes. Plant Disease 65: 107-112
Rowe RC, Farley JD and Coplin DL (1977) Airborne spore dispersal and recolonization of steamed soil by Fusarium oxysporum in tomato greenhouses. Phytopathology 67:1513-1517
Sato R and Araki T (1974) On the tomato root-rot disease occurring under vinyl-house conditions in southern Hokkaido. Annu Rep Soc Plant Protect N Japan 25:5-13
Sivan A and Chet T (1986) Possible mechanisms for control of Fusarium spp. by Trichoderma harzianum. Proc Brit Crop Prot Conf 2:865-872
Sivan A, Ucko O and Chet I (1987) Biological control of Fusarium crown rot of tomato by Trichoderma harzianum under field conditions. Plant Disease 71:587-592
Sonoda RM (1976) The occurrence of Fusarium root rot of tomatoes in South Florida. Plant Disease Reporter 60:271-274
Sonoda RM (1978) Effect of fungicide, lime and fertilizer regimes on Fusarium crown rot of tomato. Res Rep Fort Pierce, Florida, Agricultural Research Center, RL 1978-1, pp 4
Sonoda RM (1979) Observations on Fusarium crown rot of tomato in winter of 1978-79. Res Rep Fort Pierce, Florida, Agricultural Research Center, RL 1979-10, pp 6
Sonoda RM, Marois J and Augustine JJ (1978) Fusarium crown rot of tomato in Florida. Proc Florida State Hortic Soc 91:284-286
Thériault J, Dostaler D, Thibodeau PO and Gosselin A (1987) Enquête sur le pouridié fusarien (Fusarium oxysporum f. sp. radicis-lycopersici, (FORL) et les racines liégeuses Pyre-

nochaeta lycopersici) chez la tomate de serre. Phytoprotection 68:136-137 (Abstr)

Vanachter A, Van Wanleke E and Van Assche E (1983) Potential danger for infection and spread of root disease of tomatoes in hydroponics. Acta Hortic (The Hague) 133:119-128

Walker JC (1971) Fusarium wilt of tomato. Monograph 6. Am Phytopathol Soc, pp 56

Walker JC and Foster RE (1946) Plant Nutrition in relation to disease development. III. Fusarium wilt of tomato. Am J Bot 33:259-264

Yamamoto I, Kamada H, Juniyasu K, Saito M and Ezuka A (1974) A new race of *Fusarium oxysporum* f. sp. *lycopersici* inducing root rot of tomato. Proc Kanasai Plant Protect Soc 16:17-29

ON THE INFLUENCE OF ENVIRONMENTAL FACTORS ON DISEASE DEVELOPMENT

J.M.S. Martins
Departamento de Fitopatologia
Estação Agronómica Nacional
2780 Oeiras
Portugal

INTRODUCTION

Stem pith necrosis was found in several countries of Europe, in the United States of America and in New Zealand, in greenhouse-grown tomato plants, although it may also occur in the field (Jones et al 1983). Its cause was ascribed to Pseudomonas corrugata Roberts & Scarlett (Scarlett et al 1978), but several other bacteria have been found in association with this syndrome: P. viridiflava (Burkholder) Dowson, P. cichorii (Swingle) Stapp, P. syringae van Hall, and P. fluorescens (Trevisan) Migula (Alivizatos 1984; Carta et al 1981; Wilkie et al 1973; Wilkie 1974). Some of these bacteria have been referred to as "low-grade pathogens" (Scarlett et al 1978), due to their great dependence for enhancing disease on exceptional circumstances, such as unfavourable growth conditions for the host plant (excessive atmospheric humidity or nitrogen fertilization, nutritional inbalances, wide daily temperature ranges, etc.).

The most conspicuous symptom is the necrosis with laddering or complete disintegration of the pith tissues. Large and abundant adventitious roots may also grow on stems, both to the outside or into the pith cavity. In Portugal, extensive, dark patches on the stem surface, leaf and fruit spots, or a mild chlorosis of the whole plant can also occur. Affected plants usually remain alive, and their fruits can reach maturity. Occasionally, the vascular tissues are invaded, and the plant wilts and dies. These "elementary" symptoms occur in different combinations and severities in each greenhouse.

Generally, very heterogeneous populations of bacteria are present in lesions (Jacob and Martins to be published; Jacob 1987). This paper is a contribution to understanding the nature of the bacterial flora associated with the tomato pith necrosis syndrome, and its epidemiological implications.

MATERIALS AND METHODS

Diseased tomato plants (cv. Dombo, Dombito and Carmelo) were collected in plastic covered, drip irrigated greenhouses located at Muge (B), Cruz do Campo (C) and Azambuja (D), all north east of Lisbon in the valley of river Tejo, A-dos-Cunhados (A) and Obidos (E), north of Lisbon, near the coast, and Faro (F), in the coastal plain of southern Portugal. The soils were sandy loams in the first four greenhouses, and clay loams in the others.

Bacteria were isolated in plates of nutrient agar plus 5% sucrose from macerates of pieces of tissue detached from the margins of lesions. Microbiological tests were performed as described in Schaad (1980). Utilization of carbon compounds for growth was tested in the mineral base medium C of Dye (1968).

RESULTS

Each one of the isolations yielded a very diverse population of bacteria. Near 50 microbiological characteristics were determined for each of the 199 isolated Gram-negative strains. Using a principal components analysis of the 29 characteristics that were not constant in more than 90% of the strains, the bacteria were allocated to 15 groups (Jacob 1987). Some of the distinguishing characteristics of these groups are shown in Table 1. The six groups of fluorescent pseudomonads represent 43% of the studied strains.

The relationships between the bacterial populations of the six greenhouses were investigated by calculating the canonical variates of the data matrix. The first three variates account

Table 1. Frequencies (%) of microbiological characteristics in groups of strains isolated from tomato plants in six greenhouses

	a	b	c	d	e	f	g	h	i	j	k	l	m	n	o
Glucose metabolism :															
Oxidative	+	+	+	+	+	+	+	+	+	+	+				
Fermentative														+	+
Absent												+	+		
Fluor.pigments	+	+	+	+	+	+	-	-	-	-	-	-	-	-	-
Oxidase production	-	-	+	d	-	+	-	d	d	d	-	+	d	-	-
Hydrolysis of :															
Gelatine	+	+	+	d	+	-	+	d	+	d	d	+	-	+	d
Tween 80	d	d	+	-	-	-	-	-	-	-	-	d	-	-	-
Oxid.gluconate	-	-	d	+	+	+	-	-	-	-	d	-	d	-	-
Utilization of :															
Raffinose	-	+	+	-	-	-	-	-	-	-	-	-	d	+	+
L+Rhamnose	-	-	-	-	-	-	-	+	-	-	-	-	+	+	+
Maltose	-	-	-	-	-	-	-	+	-	-	-	d	+	-	+
Lactose	d	-	-	-	-	-	-	-	-	-	-	-	+	+	+
Cellobiose	+	+	d	-	-	-	d	-	d	-	-	-	+	+	+
Sorbitol	+	+	+	-	+	-	-	+	-	-	-	-	+	d	+
Erythritol	-	-	+	-	+	-	-	-	-	-	-	-	-	-	-
Lactate	-	-	+	+	+	+	+	+	+	+	+	d	+	d	+
Number of strains	23	28	13	8	8	5	8	6	34	8	4	6	16	6	26

+: more than 80% positive strains; -: less than 20% positive strains; d: between 20% and 80% positive strains.

for 87% of total variability (Fig. 1). Greenhouses became more ordered following a gradient from higher sand to higher clay content of soils, which is also the gradient of increasing soil pH, as illustrated by the minimum spanning tree of their centroids based on the Mahalanobis distance between each pair.

This analysis suggests a certain degree of uniformity of cha-

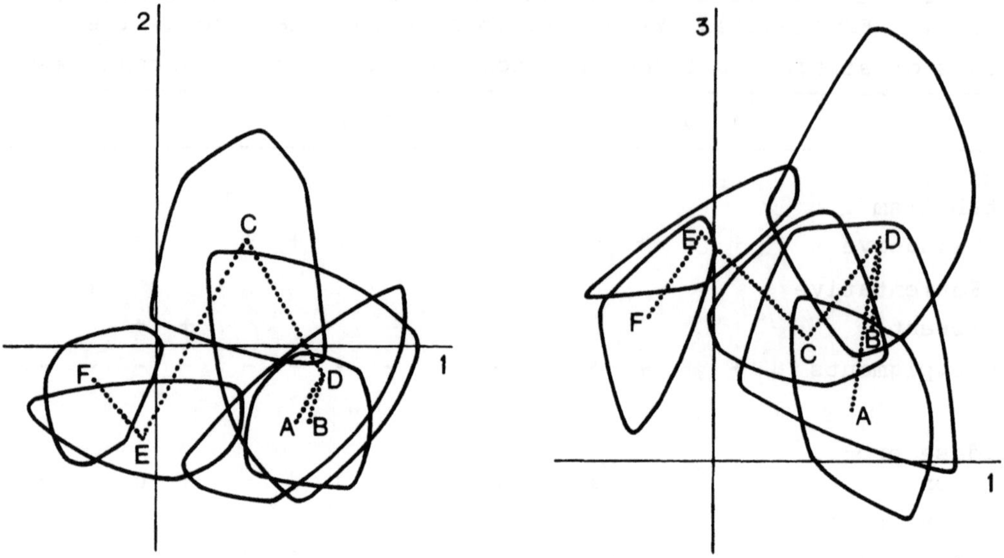

Fig. 1. Outlines of the projections of bacterial strains isolated from six greenhouses on the first three canonical variates. Dotted, minimum spanning tree of Mahalanobis generalized distances between group centroids.

racteristics in the bacterial populations isolated from the same greenhouse, possibly related to its peculiar environmental conditions, and the relative overlapping of clusters A to D indicates some similarity in their composition. The distribution of bacterial groups by greenhouses (Table 2) shows that this overlapping corresponds to the higher levels of fluorescent pseudomonads, and that the decrease in density of these bacteria follows the above mentioned gradient. A similar relationship between numbers of soil and rhizosphere pseudomonads and clay content of soils was found by Sands and Rovira (1971).

The uniformity in some characteristics of bacteria isolated in each greenhouse is revealed by comparing their distribution in the six greenhouses (Table 3) with the data in Table 2. The great majority of bacteria from greenhouse B are able to utilize raffinose, cellobiose and sorbitol, in spite of belonging to four different groups. Even more striking are the cases of greenhouses C and D, with seven and nine groups of bacterial strains, respectively.

Table 2. Frequencies (%) of groups of strains isolated from tomato plants in six greenhouses

Group	No. strains	Greenhouses					
		A	B	C	D	E	F
a	23	6.5	12.9	6.7	30.0		
b	28	64.5	12.9	10.0	2.0		
c	13	12.9	25.8			5.9	
d	8	16.1			6.0		
e	8				16.0		
f	5					29.4	
g	8			23.3			2.5
h	6				12.0		
i	34			26.7		64.7	37.5
j	8						20.0
k	4			10.0	2.0		
l	6			10.0	6.0		
m	16						40.0
n	6			13.3	4.0		
o	26		48.4		22.0		

DISCUSSION

The bacteria which cause tomato pith necrosis have not been able to establish a continuous association with their host. They were not (Crosse 1968) "stabilized through a long period of gene interaction" with it, because hybrid tomato seed is used in greenhouse crops, a procedure which continuously breaks that interaction and forces it to start again each season. This can explain the low-grade nature of the disease, and the biological variability of the pathogens, which apparently depend on environmental conditions not very common in field crops for enhancing the disease process.

Table 3. Frequencies (%) of microbiological characteristics in strains isolated from tomato plants in six greenhouses

	Greenhouses					
	A	B	C	D	E	F
Glucose metabolism :						
Oxidative	100	51.6	76.7	68.0	100	60.0
Fermentative		48.4	13.3	26.0		
Absent			10.0	6.0		40.0
Fluorescent pigments	100	51.6	16.7	54.0	35.3	
Oxidase production			13.3	16.0	94.4	
Hydrolysis of :						
Gelatine	96.8		96.7	90.2		
Tween 80			6.7	15.9	11.1	0
Oxidation of gluconate	3.2	19.4	0			
Utilization of :						
Raffinose		83.9	13.3		0	12.5
L+Rhamnose	0		13.3		5.6	
Maltose	0		0		0	
Lactose	0		13.3		0	
Cellobiose	80.6	96.8			5.6	
Sorbitol	80.6	87.1	20.0	88.2	11.1	
Erythritol	9.7		0	19.6	11.1	0
Lactate	83.9				100	97.1
Number of strains	31	31	30	50	17	40

Frequencies > 20 and < 80 suppressed from lower part of table.

It is reasonable to conclude that the relatively uniform characteristics found in strains isolated from the same greenhouse may be related to an increased fitness of the organisms to survive in that particular environment. They may have been developed in response to a common selective pressure. Only an investigation of biochemical pathways could tell if these similar phenotypes result from homology or convergence in the evolutionary process.

Having in mind the short generation time of bacteria, such an evolutionary change can take place during one season, provided that the pathogen is a plastic organism, with high nutritional versatility as a species (i.e., with a high intraspecific variability). These characteristics make it fit to survive saprotrophicaly, or to be parasitic on a wide host range. If we include P. marginalis (Brown) Stevens in P. fluorescens, of which it is perhaps a pathovar, then the four fluorescent pseudomonads isolated from plants showing pith necrosis are able to cause disease after natural infection in a great number of plant species (Bradbury 1986), but information on their occurrence as epiphytes is rather scarce. P. corrugata has been isolated from symptomless alfalfa roots (Lukezic 1979). Tomato pith necrosis pathogens are most probably recruited from populations living in the soil, the rhizosphere, or other natural habitats. Scarlett et al (1978) were able to isolate P. corrugata from irrigation water. The type of symptom produced, either in the external (stem, leaf and fruit spots) or in the inner parts of the host (pith necrosis, vascular wilting) will depend on the conditions offered by each tissue to the expression, and expansion, of the evolutionary potential of parasitic microbes.

But evolution depends on microbe reproduction, and growth rates are not under exclusive control of the host plant, as they depend mainly on temperature and water availability (water activity). Environmental conditions influence disease incidence by providing microorganisms with the necessary means of dispersion, and severity by their action on metabolic processes both of the pathogen and the host. They set the limits to the interaction of parasites with their hosts which is absolutely necessary for they becoming pathogens. With the extended possibilities of manipulating those limits in greenhouse crops, "the evolution of an epidemiologically competent pathogen, i.e., one capable of sustaining a continuous infection chain in the field, may be a comparatively rare event" (Crosse 1968).

REFERENCES

Alivizatos AS (1984) Aetiology of tomato pith necrosis in Greece. Proc 2nd Int Work Group P. syringae Pathovars. Sounion (Greece) 24-28 Apr 1984, pp 55-57

Bradbury JF (1986) Guide to plant pathogenic bacteria. C.A.B. Intern Mycol Inst, Slough

Carta C, Corda P and Fiori M (1981) La "maculatura batterica" del pomodoro (Lycopersicon esculentum Mill.) causata da Pseudomonas syringae van Hall. Studi Sassar 28-29:214-220

Crosse JE (1968) Plant pathogenic bacteria in soil. In: Gray TRG and Parkinson D (eds) The ecology of soil bacteria, an international symposium. Liverpool Univ Press, Liverpool, pp 552-572

Dye DW (1968) A taxonomic study of the genus Erwinia. 1. The "amylovora" group. N Z J Sci 11:590-607

Jacob C (1987) On the aetiology of pith necrosis of greenhouse grown tomato plants in Portugal. Proc 7th Congr Mediterr Phytopath Union Granada (Spain) 20-26 Sep 1987, p 119

Jacob C and Martins JMS (to be published) Variability of fluorescent pseudomonads isolated from diseased greenhouse-grown tomato plants. Proc 3rd Int Work Group P. syringae Pathovars. Lisbon (Portugal) 31 Aug - 4 Sep 1987

Jones JB, Jones JP, Stall RE and Miller JW (1983) Occurrence of stem necrosis in field-grown tomato incited by Pseudomonas corrugata. Plant Disease 67:425-426

Lukezic FL (1979) Pseudomonas corrugata, a pathogen of tomato, isolated from symptomless alfalfa roots. Phytopathology 69: 27-31

Sands DC and Rovira AD (1971) Pseudomonas fluorescens biotype G, the dominant fluorescent pseudomonad in South Australian soils and wheat rhizospheres. J appl Bact 34:261-275

Scarlett CM, Fletcher JT, Roberts P and Lelliott RA (1978) Tomato pith necrosis caused by Pseudomonas corrugata n. sp. Ann appl Biol 88:105-114

Schaad NW (ed) (1980) Laboratory guide for identification of plant pathogenic bacteria. Am Phytopath Soc, St Paul, Minnesota

Wilkie JP and Dye DW (1974) Pseudomonas cichorii causing tomato and celery diseases in New Zealand. N Z J Agric Res 17:123-130

Wilkie JP, Dye DW and Watson DRW (1973) Further hosts of Pseudomonas viridiflava. N Z J Agric Res 16:315-323

FACTORS AFFECTING POTATO BLACKLEG DEVELOPMENT

M.C.M. Pérombelon[1], V.M. Lumb[2], D. Zutra[3], L.J. Hyman[1] and E.M. Burnett[1]
Scottish Crop Research Institute
Invergowrie
Dundee DD2 5DA
Scotland

INTRODUCTION

Blackleg is a seedborne disease caused by soft rot erwinias. The symptoms may vary not only from one climatic region to another but also within a region according to local and annual variations in weather conditions. Regardless of temperature, a stem rot develops under wet conditions, and under dry conditions the plant is stunted and leaf desiccation may occur. Blackleg should not be confused with aerial stem rot which is initiated on the aerial stem by air-borne erwinias under wet conditions (Perombelon and Kelman 1987).

More than one erwinia is usually present on tubers, mostly in lenticels and superficial wounds. Blackleg is invariably associated with rotting of the mother tuber in which the erwinias multiply before invading the stems through the vascular system. Although most seed (mother) tubers are contaminated and rot sooner or later and most stems are latently infected (Perombelon et al 1987), usually only one stem of a relatively small proportion of plants in an average crop develops the disease. It would appear that disease develops as a result of the chance invasion of a stem by large numbers of the bacteria from the rotting mother

2. Agrolon Ltd
9-15 Neal Street
London
England
3. Volcani Center
Bet Dagan
Israel

tuber at a time when conditions are favourable for rapid multiplication in the stem. There is some evidence that independently of blackleg, latent plant infection can depress yield (Perombelon and Hyman 1987).

Since the mother tuber is the source of the pathogens, factors which favour rotting of the mother tubers and multiplication of the bacteria will also favour blackleg development. Several have been identified and the more important are examined in this paper.

SOIL MOISTURE

It is well known that blackleg in temperate regions is more prevalent in wet than in dry growing seasons. Soil moisture is the most important factor which affects disease development. In a field experiment carried out in Scotland, plots planted in April with cv. Bintje seed inoculated soon after harvest by vacuum infiltration in a suspension containing 10^5 cells/ml of one of the pathogens. Erwinia carotovora sub. sp. atroseptica (Elphinstone and Perombelon 1986) were irrigated daily for different periods in June, July and August to near field capacity as determined by calibrated soil tensiometers. Blackleg incidence c. 4 weeks after the different irrigation treatments in July and August was related to the duration the soil was maintained wet (Table 1). Disease level was not affected in June partly because resistance of the mother tubers to rotting is still high soon after emergence and partly because interruptions in the irrigation programme failed to maintain high soil moisture level continuously in the different treatments. Disease level was generally greater in plots irrigated in August than in July reflecting the greater susceptibility of mother tubers and plants with time.

Tubers will rot when covered with a water film which restricts O_2 diffusion resulting in anaerobiosis, e.g. in 2.5 hours at 22 C (Burton and Wigginton 1970). Oxygen-dependent host resistance is affected and pectolytic bacteria including soft rot erwinias in lenticels can readily multiply, invade and rot the tuber (Perombelon and Lowe 1975). The longer the soil stays wet, the more likely the mother tubers will rot rapidly with higher erwinia populations and the greater is the opportunity to invade stems to cause blackleg. Therefore, conditions which tend to maintain a high soil moisture level such as high rainfall, excessive irrigation especially surface irrigation, heavy and poor draining soils and low temperatures which restrict evaporation will all tend to favour disease development. It is of interest to note that the repeatedly higher blackleg incidence in potato crops grown in the Mediterranean region than in Northern Europe can be attributed more to the fact that irrigation is more common in the former region than to temperature which is not very different during most of the growing season because of different planting times.

Table 1. Effect of irrigation on blackleg incidence expressed as a percentage in plots planted with cv. Bintje seed inoculated with E.c. atroseptica; 50 mm/hour water applied by overhead irrigation for 5 hours/day for 0, 3, 7 and 15 consecutive days in June, July and August

Irrigation period (days)	Irrigation time		
	June	July	August
0	2.8	2.8	4.2
3	2.1	11.8	37.5
7	4.2	36.8	49.3
15	2.1	48.6	65.3
SED(df)		5.78(24)	

TEMPERATURE AND ERWINIA PATHOGENICITY

Temperature is the main factor affecting the relative virulence of the three soft rot erwinias, Erwinia carotovora sub. sp. carotovora (Ecc), E. carotovora sub. sp. atroseptica (Eca) and E. chrysanthemi (Echr) associated with potatoes and its level may determine which organism predominates in a disease situation (Perombelon and Kelman 1980). Eca is more pathogenic at low, Ecc at intermediate and Echr at high temperatures reflecting partly their different growth temperature characteristics and partly the differential effect of temperature on their pectic enzymes production (Perombelon and Hyman 1986).

In field experiments in Israel, rotting mother tubers of plants grown from seed tubers inoculated jointly with two erwinias in different combinations yielded more Eca early in the February-sown growing season when the mean weekly maximum soil temperature was < 20 C and Echr in May-June when higher temperatures prevail; Ecc was present at all times (Perombelon et al 1987). In contrast, in Scotland where Echr was not used and soil temperature was < 20 C rotting mother tubers were infected with Ecc and Eca at all times during the growing season. In plots planted with seed inoculated with Eca alone or together with Ecc, blackleg developed throughout the growing season in Scotland (Table 2). However, in plots planted in Israel with seed inoculated with Eca alone or with Ecc disease developed before May only; with seed inoculated with Echr alone or with Ecc disease developed only in May; and with seed inoculated with both Eca and Echr disease developed at all times but exhibited different symptoms.

Eca caused a stem rot more extensively in Scotland than in Israel whereas the symptoms caused by Echr in May, when temperatures were high and air humidity low, were extensive leaf desiccation and wilting with no stem rot. Echr can cause a stem rot under high humidity conditions as those found in tropical Peru (J.G. Elphinstone, personal communication). Isolations made from diseased stems showed that the temporal patterns of detection of Eca and Echr is stem rots and in plants with leaf desiccation symptoms were similar to those for expression of the two symptom types. Although Ecc can cause a rot when inoculated in stems and is the main causal agent of aerial stem rot, if failed to cause blackleg when inoculated in seed tubers in Scotland, Israel and tropical Peru.

In conclusion, the relative virulence of erwinias could be

Table 2. Blackleg incidence expressed as a percentage of plants of cv. Desiree grown from seed inoculated by vacuum infiltration with c. 10^4 cells of different erwinias alone or in different combinations in Scotland and Israel

Country	Disease assessment time	Seed inoculated with							SED
		Ecc[1]	Eca[2]	Echr[3]	Ecc+ Eca	Ecc+ Echr	Eca+ Echr	Control	
Scotland	July	0	18.3	-	15.0	-	-	0.4	6.72
	August	0	22.1	-	16.7	-	-	0.4	8.00
	September	2.5	25.8	-	17.1	-	-	0.4	8.45
Israel	April	0.7	4.7	0.7	5.3	0	8.7	0.7	1.87
	May	0.7	4.7	18.7	6.0	2.7	24.0	0.7	4.48

1 Ecc, *Erwinia carotovora* sub. sp. *carotovora*
2 Eca, *E. carotovora* sub. sp. *atroseptica*
3 Echr, *E. chrysanthemi*

tentatively explained in terms of temperature changes during the growing season, the transition points being c. 20 C for maximum soil temperature at tuber depth or c. 25 C for maximum air temperature above which Eca and below which Echr are not apparently pathogenic. However, it is not clear whether the differential temperature effect is on bacterial multiplication in the mother tuber and the dominant erwinia infecting the stem or also on the pathogenicity of the bacteria in the stems.

SEED CONTAMINATION

The relationship between seed contamination level and blackleg incidence was first demonstrated a decade ago (Aleck and Harrison 1978; Perombelon and Lowe 1978) and subsequently has been further explored following the development of methods (1) to inoculate tubers with known numbers of erwinias by vacuum infiltration in bacterial suspensions to simulate natural contamination and (2) to determine tuber contamination level accurately (Perombelon et al 1987). In field experiments, blackleg incidence in plots planted with seed inoculated with different numbers of Eca was proportional to the seed contamination level (Table 3).

Table 3. Blackleg incidence expressed as a percentage in plots planted with cvs Maris Bard and Desiree seed inoculated with different numbers of E.c. atroseptica in Scotland and Israel

Seed contamination	Scotland		Israel	
(no. Eca cells/tuber)	Maris Bard	Desiree	Maris Bard	Desiree
10^6	39.6	12.50	36.9	21.3
10^5	20.8	5.56	25.8	17.2
10^3	4.2	2.78	19.8	10.3
10^1	1.4	0.69	9.5	2.1
control ($< 10^1$)	1.4	0	2.1	0
SED (df)	2.55(20)		4.97(40)	

This relationship also holds good when seed stocks naturally contaminated with different levels of Eca were tested. (R.A. Bain and M.C.M. Perombelon, unpublished). Since blackleg development is also affected by other factors, in particular soil moisture, it is not possible to predict disease level in a crop on the basis of seed contamination level. However, more disease is likely to develop if the seed is heavily than lightly contaminated in any given situation. Moreover, recent work suggests that, regardless of cultivar resistance, there is an apparent threshold level of seed contamination (c. 10^3 Eca cells per tuber) below which little blackleg develops under average growing conditions.

Since all seed tubers in the field experiments were contaminated (in some cases with large numbers of the pathogen), it is surprising that more if not all plants were not diseased especially as most stems were probably latently infected (Perombelon et al 1987). One possible explanation is that stem resistance per se restricts growth of the bacteria, more in some stems than in others, depending on variation in the resistance between plants and in micro-ecological conditions. Alternatively, although present in all tubers, the pathogen becomes dominant in only a few because of competition with other bacteria especially pectolytic Clostridium spp. The greater the initial number of Eca present, the more likely that it will predominate when the mother tuber rots and cause blackleg when large numbers invade the stems.

Competition between Eca and other non pathogenic pectolytic bacteria in mother tubers could explain the reduced blackleg incidence found in Scotland when seed was inoculated with both Ecc and Eca compared to seed inoculated with similar numbers of Eca alone (Table 2). This effect was more evident later than earlier in the growing season; little further blackleg developed with seed inoculated with both erwinias after July when, with rising soil temperatures, Ecc and other bacteria tended to outgrow Eca in the rotting mother tubers. In contrast Eca is more likely to be the dominant organism when inoculated alone in large numbers. In Israel, high temperatures in May suppressed Eca pathogenicity in tubers and in stems but favoured that of Echr. Blackleg level was about the same in May with seed inoculated with Echr alone or Echr and Eca, but was significantly reduced with seed inoculated with Echr and Ecc. The latter organism can rot tubers as well as Echr at the prevailing temperatures and probably reduced the opportunity for Echr to predominate when both erwinias were present in mother tubers.

Although the differential effect of temperature on the relative virulence and multiplication of the different erwinias when present together in potatoes is the general rule, different results can be obtained with weakly or highly virulent bacteria. For example, Ecc will predominate instead of Eca at 15 C when inoculated together in tubers if the Ecc strain used is virulent enough to outgrow a weakly virulent Eca strain (M.C.M. Perombelon, unpublished). Similarly, when different Eca and Echr strains were inoculated in equal numbers in seed tubers, blackleg incidence was related to the virulence of the bacteria defined here by their relative pectate lyase production in vitro (Table 4). Although all the determinants of pathogenicity in soft rot erwinias have not yet been identified, recent molecular genetics studies suggest that pectic enzymes and in particular pectate lyase are important factors especially with respect to virulence (J. Hinton, personal communication). As discussed above, weakly virulent strains were probably more likely to be overgrown in a greater number of mother tubers by other more strongly pectolytic bacteria also present in the rotting mother tubers than the highly virulent strains.

ADDITIONAL FACTORS

In addition to the three factors discussed above, blackleg development is also affected by several others which are either relatively minor or about which little is known. Of the latter, cultivar resistance and interaction with other diseases are the most important. A wide range of resistance appears to exist even in the relatively narrow genetic base of present day cultivars.

Table 4. Blackleg incidence in plots of cv. Maris Bard planted with seed inoculated with E. carotovora sub. sp. atroseptica (Scotland) and E. chrysanthemi (Israel) strains of different virulence defined by their pectate lyase production in a buffered pectate broth.

E. c. atroseptica			E. chrysanthemi		
Strains	Blackleg	Enzyme units[1]	Strains	Blackleg	Enzyme units[1]
1	30.3	2.12	1	6.4	3.19
2	20.7	1.97	2	6.2	2.07
3	17.3	1.66	3	3.0	1.54
4	15.7	1.70	4	2.8	1.02
5	15.3	1.78	5	0.4	0.77
6	13.0	1.88			
7	11.7	1.70			
8	9.0	0.71			
9	3.7	0.55			
10	0.7	0.20			
SED(df)	3.24(18)	0.063(20)		0.95(16)	0.062(8)
correlation(P)	r = 0.870 (0.01)			r = 0.896(0.05)	

[1]units (Zucker and Hankin 1970) per 10^{10} cells

The ranking order of cultivar resistance to tuber soft rot and to blackleg is not always consistent, it varies for the two diseases, between years and with the method of assessment used (Lapwood and Read 1985; Bain and Perombelon 1988; Perombelon et at 1988). At best only the highly resistant or susceptible cultivars can be identified with some confidence. In the case of blackleg, two different resistance mechanisms are probably involved, one concerning the mother tuber and the other the stem. The development of a reliable and meaningful test to determine resistance will lead to a better understanding of the role of cultivar resistance in blackleg development.

Blackleg, as many other diseases, rarely develops on its own under field conditions. The weakening of the host plant by one pathogen may favour the development of another or even of opportunistic pathogens which in turn may have a synergistic affect on the primary causal agent. Blackleg development has long been associated with other diseases such as Verticillium and Fusarium wilts and Rhizoctonia solani infection. The resistance or tolerance of a cultivar to a disease may break down in the field if it is susceptible to another when dual infection occurs. Preliminary results suggest that Eca infection tends to increase V. dahliae infection of plants even in cultivars resistant to the fungus but the converse synergism does not always occur (Tsror et al 1987).

Minor factors which nevertheless can occasionally play an important role in blackleg development are treatments or conditions that damage the seed tuber or reduce its resistance: use of cut seed under conditions which favour erwinia multiplication, seed damaged by freezing or overheating (black heart) during storage and seed affected by various fungal tuber diseases (late blight, dry rot, gangrene). Blackleg is also affected by factors which have an effect on host metabolism such as soil N level with which it is inversely related (Graham and Harper 1966).

REFERENCES

Aleck JR and Harrison MD (1978) The influence of inoculum density and environment on the development of potato blackleg. Amer Potato J 55:459-494

Burton WG and Wigginton MJ (1970) The effect of a film of water upon the oxygen status of a potato tuber. Potato Res 13:150-186

Bain RA and Perombelon MCM (1988) Effect of methods of testing on the resistance of tubers of different potato cultivars to soft rot caused by Erwinia carotovora sub. sp. atroseptica. Plant Pathology (in Press)

Elphinstone JG (1986) Contamination of potatoes by Erwinia carotovora during grading. Plant Pathology 35:25-33

Graham DC and Harper PC (1966) Effect of inorganic fertilisers on the incidence of potato blackleg disease. Eur Potato J 9: 141-145

Lapwood DH and Read PJ (1985) A simplified slice method for assessing tuber susceptibility of potato cultivars to Erwinia carotovora sub. sp. atroseptica. Plant Pathology 34:284-286

Perombelon MCM and Hyman LJ (1986) A rapid method for identifying and quantifying soft rot erwinias directly from plant material based on their temperature tolerances and sensitivity to erythromycin. J appl Bact 60:61-66

Perombelon MCM and Hyman LJ (1987) Epidemiology and etiology of blackleg. Rept Scott Crop Res Inst 1986, pp 102-103

Perombelon MCM and Kelman A (1980) Ecology of the soft rot erwinias. Ann Rev Phytopathol 18:361-387

Perombelon MCM and Kelman A (1987) Blackleg and other potato diseases caused by soft rot erwinias: a proposal for a revision of terminology. Plant Disease 71:283-285

Perombelon MCM and Lowe R (1975) Studies on the initiation of bacterial soft rot in potato tubers. Potato Res 18:64-82

Perombelon MCM and Lowe R (1978) Blackleg etiology field studies. Rept Scott Crop Res Inst 1987, p 86

Perombelon MCM, Lumb VM and Zutra D (1987) Pathogenicity of soft rot erwinias to potato plants in Scotland and Israel. J appl Bacterio 63:73-84

Perombelon MCM, Lumb VM and Hyman LJ (1987) A rapid method to identify and quantify soft rot erwinias on potato seed tubers. Eur Pl Protect Org Bull 17:25-35

Perombelon MCM, Lopez MM, Carbonell J and Hyman LJ (1988) Effects of contamination by Erwinia carotovora sub. sp. carotovora and E. carotovora sub. sp. atroseptica of potato seed tubers and of cultivar resistance on blanking or non-emergence and blackleg development in Valencia, Spain. Potato Res (in Press)

Tsror L, Nachmias A and Perombelon MCM (1987) Interaction of Erwinia carotovora var. atroseptica with Verticillium dahliae pathogenesis on potatoes. In: 10th Triennial Con Eur Ass Pot Res Abs Aalborg, Denmark 1987, pp 34-35

Zucker M and Hankin L (1970) Regulation of pectate lyase synthesis in Pseudomonas fluorescence and Erwinia carotovora. J Bact 104:13-18

CHARACTERISTICS OF GREEK ISOLATES OF PSEUDOMONAS SOLANACEARUM

P.G. Psallidas
Benaki Phytopathological Institute
145 61 Kifissia, Athens
Greece

INTRODUCTION

Pseudomonas solanacearum is the causal agent of the disease known as "bacterial wilt" which is one of the most destructive bacterial diseases of plants in tropical, subtropical and warm-temperate parts of the world. It affects many plant species belonging to different families. The most important hosts are potato, tomato, tobacco, eggplant and pepper among the Solanaceae and bananas from the Musaceae families. It can also infect members of Compositae, Leguminosae and other families. Representatives of more than 35 families of plants are affected by the disease (Kelman 1953).

Pseudomonas solanacearum is not a homogeneous species. It has been established that strains adapted to different hosts or group of hosts exist. Thus strains specialized to banana (Buddenhagen and Elsasser 1962), tobacco and groundnuts (Kelman and Person 1961), mulberry (He et al 1983) have been described. According to Buddenhagen et al (1962) three races of Pseudomonas solanacearum can be designated on the base of pathogenicity on some selected hosts. Race 1 strains have a wide host range infecting tobacco, tomato,many Solanaceae and other weeds and certain diploid bananas, while strains belonging to race 2 cause bacterial wilt on triploid bananas (Moco disease) and Heliconia sp. Strains belonging to race 3 have narower host range than the race 1 strains, infecting tomato, potato and eggplant but are not highly virulent on other solanaceous hosts. Recently a 4th race has been designated (He et al 1983). This race infects mulberry and exhibits low virulence on tomato, potato, eggplant and does not infect common

NATO ASI Series, Vol. H28
Vascular Wilt Diseases of Plants

tobacco, ginger and peanuts. Based on physiological and biochemical characteristics Hayward (1964) distinguished four biotypes of the bacterium. A 5th biotype has been proposed for the strains affecting mulberry which corresponds to race 4 (He et al 1983). In Greece the disease was described for the first time by Zachos (1957) on potato tubers from the areas of Messini and on potato plants from the area of Avlona, Attiki. The origin of the inoculum probably was seed potatoes imported from Cyprus or produced in the island of Naxos where symptoms resempling those of bacterial wilt had been observed in seed potato fields during the 1953-54 growing season.

On tomato the disease was recorded for the first time by Panagopoulos and Psallidas in 1970 (unpublished data) at Pyrghi, in the district of Biotia and also reported by Malathrakis in Crete. No bacterial wilt has been reported on other hosts including tobacco and pepper.

Up to 1984 the disease had not been considered as a serious phytopathological problem neither on potato nor on tomato. In 1984 a serious incidence of the disease occurred on tomato and eggplant growing in open fields at Skala, Laconia and on tomatoes grown in plastic houses at Alfioussa in the district of Ilia. Currently the disease is considered to be endemic in the area of Skala because it appears regularly in this place.

The purpose of this work was to investigate the characteristics of the strains of Pseudomonas solanacearum which occurred in Greece and to place them into the recognized biotypes and races of the bacterium.

SYMPTOMS OF THE DISEASE

The infected plants showed the typical symptoms of bacterial wilt, that is a progressive wilting of the plant starting as light wilting of the top of the plant which was visible during the warm hours of the day. As the disease progresses the wilting becames more generalized and the plants soon collapse and die. In longitudinal sections a dark brown discolouration of the xylem vessels of the stem is obvious. On tomato the discolouration extends and covers the whole stem. On eggplant the discolouration was somehow restricted to the base of the stem. The severe and generalized wilting and the absence of cankers differentiate bacterial wilt from bacterial canker which is also a vascular disease of tomato caused by Clavibacter michiganensis subs. michiganensis.

MATERIALS AND METHODS

The bacterial strains used in this study were obtained from diseased tomato and eggplant plants from three locations, Pyrghi, Biotia District, Skala, Laconia District and Alfioussa, Ilia District. Isolations were performed by the standard technique on duplicate petri dishes with Nutrient agar and Kelman's (1954) TZC medium. Incubation was performed at 26 C. Fluidal white with pink center colonies were picked up from Kelman's medium purified and transfered to slopes containing NAG medium (Nutrient agar + 2% glycerol). The further identification of the bacterial strains was based on the following tests: Gram reaction (-), sudanophilic inclusions (+), mode of carbohydrate utilization (oxidative), oxidase activity (+), fluorescent pigment production on King's B medium (-) and hypersensitive reaction on tobacco leaves (+). After the identification, the isolates were lyophylized and kept in the freeze.

Inoculum for biochemical tests and for pathogenicity inoculations was prepared by suspending a 24-48 h growth on NAG slope culture into 6 ml of sterile distilled water.

Biotype determination

The biotype of the strains was determined according to the method described by Hayward (1964) with the exception of using the bacteriophage B1114P.

Race determination

The race determination was based on the pathogenicity of the strains on the plants suggested by Buddenhagen et al (1962) with the exeption of bananas and Helliconia sp. which were not abailable. Pepper was also included.

Pathogenicity tests

The pathogenicity of the isolates was tested on potato (Solanum tuberosum cv. Sebago), tomato (Lycopersicon esculentum cv. Marmade), eggplant (Solanum melongena cv. Black beauty), tobacco (Nicotiana tabacum cv. Berley) and pepper (Capsicum annuum). Potato plants were grown in clay pots, 20 cm in diam containing a steam sterilized soil-manure (4:1) mixture, the other plants were grown in plastic cups containing the same mixture. Inoculations were performed by placing drops of bacterial suspension on the stems, which were surface sterilized with alcohol, and then puncturing the stem with a sterile needle through the drop. After inoculation the wounds were wrapped with sellotape strips. In addition potatoes were also inoculated through the mother tubers by placing drops of bacterial suspension on a cut surface of the tuber just before planting.

Five plants per host and strain were inoculated. Potato plants were inoculated about 20 days after emergence and the other hosts two (2) weeks after transplanting. The inoculated plants were maintained in a glasshouse at 25-30 C. Observations for host response (wilting) were made up to 45 days after inoculation.

Serological tests

The indirect immunofluorescence technique was applied to investigate the serological relationship between our isolates of Pseudomonas solanacearum and the three established races. Specific antisera against these races were obtained from the serotheque of INRA Angers, France (PSS 2096, Lot 851-PSM 1842 Lot 841 and PSP 1420 Lot 841) for races 1, 2 and 3 respectively.

RESULTS

The results are summarized in Tables 1, 2 and 3. It is obvious that the Greek isolates of P. solanacearum both from tomato and eggplant can be classified into the Biotype II of Hayward, since they utilize glucose, sucrose, fructose, galactose, glycerol, inositol, lactose, maltose and cellobiose, but not mannitol, sorbitol, dulcitol and sodium malonate, they reduce nitrates to nitrites and do not produce gas from nitrates (Table 1). From the

Table 1. Characteristics of 29 greek isolates of *Pseudomonas solanacearum*-comparison with those of Hayward's biotypes

	Greek isolates from:		Hayward's Biotypes			
Character	Tomato	Eggplant	I	II	III	IV
Utilization of:						
Glucose	+	+	+	+	+	+
Sucrose	+	+	+	+	+	+
Fructose	+	+	+	+	+	+
Galactose	+	+	±	+	+	+
Arabinose	-	-	-	±	+	-
Glycerol	+	+	+	+	+	+
Meso-inositol	+	+	+	+	+	+
Maltose	+	+	-	+	+	-
Lactose	+	+	-	+	+	-
Cellobiose	+	+	-	+	+	-
Mannitol	-	-	-	-	+	+
Sorbitol	-	-	-	-	+	+
Dulcitol	-	-	-	-	+	+
Nitrite from nitrate	+	+	+	+	+	+
Gas from nitrate	-	-	-	-	+	+
Malonate ulitization	-	-	-	-	-	-

+,= positive reaction; -,= negative reaction; ±,= growth without indicator change of Buddenhagen et al (1962), since they are pathogenic to tomato, potato and eggplant but not on tobacco which is infected by race 1 strains.The pathogenicity tests on pepper were not consistent and different results were obtained with the same strain.

Table 2. Relation of greek isolates of Pseudomonas solanacearum with the races designated by Buddenhagen

Race	Tomato	Tobacco	Potato	other solanaceae	Triploid banana	Diploid banana	Heliconia sp.
1	+	+	+	+	+	-	-
2	-	-	-	-	+	+	+
3	+	-	+	-	-	-	-
Greek isolates	+	-	+	-	Ø	Ø	Ø

+,= positive reaction; -,= negative reaction; Ø= not tested

The results tabulated in Table 3 verify our conclusion that the strains which we studied belong to race 3 since they reacted positively with the antiserum prepared against the race 3 type strain but not with the antisera against the race 1 and race 2.

Table 3. Serological relationship of greek isolates with the races of Pseudomonas solanacearum

	Antiserum against race		
Races	1	2	3
1	+	-	-
2	-	+	-
3	-	-	+
Greek isolates	-	-	+

+,= positive reaction; -,= negative reaction

Although the number of strains we examined is not so large we believe that these strains represent the Pseudomonas solanacearum population in Greece. From the biochemical tests reported by Zachos in his work with strains from potato it seems that his strains also belonged to Biotype II since they utilize maltose and lactose which are not utilized by biotype I and IV strains, and consequently they considered to be race 3. Our results are in agreement with the general concept that race 3 of P. solanacearum is more wide spread than the other races and that it probably came from the strain found to infect potatoes in Peru. The homogenity of the strains we examined supports the idea expressed by Martin et al (1982) that the greatest genetic diversity of P. solanacearum as reflected by the incidence of different biotypes and pathogenicity to several hosts appears to be in the tropical zone whereas.This diversity is less at greater distances from the equator. Understanding the potential host range of strains of P. solanacearum endemic in an area and recognition of the existence of strains specialized to particular hosts, is necessary in the effort to devise rational control strategies based on crop rotation.

REFERENCES

Buddenhagen I, Sequeira L and Kelman A (1962) Designation of races in Pseudomonas solanacearum. Phytopathology 52:726 (Abstr)

Buddenhagen IW, and Elsasser TA (1962) An insect-spread bacterial wilt epiphytotic of Bluggae banana. Nature 194:164-165

Hayward AC (1964) Characteristics of Pseudomonas solanacearum. Journal of Applied Bacteriology 27:265-277

He LY, Sequeira L and Kelman A (1983) Characteristics of strains of Pseudomonas solanacearum from China. Plant Disease 67:1357-1361

Kelman A and Person LH (1961) Strains of Pseudomonas solanacearum differing in pathogenicity to tobacco and peanuts. Phytopathology 51:158-161

Kelman A (1953) The bacterial wilt caused by Pseudomonas solanacearum. North Carolina Agricultural Experimental Station Technical Bulletin 99, p 194

Kelman A (1954) The relationship of pathogenicity in Pseudomonas solanacearum to colony appearance on a tetrazolium medium. Phytopathology 44:693-695

Martin C, French ER and Nydegger U (1982) Strains of Pseudomonas

solanacearum affecting Solanaceae in the Americas. Plant Disease 66(6):458-460
Zachos DG (1957) The brown rot of potatoes in Greece. Annals Inst Phytopath Benaki 1:125-127

PROBLEMS AND PROSPECTS IN CONTROLLING VERTICILLIUM WILT

E.C. Tjamos
Agricultural University of Athens
Laboratory of Phytopathology
Votanikos 118 55, Athens
Greece

INTRODUCTION

Verticillium dahliae and V. albo-atrum are destructive soil fungi, causing great damage to many crops. This review analyses the numerous and complicated problems in controlling the diseases and focuses on the prospects for controlling these pathogens through recent advances in plant pathology, which relate directly or indirectly to the vascular wilts. The headings listed below are treated in the appropriate detail in the text.

1. Problems in controlling Verticillium wilt
   - (a) Problems originating from the nature of the pathogen
     - a_1 Wide distribution
     - a_2 Vascular parasites
     - a_3 Lack of host specialization
     - a_4 Annual and perennial host-plants
     - a_5 Resistant resting structures
   - (b) Problems related to epidemiology
     - b_1 Dissemination of the pathogen
     - b_2 Involvement of nematodes and insects
     - b_3 Susceptible weeds
   - (c) Problems associated with general difficulties of control
     - c_1 Cultural practices
     - c_2 Lack of effective fungicides
     - c_3 High cost of applying soil fumigants
2. Prospects for controlling Verticillium wilt
   - (a) Resistant cultivars or root-stocks
   - (b) New fungicides

NATO ASI Series, Vol. H28
Vascular Wilt Diseases of Plants

(c) Biological control

(d) Special target chemicals

(e) Induced resistance

(f) Soil solarization

3. Conclusions

PROBLEMS IN CONTROLLING VERTICILLIUM WILT

(a) Problems originating from the nature of the pathogens. The following are generally considered to be important in the development of Verticillium wilt:

a_1 Wide distribution

V. dahliae and *V. albo-atrum* are widely distributed all over the world; *V. albo-atrum* is dominant in temperate zones whereas *V. dahliae* is predominant in tropical and subtropical regions.

a_2 Vascular parasites

Both species become established in the vascular system where they are confined to the xylem for long periods. They emerge at late stages of disease, when the host becomes moribund, and then start colonizing neighbouring tissues. The vascular phase of the disease is difficult to control because the pathogen is inaccessible. This location in xylem is important in the survival and growth of the pathogen and provides the fungus with minerals and products of cell wall degradation.

a_3 Lack of host specialization

Physiological races such as these in *Fusarium oxysporum* are rare in the two species. But a few reports indicate host specialization of *V. dahliae* in peppermint, pepper and Brussels sprouts and for *V. albo-atrum* in lucerne (Hawksworth and Talboys 1970). Also, it is known that isolates from one host are virulent to other species or could only cause mild symptoms to others (Tjamos 1981). The cropping systems followed over large geographical regions may affect the variety and number of strains particularly those able to infect introduced crops. However, Tjamos (1981) demonstrated the lack of host specialization of a large population of *V. dahliae* and *V. albo-atrum* in Greece but emphasized the danger of an increase in numbers of strains formerly rare in areas, where certain hosts were not previously grown. This operates against crop rotation as a means for controlling Verticillium wilts.

a_4 Annual and perennial host-plants

An exceptionally wide range of annual and perennial crop plants and weeds are susceptible to Verticillium spp. Surprisingly perhaps citrus and pometrees are resistant while certain monocotyledons such as barley or wheat are susceptible hosts. Thus, Mathre (1986) isolated V. dahliae from barley plants with symptoms associated with Cephalosporium stripe, longitudinal leaf stripes with brown vascular bundlles. The barley isolate was also pathogenic to wheat, oats potato and tomato. Others have reported the pathogen in ostensibly healthy monocotyledons (Krikum et al 1987).

a_5 Resistant resting structures

A critical factor in the survival and dissemination of the pathogens and consequently epidemiology of diseases, is formation by the pathogens resting structures as dormant resistant mycelium or microsclerotia. These can withstand adverse environmental conditions up to 14 years (Wilhelm 1955). But Schnathorst (1965) has questioned the rôle of these resting structures and suggested that the principal means of the survival are the various hyaline structures produced by the fungus rather than the dark, thick walled microsclerotial cells, which are protective or function as a substrate source.

(b) Problems related to epidemiology

b_1 Dissemination of the pathogen

Verticillium spp. are rarely transmitted by seed. Some exceptions are reports for sunflower (Sackston and Martens 1959) for lupin (Parnis and Sackston 1979) and lucerne (Huang et al 1985) where it is detectable in stems, present in peducles, penticels and pods, and in seeds up to 5%.

Contaminated and infected lucerne seed introduced the pathogen to new areas e.g. new pathogenic strains in certain lucerne growing regions of the USA. Heale (1985) suggested that the 1976 outbreak of Verticillium wilt of lucerne in the USA had originated from seed-borne inoculum imported from Europe. Seed transmission is exceptionally important because chemical treatment of internally infested seed with a mixture of captan and benomyl does not prevent sebsequent disease development. Park and Kim (1986) recovered V. albo-atrum and V. dahliae from spinach seed and V. albo-atrum from pepper seeds imported into Korea from Japan. The authors concluded that seed transmission constitutes a potential danger for the development of epidemics. Van der Spek (1972) demonstrated that beet and spinach seeds are infected by V. dahliae from parent plants. Ribeiro (1972) showed that V. dahliae was transmitted by eggplant seeds (up to 25%) following inoculation of plants. Potato tubers are an efficient means of dispersal. Nachmias and Krikum (1984) suggested that serology should be used in order to detect the pathogen in imported tubers. Unfortunately there are no treatments available to exclude the pathogen from propagative material.

Infected leaves also disperse the fungus. Evidence was reported (Brinkerhoff 1969) showing production of abundant microsclerotia in dried cotton leaves. Dried particles of leaves from Verticillium diseased cotton plants which were disseminated by air and by the lint of cotton seed or machinery carry the pathogen. The pathogen could be also detected in leaves of several susceptible hosts (Robb et al 1982). Similarly Tjamos and Botseas (1987) demonstrated V. dahliae in leaves of wilted olive trees and speculated on their rôle in spreading the pathogen.

b_2 Involvement of nematodes and insects

Numerous papers refer to the involvement of nematodes, Pratylenchus, Globodera, Meloidogyne and Tylenchorynchus spp. in increasing disease incidence. Davis and Everson (1986) have recently reconfirmed that wilt severity in potatoes is positively correlated with populations of soil-borne Pratylenchus spp.Franco and Bendezú (1985) showed synergism between V. dahliae and Globodera pallida. Similarly Hide et al (1984) found that ten potato cultivars planted in soils infested with G. rostochiensis developed symptoms earlier and yielded less than did nematode free populations. Tchatchova and Sikora (1983) showed synergistic effects following inoculation of cotton plants with V. dahliae and Rotylenchulus reniformis. The effects were: reduced shoot and root dry weight and earlier symptom expression compared with single inoculations.

Involvement of insects in dispersal and survival of the pathogen has been demonstrated by Huang et al (1986). V. albo-atrum was disseminated by leaf cutter bee (Megachile rotundata) which used pieces from diseased plants to construct brood cells. Also the insect carries the pathogen on its body thus carrying the pathogen to pollen grains or stigmata which are potential sites of infection. Kalb and Miller (1986) showed that adults of Bradysia impatiens from infected lucerne plants yielded V. albo-atrum colonies on an ethanol-streptomycin-agar. Similarly infested flies transmitted the pathogen in lucerne plants. Huang and Harper (1985) demonstrated that grasshoppers (Melanoplus sanguinipes and M. bivitattus), alfalfa weevil (Hypera postica) and woolly bear (Apantesis blakei) fed on lucerne plants yielded faeces containing the pathogen. Huang et al (1983) found in a naturally infested lucerne field that Acyrthoshiphon pisum is a vector of V. albo-atrum; conidia were carried as surface contaminants of legs and antenae.

b_3 Susceptible weeds

Susceptible weeds, or symptomless carriers of the pathogen are one of the main sources of inoculum in soil and are important in epidemics of annual and perennial crops. V. dahliae occurs in several weeds in olive orchards (Thanasoulopoulos et al 1981). This has been demonstrated for several other plantations.

(c) General problems in controlling diseases

c_1 Cultural practices

Control of Verticillium wilt long been based on cultural practices aimed at diminishing the incidence and severity of the disease rather than eliminating it. Thus, in orchards and trees with superficial root systems plowing or rotovation is to be avoided because it wounds roots thus facilitating infection by the pathogen. Rotation of dicotyledons with graminaceous crops used to be used to control disease but may be used with caution in the future now that it is known that graminaceous plants may be infected by V. dahliae (Mathre 1986). Koroleva et al (1986) showed that rotation with lucerne, sorghum and barley in severely infested cotton fields reduced infection only temporarly. Furrow irrigation is currently considered to increase wilt severity (Davis and Everson 1986) and does so less than sprinkler irrigation probably because furrow irrigation disperses fungal propagules more effectively. Excess nitrogen has been positively correlated to increased symptom expression because of extensive growth which is vulnerable to the pathogen. Hafez et al (1975) found that potassium fertilizers decreased incidence and severity of Verticillium wilt of cotton. Similarly Ashworth et al (1985) found that the incidence of Verticillium wilt of pistachio in California with low densities of V. dahliae microsclerotia (0.02-0.2/g soil) increased dramatically within two years in pistachio trees with low levels of potassium in leaves during mid summer. Treatment with 1.5 kg potassium per tree reduced symptoms by 35% in the year following application of the fertilizer. They also suggested that phosphorus deficiency increases susceptibility of trees to V. dahliae.

c_2 Lack of effective fungicides

Good chemical control of Verticillium wilt still has to be achieved.Despite some success in the 1970's in the use of systemic fungicides (mainly benomyl) to control Verticillium wilt, there was the problem that these fungicides move in the xylem, occlusion of which in infected plants diverts the transpiration stream and therefore the fungicides away from the pathogen and to the leaves. Thus, the fungicides are ineffective in the vascular phase, both protectively and curatively. Similarly the pathogen is not controlled in the prevascular phase when the source of the fungicides is exhausted. Unfortunately promising data from past experiments were prematurely extra polated to use in the field and with disappointing results. Erwin (1981) showed that pre- or post- planting soil application of thiabendazole (the best effective), benomyl and thiophanate methyl (the most effective) were not considered as providing economical and practical control. Indead as much as 24.2-56 kg active ingredient/ha was necessary to achieve partial or marginal control with only slight increase in yield. Despite the evidence of Erwin et al (1979) several field trials in Greece (Tjamos et al 1981) showed that growth retardants could not be used for economic control of the pathogen. The failure of growth retardants Cycocel or Pix to mitigate Verticillium wilt symptoms in cotton was another disappointment in the attempts to control the disease in the field.

c_3 High cost of applying soil fumigants

Disinfection of soil with certain fumigants effectively controls Verticillium spp. and fumigants have been extensively used in protected crops despite the problems of their application (highly toxic to plants and mammals, pollutants of soil and the water). Although effective they are expensive to be used in the field, so their use is restricted to high income crops.

PROSPECTS FOR CONTROLLING VERTICILLIUM WILT

(a) Resistant cultivars or root-stocks

Many cultivars of lucerne have various levels of resistance and most recent research has shown the potential of symptomless cultivars and the possibility of getting resistant cultivars by using cell and tissue culture methods. Pennypacker et al (1985) tested cultivars for resistance to V. albo-atrum and found out that symptomless plants contained the fungus in stem basis and concluded that a high proportion of plants of resistant cultivars have the potential of being symptomless carriers. Heale (1985) reported that recurrent selection of commercial cultivars for resistance produced cultivar Vertus with tolerance to the disease because of limited vessel colonization. Latunda-Data et al (1983) have shown that certain plants regenerated from mesophyll protoplasts were highly tolerant and they associated this with increase in ploidy. Viands (1985) making quantitative and qualitative genetic analyses of resistance concluded that both types (Maris Kabul and Vertus progenies) were useful for developing highly resistant cultivars.

In tomato resistance based on the Ve-gene for resistance to race 1 of the pathogen, has been extensively used to control Verticillium wilt all over the world. However, race 2 in V.dahliae virulent on the Ve-gene, cvrs, is prevalent in the USA and South Europe. Bersi et al (1984) reported race 2 of V. dahliae in Marocco and Matta and Garibaldi (1984) referring to a Verticillium wilt outbreak on tomato cultivars carrying the Ve-gene in Italy assumed that it was caused by strains of race 2, originating either by selection of a rare pathotype or a previously unexpressed genotype. Couriously Tjamos (1980) found that race 2 of V. dahliae is rare in Greece. This allow the extensive cultivation of cultivars possesing the Ve-gene. Okie and Gardner (1982) in North Carolina attempted to obtain cultivars resistant to race 2 by screening for seedling resistance parental, F1F2 and backcross generations of the lines Modern Mel and Walter. Analyses of F3 progenies indicated involvement of three genes. Their results suggested that resistance to race 2 could be maintained in advanced generations by the elimination of highly susceptible plants and by progeny testing to ensure recovery of the most resistant individuals rather that escapes. O'brien (1983) in Maryland evaluating accesions and cultivars of eggplants for resistance selected 36 out of 473 USA plant introductions as potentially resistant or tolerant to the disease, and 4 out of 10 cultivars with the higher resistant ratings. Marinković et al (1984) in Yugoslavia tested resistance of Capsicum annuum and C. frutescens to find that C. frutescens line 606 with highest resistance was far removed genetically from C. annuum so, when they used instead a Capsicum sp. line 674 as mediator, obtained evidene that resistance is higher in progenies of hybrids with C. annuum as the female parent in comparison with recurrent parents.

In cotton, in California, Wilhelm (1981) underlined the potential danger from the long-practised monoculture of Acala cultivars. He suggested they should be rotated with early maturing genetically distinct cultivars with the highest tolerance to the disease and adaptable to high density planting. Wilhelm et al (1985) reported two new breeding lines with resistance from G. barbadense, or a wild line of G. hirsutum or both, and with genetic backgrounds which differ from that of the Acala SJ varieties. They pointed out that resistance is optimized by adaptability to high density planting and that the main assets of the new lines are: early maturity, suitability for once-over harvesting and with fibre quality and yield competive with those of current cotton cultivars.

In olive two resistant root-stocks have been selected in the Universities of Davis and Berkeley in California, Oblonga and

Allegra respectively (Hartman et al 1971; Wilhelm 1981). Although their resistance was acceptable and promising in California, the behaviour of Oblonga was questionable in other countries. Tjamos et al (1985) in evaluating resistance of self-pollinated selections of Oblonga in Greece, showed that the resistance, expressed as limited or no colonization of vessels and absence of foliar symptoms, could be overcome by heavy inoculation of roots with V. dahliae isolates from Greece.

(b) New fungicides

Use of "new" underlines the need for chemicals to overcome the long-lasting problem of chemical control. The main requirements are for chemicals effective in the prevascular phase and which moving in the symplast rather that the apoplast. This is because xylem-mobile, apoplastic transport of fungicides is not effective in control. Xylem-mobile fungicides are taken up along with water and minerals in the root hair zone and accumulate at the sites of high transpiration, (e.g. leaf tips and margins). Their downward movement is very limited. In contrast, phloem-mobile symplastic translocation in the sieve tubes, through mass flow from source to sink, is much more promising because fungicides applied to leaves then pass to roots and also to new growth. Unfortunately, at present phloem-mobile chemical substances controlling Verticillium wilt are not available. Effective phloem-mobile fungicides are likely to be lipophilic compounds. They are most favourable for acting in roots because of their confined movement (Jacob and Newman 1987). Generally speaking hydrophilic compounds are transferred through the soil and rapidly absorbed by the roots, whereas lipophilic chemicals undergo a retarded movement in the soil and a long-lasting uptake into the plants.

(c) Biological control

Biological control of Verticillium wilt is based on recent research work carried out in the USDA, Beltsville Maryland on Talaromyces flavus. Marois et al (1982, 1984) showed that it is a potential antagonist of V. dahliae because it is able to colonize the rhizosphere of various hosts and inhibit germination or cause death of V. dahliae microsclerotia. The initial assumption was that the antagonist produces antibiotics but the more recent evidence suggests that hydrogen hyperoxide is the inhibitor (Fravel et al 1987; Kim et al 1987). Tjamos and Paplomatas (1987) showed that T. flavus survives after solarization of soil in globe artichoke fields or olive orchards infested with V. dahliae. Survival of the antagonist was almost always correlated with increase of its population in the rhizosphere soil of treated plots. Other agent used in biological control of Verticillium wilt of tomato is Penicillium chrysogenum (Dutta 1981). Brown (1985) reported that Zygorrhynchus moelleri causes lysis of cell walls of V. albo-atrum. It produces exo and endo-β-1,-3 gluconase similar to these of Trichoderma viride. Toyoda et al (1988) using technology of gene manipulation suggested that genes for fusaric acic detoxification from Cladosporium werneckii could be incorporated into Escherichia coli and Pseudomonas syringae pv. solanacearum. An avirulent strain of P. syringae pv. solanacearum carrying the gene for fusaric acid detoxification protected tomato plants against infection by Fusarium oxysporum f. sp. lycopersici. This may be a promising approach to the control of fungal wilt pathogens.

(d) Special target chemicals

Target-specific inhibitors such as difluoromethylornithine, an inhibitor of polyamine biosynthesis, might be of some practical use in controlling Verticillium wilt (Mussel et al 1987). Because polyamine biosynthesis in plants is primarily accomplished via the arginine decarboxylase and since most fungi do not have this pathway it seems likely that fungal diseases could be controlled by specific inhibition of the ornithine decarboxylase pathway the only one existing in fungi. Also, Christias et al (1981) showed that cysteamine hydrochloride or mercamptoethanol can inhibit formation of microsclerotia of V. dahliae growing in P.D.A. and they considered this a promising approach to controlling. Christias (1987) found that in stem segments from diseased cotton plants, formation of microsclerotia was inhibited by sulphydryl or sulphur-containing compounds in infected tissue. But it is not easy to see how such substances could be developed for disease control.

(e) Induced resistance

Induction of resistance indicates that resistance is not always determined by genetic information for a resistance mechanism but may depend on whether the information is expressed soon enough to be effective against a pathogen (Kuč 1982). Induced resistance has been used to control Verticillium wilt of cotton (Schnathorst and Mathre 1966) and Tjamos (1981) showed that high molecular weight substances from V. albo-atrum culture filtrates significantly decreased colonization of vessels and mitigated symptoms in cucumber plants infected with V. albo-atrum. However, the prospects of induced resistance on a large scale seem remote because so little is known about the factors which control it.

(f) Soil solarization

Soil solarization is very different from other procedures used against soil-borne pathogens. It is particularly effective against Verticillium wilt in several parts of the world (Katan et al 1976; Pullman et al 1981). Recently in Greece (Tjamos and Paplomatas 1988) showed that it controls Verticillium wilt of globe artichokes effectively, and has long-lasting beneficial effects up to three years following one period of solarization. Its pronounced effects on wilt of perennial crops included significant decrease in numbers of diseased plants, earlier formation of primary heads and doubling of yield. Survival of V. dahliae was decreased for at least three years and this may attributed at least partially to increase of natural populations of T. flavus in the rhizosphere of artichoke plants in solarized plots. Tjamos et al (1988a) also found that solarization applied individually around Verticillium infected olive trees in an established grove effectively contributed to a long-term remission of symptoms or recovery of treated compared to the untreated trees. This beneficial effect was attributed to the substantial decrease or eradication of V. dahliae microsclerotia in the treated soil. It was also evident that T. flavus populations survived better and increased in the rhizosphere of treated than untreated trees. It is likely that soil solarization induces suppressiveness in the soil (Greenberger et al 1987). The future of solarization in areas where climatic conditions are favourable will depend on low costs of application, re-use of plastics and on the long-lasting benefits. Recent work in Central Greece on the control of Verticillium wilt of tomatoes (Tjamos et al 1988b) showed that solarization in plastic houses, which remained closed during treatment, almost eradicated microsclerotia of V. dahliae for at least two consecutive years and sharply decrease disease incidence in solarized compared to untreated plots; absence of strong air currents in closed plastic houses facilitates covering of the soil with polyethylene sheets with less need to anchor them at the edges of the plastic house with large quantities of soil. Overlapping the plastic sheets on the moisturized soil is enough to firmly keep them in position and thus trap the solar energy.

CONCLUSIONS

Most problems relating to control of Verticillium wilt remain unsolved. Prospects for effective and economic control will depend on

(a) exploitation of modern breeding technology including use of tissue and protoplast culture and protoplast fusion

(b) the discovery and development of new chemicals that act in the prevascular phase of the disease such as phloem-mobile substances which are taken up by roots and then translocated through the plant and which also could be applied to leaves

(c) advances in biological control particularly by the use of antagonists which colonise the rhizosphere and could antagonise the pathogen before and during infection

(d) inexpensive and effective techniques such as soil solarization which change the environment in ways unfavourable to the pathogen.

REFERENCES

Ashworth LJ Jr, Gaona SA and Surber E (1985) Verticillium wilt of pistachio: the influence of potassium nutrition on susceptibility to infection by Verticillium dahliae. Phytopathology 75:1091-1093

Besri M, Zrouri M and Beye I (1984) Race characteristics and comparative pathogenicity for some isolates of Verticillium dahliae (Kleb.) obtained from resistant tomatoes in Marocco.

Phytopathol Z 109:289-294

Brinkerhoff IA (1969) The influence of temperature and soil microflora on microsclerotial development of Verticillium alboatrum in absised cotton leaves. Phytopathology 50:805-808

Brown AE (1985) Ferrous complexes in fungal disease suppression. In Annual report of research and technical work of the Department of Agriculture of Northern Ireland 1985. Belfast, Northern Ireland UK (1986) 139-140

Christias C, Tjamos EC, Zioudrou C and Kornaros E (1981) In vitro inhibition of microsclerotia formation by cysteine hydrochloride in Verticillium dahliae. Proceedings of the 3rd International Verticillium Symposium, Bari, Italy, p 21

Christias C (1987) The prospect of controlling Verticillium wilt of cotton by the use of specific inhibitors of microsclerotium formation. 4th National Phytopathological Conference of the Hellenic Phytopathological Society, October 13-15, 1987 Athens Greece (Abstr) p 17

Davis JR and Everson DO (1986) Relation of Verticillium dahliae in soil and potato tissue, irrigation method and N-fertility of Verticillium wilt of potato. Phytopathology 76:730-736

Dutta BK (1981) Studies on some fungi isolated from the rhizosphere of tomato plants and the consequent prospect for the control of Verticillium wilt. Plant Soil 63:209-216

Erwin DC, Tsai SD and Khan RA (1979) Growth retardants mitigate Verticillium wilt and influence yield of cotton. Phytopathology 69:283-287

Erwin DC (1981) Chemical Control:563-594 in Fungal Wilt Diseases of Plants. Mace ME, Bell AA and Beckman CH (eds)

Franco J and Bendezú E (1985) Study of the complex Verticillium dahliae Kleb. and Globodera pallida Stone and its effect on the behaviour of some Peruvian potato cultivars. Fitopatologia 20:21-27 Abstract from Review of Plant Pathology 1987:2996

Fravel DR, Kim KK and Papavizas GC (1987) Viability of microsclerotia of Verticillium dahliae reduced by a metabolite produced by Talaromyces flavus. Phytopathology 77:616-619

Greenberger A, Yogev A and Katan J (1987) Induced suppressiveness in solarized soils. Phytopathology 77:1663-1667

Hafez AAR, Stout PR and DeVay JE (1975) Potassium uptake by cotton in relation to Verticillium wilt. Agronomy Journal 67: 359-361

Hartman H, Schnathorst WC and Whisler J (1971) Oblonga, a clonal olive root-stock resistant to Verticillium wilt. Calif Agr 25 (6):12-15

Hawksworth DL and Talboys PW (1970) C.M.I. Description of Pathogenic Fungi and Bacteria No 255, 256

Heale JB (1985) Verticillium wilt of alfalfa, background and current research. Canadian Journal of Plant Pathology 7:191-198

Hide GA, Gorbett DC and Evans K (1984) Effects of soil treatments and cultivars on "early dying"disease of potatoes caused by Globodera rostochiensis and Verticillium dahliae. Annals of Applied Biology 104:277-289

Huang HC, Harper AM, Kokko EG and Howard RJ (1983) Aphid transmission of Verticillium albo-atrum to alfalfa. Canadian Journal of Plant Pathology 5:144-147

Huang HC and Harper AM (1985) Survival of Verticillium albo-atrum from alfalfa in feces of leaf-chewing insects. Phytopathology

75:206-208
Huang HC, Hanna MR and Kokko EG (1985) Mechanisms of seed contamination by Verticillium albo-atrum in alfalfa. Phytopathology 75:482-488
Huang HC, Richards KW and Kokko EG (1986) Role of the leafcutter bee in dissemination of Verticillium albo-atrum in alfalfa. Phytopathology 76:75-79
Jacob F and Neumann S (1987) Principles of uptake and systemic transport of fungicides within plant. In Modern,Selective Fungicides - Properties, Application, Mechanisms of Action. Longman Group UK Ltd London and VEB Gustar Fisher Verlag, Jena Lyr H (ed) 13-30
Katan J,Greenberger A, Alon A and Grinstein A (1976) Solar heating by polyethelene mulching for the control of diseases caused by soil-borne pathogens. Phytopathology 76:683-688
Kalb DW and Millar RL (1986) Dispersal of Verticillium albo-atrum by the fungus gnat (Bradysia impatiens). Plant Disease 70:752-753
Koroleva NS, Kasyanenco AG and Miller VR (1986) Studies on the ecology, populations and evolution of Verticillium species. I Population density dynamics of Verticillium dahliae Kleb. Mikologiya i Fitopatologia 20:509-512. Abstract from Review of Plant Pathology 66:4830
Kim KK, Fravel DR and Papavizas GC (1987) Identification and characterization of a metabolite produced by Talaromyces flavus which mediates biocontrol of Verticillium dahliae. Phytopathology 77:1720 (Abstract)
Krikum J and Bernier CC (1987) Colonization and infestion of Verticillium dahliae on some crop species. 4th International Verticillium Symposium University at Guelph, Guelph Ontario.Canadian Journal of Plant Pathology 9:81 (Abstract)
Kuč J (1982) Induced immunity to plant disease. Bioscience 32 (11):854-860
Latunde-Dada AO and Lucas JA (1983) Somaclonal variation and reaction to Verticillium wilt in Medicago sativa L. plants regenerated from protoplasts. Plant Science Letters 32:205-211
Marinkovič N, Miladinovic Z and Aleksič Z (1984) Resistance of progenies of some interspecies hybrids of Capsicum annuum L. to Verticillium albo-atrum Reinke & Berth. Zastita Bilja 35: 363-372. Abstract from Review of Plant Pathology 1985:5279
Marois JJ, Johnston SA, Dunn MT and Papavizas GC (1982) Biological control of Verticillium wilt of eggplant in the field. Plant Disease 66:1166-1168
Marois JJ, Fravel DR and Papavizas GC (1984) Ability of Talaromyces flavus to occupy the rhizosphere and its interaction with Verticillium dahliae. Soil Biology and Biochemistry 16:387-390
Mathre DE (1986) Occurence of Verticillium dahliae on barley. Plant Disease 70:981
Matta A and Garibaldi A (1984) Strains of Verticillium dahliae virulent on tomato hybrids with the Ve-gene for resistance to race 1 in Italy. In Proceedings of the 6th Congress of the Mediterranean Phytopathological Union Cairo Egypt:283-285
Mussel H, Osmeloski J, Wettlaufer SH and Weinstein L (1987) Suppression of Verticillium wilt of tomato by difluoromethyl ornithine a suicidal inhibitor of polyamine biosynthesis. Plant Disease 71:313-316

Nachmias A and Krikum J (1984) Transmission of Verticillium dahliae in potato seeds. Phytopathology 74:535-537

O'Brien MJ (1983) Evaluation of eggplant accesions and cultivars for resistance to Verticillium wilt. Plant Disease 67:763-764

Okie WR and Gardner RG (1982) Breeding for resistance to Verticillium dahliae race 2 of tomato in North Carolina. Journal of the American Society for Horticultural Science 107:552-555

Park CI and Kim WG (1986) Fungi detected in the seeds of vegetable crops imported from Japan. Korean Journal of Mycology 14, 89-91. Abstract from Review of Plant Pathology 66:740

Parnis EM and Sackston WE (1979) Invasion of lupin seed by Verticillium albo-atrum. Can J Bot 57:597-601

Pennypacker BW, Leath KT and Hill RRJr (1985) Resistant plants as symptomless carners of Verticillium albo-atrum. Plant Disease 69:510-511

Pullman GS, DeVay JE, Garber RH and Weinhold AR (1981) Soil solarization: Effects on Verticillium wilt of cotton and soilborne populations of Verticillium dahliae, Pythium spp., Rhizoctonia solani and Thielaviopsis basicola. Phytopathology 71: 954-959

Ribeiro R de LD (1972) On the infection of eggplant seeds by isolates of Verticillium dahliae Klebahn. Arq Univ Fed Rur Rio de Janeiro 2:17-20 (From Review of Plant Pathology, 4934: 1974)

Robb J, Smith A and Busch L (1982) Wilts caused by Verticillium species. Acytological survey of vascular alterations in leaves. Can J Bot 60:825-837

Sackston WE and Martens JW (1959) Dissemination of Verticillium albo-atrum on seed of sunflower (Helianthus annuus). Can J Bot 37:759-768

Schnathorst WC (1965) Origin of new growth in dormant microsclerotial masses of Verticillium albo-atrum. Mycologia 57:343-351

Schnathorst WC and Mathre DE (1966) Cross-protection in cotton with strains of Verticillium albo-atrum. Phytopathology 56: 1204-1209

Tchatchova J and Sikora RA (1983) Alteration in susceptibility of wilt resistant cotton varieties to Verticillium dahliae induced by Rotylenchulus reniformis. Zeitschrift für Pflanzen Krankheiten und Pflanzenschutz 90:232-237. Abstract from Review of Plant Pathology 1983:4696

Thanasoulopoulos CC, Biris DA and Tjamos EC (1981) Weed hosts as inoculum source of Verticillium in olive orchards. Phytopathologia Mediterranea 20:164-168

Tjamos EC (1980) Occurence of race 2 of Verticillium dahliae in Greece. Annls Inst Phytopath Benaki (N.S.) 12:216-226

Tjamos EC (1981) Virulence of Verticillium dahliae and V. albo-atrum isolates in tomato seedlings in relation to their host of origin and the applied cropping system. Phytopathology 71: 98-100

Tjamos EC (1981) Systemic induction of protection to Verticillium wilt in cucumbers. In Active defense mechanisms in plants, p 360. Wood RKS (ed)

Tjamos EC, Chitzanides A and Kornaros E (1981) Failure of two growth retardants to suppress Verticillium wilt symptoms and increase yield in cotton field trials in Greece. In the Proceedings of the 3rd International Verticillium Symposium, Bari Italy, p 60

Tjamos EC, Biris DA and Thanasoulopoulos CC (1985) Resistance evaluation to Verticillium dahliae of olive rootstocks, pp 18-19 (Abstract). In summaries of invited and Research Papers. 3rd National Phytopathological Conference of the Hellenic Phytopathological Society, October 16-18, 1985 Volos Greece

Tjamos EC and Botseas D (1988) Occurrence of Verticillium dahliae in leaves of Verticillium wilted olive - trees. Canadian Journal of Plant Pathology 9:86 (Abstract)

Tjamos EC and Paplomatas EJ (1987) Effect of soil solarization on the survival of fungal antagonists of Verticillium dahliae. EPPO Bulletin 17:645-653

Tjamos EC, Paplomatas EJ and Biris DA (1988) Long-lasting symptom remission in Verticillium wilted olive-trees after soil solarization. 5th International Congress of Plant Pathology, Kyoto, Japan. Abstracts of papers, p 213

Tjamos EC and Paplomatas EJ (1988) Long-term effect of soil solarization in controlling Verticillium wilt of globe artichokes in Greece. Plant Pathology (in Press)

Tjamos EC, Karapapas V and Bardas D (1988) Low cost application of soil solarization in covered plastic houses for the control of Verticillium wilt of tomatoes in Greece. Acta Horticulturae (in Press)

Toyoda H, Hashimoto H, Utsumi R, Kobayashi H and Ouchi S (1988) Detixification of fusaric acid by a fusaric acid resistant mutant of Pseudomonas solanacearum and its application to biological control of Fusarium wilt of tomato. Phytopathology (in Press)

Van der Spek J (1972) Internal carriage of Verticillium dahliae by seeds and its consequences. Meded Fak Landb Wet Gent 37: 567-573

Viands DR (1985) Comparison of "Maris Kabal" with "Vertus" alfalfa for resistance to Verticillium wilt. Crop Science 25:1096-1100

Wilhelm S (1955) Longevity of the Verticillium wilt fungus in the laboratory and field. Phytopathology 45:180-181

Wilhelm S (1981) Sources and genetics of host resistance in field and fruit crops:300-369. In Fungal wilt Diseases of Plants A.P. Mace ME, Bell AA and Beckman CH (eds)

Wilhelm S, Sagen JE and Tietz H (1985) Phenotype modification in cotton for control of Verticillium wilt through dense plant population culture. Plant Disease 69:283-288

MANIPULATION OF SOIL ENVIRONMENT TO CREATE SUPPRESSIVENESS IN SOILS

C. Alabouvette
INRA - Station de Recherches sur la Flore Pathogène dans le Sol
17 rue Sully - F21034 Dijon Cédex
France

INTRODUCTION

Recognition of the occurence of soils that suppress diseases due to soil-borne plant pathogens led to the basic concept of soil suppressiveness or as we proposed in french "soil receptivity to diseases" (Alabouvette et al 1982; Corman et al 1986; Louvet in this book). It means that the soil is not a neutral medium, hosting pathogenic microorganisms freely interacting with the roots of the host-plant. On the contrary, it is well established that the soil interferes in the relationships between microorganisms and plants as it can modify the interactions between microorganisms themselves. In other words, every natural soil possesses a greater or smaller ability to control diseases. Thus, it is not unrealistic to try to manipulate the soil environment to increase this natural biological control potential to make conducive soils suppressive.

Obviously, studies have been focused on a few examples of soils showing a strong and stable suppressiveness especially to Fusarium wilts. Many review papers discuss the results of studies dealing with characteristics and mechanisms of suppressiveness (Baker and Cook 1974; Toussoun 1975; Louvet et al 1981; Schneider 1982; Cook and Baker 1983; Alabouvette 1986). Thus, I will only review the pertinent facts from which approaches to biological control arose.

From a theoritical point of view, the soil may act through its physico-chemical properties or its microbial characteristics. Therefore, it should be possible to manipulate either the abiotic or the biotic environment to create suppressiveness. In fact only a few consistent results showed the role of the physico-chemical factors such as clay content, pH or water potential in the suppressiveness. On the contrary, many evidences proved that suppressiveness is based on microbial interactions between the pathogen and all or a part of the saprophytic microflora. Soil suppressiveness can be related to "general suppression" due to the activity of the total microbial biomass acting as a nutrient sink, or to "specific suppression" due to the activity of specific groups of antagonistic microorganisms (Cook and Baker 1983). These two broad categories of mechanisms led to different approaches for biological control. In the first case, the objective is to increase the general suppression by enhancing the activity of the microbial biomass in such a way that the pathogen will suffer from an intense competition for nutrients and fail to infect the plant. In the second approach, the main objective is to screen for a specific microorganism highly antagonistic to the pathogen and to introduce it into conducive soils to make them suppressive. Finally, it would be interesting to combine these to methods of biological control to create stable suppressiveness. In fact to establish successfully an exogenous population of microorganism in raw soils, it is necessary to determine its ecological competence and its competitive ability in regard to the rest of the microflora. Thus, the two approaches described above are intimately connected; and learning how we could modify the microbial balance in soil is the real challenge.

MANIPULATION OF THE ABIOTIC ENVIRONMENT

Until the early work of Reinking and Manns (1933) establishing that the distribution of Fusarium species was not similar in all soil types examined, attention has been focused on the relationship between suppressiveness and the clay content of soils. It was established that most of the soil suppressive to Fusarium wilts showed a high content of montmorillonite type clays (Stover 1962; Stotzky and Martin 1983). Although different hypotheses have been proposed, the role of the smectite-type clays is not yet clearly understood. These clays do not have a direct inhibitory effect on the pathogen. Their great exchange capacity may control the nutrient availability and the soil pH and thus influences the microbial balance in soils. There is possibly one exception to this indirect effect of the smectite clays. In the case of the "vertisols" which content more than 50% of montmorillonite, the suppressiveness to bacterial wilt due to Pseudomonas solanacearum persists after soil desinfestation, indicating that clays inhibit some properties necessary for bacterial pathogenicity (Bereau and Messiaen 1975).

Even if clays play an important role in the mechanisms of suppressiveness, it is not possible to modify the clay content of soils in nature. It could only be considered to mix clays with peat or other growing medium used to grow vegetables or horticultural crops under greenhouses. Experiments were set out in Dijon, 10 years ago, when soil-less cultures were developed to grow tomatoes, cucumbers and carnations. Mixing montmorillonite to peat in different proportions may induce variations of the level of conduciveness of the peat. But the results were not consistent and, at the present time, there is no example of addition of clays to induce suppressiveness.

Another soil characteristic correlated with suppressiveness to Fusarium wilt is high pH. Most, if not all the suppressive soils studied have a pH higher than 7. It was demonstrated that induced modifications of the pH resulted in changes of the suppressiveness level (Scher and Baker 1980). The role of high pH can be correlated to nutrient competition for iron between saprophytic microorganisms, especially fluorescent Pseudomonads, and the pathogenic Fusarium. It is well established that pH is involved in the cation balance in soils which determine the concentration of iron available for aerobic microorganisms (Baker et al 1986).

If manipulation of pH is possible at an experimental scale, it is unrealistic to consider application of lime at a large scale to increase pH at a higher value than 7 and create suppressiveness. Moreover, increasing the pH would not result in increased suppressiveness in all soil types because pH does not inhibit directly the development of the pathogen but acts through slight changes of the microbial balance.

Some environmental factors such as temperature, water potential, O_2/CO_2 contents are known to be involved in disease incidence. But, again, it is not clearly established if these factors affect directly or indirectly the behaviour of the pathogen. Some methods based on the manipulation of these physical factors have been proposed to control Fusarium wilts. Flooding was used by the United Fruit Company in Central America to control Fusarium wilt of banana, but in this case, flooding eliminates the pathogen as would do a chemical fumigation (Stover in Cook and Baker 1983). More interesting are the biological control methods based on solar heating of soil. It seems clear that solar heating not only affects the survival of the pathogen inoculum but also induces suppressiveness (Greenberger et al 1987). The mechanisms are not clearly understood, but solar heating induces a shift in the microbial balance that favours the saprophytic microflora and renders the establishment of the pathogen more difficult. This solarization induced suppressiveness applies towards different diseases due to fungi having quite different ecological characteristics. Thus, it is probably related to "general suppression" rather than "specific suppression" as defined by Cook and Baker (1983). Soil solarization is also used to control Verticillium wilts. In this case Tjamos and Paplomatas (1987) showed that solar heating induced an increase of the specific suppression due to Talaromyces flavus a fungal antagonist of Verticillium dahliae. Soil solarization is probably the best example of a physical treatment that induces complex modifications of the biotic environment and creates suppressiveness.

At the present time, it is difficult to base biological control of wilt-diseases on a manipulation of the soil physico-chemical environment. More research is needed to better understand the relationship between the abiotic characteristics of soils and the populations of microorganisms. This lack of knowledge is of first importance because the success of biological control based on the manipulation of the biotic environment depends also on the physico-chemical characteristics of the soil.

MANIPULATION OF THE BIOTIC ENVIRONMENT

Soil suppressiveness to Fusarium wilts is essentially microbiological in nature (Louvet in this book). In relation to this characteristic, it is easy to transfer suppressiveness by mixing a small amount of suppressive soil with a peat or a previously heat-treated conducive soil.

By successive dilutions it was possible to confer suppressiveness to a soil mixture containing only 1 ppm (V/V) suppressive soil, provided that the intermediate mixtures had been cultivated over the 3-month period between successive dilutions (Alabouvette 1986). Although consistent results were obtained by several workers, it is obviously impossible to base biological control of wilts on the transfer of suppressiveness from suppressive to conducive soils at a large scale. Even in the case of horticultural crops grown in potting mixtures the volume of soil to be manipulated is too massive to consider this procedure. Moreover, mixing natural suppressive soil with a conducive medium carries the risk of transmission of other soil-borne plant pathogens (Alabouvette et al 1985). Thus, study of the suppressiveness transferrability is essentially a tool to understand the mechanisms of soil suppressiveness.

Having demonstrated the biological nature of soil suppressiveness workers tried to identify within the microflora the antagonistic microorganisms that induce suppressiveness. According to the soils studied and the techniques used many different microorganisms have been involved in mechanisms of suppressiveness: actinomycetes, <u>Arthrobacter</u> spp., <u>Clostridium</u> spp., <u>Bacillus</u> spp., fluorescent <u>Pseudomonas</u>, non-pathogenic <u>Fusarium</u>.

The easiest way to establish that these antagonists play a role in the mechanisms of suppressiveness is to grow and introduce them into conducive soils or steamed suppressive soils to establish suppressiveness. This experimental demonstration of the role of saprophytic microorganisms led directly to biological control, based on massive introduction into conducive soils of selected populations of antagonistic microorganisms.

But, it was demonstrated that competition for nutrients resulting from the activity of the total biomass was also partly responsible for suppressiveness. In fact as stated by Cook and Baker (1983) any "specific suppression operates against a background of general suppression". Figure 1 illustrates this basic concept, and shows how general suppression and specific suppression work together to create suppressiveness in a previously heat-treated suppressive soil. Either a strain of non-pathogenic Fusarium or a mixed population of microorganisms, previously isolated from the Châteaurenard suppressive soil, induced suppressiveness; but when the non-pathogenic Fusarium is introduced in the soil already colonized by the complex microflora, the resulting suppressiveness is really strong; its level is comparable to the suppressive level in the natural suppressive soil.

Thus, to propose new biological control precedures it is necessary not only to consider applications of selected antagonists but also to manipulate the soil environment to increase the intensity of general competition between microorganisms.

MANIPULATION OF THE LEVEL OF GENERAL SUPPRESSION

There is not a general agreement about the role of general suppression in the mechanisms of soil suppressiveness. For example Baker and co-workers (Sneh et al 1984; Elad and Baker 1985) wrote that only specific mechanisms based on the activity of select groups of microorganisms are responsible for soil suppressiveness. On the contrary, Alabouvette and co-workers (Alabouvette et al 1983; Alabouvette et al 1985; Lemanceau et al 1988) clearly established that general competition for nutrients especially competition for carbon plays an important role. The well known phenomenon of soil fungistasis which occurs in any soil is related to general competition for carbon and energy (Lockwood 1981). It was established that addition of glucose to a suppressive soil induces drastic changes of the level of suppressiveness and may even render the soil conducive. The increase of wilt incidence is correlated with the concentration of glucose added to the soil (Fig. 2). These results proved that providing energy

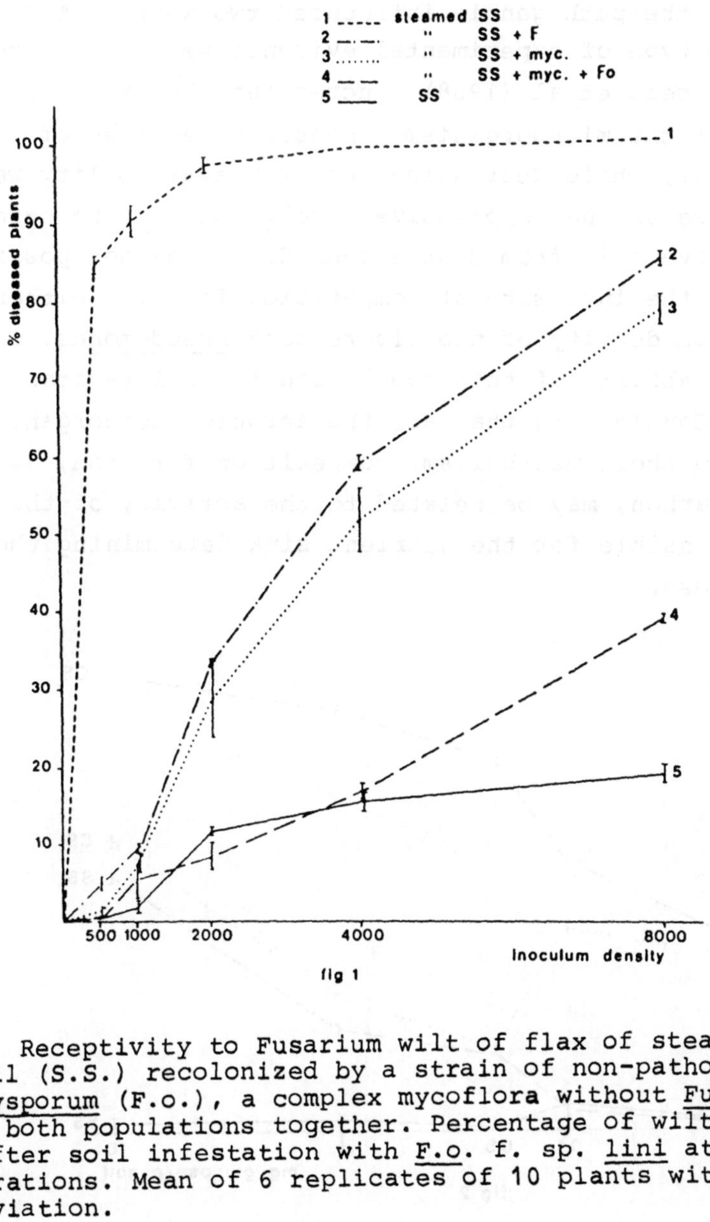

fig 1

Fig. 1. Receptivity to Fusarium wilt of flax of steamed suppressive soil (S.S.) recolonized by a strain of non-pathogenic Fusarium oxysporum (F.o.), a complex mycoflora without Fusarium(myc.) or with both populations together. Percentage of wilted plants 8 weeks after soil infestation with F.o. f. sp. lini at increasing concentrations. Mean of 6 replicates of 10 plants with its standard deviation.

at a sufficient level to counteract fungistatis allows chlamydospore germination and saprophytic development of all Fusarium including the pathogen and results in conduciveness. But, on the contrary if glucose is provided to a non infested conducive soil, it is used by the saprophytic microflora, which biomass increases resulting in an increased competition and a clear suppressive effect, when the pathogen is introduced two weeks later (Fig. 2). The same type of experimental evidence were discovered for iron by Lemanceau et al (1988). Increasing the amount of iron available for the microorganisms induces a decrease of the suppressive level, while decreasing the iron availability results in an increase of the suppressive level. But, in the case of the suppressive soil from Châteaurenard, it was not possible to correlate the intensity of competition for iron either with the population density of the fluorescent Pseudomonas, or with the specific ability of this population to chelate iron (Lemanceau 1988). Considering that all the aerobic microorganisms need iron fro their metabolism, competition for iron, as competition for carbon, may be related to the activity of the total biomass responsible for the nutrient sink determining the soil suppressiveness.

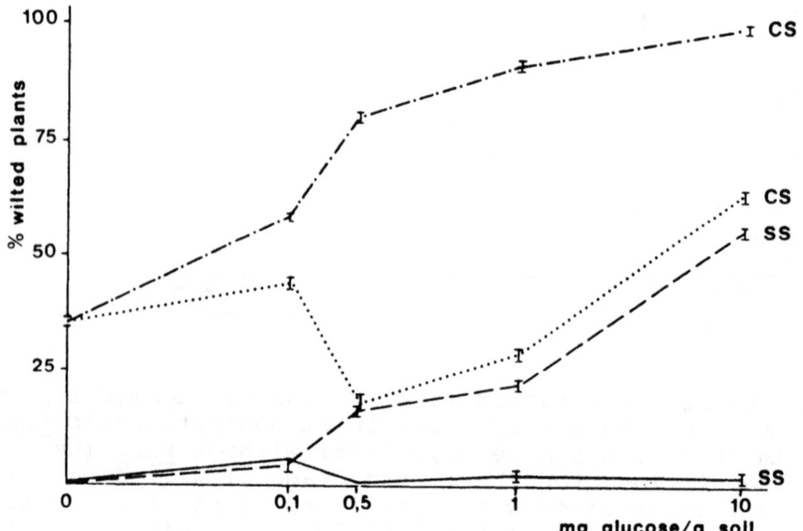

fig 2

Fig. 2. Modification of soil receptivity to Fusarium wilt induced by glucose amendments in suppressive (ss) and conducive (cs) soil.— —,—·—·= glucose amendment of soils previously infested with 2,500 CFU of Fusarium oxysporum f. sp. lini.———, ...= soils infested with 2,500 CFU of Fusarium oxysporum f. sp. lini 2 weeks after glucose amendment. Percentages of wilted plants determined after 8 weeks of cultivation.

In fact, the intensity of competition for a given nutrient

depends on both the amount of nutrient available for the microorganisms and the population density of microorganisms utilizing it at a given time. Therefore, general competition for nutrients is capable of increasing the intensity of the competition between the pathogen and antagonistic populations such as Pseudomonas especially well adapted to size upon one nutrient necessary for pathogen development.

Now the main question is to determine if it seems possible to manipulate the soil environment in such a way that the level of general competition will be increased and make the soil suppressive. It must be said that there is no clear example of such a manipulation resulting in an increased level of suppressiveness. Some of the earlier experiments performed by Papavizas and coworkers (Adams et al 1968; Lewis and Papavizas 1977) while studying the effect of organic amendements on pathogen survival and disease incidence may be interpreted in this way. They showed how organic amendements may increase the level of soil fungistasis or induce the germination and lysis of the fungal propagules. Lumsden et al (1979) studied the mechanisms of soil suppressiveness to Pythium aphanidermatum in two soils located in the same field, which differed in their organic matter content. In one case, the suppressiveness could be related to the germination lysis of the Pythium propagules due to a lack of available nutrient for the fungal growth; in the other case the high organic matter content supported an intense microbial activity resulting in fungistasis which inhibited the germination of Pythium propagules present at a high concentration in soil. In both cases, the disease incidence was low. This example, not related to wilt diseases, illustrates the complexity of the phenomena involved in disease suppressiveness and illustrates why biological control procedures based on the management of the general competition for nutrients are not used at the present time in commercial fields. The useful effects of organic amendments are difficult to monitor and may have opposite consequences on disease incidence depending on the soil type, the soil microflora, temperature and water potential, season of the year. It is why, at the present time, all the biological methods proposed to control wilt diseases are based on the activity of a selected strain of antagonist introduced into the soil or by seed treatment. But the effectiveness of such a procedure depends also on the relationships between the antagonistic population and the rest of the soil microflora. Therefore,more research is needed to better assess the interactions between populations of microorganisms in soil.

MANIPULATION OF THE LEVEL OF SPECIFIC SUPPRESSION

In most cases of soil suppressive to wilt-diseases, a specific population of antagonist has been involved. Thus, many attempts have been made to use these antagonists to control wilts.

Two different approaches are possible: increasing the biomass and/or activity of the resident population of the antagonist or introducing a selected strain as soil or seed treatment. In fact, the first proposal is not realistic. To specifically stimulate the activity of a given population of antagonist it would be necessary to know the exact requirements of this microorganism and provide a nutrient which would be metabolized by this population only. It probably will be possible in the future, especially if new molecules such as calystigine that specifically stimulate the development of a given microorganism are discovered in plant roots exudates. But at the present time, the organic amendments available do not have any specific effect, they increase the activity of various soil microorganisms and may have opposite consequences on disease incidence depending on the environmental conditions (see above).

Most of the attempts to control wilts are based on mass-production and introduction of a selected strain of antagonist. Isolated from suppressive or conducive soils many microorganisms have been described for their antagonistic activity against wilt pathogen: actinomycetes (Komada and Esuka 1970; Arjunaro 1971), Arthrobacter (Smith 1977), Clostridium and Bacillus (Tu et al 1975), fluorescent Pseudomonas (Scher and Baker 1982), Talaromyces flavus (Marois et al 1982), Trichoderma (Locke et al 1985).

Extensive studies were conducted to prove their efficacy in

microplots, but only a few of them have been tested in full-scale field experiments. A lot of problems have to be solved before being able to propose a biological control method for commercial use. I will not review all the work dealing with selection and application of antagonists to control wilts. Considering that the main problems are of the same nature and independent of the microorganism concerned, I will describe the steps necessary to achieve biological control and illustrate these aspects by examples taken from our own experience dealing with the application of non-pathogenic strains of Fusarium to control Fusarium wilts.

Screening for effective strains

The first question is to select from the population the antagonist strains with the best efficacy. Two complementary approaches may be followed. When the mode of action of the antagonist is well known it is possible to propose a screening procedure to allow the detection in vitro of strains showing specific properties related to the mode of action. For example knowing that antibiotic or siderophore production is involved in the mode of action of Trichoderma (Howell and Stipanovic 1979, 1980; Gurusiddaiah et al 1986) and fluorescent Pseudomonas (Olivier and Germain 1983), it is possible to appreciate in vitro the specific production of these metabolites by different strains of antagonist. After this first step it will be necessary to ensure that these strains are effective in the soil environment. This second procedure which consists of evaluating the efficacy of an antagonist in soil, in the presence of plant and pathogen. It is the only available approach when the mode of action is not clearly understood. This is the case with non-pathogenic Fusarium used to control Fusarium wilts. The candidate strains are grown in shake culture and mixed into a steamed soil. After two weeks of incubation, the pathogenic Fusarium oxysporum f. sp. lini is added at increasing concentrations of inoculum and a susceptible variety of flax is seeded. Disease incidence is assessed every week and the efficacy of strains is compared using a statistical method based on the calculation of the plant survival probability which is a function of time, inoculum concentrations and effectiveness of the antagonist (Corman et al 1986). The last step of this screening consists in ensuring that the selected strains effective against Fusarium wilt of flax are also effective against Fusarium wilts of tomato, carnation and muskmelon against which they will be used.

This type of screening procedure is time and space consuming but it gives consistent data. It allowed the selection of strain F.o.47 used in biological control experiments.

Mass-production of the antagonist

Having selected one or a few strains, the next problem to be solved is the production of large amounts of inoculum necessary to set out large scale experiments. Generally laboratories do not have all the technological facilities that industries possess. Thus plant pathologists use, on a large scale, the same methods developed to produce small quantities of inoculum, but they do not study all the practical problems posed by mass-production of inoculum. There are only a few papers concerning different processes to produce and formulate Trichoderma useful for biological control (Lewis and Papavizas 1985; Papavizas et al 1984) but we can assume that some results of interest for private companies have not been published.

In the case of non-pathogenic Fusarium, they are grown in fermentors, the growth medium is removed by centrifugation and the conidia are either frozen at -18 C or mixed with talc, an inert filler which, once dry, allows satisfactory storage and transport of the inoculum. This inoculum is easy to mix with soil directly as a powder or after having been suspended in water (Alabouvette et al 1987; Lemaire et al 1986).

For biological control it could be interesting to use a process of solid state fermentation to grow the antagonist (Durand and Chereau 1988). This process could facilitate the preparation of pellets including a food base already colonized by the antagonist. Introduced in the soil, the antagonist would have a greater probability to become established than when introduced as unprotected propagules.

MODES OF APPLICATION OF THE ANTAGONISTS

The mode of application of one antagonist depends on the ecological behaviour of both the antagonist and the pathogen and on the mode of action of the antagonist. The useful microorganism must be located where it has to work to be effective.

For example if the antagonist acts through a decrease of the inoculum concentration in soil it must be applied to the soil before planting; if it competes with the pathogen at the root surface, it must be located in the rhizosphere at the plant-root interface.

With respect to vascular wilts it seems necessary to protect the infection sites located at the root tips (Huisman 1982). Therefore, at the present time it is impossible to apply the proposed antagonists as seed treatment because they do not grow fast enough to colonize all the rhizosphere including the root tip. Efforts should be made to select rhizosphere competent strains as proposed by Ahmad and Baker (1987) for Trichoderma. In the case of the non-pathogenic Fusarium it was demonstrated that they must be incorporated uniformly into the soil before infestation by the pathogenic Fusarium. We showed that although the non-pathogenic strain F.o.47 does establish itself in raw soil at a level corresponding to the initial infestation concentration (Couteaudier et al unpublished), it failed to inhibit infection of the plant by the pathogen in muskmelon fields. Therefore, this non-pathogenic strain can only be used to recolonize a soil after fumigation or to colonize a soil-less growing medium such as peat mixture, rock-wool, etc. These conditions limit the interest of the biological control method to high-value crops such as vegetables and flowers grown under greenhouses.

All the experiments set out in France during the two last years were designed to control Fusarium wilts of tomato, muskmelon and carnation or Fusarium crown and root rot of tomato in soil after methyl-bromide fumigation or in different soil-less growing media.

Efficacy of biological control of Fusarium wilts

It is difficult to summarize briefly the results of a 2-year experimentation under normal growing conditions in commercial fields. With more than 2,000 kg of talc, a concentration of 2 to 5.10^8 propagules/g of F.o.47 were added either to soil-less growing medium or to soil in situ immediately after methyl-bromide fumigation, corresponding to 2 ha of experiments located in different areas in France. The establishment of the antagonist was regularly monitored and disease incidence scored. The results were variable from one site to another and did not allow any definite conclusions on the practicability of this control method.

It is interesting to illustrate some of the main problems encountered. The first one is related to the heterogeneous distribution of the pathogenic inoculum in a greehnouse. It would be necessary to carry out experiments in numerous small plots distributed all over the area of the greenhouse. But for practical reasons this design is not possible in a commercial greenhouse. For example Table 1 shows the great difference in infestation rate by F.o. f. sp. radicis lycopersici in an experiment set out in Britany in 1986. There was a positive effect of the biological control, but in these conditions it was difficult to appreciate the efficacy of the different treatments: F.o.47 added only to soil or/and to the root-ball of transplants of tomato.

Another problem is related to the efficacy of soil fumigation and to the rapidity of soil recolonization by the pathogenic Fusarium. In several experiments the disease incidence was so low in the fumigated control that it was impossible to detect any effect of biological control (Table 2).

Finally it was sometimes difficult to make biological control compatible with other cultural practices. Our biggest experiment failed for this reason. To save time and work, the grower decided to mix the talc-inoculum with an organic amendment to apply them together to the soil. This organic amendment, resulting from

Table 1. Effect of non-pathogenic strain F.o.47, added to soil and/or to the root-ball of transplants of tomato, on the % incidence of symptoms due to F.o. f. sp. radicis lycopersici in a glasshouse on a naturally infested soil. Treated rows compared with either end of the same rows as controls

	NON TREATED SOIL				TREATED SOIL			
	root balls treated		root balls non treated		root balls treated		root balls non treated	
	row 1	row 2	row 1	row 2	row 1	row 2	row 1	row 2
Control	20	20	13	27	40	67	60	80
Treatment	0	10	27	37	2	17	22	22
Control	40	40	25	60	55	45	35	40
Mean	root balls treated 5				soil and root balls treated 9		soil treated 22	
	control 30		control 31		control 52		control 54	

Table 2. Disease incidence of Fusarium wilt of muskmelon in a commercial greenhouse: the infested soil was fumigated with methyl-bromide and the non-pathogenic strain F.o.47 was introduced in the treated plot after soil fumigation at a rate of 1.10^5 CFU/g soil to a depth of 20 cm

	Percentage of wilted plants		
	06/19/87	06/26/87	06/30/87
fumigated soil	1.2	3	7
fumigated soil + treated by introduction of F.o.47	0.7	2.3	4.2

a composting process of different agricultural residues, caused a strong lytic effect on the non-pathogenic Fusarium and destroyed most of the inoculum before its introduction into the soil. The concentration was calculated to be 5.10^5 propagules of F.o.47 per g of soil to a depth of 10 cm but, microbiological analyses revealed only 5.10^3 CFU/g of soil. Moreover, in the same experiment soil recolonization by the pathogen was heterogeneous, the disease appeared in patches and depending on the localisation of the non-treated control we could prove that the biological treatment was effective or totally ineffective (Table 3).

Table 3. Population density of F. oxysporum and percentage of wilted plants in a commercial greenhouse treated with the non-pathogenic strain F.o.47

	North of the greenhouse				South of the greenhouse			
Plots	A	B	C	D	A	B	C	D
Treated	8		4,600	4.2	1		3,040	5.1
Treated	0		5,170	1.0	10		6,310	3.7
Control	53		180	7.9	9		80	14.7
Treated	10		6,210	3.2	21	+	3,510	26.9
Treated	9		5,580	5.7	156	+	4,690	17.7
Control	1	+	220	11.4	11	+	240	11.2
Treated	6		3,650	5.1	13		3,180	11.4
Treated	0		5,190	2.9	42	+	4,500	8.2
Treated	1		4,090	11.5	73	+	4,950	21.8
Control	187		300	21.5	240	+	110	14.4

A: population density of F. oxysporum (CFU /g soil) immediately after soil fumigation with methyl-bromide
B: + presence of the non-pathogenic F. oxysporum f.sp. lycopersici was detected immediately after soil fumigation
C: population density of F. oxysporum (CFU /g soil) two months after addition of the non-pathogenic strain in the treated plots
D: percentages of wilted plants at the end of the harvest period: 06/30/87 (percentage were calculated on a average of 1,080 plants per plot)

The best results were obtained in experimental plantings where the pathogen was added uniformly in all plots. Tomatoes were grown

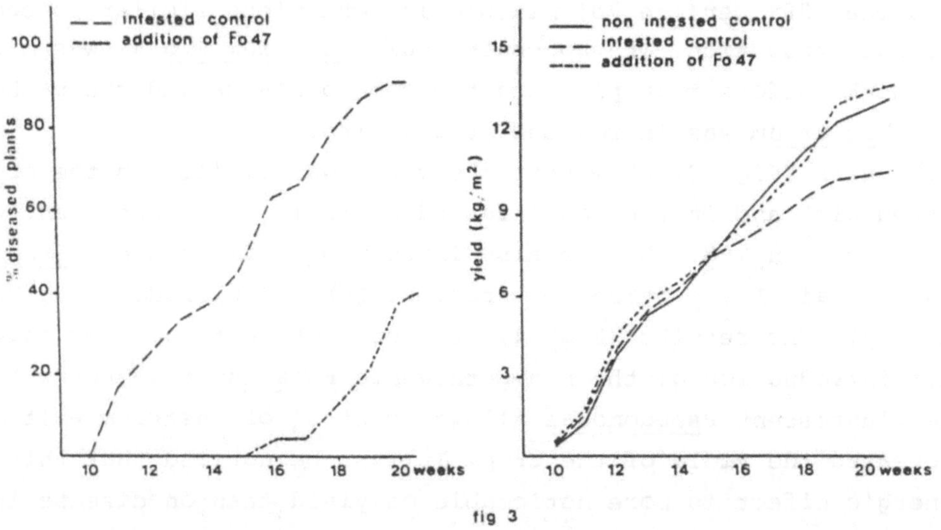

Fig. 3. Biological control of Fusarium wilt of tomato in soil-less culture (perlite-peat mixture) by introduction of non-pathogenic Fusarium strain F.o.47: percentage of diseased plants with time; yield (kg/m).

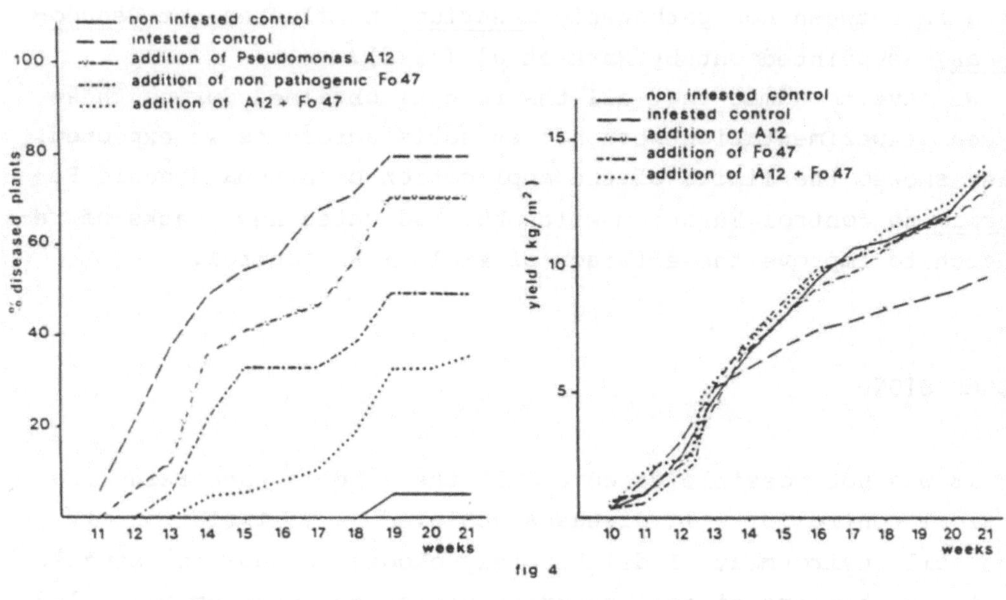

Fig. 4. Biological control of Fusarium wilt of tomato in soil-less culture (perlite-peat mixture) by introduction of non-pathogenic Fusarium strain F.o.47 and fluorescent Pseudomonas A12: percentage of diseased plants with time; yield (kg/m).

in a peat 30% perlite 70% mixture in conditions similar to commercial production. The non-pathogenic Fusarium F.o.47 was mixed two weeks before transplanting the tomato plants and the pathogenic Fusarium was introduced at planting.

Results (Fig. 3) show that the yield was similar in the protected plot and in the non infested control. This experiment was reproduced in 1987, but we also introduced a fluorescent Pseudomonas strain A12 provided by Prof. R. Baker (Colorado State University). The results (Fig. 4) are very interesting; they show that introduction of the non-pathogenic Fusarium together with the fluorescent Pseudomonas allowed control of Fusarium wilt and increased the yield of the crop. It must be noticed that this synergic effect is more noticeable on yield than on disease incidence. This illustrates another problem of this type of experiment; the yield is a more discriminant parameter than the percentage of wilted plants. But evaluation of the yield is time consuming because it is necessary to weigh the fruits every two days during 4 months. These results need to be confirmed but open the way to exciting research to study the mechanisms of synergism between non-pathogenic Fusarium and fluorescent Pseudomonas, as pointed out by Park et al (1988).

We have to admit that all the results obtained during these 2-year experimentation were not as satisfactory as we expected. They showed the limits of the application of non-pathogenic Fusarium to control Fusarium wilts but indicated new tracks of research to improve the efficacy of biological control.

CONCLUSION

It was not possible to cover all the aspects concerning biological control of wilt diseases achieved by manipulation of the soil-environment. I deliberately choose to show how studies on the mechanisms of soil suppressiveness to Fusarium wilts led to the formulation of basic concepts useful to propose new tracks towards biological control. But, we must recognize that we rediscovered the two basic approaches already studied for a long time: enhancement of the activity of the resident antagonists; introduction of selected antagonists effective against specific pathogens. However, studies of naturally suppressive soils permitted a search for effective antagonists in soils where their useful effect is obvious. Screening procedures are now mainly based on biological assays rather than on in vitro tests, and the mode of action of these antagonists is also studied in the soil environment. It is clear that progress has been made in the knowledge of the interactions between populations of saprophytic and pathogenic microorganisms but it is obvious that we are not able to really control these interactions to create suppressiveness. In regard of our own experience of studies dealing with Fusarium wilt suppressive soils, I think that we must now focus our efforts on studying the plant-microorganism relationships including both the plant-pathogen and the plant-antagonists interactions. Methodological difficulties may explain why this field of research has been neglected till today. But it is now necessary to answer some basic questions concerning the phase preceeding the vascular colonization of the plant. Do specific infection sites exist? Are they in a finite number per unit lengh of root? Are they susceptible for a finite period of time? Does the intrageneric competition between non-pathogenic Fusarium and the pathogen take place in the soil or at the root surface for the colonization of the infection sites? Does a plant stimulate specifically its pathogenic forma specialis? Do the plant exudate molecules modify the microbial balance in the rhizosphere?

The answer to these questions should provide a better understanding of the mechanisms of soil suppressiveness and open the way to an improvement of the biological control methods.

REFERENCES

Adams PB, Lewis JA and Papavizas GC (1968) Survival of root-infecting fungi in soil. IV. The nature of fungistasis in natural soil and cellulose amended soil on chlamydospore germination of Fusarium solani f. sp. phaseoli. Phytopathology 58: 378-383

Ahmad JS and Baker R (1987) Rhizosphere competence of Trichoderma

harzianum. Phytopathology 77:182-189
Alabouvette C, Couteaudier Y and Louvet J (1982) Comparaison de la réceptivité de différents sols et substrats de culture aux fusarioses vasculaires. Agronomie 2:1-6
Alabouvette C, Couteaudier Y and Louvet J (1983) Importance des phénomènes de compétition nutritive dans l'antagonisme entre microorganismes, 7-16. In: Les antagonismes microbiens. Modes d'action et application à la lutte biologique contre les maladies des plantes. 24e Coll Soc fr Phytopathol Bordeaux, 26-28 mai 1983, p 360
Alabouvette C, Couteaudier Y and Louvet J (1985) Fusarium wilt suppressive soils: mechanisms of suppression and management of suppressiveness. In Ecology and Management of Soil-borne Plants Pathogens (Parker CA, Moore KJ, Wong PTW, Rovira AD and Kollmorgen JF) (eds) pp 101-106 American Phytopathological Society, St Paul (USA)
Alabouvette C, De La Broise D, Lemanceau P, Couteaudier Y and Louvet J (1987) Utilisation de souches non pathogènes de Fusarium pour lutter contre les fusarioses: situation actuelle dans la pratique. Bulletin OEPP/EPPO 17:665-674
Arjunarao V (1971) Biological control of cotton wilt. III. In vivo effect of antagonists on the pathogen Fusarium vasinfectum. Proc Ind Acad Sci Sect B, 74:53-62
Baker KF and Cook RJ (1974) Biological control of plant pathogens. Am Phytopathol Soc St Paul, MN, pp 433
Baker R, Elad Y and Sneh B (1986) Physical, biological and host factors in iron competition in soils. pp 77-84. In Iron, Siderophores, and Plant Diseases (Swinburne TR ed) NATO ASI Series A Vol 177, pp 351
Bereau M and Messiaen CM (1975) Réceptivité comparée des sols à l'infestation par Pseudomonas solanacearum. Ann Phytopathol 7:191-193
Cook RJ and Baker KF (1983) The nature and practice of biological control of plant pathogens. Am Phytopathol Soc St Paul, MN, p 539
Corman A, Couteaudier Y, Zegerman M and Alabouvette C (1986) Réceptivité des sols aux fusarioses vasculaires: méthode statistique d'analyse des résultats. Agronomie 6:751-757
Durand A and Chereau D (1988) A new pilot reactor for solid-state fermentation: Application to the protein enrichment of sugar beet pulp. Biotechnology and Bioengineering 31:476
Elad Y and Baker R (1985) Influence of trace amounts of cations and siderophore-producing pseudomonads on germination of Fusarium oxysporum chlamydospores. Phytopathology 75:1047-1052
Greenberger A, Yogev A and Katan J (1987) Induced suppressiveness in solarized soils. Phytopathology 77:1663-1667
Gurusiddaiah S, Weller DM, Sarkar A and Cook RJ (1986) Characterization of an antibiotic produced by a strain of Pseudomonas fluorescens inhibitory to Gaeumamomyces graminis var. triticis and Pythium. Antimicrob Agents Chemothar 29:188-195
Howell CR and Stipanovic RD (1979) Control of Rhizoctonia solani on cotton seedlings with Pseudomonas fluorescens and with an antibiotic produced by the bacterium. Phytopathology 69:480-482
Howell CR and Stipanovic RD (1980) Suppression of Pythium ultimum induced damping-off of cotton seedlings by Pseudomonas fluo-

rescens and its antibiotic pyoluteorin. Phytopathology 70: 721-725
Huisman OC (1982) Interrelations of root growth dynamics to epidemiology of root-invading fungi. Ann Rev Phytopathol 20:303-327
Komada H and Ezuka A (1970) Ecological study of Fusarium diseases of vegetable crops. I. Survival of pathogenic fusaria in different soil types. Res Prog Rept Tokai-Kinki Natl Agric Exp Stn 6:1-6
Lemaire JM, Alabouvette C, Davet P and Tramier R (1986) Problems posed by the large scale application of microorganisms for biological control of soil-borne plant pathogens. Symbiosis 2:287-302
Lemanceau P, Alabouvette C and Couteaudier Y (1988a) Recherches sur la résistance des sols aux maladies. XIV. Modifications du niveau de réceptivité d'un sol résistant et d'un sol sensible aux fusarioses vasculaires en réponse à des apports de fer ou de glucose. Agronomie 8:155-162
Lemanceau P, Samson R and Alabouvette C (1988b) Recherches sur la résistance des sols aux maladies. XV. Comparaison des populations de Pseudomonas fluorescents dans un sol résistant et un sol sensible aux fusarioses vasculaires. Agronomie 8: 243-249
Lewis JA and Papavizas GC (1977) Effect of plant residues on chlamydospores germination of Fusarium solani f. sp. phaseoli and on Fusarium root rot of beans. Phytopathology 67:925-929
Lewis JA and Papavizas GC (1985) Characteristics of alginate pellets formulated with Trichoderma and Gliocladium and their effect on the proliferation of the fungi in soil. Plant Pathology 34:571-577
Locke JC, Marois JJ and Papavizas GC (1985) Biological control of Fusarium wilt of greenhouse-grown Chrysanthemums. Plant Disease 69:167-169
Lockwood JL (1981) Exploitation competition. In: Wicklow DT and Carroll G (eds) The fungal community. Its organization and role in the ecosystem M Dekker:319-350
Louvet J, Rouxel F and Alabouvette C (1977) Recherches sur la résistance des sols aux maladies. II. Incidence des traitements thermiques sur la résistance microbiologique d'un sol à la fusariose vasculaire du melon. Ann Phytopathol 9:183-192
Lumsden RD, Lewis JA, Ayers WA, Adams PB and Papavizas GC (1979) Natural biocontrol of Pythium aphanidernatum. Proc IX Int Congr Plant Bot, Washington DC, 5-12 August 1979
Marois JJ, Johnston SA, Dunn MT and Papavizas GC (1982) Biological control of Verticillium wilt of eggplant in the field. Plant Disease 66:1166-1168
Olivier JM and Germain R (1983) Etude des antibiotiques volatils des Trichoderma, 17-34. In:"les antagonismes microbiens",24e Colloque FSP Coll INRA 18 (ed INRA Publ Versailles)
Park CS, Paulitz TC and Baker R (1988) Biocontrol of Fusarium-wilt of cucumber resulting from interactions between Pseudomonas putida and non-pathogenic isolates of Fusarium oxysporum. Phytopathology 78:190-194
Reinking OA and Manns MM (1933) Parasitic and other Fusaria counted in tropical soils. Z Parasitenkunde Inf 6:23-75
Scher M and Baker R (1980) Mechanism of biological control in a

Fusarium suppressive soil. Phytopathology 70:412-417
Schneider RW (1982) (ed) Suppressive soils and plant disease. Ann Phytopathol Soc St Paul, MN, pp 88
Smith SN (1977) Comparison of germination of pathogenic *Fusarium oxysporum* chlamydospores in host rhizosphere soils conducive and suppressive to wilts. Phytopathology 67:502-510
Sneh B, Dupler M, Elad Y and Baker R (1984) Chlamydospore germination of *Fusarium oxysporum* f. sp. *cucumerinum* as affected by fluorescent and lytic bacteria from *Fusarium* suppressive soil. Phytopathology 74:1115-1124
Stotzky G and Martin RT (1963) Soil mineralogy in relation to the spread of Fusarium wilt of banana in Central America. Plant Soil 18:317-338
Stover RH (1962) Fusarial wilt of bananas and other *Musa* species. Commonwealth Mycol Inst Phytopathol Paper 4, pp 117
Tjamos EC and Paplomatas E (1987) Effect of soil solarization of the survival of fungal antagonists of *Verticillium dahliae*. Bulletin OEPP 17:645-654
Toussoun TA (1975) Fusarium-suppressive soils. In: Bruehl (ed) Biology and control of soil-borne plant pathogens. Am Phytopathol Soc St Paul, MN:145-151
Tu CC and Cheng YH (1981) Soil microbial activity in relation to Fusarium wilt-suppressive soil and conducive soil, 1-12 In: Taiwan DAIS Scientific Meeting Report

ALLELOPATHIC CONTROL OF FUSARIUM OXYSPORUM F. SP. RADICIS-LYCOPERSICI

W.R. Jarvis
Agriculture Canada
Research Station
Harrow, Ontario NOR 1G0
Canada

INTRODUCTION

There is no good conventional fungicidal control for Fusarium crown and root rot (FCRR) of tomatoes caused by Fusarium oxysporum f. sp. radicis-lycopersici Jarvis and Shoemaker (FORL), but Jarvis (1977) suggested that the disease would be amenable to control by biological means, perhaps in a similar way to the use of suppressive soils to control wilt-inducing formae speciales of Fusarium oxysporum, (Louvet et al 1981; Cook 1982), or by the augmentation of soils with substrates conducive to suppressiveness.

BIOLOGICAL CONTROL

Jarvis (1977) found that plants transplanted into a newly-steamed, reinfested groundbed had a mean disease rating of 3.64 on a 1-5 scale, those transplanted into a DD-fumigated one had a rating of 2.40 and those in an unsterilized one that had just borne a tomato crop had a rating of only 1.86. At each stage of plant propagation, leaving the medium unsterilized almost completely protected the plant from an added inoculum. Plants were also protected by the addition of non-sterile peat moss or field soil or a selected mixture of microorganisms from field soil into a sterile medium or soil, as well as by a mixture of microorganisms isolated from field soil which were highly antagonistic to FORL

NATO ASI Series, Vol. H28
Vascular Wilt Diseases of Plants
Edited by E. C. Tjamos and C. Beckman

(W.R. Jarvis and H.J. Thorpe unpublished). These findings suggested a role of microorganisms in biological control.

GREEN MANURE CROPS

Since the addition of field soil was considered likely to add other pathogens to the greenhouse groundbed, Jarvis and Thorpe (1981) investigated the use of green manure crops to enhance the population of FORL-antagonistic microorganisms in sterilized soils. Thus, plants of leaf mustard (Brassica juncea Coss.), spinach (Spinacia oleracea L.), cress (Lepidium sativum L.), and lettuce (Lactuca sativa L.) grown in sterilized FORL-reinfested soil, and chopped into the soil prior to planting tomatoes, all substantially reduced the mean disease rating. In particular, lettuce reduced disease to nil or trace levels (Table 1). On a commercial scale, the interpolation of a lettuce crop between successive tomato crops offered a cheap and even profitable means of disease control.

Table 1. Effect of preceding crop residues on incidence of Fusarium crown and root rot in 90-day-old tomato plants, cv. MR13, growing in potted soils infested with Fusarium oxysporum f. sp. radicis-lycopersici, 1979 and 1980 (Jarvis and Thorpe 1981)

	Mean disease rating[1] in	
Preceding crop	1979	1980
Lettuce	0.00 a[2]	0.35 a
Pepper cress	0.20 ab	1.81 bc
Mustard greens	0.45 bc	1.42 b
Spinach	0.68 c	2.19 c
Fallow	1.20 c	2.94 d

[1] 0 = no symptom, 5 = severe symptoms and death. Forty plants were rated in each treatment in 1979 and 48 in 1980.
[2] By Duncan's multiple range test, figures, within each year, followed by the same letter are not significantly different.

Jarvis and Thorpe (unpublished) found that dandelion (Taraxacum officinale L.) also exerts a powerful suppression of FCRR, and has been used as a companion plant in hydroponic production of tomato in sawdust in greenhouses in British Columbia.

ALLELOPATHY

It is interesting to note that Taraxacum and Lactuca are both members of the tribe Lactuceae of the Compositeae, in which very biologically-active substances are found (Gonzalez 1977). Since lettuce and dandelion are more effective in biological control than other green manure crops, it was considered possible that its effect could be an allelopathic one (Arnason and McLachlan personal communication 1983; Jarvis 1983; Jarvis and Thorpe 1981, 1983). Allelopathy in relation to plant diseases is discussed by Patrick (1986).

The mode of action of lettuce and dandelion, either as residues in soil or as companion plants with tomato, is not known. Three hypotheses are advanced: that these plants may promote a microflora that is particularly antagonistic to FORL in the rhizosphere; that chemicals secreted by them elicit a phytoalexin reaction in the host; or that allelochemicals secreted or released by T. officinale and L. sativa interfere directly with FORL.

Of these hypotheses, the first is certainly feasible; green manure crop residues stimulate intense microbial activity in soil, perhaps fairly specific anti-FORL activity, and Jarvis (1977) showed that FORL is susceptible to biological control by microorganisms. Subsequently Marois and Mitchell (1979, 1981a, 1981b), Marois et al (1981), and Sivan and Chet (1986) and Sivan et al (1987) have demonstrated the limiting effects of the soil microflora on FORL populations in soil. However, Jarvis and Thorpe (1981) found that lettuce residues controlled FCRR significantly better than residues of three other crops (Table 1), and suggested that allelopathy also played a part in control.

ALLELOCHEMICALS

Arnason, McLachlan, Morand and Kaminski (personal communication 1983) examined the constituents of extracts of root and leaves of Taraxacum officinale and found that although some filter-sterilized fractions in water were somewhat inhibitory to the germination of microconidia of FORL (Table 2), the inhibition could not be considered sufficient to account for the observed protection of tomato (Fig. 1).

Table 2. Percent germination of microconidia of Fusarium oxysporum f. sp. radicis-lycopersici in the presence of a whole extract from Taraxacum officinale and its liquid chromatograph fractions on Sephadex LH-20 after 24 hours incubation. From Arnason et al (unpublished results 1983)

	% Germination		
Fraction	Unfiltered	Filtered[1]	
	Trial 1	Trial 2	Trial 3
Whole extract	15.8	91.4	86.1
Fraction 1	23.0	94.0	87.9
2	36.2	72.6	88.9
3	77.9	100.0	82.9
4	100.0	48.2	72.4
5	82.0	85.3	81.8
6	91.8	91.0	84.6
7	47.8	68.0	81.8
Water control	----	71.0	71.0

[1] Filtered samples were passed through a 0.22 m filter to remove contaminants.

Arnason et al considered that the most likely explanation was that allelochemicals of small M.W. (87 to 238) in several fractions acted synergistically in reducing infection. They considered unlikely the hypothesis that they acted as elicitors to induce phytoalexin production in the host. The known elicitors

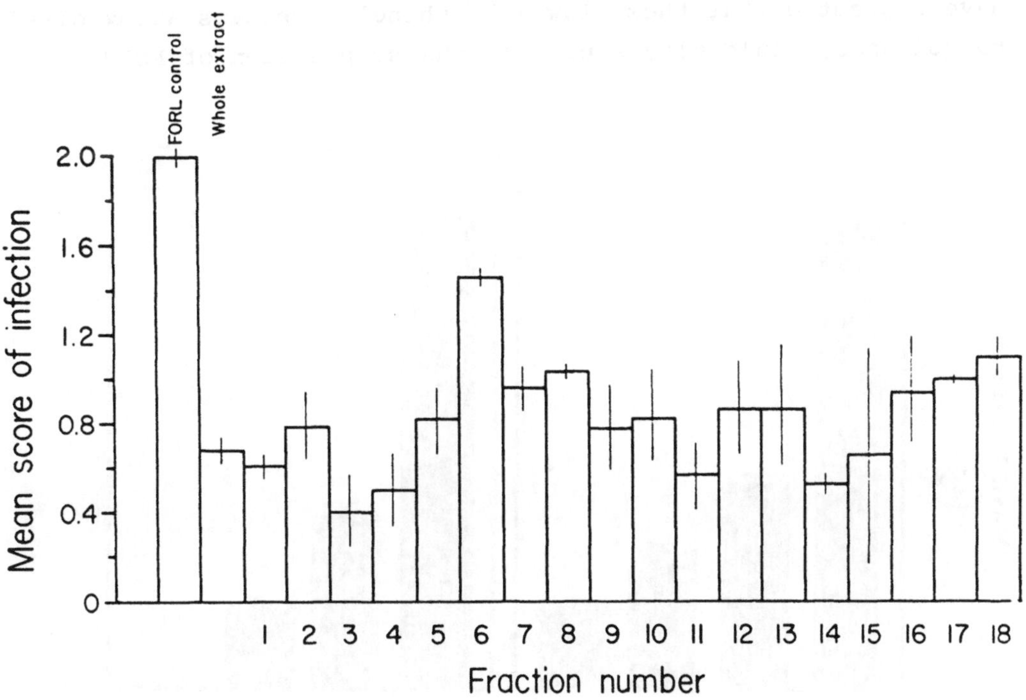

Fig. 1. Activity of different fractions of filter-sterilized lettuce extract fractions in limiting infection of tomato seedlings cv. MR13 in a petri-dish test of pathogenicity (Sanchez et al Phytopathology 65:726-727 1975). Scale: 0 = no symptoms, 3 = severe symptoms and death.

are mostly of high M.W.; those from Fulvia fulva that are responsible for the induction of the phytoalexins falcarinol and falcarindiol, and for the release of tomatine from vacuoles in leaves of tomato (Dow and Callow 1979; DeWit and Rosenboom 1980) are glycopeptides, quite unlike the substances that Arnason et al found. Arnason et al also discounted a role for polyacetylenes, very active allelochemicals also produced by the Lactuceae (Gonzales 1977; Stevens 1986).

Following the work of Arnason et al and concentrating on the phenolic components of lettuce and dandelion, Kasenberg and Traquair (1988) showed that caffeic and ferulic acids, and catechol inhibited the germination of microconidia, and reduced germ tube growth in FORL on agar (Fig. 2). Kasenberg and Traquair (1988a) have suggested that these low M.W. phenols, perhaps autoxidised to quinones, could play a part in the suppression of FCRR.

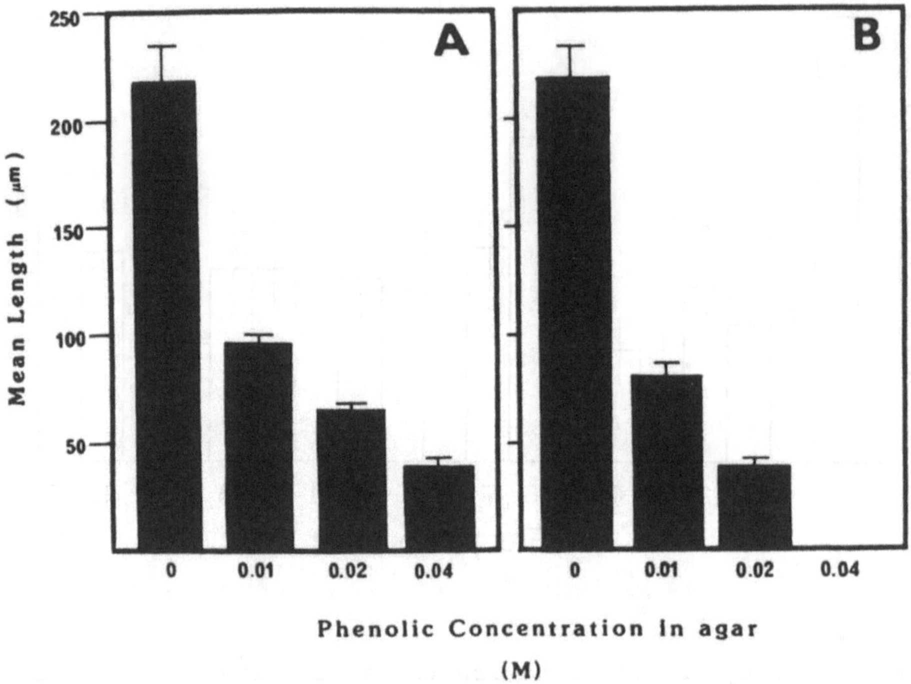

Fig. 2. The effects of A) caffeic acid and B) ferulic acid on growth of germ tubes from microconidia of *Fusarium oxysporum* f. sp. *radicis-lycopersici* on a nutrient agar (From Kasenberg and Traquair 1988).

One such mechanism is the sequestration of iron, necessary for the germination of microconidia, hyphal growth, and pathogenesis. Kasenberg and Traquair (1988b) showed that this is probably accomplished by catecholic siderophores. Thus the mechanism disease suppression by the Lactuceae is reminiscent of the suppression of Fusarium wilts mediated by *Pseudomonas putida* (Scher 1986).

Brammall (1986) indicated a possible role for phenols in lignification of the host cell wall, strengthening the mechanical barrier to colonization. He was able to break down the resistance of resistant tomato cv. Larma by treatment with low doses of glyphosate, which decreases phenolic levels in some plants and is believed to inhibit the conversion of shikimate to chorismate in the phenolic lignin pathway.

CONCLUSION

Of the three hypotheses advanced to explain the biological control of FCRR, there is no doubt that the natural soil microflora plays a part in suppressing the saprophytic competitive ability of FORL, perhaps enhanced in some specific way by residues of lettuce and dandelion; elicitation of phytoalexins by substances from lettuce and dandelion is considered unlikely; but direct action of phenolic substances from lettuce and dandelion appears to be the most likely hypothesis. It is the subject of intensive research.

REFERENCES

Cook RJ (1982) Use of pathogen-suppressive soils for disease control. In: Schneider RW (ed) Suppressive soils and plant disease. Am Phytopathol Soc, St Paul, MN pp 51-65

DeWit PJGM and Rosenboom PHM (1980) Isolation, partial characterization and specificity of glycoprotein elicitors from culture fitrates, mycelium and cell walls of *Cladosporium fulvum* (syn. *Fulvia fulva*). Physiol Plant Pathol 16:391-408

Dow JM and Callow JA (1979) Leakage of electrolytes from isolated mesophyll cells of tomato induced by glycopeptides from culture filtrates of *Fulvia fulva* (Cooke) Ciferri (syn. *Cladosporium fulvum*). Physiol Plant Pathol 15:27-34

Gonzalez AG (1977) Lactuceae-chemical review. In: Heywood VH, Harborne JB and Turner BL (eds) The biology and chemistry of the Compositeae. Academic Press, London, pp 1082-1095

Jarvis WR (1977) Biological control of Fusarium. Can Agric 22: 28-30

Jarvis WR (1983) Progress in the biological control of plant diseases. Proc 10th Int Cong Plant Protection 1095-1105

Jarvis WR and Thorpe HJ (1981) Control of Fusarium foot and root rot of tomato by soil amendment with lettuce residues. Can J Plant Pathol 3:159-162

Kasenberg TR and Traquair JA (1987) Allelopathic biocontrol of Fusarium crown and root rot in greenhouse tomatoes. Can J Plant Pathol 9:280 (Abstr)

Kasenberg TR and Traquair JA (1988a) Effects of phenolics on growth of *Fusarium oxysporum* f. sp. *radicis-lycopersici* in

vitro. Can J Bot (in Press)
Kasenberg TR and Traquair JA (1988b) Allelopathic biological control of Fusarium rot in tomatoes. Proc XIV Int Conf Groupe Polyphenols (Abstr) p 63
Louvet J, Alabouvette C and Rouxel F (1981) Microbiological suppressiveness of some soils in Fusarium wilts. In: Nelson PE, Tousson TA and Cook RJ (eds) Fusarium diseases, biology and taxonomy. PSU Press, University Park, PA pp 261-275
Marois JJ and Mitchell DJ (1979) The relationship of density of chlamydospores of Fusarium oxysporum to infection of tomatoes in fumigated and nonfumigated soils. Phytopathology 69:1037
Marois JJ and Mitchell DJ (1981a) Effects of fumigation and fungal antagonists on the relationships of inoculum density to infection incidence and disease severity in Fusarium crown rot of tomato. Phytopathology 71:167-170
Marois JJ and Mitchell DJ (1981b) Effects of fungal communities on the pathogenic and saprophytic activities of Fusarium oxysporum f. sp. radicis-lycopersici. Phytopathology 71:1251-1256
Marois JJ, Mitchell DJ and Sonoda RM (1981) Biological control of Fusarium crown rot of tomato under field conditions. Phytopathology 71:1257-1260
Patrick ZA (1986) Allelopathic mechanisms and their exploitation for biological control. Can J Plant Pathol 8:225-228
Scher F (1986) Biological control of Fusarium wilts by Pseudomonas putida. In: Swinburne TR (ed) Iron, siderophores, and plant diseases. Plenum Publishing Corporation, London, pp 109-117
Sivan A and Chet T (1986) Possible mechanisms for control of Fusarium spp. by Trichoderma harzianum. Proc Brit Crop Prot Conf 2:865-872
Sivan A, Ucko O and Chet I (1987) Biological control of Fusarium crown rot of tomato by Trichoderma harzianum under field conditions. Plant Disease 71:587-592
Stevens KL (1986) Polyacetylenes as allelochemicals. In: Putnam AR and Tang CS (eds) The science of allelopathy. John Wiley & Sons, New York pp 219-228

BIOCONTROL OF VERTICILLIUM WILT OF EGGPLANT AND POTATO

Debora R. Fravel
Biocontrol of Plant Diseases Laboratory
Plant Sciences Institute
United States Department of Agriculture
Agricultural Research Service
Beltsville, MD 20705
USA

INTRODUCTION

Verticillium dahliae is a soil inhabiting fungus that causes wilt diseases on over 250 different plant species worldwide. In most hosts there is no agronomically acceptable resistance to this pathogen, and with the exception of the broad spectrum biocides such as metham-sodium and methyl bromide, there are no chemicals which will control this pathogen. Moreover, use of these chemicals often is not economically feasible and these chemicals may have adverse ecological consequences. For these reasons, a program to investigate alternative control measures for V. dahliae, such as biocontrol, was initiated.

SUPPRESSION OF VERTICILLIUM WILT OF EGGPLANT BY TALAROMYCES FLAVUS

Eggplant - V. dahliae was selected as a model disease system for a study because of the ease of integration of biocontrol agents into current grower practices (Marois et al 1982). Eggplants are usually started in the greehnouse and transplanted, along with the soilless potting mix in which they are grown, to the field after 6-8 weeks. The 6-8 weeks period may permit establishment of the antagonist in the rhizosphere under conditions of relatively low competition, before going to the field under a situation of relatively high competition. The greenhouse envi-

NATO ASI Series, Vol. H28
Vascular Wilt Diseases of Plants
Edited by E. C. Tjamos and C. Beckman

ronment also facilitates introduction of the antagonist, since the propagules can be added during regular watering. Marois et al (1982) demonstrated that preplant ascospore drenches of the soil - inhabiting fungus Talaromyces flavus (Klöcker) Stolk and Sampson (anamorph: Penicillium dangeardii Pitt), isolate Tf1, suppressed incidence of Verticillium wilt and resulted in a yield increase comparable to chemical control. T. flavus is an ascomycete which is widely distributed throughout temperate regions (Fravel and Adams 1986).

STUDIES ON BIOCONTROL OF VERTICILLIUM WILT OF POTATO BY TALAROMYCES FLAVUS

Studies on potato were initiated to extend the usefulness of T. flavus to other crops. Potatoes were selected because (1) the production conditions associated with potato make it ecologically simpler than other hosts (eg. pistachio, olive), (2) the large size of the seedpiece provides a great deal of surface area on which to introduce the antagonist, and (3) potatoes are in the same plant family as eggplants and thus it is more likely that Verticillium wilt will be controlled with greater success in plants of the same family than in plants of other families.

Verticillium wilt limits potato production in many areas of the United States. On potato, the disease is usually called Early Dying. In some areas there are other components involved in early dying such as nematodes but while other components may be involved in some cases, the Verticillium component is always present.

Potato seedpieces are normally treated with fungicides to prevent seedpiece decay caused by Fusarium and other fungi. Therefore, studies were initiated to determine whether T. flavus is compatible with these fungicides (Fravel et al 1985). There are five fungicides labeled for use on potato seedpieces with thiabendazole (TBZ) being most commonly used. Either ascospores or conidia of T. flavus were mixed with a talc-like clay carrier and combined with the fungicides. These preparations were dusted onto potato seedpieces. The seedpieces were planted in agricultural fields and recovered after various lenghts of time. Populations of T. flavus in the soil adhering to the seedpieces were determined. T. flavus survived better in the field and was more tolerant of fungicides when introduced as ascospores than as conidia. T. flavus was most tolerant of TBZ and metiram and least tolerant of captan.

Field testing of T. flavus for control of Verticillium wilt of potato was initiated in cooperation with J.R. Davis, University of Idaho, USA. Seedpieces of Russet Burbank potato were dusted with TBZ and ascospores of T. flavus, and planted in a field in Aberdeen, Idaho naturally infested with V. dahliae. Although there were no yield differences among the treatments used, T. flavus suppressed incidence of Verticillium wilt measured both by visual assessment and by stem isolation in the presence of the fumigant metham-sodium (Davis et al 1986).

Since limited distribution of the antagonist may have contributed to the marginal level of control observed, the following year T. flavus was introduced as ascospores in alginate clay granules (Fravel et al 1985; Papavizas et al 1987). The granules were broadcast at a rate of 36.2 kg/ha and rotovated to a depth of 15 cm. T. flavus was applied to plots which had or had not been fumigated with metham-sodium (Vapam) at 468 l/ha. Russet Burbank potatoes were planted in this field which was naturally infested with V. dahliae. There were no yield differences in any treatment; however, there was a trend ($P = 0.10$) toward suppression of disease incidence with T. flavus. The following year Russet Burbank and Norgold potatoes were planted in this same field without additional treatment. Following the second growing season, disease incidence in plots having received T. flavus the previous year was significantly lower than in other plots ($P = 0.05$) for both potato varieties (Fravel et al 1986). Further, populations of T. flavus were significantly greater in plots having received the antagonist 16 months earlier than in plots not treated with T. flavus.

In additional field tests during a 3-year period there was no yield or with incidence difference in plots receiving T. flavus. Although edaphic parameters including cation exchange capacity, and potassium, sodium, and zinc concentrations were implicated in survival of T. flavus (Fravel and Marois 1986), these parameters are not thought to be involved in the lack of biocontrol during these 3 years.

MECHANISMS OF BIOCONTROL OF VERTICILLIUM DAHLIAE BY T. FLAVUS

Mechanisms of biocontrol include competition, parasitism and antibiosis. T. flavus competes well in soil. As discussed earlier, T. flavus survived for 16 months when introduced into an agricultural field in Idaho (Fravel et al 1986). It was the only one of eight species tested that consistently increased soil respiration in natural soils (Marois and Fravel 1983). T. flavus also survived when added to 25 different natural soils (Fravel and Marois 1986). To date, T. flavus isolate Tf1 has not been found to be hyperparasitic as have other T. flavus isolates (Mc Laren et al 1986). T. flavus may suppress wilt incidence by killing V. dahliae microsclerotia in soil (Marois et al 1984), although this was not substantiated by Spink and Rowe (R. Rowe, personal communication).

Isolates of T. flavus have been reported to produce four antibiotics (Fuska et al 1972, 1979a,b; Mizuno et al 1974). Our research indicated that T. flavus Tf1 produced a compound of lower molecular weight than other reported antibiotics which killed microsclerotia of V. dahliae in vitro and in soils (Fravel et al 1987). The compound was subsequently purified and identified as glucose oxidase (Kim et al 1988). Glucose oxidase catalyzes the reaction of glucose + water to produce hydrogen peroxide and gluconic acid. Gluconic acid is not toxic to V. dahliae and can even be utilized as a carbon source by the pathogen. However, hydrogen peroxide is toxic to V. dahliae at concentrations as low as 12 ppm in vitro. Hydrogen peroxide is rapidly inactivated in soil and can not kill microsclerotia unless it is added at concentration greater than 60 ppm. Our results indicated that, microsclerotia are killed in soils by the simultaneous addition of glucose and glucose oxidase.

An experiment was designed to determine whether roots can supply glucose in sufficient quantity to produce enough hydrogen peroxide to kill microsclerotia of V. dahliae. Microsclerotia embedded in nylon meshes (Lumsden 1981) were places on a 2 cm tall (3 cm wide) column of washed, sterile sand. The meshes were covered with an additional 4 cm of sand. In half of the columns, sterile germinated pea seeds were planted in the sand with the cotyledons not touching the surface of the sand. After 12 days, columns were treated with water, sodium acetate buffer (pH 5), hydrogen peroxide, glucose oxidase in buffer, glucose, or glucose oxidase + glucose. The meshes were retrieved and survival of V. dahliae was assessed. As in previous experiments (Kim et al 1987), glucose or hydrogen peroxide alone did not kill microsclerotia but glucose + glucose oxidase killed 100% of the microsclerotia. In the absence of plant roots glucose oxidase had no effect of V. dahliae while in the presence of the plant, addition of glucose oxidase resulted in a significant reduction in survival of microsclerotia. Additional experiments are in progress to asses the relative importance of glucose oxidase in the biocontrol of V. dahliae by T. flavus in natural soils. Selections from the wild-type Tf1 which produce greater or smaller quantities of glucose oxidase than Tf1 are being compared to Tf1 for biocontrol ability.

Although field results have been inconsistent, T. flavus is a promising biocontrol for V. dahliae. Greater understanding of the mechanism and of the biology and ecology of T. flavus may result in improved field control.

Currently studies are in progress to search for additional antagonists of V. dahliae.

REFERENCES

Davis JR, Fravel DR, Marois JJ and Sorensen LH (1986) Effect of soil fumigation and seedpiece treatment with Talaromyces flavus on wilt incidence and yield 1983. Biol Cult Tests 1:18

Fravel DR and Adams PB (1986) Estimation of United States and world distribution of Talaromyces flavus. Mycologia 78:684-686

Fravel DR, Davis JR and Sorensen LH (1986) Effect of Talaromyces flavus and metham on Verticillium wilt incidence and potato yield 1984-1985. Biol Cult Tests 1:17

Fravel DR, Kim KK and Papavizas GC (1987) Viability of microsclerotia of Verticillium dahliae reduced by a metabolite produced by Talaromyces flavus. Phytopathology 77:616-619

Fravel DR and Marois JJ (1986) Edaphic parameters associated with establishment of the biocontrol agent Talaromyces flavus. Phytopathology 76:643-646

Fravel DR, Marois JJ, Dunn MT and Papavizas GC (1985) Compatibility of Talaromyces flavus with potato seedpiece fungicides. Soil Biol Biochem 17:163-166

Fravel DR, Marois JJ, Lumsden RD and Connick WJ Jr (1985) Encapsulation of potential biocontrol agents in an alginate-clay matrix. Phytopathology 75:774-777

Fuska J, Fuskova A and Nemec P (1979) Vermistatin, an antibiotic with cytotoxic effects, produced from Penicillium vermiculatum. Biologia (Bratislava) 34:735-739

Fuska J, Nemec P and Fuskova A (1979) Vermicillin, a new metabolite from Penicillium vermiculatum inhibiting tumor cells in in vitro. J Antibiot 32:667-669

Fuska J, Nemec P and Kuhr I (1972) Vermiculine, a new antiprotozoal antibiotic from Penicillium vermiculatum. J Antibiot 25: 208-211

Kim KK, Fravel DR and Papavizas GC (1987) Biocontrol of Verticillium dahliae by Talaromyces flavus mediated by glucose oxidase. In: Proc 10th Korean Symp Sci Technol,vol 1. Korean Scientists and Engineers Assoc, Seoul, p 600

Kim KK, Fravel DR and Papavizas GC (1988) Identification of a metabolite produced by Talaromyces flavus as glucose oxidase and its role in the biocontrol of Verticillium dahliae. Phytopathology 78:488-492

Lumsden RD (1981) A nylon fabric technique for studying the ecology of Pythium aphanidermatum and other fungi in soil. Phytopathology 71:282-285

Marois JJ, Fravel DR and Papavizas GC (1984) Ability of Talaromyces flavus to occupy the rhizosphere and its interaction with Verticillium dahliae. Soil Biol Biochem 16:387-390

Marois JJ, Johnston SA, Dunn MT and Papavizas GC (1982) Biological control of Verticillium wilt of eggplant in the field. Plant Disease 66:1166-1168

McLaren DL, Huang HC and Rimmer SR (1986) Hyperparasitism of Sclerotinia sclerotiorum by Talaromyces flavus. Can J Plant Pathol 8:43-47

Mizuno K, Yagi A, Tanaka M, Matsuura K, Yamaguchi K and Asano K (1974) A new antibiotic, talaron. J Antibiot 7:560-563

Papavizas GC, Fravel DR and Lewis JA (1987) Proliferation of Talaromyces flavus in soil and survival in alginate pellets. Phytopathology 77:131-136

THE BIOLOGICAL CONTROL INDUCED BY SOIL SOLARIZATION

J. Katan[1], J.E. DeVay[2] and A. Greenberger[3]
Department of Plant Pathology
and Microbiology
The Hebrew University
Rehovot 76100
Israel

INTRODUCTION

Soil solarization (solar heating) is a nonchemical method for disease control by tarping the soil with transparent polyethylene for one month or more during the hot season, thereby heating the soil and killing the pathogens. An increase in soil temperatures is the most apparent pronounced effect of solarization. The temperatures in the upper layers of solarized soils may reach 44-50 C. Thus, heating can be responsible for the drastic reduction in pathogen populations observed in solarized soils (Katan 1981; Stapleton and DeVay 1986). However, there are cases in which pathogen populations were reduced in solarized soils also at marginal temperatures e.g. at deep soil layers or at marginal climatic conditions. For example, an effective killing of pathogens to the depths of 50-70 cm (where temperature of solarized are only 1-4 C higher than the nonsolarized) was observed with _Dematophora_, _Verticillium dahliae_ and _Phytophthora cinnamomi_ (Pinkas et al 1984; Pullman et al 1981; Sztenjberg et al 1987). Another phenomenon frequently observed in sola-

2. Department of Plant Pathology
University of California
Davis, CA 95616
USA
3. Department of Vegetable
Crops, ARO
The Volcani Research
Center, Bet Dagan 50250
Israel

NATO ASI Series, Vol. H28
Vascular Wilt Diseases of Plants

rized soils is the long-term effect under field conditions i.e., an effective control which lasted for two or even three seasons after solarization in spite of the fact that recontamination cannot be avoided (Abdel-Rahim et al 1988; Katan et al 1983; Pullman et al 1981). The above findings strongly suggest that biological control processes are induced or stimulated in the soil, during or after solarization, thus contributing to the killing of pathogens. This was suggested already in the earlier works on this newly established technique (Katan et al 1976). Data from various regions and soils and with different pathogens, which accumulated in the following years, further verified this concept.

MECHANISMS OF BIOLOGICAL CONTROL IN SOLARIZED SOILS

Soil solarization, as any control method, may affect one or more of the living components involved in plant disease caused by a soil-borne pathogen, i.e. the pathogen, the host and the surrounding soil microorganisms. This may affect inoculum density (ID) or inoculum potential (IP) or both. According to Baker (1968), disease severity = Inoculum potential x disease potential. More specifically, the equation becomes = (inoculum density x capacity) x (proneness x susceptibility), where capacity is the effect of the environment on energy for colonization and proneness is the effect of the environment on the host. Of these four components, inoculum density is the one most affected by solarization, either through the direct physical effect of the heat or by microbial processes induced in the soil. The other components, however (except for susceptibility which is genetically determined), might also be affected.

The possible mechanisms of biological control in solarized soils can be classified according to the component which they affect (Katan 1981). This can be summarized as follows:

I. The effect on the inoculum existing in the soil.

A. Reduction in ID (in the dormant stage or during penetration of the host) through the following:

1. microbial killing of the pathogen, already weakened by

sublethal heat;

2. partial or complete annulment of fungistasis and subsequent lysis of the germinating propagule;
3. parasitism or lysis by antagonists stimulated by solarization.

B. Reduced inoculum potential due to antibiosis or competition enhanced by solarization.

C. Diminished competitive saprophytic ability of the pathogen, in the absence of the host, due to antibiosis or competition.

II. Suppress on of inoculum introduced in to soil after solarization, from deeper soil layers or adjacent nontreated plots, i.e. preventing reinfestation through activity of microorganisms possessing mechanisms A_2, A_3, B and C.

III. The effect on the host due to cross-protection.

The mechanisms of biological control can be also classified according to the time of their action and the source of inoculum (Katan 1987):

1. Affecting, ID, IP, or saprophytic activity of the pathogen existing in the infested soil, during or after solarization.
2. Affecting inoculum introduced into the soil after solarization, from deeper soil layers, adjacent infested soil, or external sources.

Most studies on mechanisms of conventional biological control, soil disinfestation, or disease control of soilborne pathogens emphasize stage 1. Though eradicating the primary inoculum is a prerequisite for disease control, studying the fate and behavior of inoculum introduced in previously treated soil is crucial for the evaluation of disease control during the growing season, since under field conditions reinfestation can be restructed but not totally avoided. Studies on soil suppressiveness are especially relevant to the mechanisms of stage 2.

a. Biological Control of the Existing Population of a Pathogen by Solarization

Populations of antagonistic flora may be stimulated into attack pathogens propagules and reduce ID following solarization.

Antagonism might by further improved if propagules of the pathogens are weakened by sublethal heating and become vulnerable to microbial acivity. Sublethal heating increased the leakage of water-soluble organic compounds from S. rolfsii sclerotia, increased their microbial colonization and the frequency of surface cracks, and resulted in decreased disease incidence (Lifshitz et al 1983). Sublethal heating affected the survival of Armillaria mellea (Munneke et al 1976), since less heating was required for indirect killing of the pathogen. Trichoderma was the dominant colonizer of the heated infected roots. A better picture regarding this phenomenon might be obtained by assessing not only the number but also the growth capacity of the propagules, as well as ID and IP analyzed at various periods after the termination of solarization (or disinfestation). For example, the population of the nematode Helicotylenchus dignonicus was significantly reduced 3 months after solarization, but not immediately after treatment (Stapleton and DeVay 1983). Diminished fungistasis in solarized or heated soils (Greenberger et al 1987; Katan et al 1976) is regarded as harmful to resistant resting structures, which, in such a situation, are rendered as germinating propagules, vulnerable to microbial activity.

The mulch cover seals the soil and causes an accumulation of volatiles such as CO2, ethylene, and possibly other substances (Greenberger et al 1984; Rubin and Benjamin 1984). Volatile compounds toxic to S. rolfsii and Fusarium, accumulated in heated soils, and microorganisms producing antifungal toxic volatiles were isolated from solarized and heated soils and sclerotia (Greenberger et al 1984). Decreased nematode populations were observed in moist, plastic-covered but shaded (nonheated) soil; hence both the high soil moisture and the accumulation of volatiles may contribute to nematode control (Stapleton and DeVay 1983). This aspect offers many options for improving solarization, such as using mulch that is less permeable to volatiles or amending with suitable organic matter before mulching. Indeed, Ramirez - Villapudua and Munnecke (1988) found that combining cruciferous residues with solar heating improved control of Fusarium in cabbage. A shift in population of soil microorganisms in solarized soils in favour of antagonists, was reported from different regions. Such potential antagonists include species of Trichoderma, Talaromyces, and lytic bacteria (Elad et al 1980; Greenberger et al 1987; Tjamos and Paplomatas 1987). Studies on microbial changes in the soil, during and after solarization, by Stapleton and DeVay (1982, 1984) show that while populations of many microorganisms were reduced immediately following solarization, thermophilic and thermotolerant fungi, and Bacillus spp. were less affected or even increased. Fluorescent pseudomonads (known as beneficial root colonizers) rapidly colonized solarized soil and plant roots. The percentage of gram-positive bacteria exhibiting in vitro antibiosis increased 20-fold in solarized soil. Speculation was raised regarding the reasons for the frequent stimulation of beneficial microorganisms in heated soils. Temperatures in solarized soils are not too high, allowing many saprophytes, some of which are potential antagonists, to survive (Stapleton and DeVay, 1982). It appears that soil solarization also causes changes in soil biota and substrate in favour of organisms with greater saprophytic ability, usually saprophytes, rather than phytopathogens, which have more specialized growth requirements (Stapleton and DeVay 1986).

b. Effect of solarization on inoculum introduced in soils

Studies have frequently shown that solarized soils become hostile and less receptive to pathogen reinfestation. This can be regarded as an induced suppressiveness (Greenberger et al 1987), similar to the natural one. The incidence of Fusarium, S. rolfsii, and Verticillium diseases was lower in solarized and subsequently inoculated soils than in untreated inoculated ones, while populations of lytic and antibiotic microorganisms were higher. The suppression of chlamydospore formation was frequently observed (Greenberger et al 1987). A similar phenomenon of induced suppressiveness in solarized soils was found with Fusarium wilt of carnation (Hardy and Sivasithamparam 1985) and Phytophthora cinnamomi (Pinkas et al 1984). The widespread solarization-induced suppressiveness, somewhat similar to the natural one but less specific, should not be regarded as a universal phenomenon. The possibility that in certain soils a conduciveness might be induced should not be excluded. Populations of the nematode *Pratylenchus hamatus* were drastically reduced immediately after solarization,and much increased seven months later (Stapleton and DeVay 1983).

CONCLUSIONS

Soil disinfestation is a drastic soil treatment and therefore both harmful and beneficial microbial shifts should be expected, looked for and determined. The harmful phenomena should be avoided or controlled while the beneficial ones should be stimulated. Combining with biocontrol agents is especially recommended for this purpose (Elad et al 1980). The findings from solarized soils have many implications to other control methods e.g. fumigation. Thus, reducing dosage of fumigation may enable us to avoid biological vacuum, and ascertain a longer lasting effect.

REFERENCES

Abdel-Rahim, Satour MM, Mickail KY, El-Eraqi SA, Grinstein A, Chen Y and Katan J (1988) Effectiveness of soil solarization in furrow-irrigated Egyptian soils. Plant Disease 72:143-146

Baker KF and Cook RJ (1974) Biological Control of Plant Pathogens, Freeman, San Francisco

Baker R (1968) Mechanisms of biological control of soil-borne pathogens. Ann Rev Phytopathol 6:263-294

Elad E, Katan J and Chet I (1980) Physical, biological and chemical control integrated for soilborne diseases in potatoes. Phytopathology 70:418-422

Greenberger A, Yogev A and Katan J (1984) Biological control in solarized soils. Proc Sixth Cong Mediter Phytopath Union, Cairo, p 112

Greenberger A, Yogev A and Katan J (1987) Induced suppressiveness in solarized soils. Phytopathology 77:1663-1667

Hardy GE St and Sivasithamparam (1985) Soil solarization effects on Fusarium wilt of carnation and Verticillium wilt of eggplants. In Parker CA (ed) Biology and Management of Soilborne Plant Pathogens, APS, St Paul

Katan J (1981) Solar heating (solarization) of soil for control of soilborne pests. Ann Rev Phytopathol 19:211-236

Katan J (1987) Soil solarization. In Chet I (ed) Innovative approaches to plant disease control, J Willey, p 77

Katan J, Fishler G and Grinstein A (1983) Short- and long-term effects of soil solarization and crop sequence on Fusarium wilt and yield of cotton in Israel. Phytopathology 73:1215-1219

Katan J, Greenberger A, Alon H and Grinstein A (1976) Solar heating by polyethylene mulching for the control of diseases caused by soilborne pathogens. Phytopathology 66:683-688

Lifshitz R, Tabachnik M, Katan J and Chet I (1983) The effect of sublethal heating on sclerotia of Sclerotium rolfsii. Can J Microbiol 29:1607-1610

Munnecke DE, Wilbur W and Darrley DF (1976) Effects of heating or drying on Armillaria mellea or Trichoderma viride and the relation to survival of A. mellea in soil. Phytopathology 66:1363-1368

Pinkas Y, Kariv A and Katan J (1984) Soil solarization for the control of Phytophthora cinnamomi: thermal and biological effects. Phytopathology 74:796 (Abstr)

Pullman GS, DeVay JE, Garber RH and Weinhold AR (1981) Soil solarization effects on Verticillium wilt of cotton and soilborne populations of Verticillium dahliae, Pythium spp. Rhizoctonia solani and Thielaviopsis basicola. Phytopathology 71:954-959

Ramirez-Villapudua J and Munnecke DE (1988) Effect of solar heating and soil amendments of cruciferous residues on Fusarium oxysporum f.sp. conglutinans and other organisms. Phytopathology 78:289-295

Rubin B and Benjamin A (1984) Solar heating of the soil: involvement of environmental factors in the weed control process. Weed Sci 31:819-825

Stapleton JJ and DeVay JE (1982) Effect of soil solarization on populations of selected soilborne microorganisms and growth of deciduous fruit tree seedlings. Phytopathology 72:323-326

Stapleton JJ and DeVay JE (1983) Response of phytoparasitic and free-living nematodes to soil solarization and 1,3-dichloropropene in California. Phytopathology 73:1429-1436

Stapleton JJ and DeVay JE (1986) Soil solarization: a nonchemical approach for management of plant pathogens and pests. Crop Prot 5:190-198

Sztejnberg A, Freeman S, Chet I and Katan J (1987) Control of Rosellinia necatrix in soil and in apple orchard by solarization and Trichoderma harzianum. Plant Disease 71:365-369

Tjamos EC and Paplomatas EJ (1987) Effect of soil solarization on the survival of fungal antagonists of Verticillium dahliae. Bull OEPP/EPPO 17:645-653

STRATEGIES FOR THE CONTROL OF BAYOUD DISEASE OF THE DATE PALM CAUSED BY *FUSARIUM OXYSPORUM* F. SP. *ALBEDINIS*

A. Vanachter
Laboratory of Phytopathology and Plant Protection
Katholieke Universiteit Leuven
Kardinaal Mercierlaan 92
3030 Leuven (Heverlee)
Belgium

INTRODUCTION

"Le Bayoud" (the name Bayoud comes from the Arabic word "abiadh", meaning white, which refers to the whitening of the fonds of diseased trees - Djerbi, 1983) is the most serious cryptogamic disease of the date palm (*Phoenix dactylifera* L.). The problem occurred for the first time in the Drâa Valley in Marocco, and was already recognized in a few foci before 1870. Only in 1921, the causal relation with a fungal infection was made, and the isolated fungus was identified as *Neocosmospora vasinfecta*. In 1930 the fungus was described as *Cylindrophora albedinis*, and after the detection of macroconidia and chlamydospores (1934), a new classification was made as *Fusarium albedinis*. The current designation of the fungus is:*Fusarium oxysporum* f. sp. *albedinis* (Malancon) Snyd. et Hans (Kada and Dubost 1975; Saadi 1979; Djerbi 1983).

Since its appearance, the problem has developed as an epidemic. In one century the whole area of date palm plantations in Marocco was affected and only a fraction of the initial number of trees is still productive. The spread of the disease in Marocco can be explained by the advancement from one oasis to another along the river valleys. From Marocco, Bayoud entered Algeria via oases close to the border. From there the fungus travelled via the caravan-routes in infested wood and offshoots to the oases of central Algeria.

NATO ASI Series, Vol. H28
Vascular Wilt Diseases of Plants

Actually, the oases of Metlili (1950), Ghardaia (1965) and El Goléa (1978) in North Central Algeria are the most distant areas where F. oxysporum f. sp. albedinis (FOA) has been detected (Djerbi 1987).

The symptoms are very typical, with initial wilting and discolouration of one or a few leaves. The discolouration starts at the base of one side, developing progressively to the top and returning to the base at the other side. Affected leaves will die in a few days to several weeks. Successively also other leaves show symptoms and the disease progresses to the central leaf cluster. When the terminal bud is affected, the tree will die. This process can take from 6 months up to 2 years. At that time, only a part of the roots and of the vascular bundles of the stem show a reddish-brown discolouration.

FOA can be isolated from all plant parts showing symptoms. The resting structures in dead plant material are chlamydospores. This means that all diseased plant material is a potential source for spread of the Bayoud. When thinking about eradication, this should be taken into consideration.

MAIN FACTORS, RESPONSIBLE FOR SPREAD AND DEVELOPMENT OF BAYOUD

For the spread of the disease, we can state some important ways, which can differ according to the situation.

Spread between oases occurs mostly by exchange of offshoots, originating from infested mother plants, but also by the trade in woody materials, made from infested wood. It is proven by the history of the epidemic that spread from one oasis to another is mostly correlated to caravan-routes of nomads, who take diseased material with them. In case there is a river connection between oases (as is the case in the Drâa valley in Marocco), spread can happen via the river water too.

Spread between plantations in the same oasis can, of course, occur in the same way. Irrigation water is here very important. Another important factor is the wind which can transport infested soil and spores all around the oasis, as well as the transportation of infested wood by man (e.g. to an incineration place).

Spread within the plantation is mostly seen starting by one of a few infested trees and progressing to many foci. This is due to transport of infested soil and irrigation water. Development of the individual foci is mostly circular, as a result of the root contact between diseased and healthy trees.

Development of disease symptoms after infection of a tree is strongly dependent on the cultural practices. Intensive irrigation results in a better development of trees and a higher production of quality dates, but accelarated also the development of Bayoud. Cropping of other plants in the shade of date palms increases the biological activity of the soil and can also increase the saprophytic growth of FOA. Especially when crops are grown which can act as symptomless carriers, such as lucerne (*Medicago sativa* L.) or henna (*Lowsonia inermis* L.), FOA growth will be influenced significantly.

FAO PROGRAM FOR CONTROL OF BAYOUD

Because the date palm is of major importance for the Sahara communities and for the ecosystem, control of the disease will reduce desertification and increase food supply and income of the local inhabitants. For that reason, a regional project for control of Bayoud disease of date palm (RAB/84/018) was initiated as a UNDP project, with the collaboration of Algeria, Marocco and Tunesia, and executed by the FAO (Djerbi 1987). The international coordinator of this project is Dr. M. Djerbi.

The project will deal with scientific research in different fields, to come to a better understanding and development of control methods for the Bayoud disease. Main research topics are:

a) fundamental studies on disease epidemiology, pathogen detection and rapid identification;
b) selection and breeding for resistance and high quality;
c) in vitro propagation;
d) eradication of infested foci.

The aim of the study reported here is to find methods for stopping, or at least for delaying the spread of Bayoud disease.

ERADICATION PROGRAM

The specific aim of the eradication program is the prevention or the limitation of disease spread within plantations and within oases. But the major goal is to prevent the spread of Bayoud disease into other oases. As stated earlier, the Ghardaia, Metlili and El-Goléa oases in North-Central Algeria are the most advanced areas were Bayoud is reported. The intention of the eradication program is to stop spread here and protect the high quality 'Deglet Nour' plantations in North-East Algeria and in Tunesia.

For that purpose, a total destruction of all infested trees, as sources of inoculum, is necessary. This means cutting down the trees and uprooting of the bulbs, leaving in the soil as little plant material as possible. This procedure is followed by incineration of all plant parts on the site. Destruction of the inoculum that remains in the soil (in infested roots and as chlamydospores) is also very important. For that purpose, a soil disinfestation is the only possibility.

When considering soil disinfestation, one has to take into account the very specific situation. Because FOA propagules are present in infested roots to a depth of 1 m (Nash-Smith 1978; Djerbi 1987), disinfestation should penetrate the woody material to 1 m depth and kill almost 100% of the propagules present to achieve a satisfactory effect. After such a drastic treatment a protection against reinfestation and/or recolonisation is necessary.

SOIL DISINFESTATION IN THE FRAME OF THE ERADICATION PROGRAM

Possible methods

Steam sterilization: Experiments have shown that FOA chlamydospores, encysted in sand, can be destroyed in 3 hours at 65 C, in 75 minutes at 70 C and in 20 minutes at 80 C (Djerbi 1987). This means that steam sterilization can be considered as an eradication treatment. However, a satisfactory penetration of a sufficiently high temperature to 1 m deep will be a problem. Nevertheless, in some situations, steam sterilization can be a useful method.

Soil solarization: During the summer period in the Sahara-region, conditions are ideal for soil solarization. Unfortunately, with this treatment only a diminution of the inoculum can be reached, and not a total control. However, in combination with a chemical treatment, soil solarization can give an additional effect (Katan 1984; Tjamos 1984).

Chemical soil disinfestation: In this field only treatments with methyl bromide and/or chloropicrin can be considered, due to the high fungicidal value (chloropicrin) and the penetrating ability (methyl bromide) of these chemicals. It is known from our own experiments as well as from literature (Kolbezen et al 1974) that methyl bromide can penetrate deeply into the soil, giving excellent control of pathogens that are encysted in root tissue. Because of their particular properties, both methyl bromide and chloropicrin were choosen for experiments to develop a standard method of soil treatment in the eradication program.

When considering soil fumigation, one should know that success is dependent on a lot of factors, that influence the result of a treatment in a positive as well as in a negative way (Vanachter 1984).

Factors affecting the result of a soil fumigation in a positive way

The soil type: The date palm is a typical tree for desert areas, where the soil is sandy, with a very low organic matter content. This means a high porosity, even in undisturbed conditions. This guarantees an optimal aeration and thus an easy spread of fumigants, even to deeper soil layers. This can be achieved without soil preparation, which avoids also a potential risk for spread of *Fusarium* propagules. In these soils the adsorption of the fumigants will be negligable, and the whole applied dose can work actively against the pathogen.

The soil humidity: A dry (Sahara) soil means also that the initial porosity remains unchanged, keeping the possibilities for gas diffusion at maximal level.

The temperature: The high summer temperature of Sahara areas have some important advantages. The vapour pressure of the applied fumigant will be high and gas distribution can go fast and deep into the soil. This temperature also increases strongly the sensitivity of the fungi in the soil. When fumigating during the summer period, temperature under the plastic cover can be so high that a combined effect of soil fumigation and soil solarization can be obtained (Katan 1984; Tjamos 1984).

The dosage: The choice of the dose can be made independent of the risk of residues, with a long after effect, or other negative influences. So, high dosages can be used to function with maximal effect in CTP over the whole treated zone, giving a maximal penetration and efficiency in killing propagules inside the plant remnants.

The plastic cover: To reach a sufficient dose in the whole soil profile, a plastic cover is necessary in any method of treatment. The formerly used LDPE plastics allow escape of fumigants to the air to an important degree. The introduction of impervious covers is the most interesting innovation in soil fumigation in the last years. Different types of tarps are available now, some of them barrier properties of almost 100% (Kolbezen and Abu-El-Hay 1978; de Heer et al 1984; Van Wambeke 1984). Until now, most available data are from experiments with Saranex. Kolbezen et al (1974) could maintain a methyl bromide concentration between 1,000 and 1,300 ppm at a depth of 2.50 m for 50 days when they applied 250 g/m^2 to a dry soil, using Saranex as tarping material. In these circumstances almost the whole amount of product must be decomposed in the soil (with a very long antiparasitic action), while in the case of covering with LDPE, about 80% of the applied fumigant escaped to the environment.

Factors affecting the result of a soil fumigation in a negative way

The root system: As mentioned earlier, studies of Nash-Smith (1978) and of Djerbi (1987) have shown that FOA propagules are present in the roots of infested trees to a depth of about 80-100 cm. The concentration of propagules is much higher in the root than in the soil at that depth.

For a successful fumigation it is necessary to penetrate these infested roots. When trees are pulled up in an early stage after infection is started (which is necessary to prevent spread of spores from that tree), the roots are very hard and everything possible must be done to achieve sufficient penetration of the fumigants in that material.

The causal agent is a vascular *Fusarium*: *Fusarium oxysporum*, and especially the chlamydospores, are less sensitive to most fumigants than is the case with other fungi. Moreover, a successful infection of a plant vessel can occur from only one resting propagule and plant death will follow. This means that the objective must be to reach the highest possible efficiency.

The risk for reinfestation and repopulation is extremely high: Because date palms are productive for more than 50 years and inoculum pressure can be extremely high, when sufficient precautions are not taken, there is a high risk for reinfestation. This is particularly the case because most *F.oxysporum* strains can develop very quickly as saprophytes after soil disinfestation.

SOIL FUMIGATION EXPERIMENTS

Experimental conditions

During the period June 1987 - February 1988, soil fumigation experiments were carried out at the "Station Régional de Protection des Végétaux" at Ghardaia-Algeria, by the FAO experts M. Fredrix and K. den Braber, under the consultancy of the author. The general set-up of all experiments was:

a) a deep placement (50 or 100 cm) of methyl bromide cans with 620 g pure product;

b) in the case of mixed application of methyl bromide (MB) and chloropicrin (CP), only half the MB amount was used. CP was then applied first, by injection of 5.5 ml product every 50 cm, at a depth of 35 cm with a pal-injector. This was an equivalent dose, calculated as gram per square meter, as for MB (35.5 g/m^2).

c) test material was placed in holes into the soil at depths of 25, 50, 75 and 100 cm respectively, and at the surface; distance from the MB application point was 50, 100, 125, 175 and 215 cm respectively;
d) as test material either a talc + FOA preparation containing about 5.9 x 10^6 propagules per gram, or 3 cm long pieces of rachis cut from an infested plant were used. This was done to measure the penetration of the chemicals into woody material;
e) a minimal soil preparation was done, only at the surface soil below 25 cm was undisturbed;
f) covering was done for 3 days either by Saranex (0.04 mm), as an impervious film, or by LDPE (0.1 mm) from Algerian production.

RESULTS AND DISCUSSION

An important difference was observed between the results obtained during the warm summer and fall period, and those from the winter experiments. As an example, results of a fumigation experiment of October 1987 (Table 1) are compared with those of a fumigation of February 1988 (Table 2).

In Table 1, the results of a fumigation with MB or with MB + CP, both covered with Saranex film, are illustrated. We can report very good results for both products to a depth of 100 cm and a lateral distance of 125 cm from the MB-application point. At a lateral distance of 175 cm, which was the overlap zone between two application points, results were also very good. At distances of 150 cm and of 215 cm, the result of the MB + CP treatment was better than that of the pure MB treatment.

In a second part of the experiment, Algerian LDPE was used as covering material. Both MB + CP treatments gave the same results as with the Saranex cover. In this experiment, talc + FOA was also used as a test material. In a fumigation experiment in September 1987, equal results were obtained when this preparation

Table 1. Results of a fumigation experiment with methyl bromide (MB) and methyl bromide + chloropicrin (CP), using a talc preparation of <u>Fusarium oxysporum</u> f. sp. <u>albedinis</u> (<u>FOA</u>) as test material, carried out at the SRPV Station of Ghardaia - Algeria, on October 26th, 1987. Covering material was Saranex (0.04 mm).

	Fumigant	MB + CP					MB				
	Covering	Saranex					Saranex				
	Depth (cm)	0	25	50	75	100	0	25	50	75	100
Dist. (cm)	Replication										
50	A	--	-	-	-	-	--	-	-	-	-
	B	--	-	-	-	-	--	-	-	-	-
100	A	--	-	-	-	-	--	-	-	-	-
	B	--	-	-	-	-	--	-	-	-	-
125	A	--	-	-	-	-	--	-	-	-	-
	B	--	-	-	-	-	--	-	-	-	-
150	A	--	-	-	-	-	--	-	+	-	-
	B	--	-	-	-	-	--	+	+	-	+
175		--	-	-	-	-	--	-	-	-	-
215	A	0	0	-	0	0	0	0	+	0	0
	B	0	0	-	0	0	0	0	+	0	0

+ Normal growth of <u>FOA</u> from soil seeded onto agar medium after fumigation.
- No growth of <u>FOA</u> from soil seeded onto agar medium after fumigation.
0 No samples.

Table 2. Results of a fumigation experiment with methyl bromide, using a talc preparation of Fusarium oxysporum f. sp. albedinis (FOA) as test material, carried out in an infested palm garden in Ghardaia - Algeria on February 10th, 1988. Covering material was Saranex (0.04 mm) and Algerian LDPE (0.1 mm).

Covering	Algerian LDPE					Saranex				
Depth (cm)	0	25	50	75	100	0	25	50	75	100
Distance (cm)										
50	-	-	-	-	-	-	-	-	-	-
100	+	-	+	-	+	-	-	-	-	-
125	+	+	+	+	+	+	+	-	+	+
150	+	+	+	+	+	+	+	+	+	+

+ Normal growth of FOA from soil seeded onto agar medium after fumigation.
- No growth of FOA from soil seeded onto agar medium after fumigation.

or infested rachis pieces were used, showing that penetration in that woody material did not create problems. Fumigation experiments at the same location, but during winter time, showed entirely different results; an example of a MB fumigation is illustrated in Table 2. Here the effective lateral spread of the fumigant is only 50 cm under Algerian LDPE cover, and 100 cm under Saranex, showing the difference in barrier capacity and thus in fumigant concentration into the soil. Compared with the fall experiment, differences are striking. Equal differences are observed also in a second experiment, where penetration into rachis pieces was strongly decreased, giving a great difference in the results, obtained with the talc + FOA preparation, and an even greater difference in the summer experiments.

From these results it is proven that successful penetration of fumigants 1 m deep into the soil and even into woody material, such as rachis and roots, is possible. However we have stated the enormous influence of soil temperature. It is also clear that plastic covers with a high barrier capacity maintain a higher fumigant concentration in the soil, resulting in a better control of the pathogen. The combined application of MB + CP combines the high penetration ability of MB with the excellent fungicidal value of CP and guarantees a more efficient killing action.

CONCLUSIONS

To stop spread of Bayoud disease, it is necessary to destroy the inoculum in infested foci as efficiently as possible, before any other measure can be taken into account. For that, a careful clearing-up of infested foci, with incineration of all woody material on the spot, is absolutely necessary. In connection with this, the killing of inoculum, resting in the soil, is also necessary. The possibilities of soil disinfestation for this particular purpose are under study as a part of the work done by the UNDP project RAB/84/018.

Fumigation with methyl bromide and with a combination methyl bromide and chloropicrin have shown good results, when applied in the summer. Winter fumigations were inefficient. However, it is proven from these experiments that practical application methods can be worked out by defining the limits of effective action. A possibility for extending the zone of effective action is the use of a much longer covering time to achieve a combined effect of soil fumigation and soil solarization.

A very important danger in this procedure is the real risk for reinfestation of the soil by the FOA pathogen. Precautions should be taken to avoid this. Some possible measures can be:

a) to maintain the plastic cover on the soil as long as possible, giving a protection against air-borne spores and contaminated sand, until the natural soil microbial population is recovered;
b) to keep soil dry will decrease the biological activity and also the possible saprophytic development of FOA;
c) to inoculate with antagonistic populations originating from suppressive soils (e.g. Marrakech), to give a long term protection. This problem is also in study by the project;
d) to replant treated areas with non-host plants, e.g. fruit trees or resistant date palm varieties.

REFERENCES

de Heer H, Hamaker PH, Tuinstra LGMTH and van der Burg AMM (1984) Use of gas-tight plastic films during fumigation of glasshouse soils with methyl bromide. I. Significance of permeation and leakage for the emission into the outside air. Acta Horticulturae 152:105-126

Djerbi M (1983) Diseases of the Date Palm (Phoenix dactylifera L.). FAO - Regional Project for Palm and Dates Research Center in the Near East and North Africa, Bagdad, pp 106

Djerbi M (1987) Report on work on Bayoud - September 1985 - September 1987. FAO/PNUD, Control of Date Palm Bayoud disease - RAB/84/018, pp 1-14

Kada Annick and Dubost D (1975) Le Bayoud à Ghardaia. Bull Agr Sahar, 1:29-61

Katan J (1984) Soil solarization. Acta Horticulturae 152:227-236

Kolbezen MJ and Abu-El-Hay FJ (1978) Permeability of plastic films to fumigants. Proc Int Agric Plastic Congress, San Diego:476-481

Kolbezen MJ, Munnecke DE, Wilbur WD, Stolzy CH, Abu-El-Hay FJ and Szuszkiewisz TE (1974) Factors that affect deep penetration of field soils by methyl bromide. Hilgardia 42:465-492

Nash-Smith Shirley (1978) Différents points étudiés au cours

de la mission dans les oasis Algeriennes - Novembre 1977. Bull Agr Sahar 1:63-71
Saadi M (1979) Contribution à la lutte contre le Bayoud, Fusariose vasculaire du palmier dattier. Thèse, Faculté des Sciences de la Vie et de l'Environment, Dijon, France, pp 140
Tjamos EC (1984) Control of Pyrenochaeta lycopersici by combined soil solarization and low dose of methyl bromide in Greece. Acta Horticulturae 152:253-258
Vanachter A (1984) The possible role of soil disinfestation in the total complex of control measures against Fusarium oxysporum f. sp. albedinis, Le Bayoud of date palms. Acta Horticulturae 152:179-185
Van Wambeke E (1984) Efficiency increase of methyl bromide soil fumigation by admixture with methyl chloride or ameliorated tarps. Acta Horticulturae 152:137-141

RACES OF FUSARIUM OXYSPORUM F. SP. CICERI INFECTING CHICKPEAS IN SOUTHERN SPAIN

R.M. Jiménez-Díaz, A. Trapero-Casas and J. Cabrera de la Colina
Departamento de Agronomia
Universidad de Córdoba
Apartado 3048
14080 Córdoba
Spain

INTRODUCTION

About 60,000 ha are sown annually to chickpea (Cicer arietinum L.) in Andalucia, southern Spain, approximately 66% of the total national acreage of the crop. Wilt induced by Fusarium oxysporum Schlecht. emend. Snyd. & Hans. f. sp. ciceri (Padwick) Snyd. & Hans. (Foc) severely affects chickpeas in that region causing an average 10% yield loss (Trapero-Casas and Jiménez-Díaz 1985). Control of the disease is mainly by the use of resistant cultivars (Haware and Nene 1980a; Nene and Haware 1980; Nene and Reddy 1987), the efficacy of which is curtailed by the occurrence of pathogen races. Haware and Nene (1982) were the first to show the existence of races of Foc, namely races 1,2,3 and 4, based on the differential interaction among 10 chickpea cultivars and isolates of the pathogen from India. Recent work indicates that resistance to race 1 is conferred by recessive alleles from at least two independent loci (Upadhyaya et al 1983a, 1983b; Singh et al 1987). Cultivars WR 315, CPS 1, BG 212 and JG 74, carrying recessive alleles at both loci, are completely resistant, while cultivars C 104 and K 805, which are homozygous recessive at one of the loci, show a delayed wilt (late-wilting, Haware and Nene 1980b). Cultivar JG 62, which is susceptible to all four races of Foc in India, carries neither of the recessive alleles and shows wilt symptoms within 20 days after sowing (early-wilting).

Research in our laboratory first indicated that isolates of

NATO ASI Series, Vol. H28
Vascular Wilt Diseases of Plants
Edited by E. C. Tjamos and C. Beckman

Foc from southern Spain induce either vascular wilt or vascular yellowing syndromes in local kabuli cultivars (Trapero-Casas and Jiménez-Díaz 1985). The wilt syndrome, which develops 15-20 days after inoculation, is characterized by chlorosis, flaccidity and plant death. The yellowing syndrome is characterized by a progressive foliar yellowing which develops 30-40 days after inoculation and late death of the plant. While wilt-inducing isolates are highly virulent to most chickpea cultivars, yellowing-inducing isolates are not pathogenic to cultivar JG 62 but virulent mainly to local 'kabuli' cultivars (Trapero-Casas and Jiménez-Díaz 1985). This suggested the existence of pathogenic races of Foc in Andalucia. This paper presents the results of research to determine the identity and distribution of Foc races in that region.

MATERIALS AND METHODS

Sixty-one isolates of Foc were obtained from affected plants collected in the major chickpea-growing areas of Andalucia. Pathogenicity of these isolates to kabuli cultivars PV 24 (local land race) and PV 2245 was tested first with mass cultures. Thereafter, all wilt-inducing isolates and 11 selected yellowing-inducing isolates representing a geographical range were single-spored. Single-spore isolates and races 1,2,3 and 4 from India were then used in further inoculation experiments for disease reaction on race differential cultivars. All cultivars except PV 24 and cultures of races 1,2,3 and 4 were obtained from Y.L. Nene, ICRISAT, Hyderabad, India.

Isolates were grown on PDA and incubated at 23 ± 3 C and 12 hours photoperiod of fluorescent and near UV light at 2 klux. Isolates were maintained in autoclaved soil in test tubes. Inoculum was increased in a cornmeal-sand (CMS) mixture (Nene and Haware 1980) and incubated under the same conditions as isolates for 2 weeks. Infested CMS was mixed with an autoclaved soil mixture (1:12, w/w) and inoculations were accomplished by sowing germinated seeds in pots filled with that mixture (Nene and

Haware 1980; Trapero-Casas and Jiménez-Díaz 1985). Control plants were grown in a comparable mixture of uninfested CMS and autoclaved soil. Disease reaction was assessed by severity of symptoms on a 1-5 scale (1 = no symptoms, 5 = dead plant) according to the percentage of the foliage affected at 20 and 40 days after inoculation. Plants were grown in a growth chamber adjusted to 26 ± 2 C/21 ± 2 C (day/night), 60-90% relative humidity, and a 14-hours photoperiod of fluorescent light at about 14 klux. Experiments were designed as randomized complete blocks with three replicated pots (four plants/pot) for each isolate-cultivar combination.

RESULTS

Of 61 mass isolates of Foc, six induced the wilt syndrome and 55 induced the yellowing syndrome. Disease reaction of race-differential cultivars to the 17 selected single-spore isolates indicated that three races of the pathogen, namely races 0,5 and 6, occur in Andalucia. All three appear to differ from the descriptions of races 1,2,3 and 4 in India (Table 1).

All 11 isolates inducing the yellowing syndrome are not pathogenic to 'desi' cultivar JG 62, which is susceptible to all other races (Haware and Nene 1982, Table 1). These 11 isolates were designated as race 0 (Cabrera de la Colina et al 1985). Race 0 is mildly to highly virulent to 'kabuli' cultivars C 104, PV 24 and P 2245 and to 'desi' cultivar 12-071/10054. Race 5 was proposed to designate five wilt-inducing isolates pathogenic to 'desi' cultivars JG 74 and CPS 1 which are highly resistant to race 1 (Cabrera de la Colina et al 1985), and nonpathogenic to cultivar BG 212 which together with JG 74 and CPS 1 is susceptible to race 2 (Haware and Nene 1982). Furthermore, race 5 is not pathogenic to cultivar 12-071/10054 but highly virulent to all 'kabuli' cultivars inoculated. Race 5 isolates induce early wilt in all susceptible cultivars except JG 74 and CPS 1. In these cultivars, symptoms consisted of a foliar yellowing that started at the leaflet margin and affected the entire plant 40

Table 1. Differential interactions of races of *Fusarium oxysporum* f. sp. *ciceri* with chickpea cultivars[1]

	Race origin and designation						
	Spain			India			
Cultivar	0	5	6	1	2	3	4
12-071/10054	S[2]	R	R	M	R	R	R
JG 62	R	S	S	S	S	S	S
C 104	M	S	M	M	S	S	S
JG 74	R	M	R	R	S	R	R
CPS 1	R	M	R	R	S	M	M
BG 212	R	R	R	R	S	M	M
WR 315	R	R	R	R	R	S	R
ICCV 2	R	S	M	R	M	M	M
ICCV 4	R	S	M	R	S	S	S
PV 24	S	S	S	S	S	S	S
P 2245	S	S	S	S	S	S	S

[1]Pooled results of six experiments that shared several race-cultivar combinations. Disease reactions for those combinations were consistent over the experiments.

[2]Assessed on a 1-5 scale according to percentage of foliage with yellowing or necrosis in acropetal progression (1 = 0%, 2 = 1-33%, 3 = 34-66%, 4 = 67-100%, 5 = dead plant) 40 days after inoculation. Scores < 2 and > 4 were considered as resistant (R) and susceptible (S) reactions, respectively. Scores in between were considered as moderately susceptible reaction (M).

days after inoculation. Race 6 is proposed to designate a single wilt inducing isolate that is pathogenic to cultivars JG 62, C 104 and to 'kabuli' ICCV 2 and ICCV 4, and nonpathogenic to cultivars 12-071/10054, JG 74, CPS 1 and BG 212.

Previously in our studies the race 6 isolate had been identified as race 1 (Cabrera de la Colina et al 1985). That occurred because tests were carried out in the absence of cultivars 12-071/10054, ICCV 2 and ICCV 4, and without a control for the differential reaction induced by race 1 from India. When the cultivars mentioned above as well as others were inoculated with the two isolates, our race 6 isolate differed from race 1 in that it is pathogenic to 12-071/10054 as well as to ICCV 2 and ICCV 4 which are resistant to race 1 (Kumar et al 1985).

DISCUSSION

Three races of Foc occur in southern Spain which are different from those hitherto described in India (Haware and Nene 1982). As criterion for race identification we have used severity of symptoms rather than disease incidence as Haware and Nene (1982) did. However, our results with cultures of the races from India reproduced the expected disease reactions and confirm differences between the two groups of races. Our results ilustrate that to identify races of Foc, experimental comparison of differential interactions may be more reliable than the use of published description of the races, particularly when new plant genotypes can be tested.

Races 0, 5 and 6 differ very much in virulence to chickpea cultivars and in geographical distribution. Race 0, the least virulent one, is widespread in Andalucia, while races 5 and 6 are found only within a restricted area. Race 0 also appears to be the simplest race since it is not pathogenic to cultivar JG 62 which carries none of the recessive alleles for resistance to race 1 (Upadhyaya et al 1983a, 1983b). Furthermore, througthout our studies race 0 has proven pathogenic only to 'kabuli' cultivars, with the exception of the 'desi' cultivar 12-071/10054. 'Kabuli' chickpeas are the only ones grown in Andalucia and one might speculate that race 0 has become pathogenically adapted to these cultivars. On the other hand, races 5 and 6 carry genes for pathogenicity to recessive resistant alleles that to our knowledge Foc has not been exposed to. It would be of interest to determine whether those genes for pathogenicity become more widely spread in Andalucia.

ACKNOWLEDGMENTS

This research was supported by grant C 2-0238/81 from Comisión Asesora de Investigación Científica y Técnica, Ministerio de Educatión y Ciencia.

REFERENCES

Cabrera de la Colina J, Trapero-Casas A and Jiménez-Díaz RM (1985) Races of Fusarium oxysporum f. sp. ciceri in Andalucia, southern Spain. Int Chickpea Newsletter 13:24-26

Haware MP and Nene YL (1980a) Sources of resistance to wilt and root rots of chickpea. Int Chickpea Newsletter 3:11-12

Haware MP and Nene YL (1980b) Influence of wilt at different growth stages on yield loss in chickpea. Tropical Grain Legumes Bulletin 19:38-44

Haware MP and Nene YL (1982) Races of Fusarium oxysporum f. sp. ciceri. Plant Disease 66:809-810

Kumar J, Haware MP and Smithson JB (1985) Registration of four short duration Fusarium wilt-resistant kabuli (garbanzo) chickpea germplasms. Crop Sci 25:576-577

Nene YL and Haware MP (1980) Screening chickpea for resistance to wilt. Plant Disease 64:379-380

Nene YL and Reddy MV (1987) Chickpea diseases and their control. In: Saxena MC and Singh KB (eds) The Chickpea. CAB International, Wallingford, pp 233-270

Singh H, Kumar J, Smithson JB and Haware MP (1987) Complementation between genes for resistance to race 1 of Fusarium oxysporum f. sp. ciceri in chickpea. Plant Pathology 36:539-543

Trapero-Casas A and Jiménez-Díaz RM (1985) Fungal wilt and root rot diseases of chickpea in southern Spain. Phytopathology 75:1146-1151

Upadhyaya HD, Haware MP, Kumar J and Smithson JB (1983a) Resistance to wilt in chickpea. I. Inheritance of late-wilting in response to race 1. Euphytica 32:447-452

Upadhyaya HD, Smithson JB, Haware MP and Kumar J (1983b) Resistance to wilt in chickpea. II. Further evidence for two genes for resistance to race 1. Euphytica 32:749-755

INHIBITION OF MICROSCLEROTIUM FORMATION IN VERTICILLIUM DAHLIAE BY THIOGLYCOLIC ACID IN INFECTED COTTON STEMS

C. Christias
Department of Biology
National Research Center
"DEMOCRITOS"
153 10 Aghia Paraskevi
Greece

INTRODUCTION

Sclerotia are small fungal reproductive bodies produced asexually from mycelial hyphae. They play a very important role in the life cycle of the fungi which produce them because they are the structures which survive through very long periods of unfavourable conditions in nature (Coley-Smich and Cooke 1971). The survival of many phytopathogenic soil fungi would be problematic if they were deprived of the ability to produce sclerotia. Sclerotial biogenesis can be inhibited by the use of certain sulfhydryl compounds such as mercaptoethanol, propyl- and butyl mercaptan, dithiothreitol, dithioerythritol, cysteamine hydrochloride (Christias 1975; Christias et al 1981; Christias 1978; Christias and Zioudrou 1983, 1984). This paper presents experimental data showing a strong inhibitory effect of thioglycolic acid on microsclerotium formation in pure culture as well as in the host plant.

MATERIALS AND METHODS

The strain of V. dahliae used in this study was isolated from infected cotton stems taken from cotton fields near Lamia, Greece. The fungus was grown in potato-dextrose-agar medium (PDA) at 21 C

NATO ASI Series, Vol. H28
Vascular Wilt Diseases of Plants
Edited by E. C. Tjamos and C. Beckman

in standard 9-cm Petri dishes. Thioglycolic acid was incorporated into the growth medium to give final concentrations of 1,2 and 4 mM. At the same time naturally infected cotton stems were incubated with thioglycolic acid as follows: In standard 9 cm Petri dishes, 5-7 pieces of infected cotton stems were placed evenly on the dish bottom. The stem pieces were 3 cm long and had a diameter of 5-8 cm. In all Petri dishes a small glass cylindrical vessel with base diameter 25 mm and height 6 mm was placed. In 10 Petri dishes one ml of a stock solution of 240 mM thioclycolic acid was placed in the glass vessel. In the remaining Petri dishes one ml of distilled water was placed in the glass vessel to serve as the controls. All Petri dishes were incubated at 21 C for 5 months. At the end of the incubation period the stem pieces were examined microscopically for the presence or absence of microsclerotia.

RESULTS

Fungal growth was normal in agar cultures (Fig. 1). Thioclycolic acid inhibited the formation of microsclerotia completely at concentrations higher than 1 mM (Fig. 2), while at 4 mM the mycelial growth was completely inhibited (Fig. 3). Microscopic examination of control cultures showed abundant microsclerotia typical of this fungus. (Fig. 4). The presence of microsclerotia in great numbers gives the culture a black colouration as shown in Fig. 1. Similar microscopic examination of cultures grown in the presence of 2 mM thioglycolic acid showed complete absence of microsclerotia (Fig. 5).

At the end of the incubation period the cotton stem pieces incubated without thioglycolic acid (controls) had an intense dark brown discolouration (Fig. 6). Microscopic examination of these stem pieces showed the presence of abundant microsclerotia (Fig. 7). The microsclerotia were formed mainly on the inner surface of the bark which also had a brown discolouration. An extensive mycelial development was also observed on these stem cuttings, especially at the two cut ends. Typical verticillate conidiophores bearing conidia were present. In contrast, the stem cuttings incubated with thioglycolic acid were very clean and had the characteristic colour of healthy wood (Fig. 8). Microscopic examination of these stem pieces showed complete absence of microsclerotia (Fig. 9). The inner bark surface was clean, without any indication of infection. Fungal mycelium was never detected on these stem pieces.

DISCUSSION

The Verticillium wilt of cotton can be one of the most serious problems in cotton production. In spite of extensive research work there are not effective chemical means available for the control of this disease. Even the modern systemic fungicides are not effective for the control of this pathogen (Erwin 1973). The problem is faced with the use of resistant varieties and sound cultural practices like disposal and destruction of infected plant residues, weed control, disinfection, soil drainage, crop rotation etc. It is, therefore, imperative to develop alternative ways to control the disease. Thioglycolic acid inhibited the formation of microsclerotia of V. dahliae at relatively low concentrations. This inhibition was complete and permanent. In infected cotton stems this compound inhibited completely the formation of sclerotia as well as the mycelial growth. At the end of the incubation period the cuttings had the appearance of healthy wood.

The possibility of inhibition of microsclerotium formation in nature by chemical means would be detrimental for the pathogen. Without the ability to form microsclerotia it would be difficult to withstand the pressure, antagonism and parasitism exerted on it by the indigenous soil microflora. However, it is doubtful if thioglycolic acid could be used effectively under field conditions because, like mercaptoethanol, it is a volatile compound. Inhibition of microsclerotium formation under field conditions could possibly be achieved by the use of non volatile analogues of the acid. Such an analogue with greatly reduced volatility has been synthesised and found to be very effective against the fungus

Sclerotium rolfsii (Christias and Kallias 1980).

It is obvious that any control measures based on selective inhibition of sclerotium formation rather than direct killing of phytopathogenic fungi by the use of traditional fungicides may have the following advantages: (1) Considerable specificity of action. Only the pathogen or few very closedly related species are affected. (2) There is no direct killing of the target organisms. Only a vital physiological process is inhibited in such a way that these organisms cannot survive in nature. (3) The displacement is gradual so that the danger of creating a biological vacuum is greatly diminished. (4) It is anticipated that pollution problems resulting from the use of inhibitors of sclerotium formation will not be as serious as those known to occur as a result of the application of traditional fungicides. (5) The non-target soil microbial flora are not apparently affected. A recent study has shown that mercaptoethanol, another inhibitor of sclerotium formation, did not affect the mycelial development and conidiogenesis of at least 75 different species and isolates of soil fungi and actinomycetes. (Christias, unpublished data).

It is recognized that this work is only preliminary. A lot of research has to be carried out towards the development of especially active, non-volatile inhibitors of sclerotium formation. Of special interest would be the development of such inhibitors bearing systemic movement in the host plants. Research efforts directed towards the above goals are currently under way.

REFERENCES

Christias C (1975) Specific inhibition of sclerotium formation by 2-mercaptoethanol and related sulfhydryl compounds in Sclerotium rolfsii. Can J Microbiol 21:1541-1547

Christias C (1978) Response of 5 sclerotium-forming fungi to an inhibitor of sclerotium formation. Proc 3rd Int Congr Plant Pathology, p 115

Christias C and Kallias D (1980) Specific inhibition of Sclerotium formation by a thioglycolic acid analogue in Sclerotium rolfsii. Proc 5th Congr Un Phytopath Mediterr p 161-163

Christias C, Tjamos EC, Zioudrou C and Kornaros E (1981) In vitro inhibition of microsclerotia formation by cysteamine hydrochloride in Verticillium dahliae. Abstr 3rd Int Symp on Verticillium, Bari Italy, p 21
Christias C and Zioudrou C (1983) Inhbition of sclerotium formation by dithiothreitol and dithioerythritol in Sclerotium rolfsii. Proc 5th Conf Hellenic Soc Biol Sci,pp 41-42
Christias C and Zioudrou C (1984) A potent inhibitory action of cysteamine hydrochloride on sclerotial morphogenesis in Sclerotium rolfsii. Proc 6th Conf Hellenic Soc Biol Sci, pp 39-40
Coley-Smith JR and Cooke RC (1971) Survival and germination of fungal sclerotia. Ann Rev Phytopath 9:65-92
Erwin DC (1973) Systemic fungicides: Disease control, translocation and mode of action Ann Rev Phytopath 11:389-422
Willetts HJ (1971) The survival of fungal sclerotia under adverse environmental conditions. Biol Rev (Camb.) 46:387-407

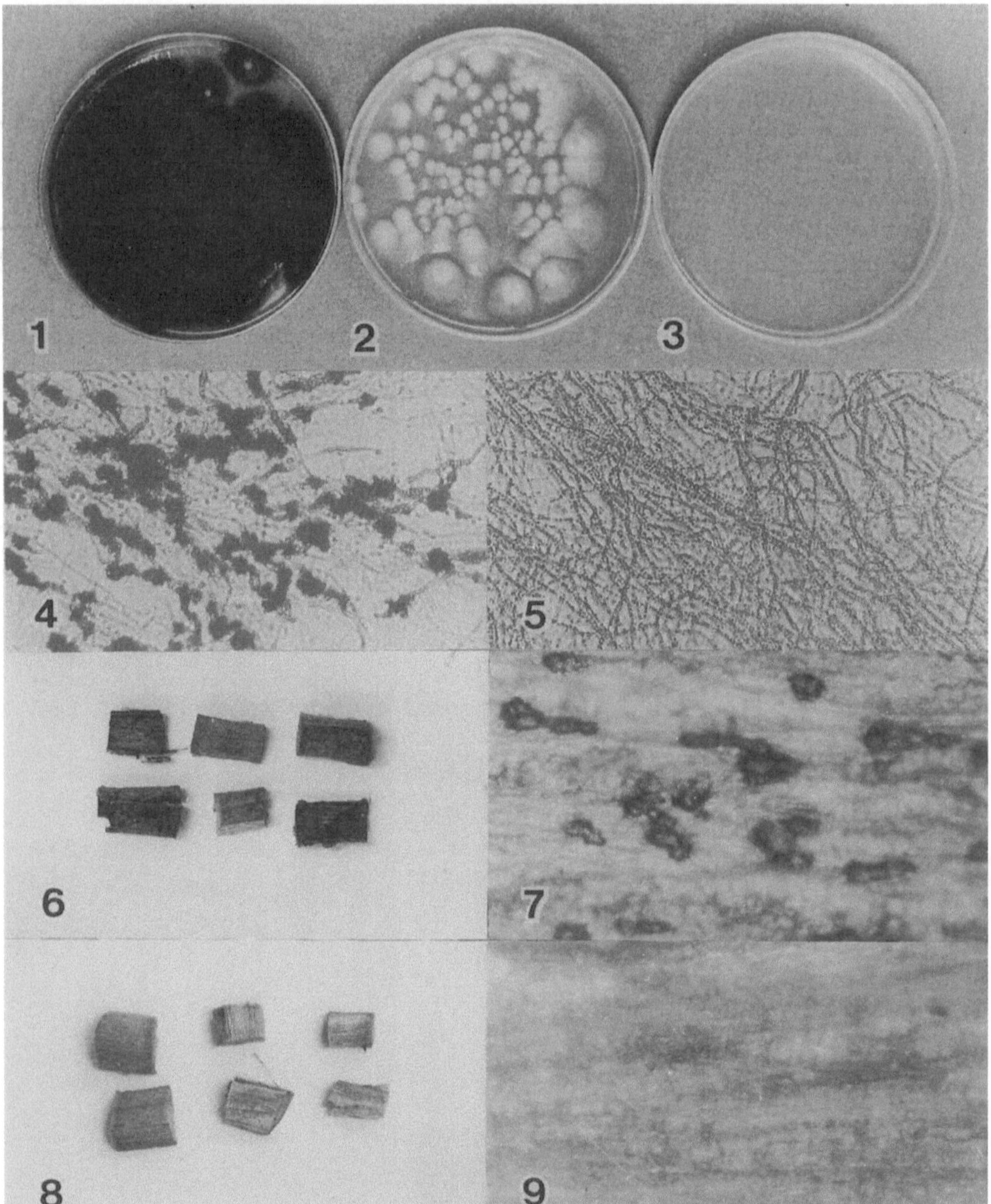
1
2
3
4
5
6
7
8
9

Fig. 1. Normal growth of V. dahliae on PDA agar.
Fig. 2. Complete inhibition of microsclerotium formation in cultures grown in 2 mM thioglycolic acid.
Fig. 3. Complete inhibition of mycelial growth at 4 mM.
Fig. 4. Abundant microsclerotia produced in control plates.
Fig. 5. Complete absence of microsclerotia in cultures grown at 2 mM.
Fig. 6. Infected cotton stems incubated without inhibitor. Notice the intense dark-brown discolouration.
Fig. 7. Abundant microsclerotia formed on the inner bark surface of the cotton stems shown in Fig. 6.
Fig. 8. Infected cotton stems incubated in the presence of the inhibitor. Notice the normal, healthy colour of the wood.
Fig. 9. Complete absence of microsclerotia in the cotton stems shown in Fig. 8.

STUDIES ON SELECTIVITY AND SYSTEMICITY OF STEROL BIOSYNTHESIS INHIBITORS IN TRANSFORMED ROOTS INOCULATED WITH FUSARIUM OXYSPORUM

J. Mugnier
Rhône-Poulenc Agrochimie
69263 Lyon
Cedex 09
France

INTRODUCTION

The sterol biosynthesis inhibitors (SBIs) are systemic fungicides with strong fungitoxicity to genera such as Fusarium. SBIs inhibit sterol metabolism in fungi and also have plant growth retardant activity (Lyr 1987). It has been shown that growth retardation caused by SBIs can be reversed by addition of gibberellic acid (GA). SBIs also interfere at high concentrations in sterol synthesis of plants (Buchenauer and Röhner 1981; Lyr 1987).

Various plant tissues, including seedlings, cell and callus cultures have been used for SBI studies. We suggest here the use of root cultures transformed with Agrobacterium rhizogenes. When A. rhizogenes infects a plant, a part of the bacterial DNA (T-DNA) from the Ri (root-inducing) plasmid is transferred to plant cells (White et al 1985). Plant cells that receive the T-DNA and incorporate it into their genome differentiate into root meristems and the roots grow rapidly. The transformed roots are indefinitely cultured in liquid or on solid media without phytohormones (Mugnier 1988).

Various SBIs were applied to different transformed roots over a range of concentrations to determine any phytotoxic effects, and were also tested for their systemic activity against Fusarium. The method, previously reported (Mugnier and Mosse 1987), for testing the transport process involved the use of bicompartmental cultures. Roots growing from a Murashige and Skoog (Murashige

NATO ASI Series, Vol. H28
Vascular Wilt Diseases of Plants
Edited by E. C. Tjamos and C. Beckman

and Skoog 1962) medium into a compartment containing water were infected with spores of Fusarium. The use of such a system for studying the infection process by Fusarium is discussed.

MATERIALS AND METHODS

Transformed root organ cultures

Roots induced by A. rhizogenes according to the procedures previously described (Mugnier 1988) were established for flax (Linum grandiflorum Desf.), tomato (Lycopersicon esculentum Mill), carnation (Dianthus caryophyllus L.), marigold (Tagetes erecta L.) and cucumber (Cucumis sativus L.). Thereafter, the transformed roots were maintained in a modified Murashige and Skoog (M&S) liquid medium (Murashige and Skoog 1962) containing 0.33 g/liter NH_4NO_3 and 0.38 g/lit KNO_3.

The lowest concentration of SBIs that was sublethal, but completely inhibitory to roots was determined. The roots were incubated in the liquid M&S medium containing the fungicides (active ingredient) at different concentrations. Complete growth inhibition was estimated visually after 5 days incubation at 25 C in the dark. Toxicity is expressed as the minimum inhibitory concentration, being the lowest concentration giving no visible growth, and is illustrated in Fig. 2 and 3. Histograms in Fig. 3 correspond to the fungicidal concentration range where the roots grew at the same rate and manner as the controls.

Bicompartmental culture system

The transformed roots were cultured in divided Petri dishes. The method is illustrated in Fig. 1. Compartment a contained the M&S agar (Difco 1%) and compartment b contained water agar. Roots were placed on the nutrient agar and grew over the barrier that divided the Petri dish and into the water compartment.

Fungicides (a.i.), in liquid agar at 40 C, were placed in compartment a. The root cultures used for these experiments were well established to avoid the phytotoxic effect on root growth.

The cultures were then incubated for 4 days and after this period of fungicide uptake, spore suspensions (c.a. 1,000 spores/ Petri dish) were added to compartment b. Controls were made to investigate the vapor phase activity by adding the compounds and inoculum of Fusarium to separate compartments of the Petri dish in the absence of roots.

Fusarium strains

Fusarium oxysporum (Schlecht.) Snyd. & Hans belonging to the formae specialis dianthi, melonis, lycopersici, callistephi, lini and Fusarium solani (Mart.) Snyd. & Hans. (ex Institut de la Recherche Agronomique, France) were used for the experiments. The formae specialis were associated with their respective host root species listed below.

Fungicides tested

The following compounds were tested: fenarimol, propiconazole, flutriafol, flusilazol, prochloraz, triadimefon, triadimenol, tridemorph, bitertanol, fenpropimorph.

RESULTS

Selectivity

The effect of SBIs on the root growth is illustrated in Fig. 2. Histograms in Fig. 3 show that growth of roots in the liquid M&S medium was reduced after fenarimol (0.5 mg a.i./liter) treatment. Other SBI-fungicides appeared more selective; for example, roots grew in the presence of 5 mg/liter flutriafol or tridemorph. Treatment with higher concentrations of SBIs altered the root morphology. The fungicides produced swelling of the root tips in the region of maximum cell division.

GA reversal experiments

We reproduced the GA reversal experiments (Buchenauer and Röhner 1981) with the different SBIs by growing the roots in the liquid M&S medium in the presence of 10,20 and 50 mg/liter GA. Above 20 mg/liter, the hormone became phytotoxic. Of the different roots treated with sublethal concentrations of SBIs, it was found that the addition of GA did not prevent growth inhibition caused by the fungicides.

Systemicity

Propiconazole, flutriafol, prochloraz, triadimefon, triadimenol, fenpropimorph and, fluzilazol applied to the bicompartmental Petri dishes moved into the roots. The compounds were absorbed by the roots in compartment a and translocated throughout the root tissues into compartment b, and exuded into the water. The fungicides present in the root exudated inhibited spore germination and hyphal growth of all the formae specialis tested. In Petri dishes not-treated with the fungicides, the spores of F. oxysporum germinated and the vegetative mycelia invaded the medium surrounding the roots.

Infection process.

The experiments were carried out with the different transformed root species and their specialized formae specialis of F. oxysporum. We studied the development of the germ tubes and hyphae near the root hairs and epidermal cells behind the root tips, or within the zone of elongation. Unfortunately, none of the germ tubes or hyphae of the different formae specialis was capable of infecting the root hairs or epidermal cells of their host-roots, and therefore the colonization of the xylem vessels could not be observed.

Various modifications of the M&S medium were tested (e.g. nitrogen concentration and form), but even when the nutrient medium contained lower nitrogen and sugar concentrations, the various stages of the infection process were not observed.

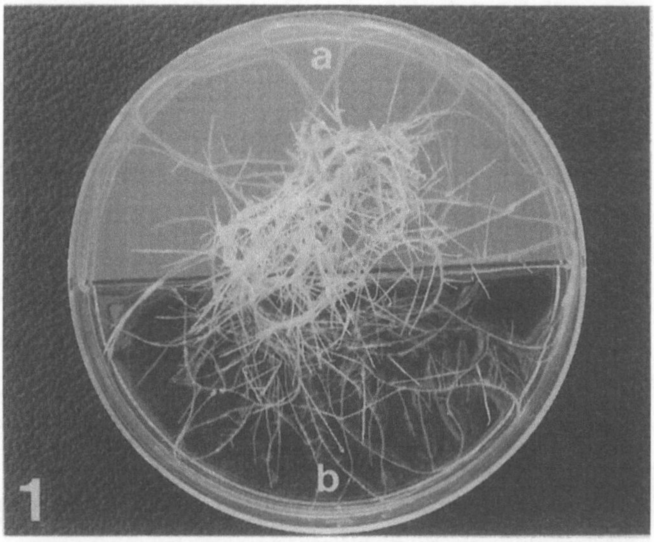

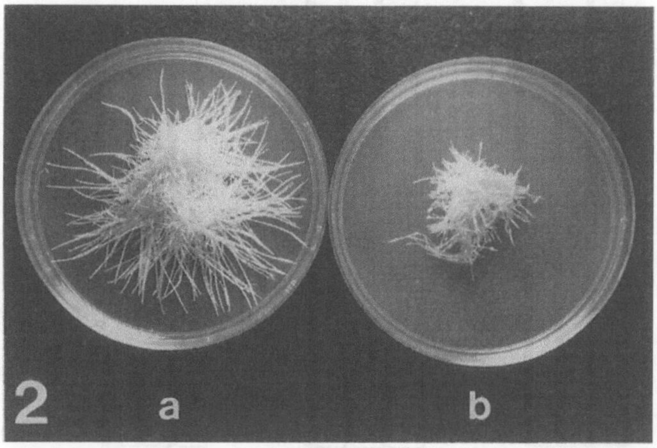

Fig. 1. Ten day transformed roots of flax (Linum grandiflorum) growing in a 9 cm bicompartmental Petri dish. Compartment a contains the Murashige & Skoog agar medium, compartment b contains water.

Fig. 2. Effect of triadimenol (5 mg/liter) on the growth of roots of flax (Linum grandiflorum) on the Murashige & Skoog agar medium after 5 days of incubation. a. control; b. treated.

DISCUSSION

Specific information can be obtained by studying the effects caused by SBI-fungicides on the growth of transformed roots. The range of concentrations over which one could obtain activity without phytotoxicity in the M&S liquid medium was in the order of 5-10 mg/litre. We have not reproduced the GA reversal experiment carried out on seedlings (Buchenauer and Röhner 1981). However it was suggested (Lyr 1987) that growth inhibition by SBIs such as pyrimidine and triazole derivatives might involve other sites than those affecting GA biosynthesis directly. Bicompartmental root cultures are easily adapted for radioisotope bioassays (Mugnier and Mosse 1987) for studying root translocation of fungicides. Unfortunately the study is limited to dicotyledonous roots because it involves the use of roots which have been genetically transformed by Agrobacterium rhizogenes.

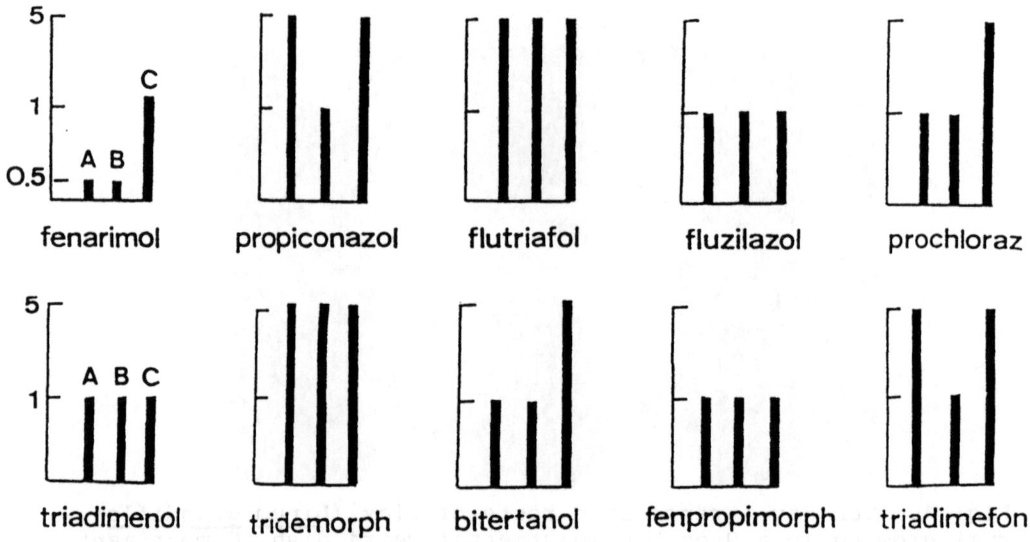

Fig. 3. Minimum inhibitory concentrations of sterol biosynthesis inhibitors on transformed roots growing in the Murashige & Skoog liquid medium (in Y-axis, concentration in mg/liter). A. flax; B. tomato; C. sugarbeet.

Studies with other soil-borne pathogens such as Pythium Phytophthora or Rhizoctonia showed normal features in the infection process, including responses of zoospores or appressorial development followed by inter- and intracellular colonization of the transformed root tissues. These pathogens depend little on host nutrition. That emphasizes the possibly central role of root nutrition in the development of Fusarium infection. The nitrogen source in the M&S medium might be one explanation for the transformed root resistance. The bicompartmental root system can supply separate nutrition for the roots and Fusarium, and can thus provide a relatively normal nutritional environment for Fusarium. The method may be effective and merits further study.

REFERENCES

Buchenauer H and Röhner E (1981) Effect of triadimefon and triadimenol on growth of various plant species as well as gibberellin content and sterol metabolism in shoots of barley seedlings. Pesticide Biochemistry and Physiology 15:58-70

Lyr H (1987) Modern selective fungicides. John Wiley & Sons. New York, p 383

Mugnier J and Mosse B (1987) Vesicular-arbuscular mycorrhizal infection in transformed root-inducing T-DNA roots grown axenically. Phytopathology 77:1045-1050

Mugnier J (1988) Establishment of new axenic hairy root lines by inoculation with Agrobacterium rhizogenes. Plant Cell Reports 7:9-12

Murashige T and Skoog F (1962) A revised medium for rapid growth and bioassays with tobacco tissue culture. Physiologia Plantarum 15:457-497

White FF, Taylor BH, Huffman GA, Gordon MP and Nester EW (1985) Molecular and genetic analysis of the transferred DNA regions of the root inducing plasmid of Agrobacterium rhizogenes. Journal of Bacteriology 164:33-44

CULTURAL PRACTICES: THE EFFECT OF PLANT DENSITY AND IRRIGATION REGIMES ON VERTICILLIUM WILT OF COTTON

O.C. Huisman[1] and D.W. Grimes[2]
Department of Plant Pathology
University of California
Berkeley, CA 94720
USA

INTRODUCTION

Verticillium wilt caused by the soil-borne fungus Verticillium dahliae, affects a wide variety of cultivated and ornamental plants. In California, as in many other parts of the world, the disease is of major economic importance on cotton, pistachio, olive, tomato, strawberry, and potato. At present, control of the disease is limited to planting resistant or tolerant cultivars when available and to reduction of inoculum density in soil by fumigation or solarization, which, with crops such as cotton, is often economically unfeasible. Alternative approaches involve the use of cultural practices which minimize disease incidence and/or disease severity. This report summarizes the findings of our work on the use of cultural practices for minimizing yield losses from Verticillium wilt in cotton.

BASIC METHODS

Over a three year period, cotton cultivar SJ-2 was planted, in mid April, in field plots severely infested with V. dahliae (40-80 propagules per gram soil) in 8 row plots (8 m x 100 m).

2. Department of Land, Air and Water Resources
University of California
Davis, CA 95163
USA

NATO ASI Series, Vol. H28
Vascular Wilt Diseases of Plants
Edited by E. C. Tjamos and C. Beckman

The field plots, a deep clay loam soil, were located at the West Side Field Station in the San Joaquin valley CA. All plots received a pre-plant irrigation of 45 cm water. Post-plant precipitation was negligable, being less than 1 cm. Treatments, replicated three times, consisted of varying post-plant furrow irrigation regimes and varying plant densities (25,000 to 200,000 plants per hectare). The irrigation regimes varied in (1) number of post-plant irrigations (0 to 5, ranging from 12 to 20 cm water each), (2) timing of each irrigation relative to plant water stress (mild to severe as measured by leaf water potential at mid-day, Grimes and Yamada 1982), and (3) timing of the first irrigation. Data were collected on the incidence of systemic infection (bases of foliar symptoms), incidence of disease severity (bases of percent plants with 80% leaf loss due to Verticillium wilt), and lint yields at the end of the season.

EFFECTS ON DISEASE INCIDENCE

Percent of infected plants at any given date was not a simple function of soil inoculum density, being linear only at low levels of disease incidence (Ashworth et al 1979). However, a linear relationship between inoculum density and disease incidence was observed at all levels of soil inoculum density and disease incidence when the fraction of infected plants was corrected for random multiple infections: mean number of infections = ln (1/(1-X), where X = fraction of plants infected.

Disease incidence was found to be inversely related to plant densities. Plots of mean number of infections verses 1/(plants/ha) were linear. These results are in good agreement with those reported by Minton et al (1972), particularly when their infection data are corrected for random multiple infections. The results are consistent with an interpretation that the expected disease incidence is proportional to the size of the plant root system. This is derived from the following considerations. Excluding extremely low plant densities which preclude inter- and intra-row closure by the canopy, root densities and canopy densities (on a

m^2 bases) are only minimally influenced by plant densities for most of the season, particularly so after row closure by the canopy. If the infection rate per unit root density is constant, that is, infections are proportional to number of encounters between roots and soil-borne inoculum (Huisman 1982) at any given time, then a linear relationship between disease incidence and the inverse of plant density would be expected.

The irrigation regime had only a minor influence on disease incidence. Disease progression was similar for most irrigation treatments. However, disease progression slowed slightly in plots during periods of water stress as compared to plots with adequate soil moisture.

EFFECTS ON DISEASE SEVERITY

The relationship between disease severity and plant density was more complex. With low quantities of applied water (0 to 1 post-plant irrigations), defoliation decreased with increasing plant densities, in line with the observed effects on disease incidence. In treatments with only a single post-plant irrigation, the lowest disease severity occurred in plots with the longest delay in the irrigation. In contrast, under high soil moisture regimes (3 to 5 post-plant irrigations), the percentage of plants defoliated at the end of the season was greater in plots with high plant densities than in those with low plant densities. When taking into consideration the lower level of disease incidence in the plots with higher plant densities, the fraction of defoliated plants was a far higher proportion of infected plants in plots with high plant densities than in plots with low plant densities.

Delays in the timing of the first irrigation had the greatest influence on disease severity. Percentage of defoliated plants at the end of the season had a linear, inverse relationship to the post-plant day of the first irrigation (values ranged from 44 to 7% defoliation for plots first irrigated, respectively, on 19 June and 7 August). In contrast, total number of irrigations had only a minor effect on disease severity. For plots receiving a single irrigation, the lowest disease severity (3-5% defoliation) was observed in those with high plant densities (> 150,000 plants per hectare) receiving a late (early August) irrigation.

EFFECTS ON YIELD

Effects of irrigation regimes and plant density on yields were most closely correlated (negatively) with the effects of these variables on disease severity and water stress. Yield losses attributable to Verticillium wilt had a high negative correlation with the fraction of defoliated plants at the end of the season. Delayed irrigations drastically reduced the yield losses induced by Verticillium wilt: with negligable losses occurring with no post-plant irrigation. At the same time, delays in the first irrigation induced increasing yield losses directly attributable to water stress (Grimes 1977). With high plant densities and a normal irrigation regime, yield losses were around 50% relative to those expected in the absence of Verticillium wilt. With high plant densities and a single late irrigation, yield losses were around 20 to 25%. In the latter case most of the yield loss (20%) could be attributed to yield reductions due to water stress with only a minor component of the loss (3 to 5%) attributable to Verticillium wilt. Yields were linearly related to the timing of the first post-plant irrigation (ranging from 860 to 1250 kg lint/ha for plots with high plant densities first irrigated, respectively, on 19 June and 7 August).

Our results show that minimizing yield losses due to Verticillium wilt in cotton involves the balancing of two opposing trends. Delaying the first post-plant irrigation progressively decreased disease severity (from around 50% to negligable defoliation for, respectively, normal irrigation regimes to no irrigation) and suppresses yield loss attributable to Verticillium wilt to very low levels. At the same time, delayed irrigation progressively decreased yields due to water stress in the plants (Grimes 1977).

In clay loam soils heavily infested with V. dahliae, optimum yields were achieved by using high plant densities (> 150,000 plants/ha) and a single irrigation applied in early August. The cultural practices resulting in the highest yield losses were those involving the use of high plant densities and a normal irrigation regime (3 to 5 post-plant irrigations). Under a normal irrigation regime, low plant densities (< 40,000 plants/ha) were preferable to high plant densities.

REFERENCES

Ashworth LJ Jr, Huisman OC, Harper DM, Stromberg LK and Basset D (1979) Verticillium wilt disease of cotton: influence of inoculum density in the field. Phytopathology 69:483-489

Grimes DW (1977) Physiological response of cotton to water and its impact on economical production. Western Cotton Prod Conf Summary Proc 1977:26-29

Grimes DW and Yamada H (1982) Relation of cotton growth and yield to minimum leaf water potential. Crop Sci 22:134-139

Huisman OC (1982) Interrelations of root growth dynamics to epidemiology of root-invading fungi. Ann Rev Phytopathol 20:303-327

Minton EB, Brashears AD, Kirk IW and Hudspeth EB Jr (1972) Effects of row and plant spacings in Verticillium wilt of cotton. Crop Sci 12:764-767

EFFECT OF CROP ROTATION AND METHAM-SODIUM ON *VERTICILLIUM DAHLIAE*

Y. Ben-Yephet[1], Z.R. Frank[1], J.M. Malero-Vera[2] and J.E. DeVay[3]
Dept. of Plant Pathology
ARO
The Volcani Center
Bet Dagan 50250
Israel*

INTRODUCTION

The fungus *Verticillium dahliae* Kleb. survives in soil for long periods, in the form of microsclerotia (MS) (Green 1969; Wilhelm 1955). In Israel the fungus was introduced into the Negev area with infected potato tubers (Krikum et al 1974). Following expansion of potato growing in the Negev, this pathogen became established in many fields in the region. The ordinary practice there is to grow potatoes only once every 3-4 years. Nevertheless, the fungal population increased and induced serious losses in potato yields although potato cultivars tolerant to *V. dahliae* ('Désirée'and'Spunta') have being used. In addition to the use of potato cultivars tolerant to *V. dahliae*, treatment of soil with metham-sodium (MES) was found to improve potato yields (Krikum et al 1974). This fungicide is used against *V. dahliae* because of its effectiveness, low cost, and ease of ap-

2. Departmento de
Protection Vegetal
Apartado Correos 240
14071 Cordoba
Spain
3. Dept. of Plant Pathology
Univ. of California
Davis, CA 95616
USA

*Contribution No.2421-E, 1988 series, from the ARO Volcani Center, Bet Dagan, Israel

plication to soil via the sprinkler irrigation system (Ben-Yephet et al 1983).

In water solution MES changes rapidly to methylisothiocyanate (MIT), which possesses fungicidal activity (Munnecke et al 1962). About 99% of MIT is soluble in water and about 1% in the vapor phase, depending upon the type of soil and its moisture content (Gerstl et al 1977). A recent study has shown that the use of MES coupled with soil solarization (S—S), causes a synergistic killing action against some plant pests (Frank et al 1986).

The persistent structures of the fungus are microsclerotia (MS) widespread in the soil profile of the upper 40 cm. Even a very low population of microsclerotia can cause an epidemic. Thus for efficient control, microsclerotia should be killed throughout the soil profile (Ben-Yephet et al 1980).

The purposes of this study were to follow the inoculum density of V. dahliae, in infested soil, under non-host crop rotation and the effect of application of MES through the irrigation system, rate of water application, soil moisture and soil temperature on MES efficiency in controlling V. dahliae.

MATERIALS AND METHODS

Soil

Field and laboratory experiments were conducted in silty loam loessial soils (20-25% clay, field capacity ca. 21%, pH 8.4).

Crop sequences and cultivations

Potatoes were grown in the autumn at two different locations in the Negev. Field A was planted with cv. Spunta and field B with cv. Désirée. The plants in both fields suffered from severe wilt induced by V. dahliae. Their residues were incorporated into the soil in January. The soil was cultivated with a disc plow to 15-20 cm and prepared for a cotton crop. Following cotton, the soil was plowed to a depth of 40 cm. In the successive years the soil was cultivated to a depth of 15-20 cm before seeding. In field A, 3 years of wheat followed the cotton (San Joaquin SJ-2, tolerant to V. dahliae). In field B, the cotton was followed by peas and wheat (in 1 year), 2 years of wheat, 1 year of vetches, and again 2 years of wheat.

Estimation of V. dahliae viable units in soil

In both fields inoculum levels were first estimated in April following the potato crop and again in February of other years: in field A in each of the next 4 years and in field B in the sixth and seventh years. From each of the two fields a composite soil sample comprising 25 subsamples was collected along a diagonal transect. Separate samples were collected from four soil layers: 0-10, 10-20, 20-30 and 30-40 cm deep, using a core-type auger (7.5 cm diam). The soil was air dried for 4 weeks, gently ground with mortar and pestle, and passed through a 250-µm sieve. From each of the eight samples (four layers from each of two fields), five 5-g aliquots were processed with the cesium chloride floatation technique (Ben-Yephet and Pinkas 1976) to determine the number of viable V. dahliae units. Organic particles floating on the cesium chloride solution were seeded onto a synthetic growth medium (Ausher et al 1975); 24 days after plating, V. dahliae colonies were counted.

Inoculum and fungicide

The inoculum used in all experiments with MES was MS (< 250 µm in diameter) of V.dahliae from potato stems (Ben-Yephet et al 1983). The MS were mixed with air-dried loessial soil at 10^4/g and 0.5- to 1.0-g aliquots of this mixture were placed in 20-µm-mesh nylon bags. In microplot and field experiments, these bags were then tied along a string at intervals of 5, 10, 20, 30 and 40 cm from the upper end and buried at those depths from the soil surface following seedbed preparation. After soil treatments, samples were suspended in 50 ml of sterile distilled water, poured onto a filter paper, and rinsed three times with sterile distilled water. To determine viability, 50 MS from each sample were identified under low-power magnification with a dissecting microscope, transferred singly with a botanical needle, and planted on a synthetic medium containing PCNB and sucrose (Ausher et al 1975).

Commercial-grade MES (sodium-methyldithiocarbamate) was used in an aqueous solution (pH 9.1) containing 32.7% active ingredient (a.i.). The fungicide dose was calculated as milliliters of product per liter of water per square meter, or as the amount of solution at a particular concentration of MES.

Penetration of metham-sodium in loessial soil

Microsclerotia inserted to a depth of 40 cm in the middle of 1-m^2 microplots were treated with the fungicide using a watering can. The irrigation volume necessary to reach field capacity of the soil down to 40 cm was 80 l/m^2. The same dose (60 ml/m^2) of MES diluted in different volumes of water (4,8,16,32,64 and 80 l) was applied and then supplemented with water to a total of 80 l/m^2. Treatments were replicated in randomized blocks at least three times in each of the two experiments.

Effect of incubation temperatures on metham-sodium concentrations required to kill <u>V. dahliae</u>

Soil and columns were equilibrated at the experimental temperatures. Then columns 4.4 cm in diameter and 10 cm long were packed with air-dried loessial soil, 180 g per column. <u>V. dahliae</u> microsclerotia were placed at mid-depth in the columns and the soil was irrigated with 25 ml of either water or fumigant solutions at 25,50,100,200,300 or 400 mg of MES a.i./l water. Water temperatures were adjusted before the columns were irrigated. Subsequently, the incubation periods were adjusted to the half-life time of MIT at each temperature to obtain toxicity data of the different concentrations on a comparable time-temperature basis. The incubation periods were 8,4,3 and 2 days at 15,20, 25 and 35 C, respectively. All treatments were replicated three times in each of three experiments.

In addition, a field experiment in microplots was conducted

in mid-summer to test the effect of combining solarization, which increases soil temperatures (Katan 1981), with MES at two sub-lethal doses. Treatments were replicated five times in randomized blocks for each of the following treatments: soil solarization (S—S); MES 12.5 ml/m^2; MES, 25 ml/m^2; S—S plus MES 12.5 ml/m^2 and S—S plus MES, 25 ml/m^2. The MES dosages selected were based on data from preliminary experiments in columns. The site selected for the solarization studies was on Reiff loam (sand 46%, silt 37%, clay 16%, organic matter 1%), with a field capacity of 20%, and pH 6.8.MES was applied uniformly at the beginning of the irrigation by sprinkler in the first 10% of the irrigation water. The transparent polyethylene tarps (50 µm thick) were applied for solarization treatments 30 hours after the irrigation. During the solarization treatments, soil temperatures were measured at 25 cm depth and recorded by inserting thermograph sensors in the solarized and control soils. Before the irrigation, multiple soil samples, naturally infested with known populations of fungal propagules were buried at depth of 25 cm, in each plot, and sampled 1,2,4 and 6 weeks after the beginning of the solarization process.

Separate factorial analyses of variance were carried out for each time of sampling. Logarithmic transformation was done prior to analysis to stabilize the variance. Duncan's multiple range tests were carried out on the treatment combinations rather than on the main effects because of the significant interactions which were found.

RESULTS

Survivability of V. dahliae in the field

In field A, populations of the fungus decreased slightly in the first year (with Verticillium-resistant cotton) and more rapidly in the next 3 years (with wheat every year). In field B, where non-susceptible crops were grown for 7 years, the inoculum of V. dahliae decreased to 4% of the original density of viable units. Following disc incorporation of the infested potato plant residues in field A, 98.5% of the fungal units counted were found in the top 20 cm of soil. After another year, following deep plowing, the inoculum density of viable units decreased in the upper layers and increased in the lower layers. After a second year of cropping, the inoculum density was similar throughout the 40 cm of soil sampled. By the end of the third and fourth years, however, the highest inoculum levels were found in the top 10 cm of soil and decreased with increasing soil depth. In field B, after 6 and 7 years of cropping, inoculum density also decreased with increasing soil depth (Table 1).

Table 1. Distribution of viable Verticillium dahliae units, at various depths in infested fields under crop rotation

Soil depth (cm)[z]	Number of viable V. dahliae units per 5 g soil							
	Field A					Field B		
	Time (years)[z]							
	0	1	2	3	4	0	6	7
0-10	143^{aA}	65^{aB}	36^{aC}	9^{aD}	13.6^{aD}	209^{aA}	10.2^{aB}	6.2^{aC}
10-20	65^{bA}	50^{bB}	31^{bC}	5.2^{bD}	5.5^{bD}	104^{bA}	6.8^{bB}	4.4^{bC}
20-30	1.8^{cB}	33^{cA}	29^{bA}	3.4^{bcB}	1.0^{cC}	--	3.0^{cA}	1.0^{cB}
30-40	1.2^{cC}	19^{dA}	14^{cA}	2.4^{cB}	1.0^{cC}	--	1.2^{dA}	1.0^{cA}

[z]Following incorporation of infested potato plants in January soil samples were first collected in the fields in early April (=zero time). In the other years the fields were sampled manually in February, when winter crops were grown.
Figures followed by different letters within a field differ significantly ($P<0.05$) according to Duncan´s Multiple Range Test. Capital letters describe horizontal comparison and lower case letters describe vertical comparison. Dashes symbolize unassembled data.

Penetration of a constant dose of MES in loessial soil when diluted in different volumes of water

The chemical killed the MS throughout the 40 cm when applied in the first 5% or 10% of water volume (concentrate mode) and only to 20 cm when diluted in the total water volume (dilute mode) (Table 2).

Table 2. Penetration of a constant dose of metham-sodium (MES), in loessial soil, when diluted in different volumes of water

Solvent water (%)	Supplementary water (%)	Penetration of MES (cm)[z]
5	95	40
10	90	40
20	80	35
50	50	24
100	0	20

[z]Penetration to soil depth at which 90-100% of inserted microsclerotia were killed

Effect of the rate of water application on efficiency of MES in controlling V. dahliae

Two rates of water application were tested in a field experiment in loessial soil when the chemical was applied in the concentrate mode. The low rate of water application is the one used by the growers. At this rate the fumigant killed MS of the fungus throughout the 40 cm. At the higher rate, the fumigant leached faster from the top and left some of the propagules viable (Table 3).

MES penetration in moist soils

When the fumigant is applied in the concentrate mode to dry loessial soils at a low rate of water application, the chemical moves with the irrigation water and kills the fungus uniformly throughout the 40 cm. However when soil moist, control of the pathogen is not uniform. Penetration of the chemical in moist soil was demonstrated in a field experiment conducted in moist alluvial soil. In this experiment, MS were controlled variably at difference depths. Thus, in 3, 1, 3 and 4 replicates, MS were killed down to the 40, 30, 20 and 10 cm depth, respectively.

Table 3. Effect of rate of water application on efficiency of MES applied in a cencentrate mode, in controlling Verticillium dahliae

	Percent V. dahliae killed by MES[z]	
Soil depth	Rate of water application	
(cm)	26 m^3/hour	9 m^3/hour
5	97	100
10	84	100
20	100	100
30	100	100
40	100	100

[z]MES, 800 l/ha, was injected into the first 80 m^3 water/ha of the total 800 m^3 water/ha applied.

Effect of soil temperatures on toxicity of MES to V. dahliae

Although the half-life period of MIT shortens as soil temperature rises, toxicity of MIT to V. dahliae increased with higher temperatures. Thus, the fumigant dose required to kill 50% of MS at 25 C was four times higher than required at 35 C (Table 4).

In the field experiment, soil temperatures in the control (nonsolarized) plots at 25-cm depth ranged from 25 C minimum to 26 C maximum; whereas in the solarized plots temperature ranged from 31 C to 35 C. Solarization gradually reduced the viability of V. dahliae and after 6 weeks viability was 18% of the control (nonsolarized) treatment. MES alone at the lower rate had no effect on viability of propagules; however, solarization plus MES at this rate reduced viability of V. dahliae to 12% of the control after only 1 week. MES alone at the higher rate reduced viability of V. dahliae to 30% of the control after 1 week. In contrast, solarization plus metham-sodium at the higher rate killed 100% of the propagules within 1 week. The P values confirm the significance of the synergistic interactions which exist between S—S and MES, resulting in faster and more effective control of V. dahliae than by solarization alone (Table 5).

Table 4. Toxicity effect of metham-sodium (MES) to Verticillium dahliae microsclerotia in columns of loessial soil at four temperatures

Temperature (C)	Percent V. dahliae killed by MES at (mg a.i./1 of water) 25	50	100	200	300	400
15	0	0	24	78	97.5	100
25	0	0	42	99	100	100
30	0	4	89	100	100	100
35	50	94	100	100	100	100

Each column was filled with 180 g of loessial soil and irrigated with 25 ml of water or fumigant solution.
The incubation periods were adjusted to obtain the half-life time of the fumigant at each temperature, and were 8,4,3, and 2 days at 15,25,30 and 35 C, respectively.

DISCUSSION

In general, the inoculum density decreased from the upper to the lower soil levels tested and also from one year to the next. Nevertheless, in the inspected fields 12% (field A) and 4% (field B) of the original population density remained viable after 4 and 7 years, respectively, of crop rotation with non-suscepts hosts. The existence of the fungus for long periods in the field was probably due to the ability of the fungus to colonize and produce new MS on the root system of non-hosts, including wheat (Krikum 1976; Martinson and Horner 1962).

Crop rotation is not an effective practice for controlling Verticillium-incited disease (Nnodu and Harrison 1979; Sewell and Wilson 1966; Wilhelm 1955). Ashworth et al (1972), who studied the relationship between inoculum density of V. albo-atrum (microsclerotial type) and infection of cotton, concluded that crop rotation may be effective if levels are reduced below a certain threshold. Nnodu and Harrison (1979) reported that a minimum of approximately 100 units of V. albo-atrum (MS type) per 5 g soil was required to cause significant potato yield reductions. In the Negev area, however, we found (Ben Yephet et al 1980) that severe potato wilt in the autumn season may occur when the inoculum density is greater than one unit of V. dahliae per 5 g of soil. In the present study, 7 years of crop rotation reduced V. dahliae inoculum to 3.15 propagules per 5 g soil, which is still high enough to cause severe wilt in autumn-planted potatoes. Thus, in order to minimize the risk of wilt in potato regrown in a field which previously was heavily infested with V. dahliae, it seems necessary to supplement an inoculum-reducing crop rotation by chemical control.

Table 5. Effect of soil solarization (S—S) and MES (MES) on the ability of Verticillium dahliae propagules in July-August

	V. dahliae (propagules/g soil) July-August			
	Time (weeks)			
Treatment	1	2	4	6
Control	14.8	16.4	12.4	16.0
MES_I	16.4	16.8	22.4	17.6
MES_{II}	4.4	2.0	1.6	2.0
S—S	14.0	13.2	4.8	2.8
S—S + MES_I	2.8	5.2	2.8	3.2
S—S + MES_{II}	0.0	0.0	0.0	0.0
LSR (5%)	1.57	2.35	2.55	2.56
P values for interaction	0.0001	0.05	0.09	0.32

MES_I and MES_{II}-MES applied at 12.5 and 25 ml/m^2, respectively.
LSR-least significant ratio.

In loessial soil, MES killed MS throughout the 40 cm when applied in the first 5-15% of water volume and to only 20 cm when diluted in the total water volume. In this soil, water percolation is slow and it takes 60 hours for water to percolate down to 40 cm. As the time required for penetration of the fumigant increases, more MIT molecules are absorbed to clay particles and decompose; thus fewer MIT molecules arrive at deeper layers. When MES is applied in the first 5-15% of the water volume more MIT is leached to the lower layers by the complementary water as compared with dilution of MES in the total water volume.

When the fumigant is applied in a concentrate mode to dry loessial soil at a low rate of water application, the chemical moves with the irrigation water and kills the fungus uniformly throughout the 40 cm. However, when soil is moist, control of the pathogen is non uniform probably for two reasons:(a) unequal penetration of the irrigated water in the soil profile (like fingers rather than as a piston), and (b) an uneven fumigant diffusion into aggregates and in voids to various sites filled with previously stored water.

Toxicity of MES was reported to be greater at 20 C than at other temperatures that were tested (Lambe 1961). In a previous study, however, toxicity of MES to V. dahliae in soil columns was higher at 35 C than at lower temperatures (Ben-Yephet 1979). This trend was verified in the present study: several MES concentrations were employed for incubation periods that were deliberately adjusted to the half-life of MIT at each temperature. These comparable incubation periods were used to enable similar active concentrations to affect the pathogen tested at compared temperatures. Smaller MES concentrations were required to kill the pathogen at the higher temperatures than at the lower ones, but the relation was not linear.

Combining solarization — which increases soil temperatures — with soil fumigants increases the effectiveness of the combined treatments as compared with each single treatment (Frank et al 1986; Sharoni 1985; Stapleton and DeVay 1983). A synergistic interaction was recently reported between solarization and MES in controlling delimited shell spot of peanut pods, a disease of uncertain etiology (Frank et al 1986). The combination of MES at low doses with S—S resulted in a synergistic interaction between the treatments in controlling V. dahliae propagules

(Table 5). The synergism of solarization with MES not only increased pathogen destruction compared with solarization alone, but also reduced the time necessary to kill the pathogen.

REFERENCES

Ashworth LJ Jr, McCutcheon OD and George AG (1972) Verticillium albo-atrum: the quantitative relationship between inoculum density and infection of cotton. Phytopathology 62:901-903

Ausher R, Katan J and Ovadia S (1975) An improved selective medium for the isolation of Verticillium dahliae. Phytoparasitica 3:133-137

Ben-Yephet Y (1979) Penetration of metham and mylone into soil columns as measured by their effect on viability of microsclerotia of Verticillium dahliae. Plant and Soil 53:341-349

Ben-Yephet Y, Gurevich L and Frank Z (1980) The relationship between inoculum density of Verticillium dahliae microsclerotia in the field and disease incidence in potato plants. Hassadeh 60:1277-1280 (in Hebrew)

Ben-Yephet Y and Pinkas Y (1976) A cesium chloride flotation technique for the isolation of Verticillium dahliae microsclerotia from soil. Phytopathology 66:1252-1254

Ben-Yephet Y, Siti E and Frank ZR (1983) Control of Verticillium dahliae by metham-sodium in loessial soil and effect on potato tuber yields. Plant Disease 67:1223-1225

Frank ZR, Katan J and Ben-Yephet Y (1986) Synergistic effect of metham and solarization in controlling delimited shell spots of peanut-pods. Crop Protection 5(3):199-202

Gerstl Z, Mingelgrin U and Yaron B (1977) Behaviour of Vapam and methylisothiocyanate in soils. J Soil Sci Soc Am 41:545-548

Green RJ Jr (1969) Survival and inoculum potential of conidia and microsclerotia of Verticillium albo-atrum in soil. Phytopathology 59:874-876

Katan J (1981) Solar heating (solarization) of soil for control of soilborne pests. Ann Rev Phytopath 19:211-236

Krikum J (1976) Verticillium and wheat. Proc 2nd Int Verticillium Symp (Univ of California, Berkeley) p 28

Krikum J, Netzer D and Sofer M (1974) The role of soil fumigation under conditions of intensive agriculture. Agro-Ecosystems 1:117-122

Lambe RC (1961) The influence of temperatures, moisture and method of application on the fungitoxicity of mylone and vapam. Diss Abstr 21:418-419

Martinson CA and Horner CE (1962) Importance of nonhosts in maintaining the inoculum potential of Verticillium. Phytopathology 52: 742 (Abstr)

Munnecke DE, Domsch KH and Eckert JW (1962) Fungicidal activity of air passed through columns of soil treated with fungicides. Phytopathology 52:1298-1306

Nnodu EC and Harrison MD (1979) The relationship between Verti-

cillium albo-atrum inoculum density and potato yield. Am Pot J 56:11-25
Sewell GWF and Wilson JF (1966) Verticillium wilt of the hop: the survival of V. albo-atrum in soil. Ann appl Biol 58:241-249
Sharoni Y (1985) The effect of heat, solarization and fungicides, separately or combined,on viability of sclerotia of Sclerotinia sclerotiorum. MSc thesis 59 p. Faculty of Agriculture, Hebrew University of Jerusalem, Rehovot, Israel (summary in English)
Stapleton JJ and DeVay JE (1983) Response of phytoparasitic and free-living nematodes to soil solarization and 1,3-dichloropropene in California. Phytopathology 73:1429-1436
Wilhelm S (1955) Longevity of the Verticillium wilt fungus in laboratory and field. Phytopathology 45:180-181

PRELIMINARY STUDIES ON THE CONTROL OF POTATO BLACKLEG BY A HOT WATER TREATMENT OF SEED TUBERS

M.C.M. Pérombelon[1], E.M. Burnett[1], J.S. Melvin[2] and S. Black[2]
Scottish Crop Research Institute
Invergowrie
Dundee DD2 5DA
Scotland

INTRODUCTION

Blackleg is a seed-borne disease affecting potato plants and in temperate climates is caused by the soft rot bacterium Erwinia carotovora sub. sp. atroseptica (Eca) which together with E. carotovora sub. sp. carotovora (Ecc) is also responsible for tuber soft rot in the stored crop. The bacteria are mostly found superficially on the tubers, in lenticels and in wounds. Although most tubers in most seed stocks are usually contaminated by both bacteria, blackleg and soft rot occur only when environmental conditions favour growth of the bacteria (Perombelon and Kelman 1980). Blackleg development is favoured by wet growing conditions which induce the mother tuber to rot rapidly and the pathogen to multiply before infecting the stem. Recent work has demonstrated that blackleg incidence is related to the level of seed contamination by the pathogen and that little disease develops below a threshold level of seed contamination under average conditions, 10^3 cells per tuber (Bain and Perombelon 1987).

Current disease control strategy in many seed producing countries is based on the production of erwinia-free seed tubers from micropropagated plants as initial vegetative reproduction material and seed certification which relies on field inspection for

2. Aberdeen Biotechnology Co Ltd
50 Skene Terrace
Aberdeen
Scotland

blackleg. However, disease control has not been achieved in practice because extensive recontamination of stocks takes place within a few generations even in the absence of blackleg (Perombelon et al 1980).

Attempts to reduce tuber contamination by chemotherapy have proved unsuccessful (Wyatt and Lund 1981; Storey et al 1984) as the bacteria are well protected in lenticels and in suberised wounds. In fact, more harm than good can be done since wetting of the tuber surface without adequate rapid drying tends to induce the development of anaerobic conditions within the tuber which in turn favour multiplication of the bacteria. In contrast, thermotherapy of tubers using hot water has been shown to be promising in reducing tuber infection by fungal pathogens (Hide 1975) and more recently tuber contamination level by erwinias (Mackay and Shipton 1983). Heat is transferred to a solid more efficiently from water than air and provided the thermal death point of the pathogen is reached without affecting the viability of the eyes, disease free seed tubers can be produced.

Wale and Robertson (1986) have reported on an attempt at controlling blackleg on a commercial scale by dipping one tonne boxes of seed tubers at 45 C for 30 min. Although initial results appeared promising, certain problems inherent perhaps in large scale bulk dipping systems have become evident; (a) difficulties are experienced in restoring water temperatures to the desired level uniformly within the tuber mass. Uneven treatment of tubers can affect both erwinia control and sprouting potential, and (b) more importantly, uniform and rapid drying of treated tubers is not readily achieved in a bulk system even when forced ventilation is used, especially so when the tubers are encased in soil.

A patented continuous flow hot water treatment system has been under development since 1983 by Aberdeen Biotechnology Ltd and recently in collaboration with SCRI. This process does not suffer from the above mentioned problems associated with a bulk dipping system and produces soil free dry tubers with a uniformly low contamination level and high sprouting ability. This paper describes the treatment and discusses its potential and limitations.

THE CONTINUOUS FLOW HOT WATER TREATMENT MACHINE

Treatment of tubers occurs continuously and the tubers pass sequentially through the following four stages: (a) tubers are unloaded usually from one tonne boxes into a precleaning hopper which feeds a pre-washer in which soil from the tuber surface is removed as the tubers move forward; (b) the washed tubers then pass over a roller inspection table where any rotting tubers present can be picked out; (c) next they move in to the hot water plant which consists of a revolving wheel with perforated baskets which can hold small lots of tubers each on its periphery. As the wheel rotates, the tubers in the baskets are immersed for a predetermined time in a circulating hot water tank maintained at the desired temperature. Setting of the thermostat has to take into account the fact that the necessary heat input is dependent on the initial temperature, cultivar, size and flow rate of the tubers and the ambient temperature and certain characteristics of the machine; (d) treated tubers are dried by blasting air from a bank of air knives before boxing or bagging. Heat absorbed by the tubers ensures that any residual moisture on the surface evaporates within about a minute after leaving the machine. The machine is mobile and the present model can treat five to seven tonnes per hour.

ERWINIA THERMAL DEATH POINT

Several workers have reported that thermal death of Ecc and Eca falls in the range 46.5 to 54 C (Morse 1917; Shapovalov and Edson 1921; Stapp 1961; Robinson and Foster 1987) but the methodology used varied and was sometimes questionable. Overnight cultures of six strains of Ecc and Eca in nutrient broth at 27 C were diluted in sterile water and 0.05 ml of suspensions containing 10^7 cells/ml was pipetted in sterile glass Pasteur pipettes with the tips sealed. The pipettes in triplicates were then immersed in a circulating water bath (± 0.1 C) for different periods of time at different temperatures. The tips were then broken and the suspension plated on nutrient agar plates at 27 C for 48 hours. The thermal death point was defined as the temperatures required to kill all cells heated for a given exposure time. The thin wall narrow bore of Pasteur pipette tips allows for rapid heating and cooling of the bacteria. Results given in Table 1 suggest that the appropriate thermal death points for Eca and Ecc were 51 and 52 C for 2 min and 50 and 51 C for 4 min respectively. Quantitative studies showed that 90% of the bacteria were dead at temperatures 1 C below the thermal death point.

Table 1. Number of strains of *Erwinia carotovora* sub. sp. *carotovora* (Ecc) and *E. carotovora* sub. sp. *atroseptica* (Eca) out of six not killed after different exposure time-temperature combination treatments

Exposure time (min)	Temperature C						
	45	47	48	49	50	51	52
2	6/6[1]	6/6	6/6	6/5	6/4	4/0	0/0
4	6/6	6/6	6/6	6/1	5/0	0/0	0/0
6	6/6	6/6	6/6	1/0	0/0	0/0	0/0
10	6/1	6/5	0/0	0/0	0/0	0/0	0/0
20	2/0	1/0	0/0	0/0	0/0	0/0	0/0
30	0/0	0/0	0/0	0/0	0/0	0/0	0/0

[1] Ecc/Eca

HEAT TRANSFER IN TUBER TISSUE

Since heat transfer in tuber tissue is likely to be slower than through glass and water and since a reservoir of cold tissue in interior of the tuber exists, longer tuber treatment times would be necessary to inactivate the bacteria. But since the bacteria are sited mostly not deeper than 2-3 mm from the surface in lenticels and most wounds, treatment time need not be excessively long for the temperature to reach the desired level at that depth.

When the temperature at the center of a core (radius 5 mm and 30 mm long) taken from a cv. Maris Bard tuber four fifths immersed in water at 49.7 C was monitored with a thermistor probe inserted axially to c. 5 mm from the bottom end,it took 2.5 min to reach 48.9 C and another 1.5 min to the water temperature from an initial core temperature of 22.8 C. When the probe was inserted at the cent of a tuber 5.0 cm in diameter and the tuber immersed just below the surface in a water bath at 60.2 C, the temperature rose from 5.5 C at the start to only 12.8 C after 5 min and to the water temperature in 68 min. Because some heat could have been conducted along the probe, these data should be interpreted with caution. Nevertheless they suggest that sufficient heat transfer within the 3 mm superficial tuber layer can occur in a short time to raise the temperature above the thermal death point of erwinias without having to use excessively hot water. The final choice of suitable temperature-exposure time combinations can be arrived at by experimentally determining the efficiency of a selected range of treatments on tubers.

HOT WATER BATH TUBER TREATMENT

The effect of a range of temperature-time combination treatments on tuber contamination level was examined by immersing naturally contaminated 35-45 mm tubers, four at a time, in a large water bath. Tuber contamination level of the tubers was determined before and after treatment by peeling duplicate lots of eight tubers to 3-4 mm depth by dry abrasion, expressing the sap from the peel pulp and estimating the number of erwinias present by dilution-plating on a selective medium (Perombelon et al 1982). Selected heavily contaminated (Ecc and Eca) stocks of several cultivars were examined and 35-55 mm (seed size) washed tubers were used straight from cold storage at 5 C. Data obtained with cv. Estima (Table 2) are representative of the results obtained.

Table 2. Effect of different water temperature and exposure time combination treatments applied in a water bath on the level of contamination by <u>Erwinia carotovora</u> expressed as number (and $\log_n$ number) cells per tuber of cv. Estima

Water temperature C	Exposure time (min) 3	5	7
49	1.08×10^4 (9.286)	7.75×10^3 (8.887)	4.56×10^3 (8.406)
51	7.15×10^3 (8.871)	2.51×10^3 (7.781)	4.05×10^3 (5.952)
53	1.50×10^3 (7.000)	8.60×10^1 (4.223)	1.27×10^1 (2.641)
55	9.10×10^1 (4.493)	7.40×10^1 (4.255)	5.60×10^1 (4.022)
57	6.25×10^2 (6.403)	5.10×10^1 (3.422)	$< 3.30\times10^1$ (3.409)
59	$< 5.60\times10^1$ (3.907)	$< 5.30\times10^1$ (3.967)	$< 4.90\times10^1$ (3.811)
Untreated		1.04×10^5 (11.523)	
SED $\log_n$ no. (df)			0.5829 (24)

Tuber contamination was affected by all treatments but only those at 59 C for 3, 5 and 7 min and 57 C for 7 min reduced it to an undetectable level. However, contamination was lowered to a barely detectable level by some treatments, e.g. 53 C for 7 min,

55 C for 5 and 7 min and 57 C for 5 min. As expected 3 min treatment at all temperatures except 59 C failed to reduce contamination to a low level because the exposure time was too short for sufficient heat transfer. Results obtained with four other cultivars/stocks tested indicated that the effective temperature-time combinations shifted sometime downwards. This could be explained partly in terms of the initial contamination level, partly by the dominant erwinia present and partly by the cultivar. Low initial numbers of erwinia, Eca more than Ecc, are more readily reduced than high numbers and heat transfer in tuber tissue may be greater in some cultivars than in others and may vary at different times during the storage period.

The choice of a suitable temperature-time combination treatment rests not only on its effect on tuber contamination which must be reduced to below the threshold level for blackleg development as discussed before but also on its effect on the viability of the tuber eyes. Preliminary work has shown that sprouting of treated tubers was affected by both the heat input given, the maturity characteristics of the cultivar, and the physiological age of the tubers. In general, emergence tends to be delayed increasingly and may even fail if the temperature exceeds 55 C. This effect is more pronounced with early maturing cultivars, especially when tuber dormancy is already broken prior to treatment than with late maturing cultivars even when such tubers are already well sprouted. With the latter, although the sprouts may be killed, new ones will emerge.

FIELD TRIALS

The results obtained with the water bath cannot be applied directly to the continuous flow hot water machine because of the different heating characteristics inherent in the system. However, they can be a guide when calibrating the equipment to obtain blackleg control while retaining the agronomic characteristics of the crop. The choice of the time-temperature combination will depend on what is considered more important, blackleg control or high yield in the case of seed or ware crops respectively.

In a field trial, the effect of treating cv. Desiree seed tubers heavily contaminated with Eca and taken after loss of dormancy in April directly from cold storage at 5 C and treated at different temperatures for 5 min on blackleg control and crop growth and yield was examined. The temperatures given in Table 3 are arbitrary values obtained by averaging temperatures taken at different points. The prevailing wet growing season was favourable for blackleg development and magnified the effect of the different treatments.

The results summarised in Table 3 show that (a) seed contamination level was reduced by all the treatments, less at 50 C than at the other temperatures, in line with the results obtained with the water bath; (b) blackleg incidence in July was high only in plots planted with untreated seed but increased significantly in the 50 C and to a lesser extent in the 53 C treatments in August and September. Disease level also increased with time in the 55, 57 and even 60 C treatments indicating that under favourable conditions low seed contamination levels can still give rise to blackleg albeit at a low incidence and late in the growing season; (c) the effect of the treatments on yield was more complex, on one hand, blackleg control tended to increase yield and, on the other hand, delayed emergence retarded crop maturity hence decreased yield. Thus flowering occurred progressively later as the treatment temperature increased. Any effect on yield of non emergence or blanking caused by tuber eye death which occurred at a relatively low level was probably negated by compensatory growth of neighbouring plants.

In this experiment the 53 C and 55 C treatments significantly increased yield and satisfactorily controlled blackleg. Under better growing conditions less favourable for disease development and/or with less contaminated seed, treatment at lower temperatures could have been adequate. It is likely that in practice when the technique has been fully developed more than one temperature regime would be used when dealing with seed destined for seed and ware crops and to take into account different cultivar eye susceptibility to heat treatment especially on non dormant tubers.

Table 3. Effect of heat treatment for 5 min at different temperatures done in the continuous flow hot water treatment machine on *Erwinia carotovora* sub. sp. *atroseptica* (*Eca*) contamination of cv. Desiree seed, plant growth, blackleg development yield

Mean Water temperature C	Seed contamination (*Eca*/tuber)	Non-Emergence %	Blackleg % July	Blackleg % Aug	Blackleg % Sept	Yield kg/plot
50	3.40×10^3	0	1.0	15.1	29.7	75.4
53	4.10×10^2	0	0.5	6.3	13.5	84.0
55	2.07×10^2	1.0	0	1.6	6.8	77.1
57	1.00×10^2	1.6	0	0.5	4.7	72.6
60	6.00×10^1	9.9	0	0	2.1	57.5
Untreated	1.75×10^5	2.1	32.8	51.6	63.0	65.4
SED(df)		1.72(18)		2.92(54)		5.27(18)

Work is also in progress to extend the hot water treatment to control certain tuber-borne fungal diseases, skin spot (*Polyscytalum pustulans*), gangrene (*Phoma foveata*), black scurf (*Rhizoctonia solani*), silver scurf (*Helminthosporium solani*), dry rot (*Fusarium solani*, *F. coeruleum*) and possibly powdery scab (*Spongospora subterranea*), if necessary by the incorporation of a fungicidal treatment in the process. Preliminary results have been promising especially when the infection is still latent and hence sited mostly superficially. Lastly, hot water treatment can ... ting in ware tubers stored for a long time for industrial processing purposes by raising the water temperature high enough to kill the eyes but not damage the tubers.

REFERENCES

Bain R, Perombelon MCM, Tsror L and Nachmias A (1987) Epidemiology and etiology of blackleg. Rept Scot Crop Res Inst 1986, p 103

Hide GA (1975) Effect of heat treatment of potato tubers on Oospora pustulans. Plant Pathology 24:233-236

Mackay JM and Shipton PJ (1983) Heat treatment of seed tubers for control of potato blackleg (Erwinia carotovora sub. sp. atroseptica) and other diseases. Plant Pathology 32:385-393

Morse WJ (1917) Studies upon the blackleg disease of the potato with special reference to the relationship of the causual organisms. J agr Res 8:79-126

Perombelon MCM and Kelman A (1980) Ecology of the soft rot erwinias. Ann Rev Phytopathol 18:361-387

Perombelon MCM, Lowe R, Quinn C and Sells A (1980) Contamination of pathogen free and seed potato stocks by Erwinia carotovora during multiplication: Results of a six year monitoring survey. Pot Res 23:413-425

Perombelon MCM, Lumb VM and Hyman LJ (1987) A rapid method to identify and quantify soft rot erwinias on seed potato tubers. EPPO Bul 17:25-35

Robinson K and Foster G (1987) Control of potato blackleg by tuber pasteurisation: the determination of time-temperature combinations for the inactivation of pectolytic erwinia. Pot Res 30:121-125

Shapovalov M and Edson HA (1921) Blackleg potato tuber-rot under irrigation. J agr Res 22:81-92

Stapp C (1961) Bacterial Plant Pathogens. (Trans. A. Schoenfeld) Oxford. Oxford Univ Press

Storey MMJ, Dent TJ, Ginger MW and Hampson CP (1984) Comparison of 8-hydroxyquinoline/thiabendazole as a seed and/or progeny treatment to control bacterial soft rot in potato stores. Tests Agrochem cvs 5:64-65

Wale SJ and Robinson K (1986) Evaluation of large scale hot water dipping and forced ventilation of seed potatoes to reduce tuber contamination with blackleg bacteria (Erwinia spp.). In Br Crop Protect Conf - Pests Dis 1137-1143

Wyatt GM and Lund BM (1981) The effect of antibacterial products on bacterial soft rot of potatoes. Pot Res 24:315-329

BIOLOGICAL CONTROL OF PSEUDOMONAS SOLANACEARUM

A. Trigalet
Laboratoire de Biologie Moléculaire
INRA - CNRS
BP 27
31326 Castanet-Tolosan Cedex
France

INTRODUCTION

Bacterial wilt caused by Pseudomonas solanacearum is a soil-borne plant disease of economic and international importance. The most successful method used to control this disease has been the use of resistant varieties. Because of difficulties with this approach, alternative, control measures such as biological control have been investigated.

Pseudomonas solanacearum spontaneously segregates colony morphology variants which differ from the wild-type strain by many characters. In particular, these pleiotropic mutants have lost the ability to cause disease and instead induce a hypersensitive reaction once infiltrated into tobacco leaves. As a result these bacteria are no longer able to multiply within their originally susceptible host (Sequeira et al 1977). Because of their rough colony aspect on agar plates, these spontaneous avirulent mutants have been referred as Nfl, i.e. non fluidal mutants.

There have been some reports of biological control of Pseudomonas solanacearum based on prior inoculation of the susceptible host by heat-killed, wild-type pathogens and/or by saprophytic bacteria, and/or by spontaneous Nfl avirulent mutants (Averre and Kelman 1964; Kempe and Sequeira 1983; Tanaka 1985a and b). These experiments, which were done under laboratory conditions, showed a reduction in disease severity with control rate (% protection) up to 80% in the most favourable instances. Some field experiments have been reported with less successful control

NATO ASI Series, Vol. H28
Vascular Wilt Diseases of Plants

rates (Tanaka 1985b).

Random mutagenesis of the wild-type strain with transposon Tn5 led to the isolation of thirteen mutants which do not produce any symptoms on susceptible tomato (Boucher et al 1985) although most of them have retained, to a certain extent, the ability to colonize and to multiply within the plant after root inoculation (Trigalet and Demery 1986). The latter property suggested that these mutants may durably exclude the wild-type within the host plant.

In this communication, we report that, under laboratory conditions, some Tn5-induced avirulent mutants are effective in colonizing tomato plants and in preventing subsequent colonization by the wild-type strain, thus leading to stable, high, biological control.

MATERIALS AND METHODS

The origin ot strains of Pseudomonas solanacearum, the preparation of bacterial inoculum, plant inoculation and bacterial isolation and numerical procedures have been published elsewhere (Boucher et al 1985; Trigalet and Demery 1986). In addition to the H63-5 tomato cultivar, the susceptible super Marmande cv has also been used. Induction of protection against the virulent wild-type Pseudomonas solanacearum has been assayed by making simultaneous and sequential inoculations with Tn5-induced avirulent mutants and with the virulent strain. Different inoculum concentration ratios between virulent vs avirulent bacteria, within the range 1 to 0 01 respectively, have also been tested. Disease ratings for individual plants were recorded at different times (30 to 50 days) after the challenge inoculation, according to the standard disease index (DI) scale ranging from 0 to 5 (Winstead and Kelman 1952). The control rate, i.e. percentage of protection (% P) was estimated according to Tanaka (1985) namely:

$$\% P = \left[1 - \frac{\text{average DI in the avirulent/virulent combination}}{\text{average DI in the virulent control}} \right] \times 100$$

RESULTS

Tomato plants were inoculated at the root level without any artificial injury and kept under temperature and relative humidity conditions most favourable to disease development as previously described (Trigalet and Demery 1986). We reported only results from experiments which led to 100% wilted plants within 12 days in the control inoculation with the wild-type strain derivative GMI 1229. Thus, the protective effect of a given mutant has been estimated under the highest disease severity. Results which are presented in this communication have purposely been presented in a summarized form, since they will be extensively published elsewhere.

Simultaneous inoculations of each of the Tn5-induced avirulent mutants with the wild-type derivative GMI 1229 in mixed inocula at comparable relative ratios, revealed that only three mutants had some protective effect, namely GMI 1350, 1351 and 1353. These mutants were obtained by transformation of the parental strain GMI 1229 with genomic DNA from each of the mutants GMI 1229, 1301 and 1305, respectively (Boucher et al 1985). These three mutants exhibited relatively good invasiveness in tomato, when compared with the other Tn5 induced avirulent mutants (Trigalet and Demery 1986). Two of these mutants, GMI 1350 and GMI 1353, were further assayed for variations in the relative ratio of wild-type to mutant and for sequential inoculations. Changing the proportions of the virulent GMI 1229 and of the avirulent mutant from 1 to 0.1 resulted in increased percentages of protection in simultaneous inoculations from 24% to 45% for the GMI 1353 mutant and from 45% to 90% for the GMI 1350 mutant. Keeping the same relative ratio (wild-type/mutant) of 0.1, but changing the inoculation time schedule, i.e. using delayed challenge inoculations 1 or 8 days after prior inoculation with the avirulent GMI 1353 mutant, resulted in increased percentages of protection as compared with simultaneous inoculations. This protection was even greater when the ratio wild-type/avirulent mutant was changed from 0.1 to 0.01. Thus, up to 80% or even more protection could be obtained with GMI 1353 in sequential inoculations with a ratio (wild-type/avirulent) of 0.01 and an 8 days delay between mutant and challenge inoculations. One must remain very cautious with interpreting these percentages of protection since they can vary within some limits for the same wild-type/mutant combination in repeated experiments. This is probably due to variations in factors which have not previously been taken into account. In fact, the root system development has hardly been taken into consideration, and could be a cause of non-reproductible results that render any comparison between simultaneous and sequential inoculations somewhat difficult.

Isolation and bacterial counts from stem pieces at the cotyledon level provided additional information about the protection mechanism. In a sequential inoculation 45 days after the challenge inoculation, symptomless plants which gave positive isolations turned out to be contaminated by the wild-type derivative (8% of the plants) and by the GMI 1358 mutant (more than 74%). In another sequential inoculation, isolations from the cotyledon level revealed 3% contamination by the wild-type strain to more than 83% by the GMI 1353 mutant. In 75% of the cases so far tested, symptomless plants appeared to be invaded at the cotyledon level whereas 25% of the plants were free of any bacteria. The root system was not checked (for an explanation see Trigalet and Demery 1986). A majority of the symptomless plants, from which positive isolations were made, were colonized exclusively by the mutant which seemed to exclude the wild-type strain. In plate culture, however, no antagonism was detected between the wild-type GMI 1229 and either GMI 1353 or GMI 1350. Bacterial counts at the cotyledon level in symptomless plants showed a great variability in bacterial concentration ranging from 10^2 up to more than 10^7 CFU in individual 5 mm length stem pieces.

Higher in the stem (30 cm above soil level), the bacterial mutant population was very weak, however, and wild-type bacteria were never isolated.

DISCUSSION AND PROSPECTS

Our results indicate that even under conditions which are most favourable to disease development, some Tn5-induced avirulent mutants are able to infect from within the soil and to spread extensively through the host without causing any disease and, in addition, are able to exclude the wild-type virulent strain from subsequent invasion. Although these mutants are invasive, they show reduced colonization and multiplication as compared to the wild-type strain. In addition, their limited, permanent invasion is advantageous provided that the plant remains healthy and productive. Whatever the nature of the involved antagonism between virulent wild-type and Tn5-induced avirulent mutant, it ensures permanent control of the disease. The two protective Tn5-induced avirulent mutants, GMI 1353 and GMI 1350 are different with respect to the Tn5 insertion. Both insertions are located on the megaplasmid of Pseudomonas solanacearum (Boucher et al 1987), in a hrp gene cluster for GMI 1353 and in a dsp gene cluster for GMI 1350 (Arlat and Boucher 1988). Within the hrp gene cluster some Tn5 induced mutants are neither invasive nor protective. Thus seems to be a correlation between invasiveness and protective ability. Work is in progress to characterize the genomic regions involved in both properties.

Sequences homologous to the hrp gene cluster have been found in all different wild-type strains of Pseudomonas solanacearum so far tested. Therefore, site-directed mutagenesis (Ruvkun and Ausubel 1981) should help in constructing ecologically acceptable mutants. Such mutants must be invasive in the host plant, competitive against the virulent wild-type strain and devoid of the virulent trait and of any antibiotic marker. Such mutants could be used in field trials to determine whether the level of [illegible] induced in young tomato plants is effective in older plants without affecting fruit production.

REFERENCES

Arlat M, Barberis P and Boucher C (1988) Additional genes controlling pathogenicity in Pseudomonas solanacearum. In: Proceed 4th Int Symp Molecular genetics Plant Microbe Interactions Acapulco Mexico May 15-20 (in Press)

Averre C and Kelman A (1964) Severity of bacterial wilt as influenced by ratio of virulent to avirulent cells of Pseudomonas solanacearum in inoculum. Phytopathology 54:779-783

Boucher C, Barberis P, Trigalet A and Demery D (1985) Transposon mutagenesis of Pseudomonas solanacearum : isolation of Tn5-induced avirulent mutants. J Gen Microbiol 131:2249-2257

Boucher C, Van Gijsegem F, Barberis P, Arlat M and Zischek C (1987) Pseudomonas solanacearum genes controlling both pathogenicity on tomato and hypersensibility on tobacco are clustered. J Bact 169:5626-5632

Kempe J and Sequeira L (1983) Biological control of bacterial wilt of potatoes : attempts to induce resistance by treating tubers with bacteria. Plant Disease 67:499-503

Ruvkun G and Ausubel F (1981) A general method for site-directed mutagenesis in prokaryotes. Nature 289:85-88

Sequeira L, Gaard G and de Zoeten GA (1977) Attachment of bacteria to host cell walls its relation to mechanisms of induced resistance. Physiol Plant Pathol 10:43-50

Sequeira L and Hill L (1974) Induced resistance in tobacco leaves: the growth of Pseudomonas solanacearum in protected tissues. Physiol Plant Pathol 4:447-455

Tanaka H (1985a) Induced resistance in tobacco against bacterial wilt and its possible mechanism (in Japanese). Bull Utsunomiya Tobacco Exp Station 21:1-64

Tanaka H (1985b) Pseudomonas solanacearum M4S strain, composition containing the same for controlling soil-borne diseases of solanaceous plants, and method for controlling said soil-borne diseases with the same. Europ Pat Appl n° 85900771.8

Trigalet A and Demery D (1986) Invasiveness in tomato plants of Tn5-induced avirulent mutants of Pseudomonas solanacearum. Physiol Mol Plant Pathol 28:423-430

Winstead N and Kelman A (1952) Inoculation techniques for evaluating resistance to Pseudomonas solanacearum. Phytopathology 42:628-634

PARTICIPANTS

Alabouvette, C.
Station De Recherches Sur La Flore Pathogene, Dans le sol INRA,17 Rue Sully, 21034 Dijon, Cedex, France

Alivizatos, A.
Benaki Phytopathological Institute, Kiphissia 14561, Athens, Greece

Beckman, C.H.
Department of Plant Sciences, College of Resource Development, The University of Rhode Island, Kingston, RI 02881-0804, U.S.A.

Biris, D.
Plant Protection Institute of Volos, 38001 Volos, Greece

Ben-Yephet, Y.
Institute of Plant Protection, Volcani Center, P.O. Box 6, Bet Dagan 50-250, Israel

Blanco Lõpez, M.A.
Department of Agronomia, Universidad de Cordoba, Apartado de Correos 3048, 14080 Cordoba, Spain

Christias, C.
Biology Department, Nuclear Research Center "Demokritos", Aghia Paraskevi 15310, Athens, Greece

Cohen, Y.
Department of Life Sciences, Bar-Ilan University, Ramat-Gan 52100, Israel

Cooper, R.
School of Biological Sciences, University of Bath, Claverton Down, Bath BA2 7AY, England

Couvaraki, C.
Agricultural University of Athens, Laboratory of Plant Pathology, Votanikos 11855, Athens, Greece

Davis, M.
University of Florida, Institute of Food and Agricultural Sciences, 3205 College Avenue, Fort Lauderdale, Florida 33314, U.S.A.

Demery, D.
Laboratoire Moleculaire des Relations Plantes - Micro-organisms, INRA-CNRS, Auzeville, BP-27, 31326 Castanet-Tolosan, France

DeVay, J.
Department of Plant Pathology, College of Agricultural and Environmental Sciences, University of California, Davis, California 95616 U.S.A.

Elgersma, D.
Willie Commelin Scholten, Phytopathological Laboratory, Javalaan 20, 3742 CP BAARN, Netherlands

Engelhard, A.
Agricultural Research and Education Center, University of Florida, 5007 60th Street East, Bradenton, FL 34203 U.S.A.

Fravel, D.
Biocontrol of Plant Diseases Laboratory, USDA-ARS, Room 275, Building 011A, Barc W, Beltsville, Maryland 20705, U.S.A.

Georgopoulos, S.
Lapafarm Inc., 73 Menandrou Str., 10437 Athens, Greece

Hastie, A.C.
Department of Biological Sciences, The University, Dundee DDI 411N, Scotland

Heale, J.

Department of Biology, King's College London, Campden Hill Road, Kensington, London W 8 7AH, U.K.

Huisman, O.

Department of Plant Pathology, University of California, Berkeley, CA 94720, U.S.A.

Jarvis, W.R.

Plant Pathology Section, Research Station, Harrow Ontario, NOR IGO, Canada

Jiménez-Díaz, R.

Departmento de Agronomia, Universidad de Cordoba, Apartado de Correos 3048, 14080 Cordoba, Spain

Katan, J.

The Levi Eshkol School of Agriculture, The Hebrew University of Jerusalen, Rehovot 76-100, P.O. Box 12, Israel

Lemanceau, P.

Station De Recherches Sur La Flore Pathogene, Dans Le Sol INRA, 17 Rue Sully, 21034 Dijon,Cedex, France

Louvet, J.

Station De Recherches Sur La Flore Pathogene, Dans Le Sol INRA, 17 Rue Sully, 21034 Dijon, Cedex, France

Mace, M.E.

Cotton Pathology Research Unit, United States Department of Agriculture, P.O. Drawer JF, College Station, TX 77841,U.S.A.

Martins, J.M.S.

Estacao Agronomica Nacional, 2780 Oeiras, Portugal

Matta, A.

Istituto di Patologia Vegetale, Università di Torino, Via P. Giuria 15, 10126 Torino, Italy

Mugnier, J.
Rhône-Poulenc Agrochimie, 14-20 Rue Pierre Baizet B.P. 9163, 69263 Lyon, Cedex 09, France

Panagopoulos, C.
Agricultural University of Athens, Laboratory of Phytopathology, Votanikos 11855, Athens, Greece

Panopoulos, N.
Department of Plant Pathology, University of California, Berkeley, CA 94720, U.S.A.

Pegg, G.
Department of Horticulture, University of Reading, P.O. Box 239, Earley Gate, Reading RG6 2AU, U.K.

Perombelon, M.
Scottish Crop Research Institute, Invergowrie, Dundee, DD2 5DA, Scotland

Psallidas, P.
Benaki Phytopathological Institute, Kiphissia 14561, Athens, Greece

Sackston, W.
Department of Plant Science, MacDonald College, McGill University, Quebec H9X IC0, Canada

Sletten, A.
Department of Plant Pathology, Norwegian Plant Protection Institute, BOKS 70, 1432 AS-NLH, Norway

Tjamos, E.C.
Laboratory of Plant Pathology, Agricultural University of Athens, Votanikos 11855, Athens, Greece

Trigalet, A.
Laboratoire Moleculaire des Relations Plantes - Micro-organismes, INRA-CNRS, Auzeville, BP-27, 31326 Castanet-Tolosan, France

Van Alfen, N.

Department of Biology, College of Science, Utah State University, Logan, Utah 84322-4500, U.S.A.

Vanachter, A.

Laboratory of Phytopathology and Plant Protection, Katholieke Universiteit Leuven, Kardinaal Mercierlaan 92, B-3030 Leuven, Belgium

INDEX

NATO ASI Series H

Vol. 1: **Biology and Molecular Biology of Plant-Pathogen Interactions.** Edited by J. A. Bailey. 415 pages. 1986.

Vol. 2: **Glial-Neuronal Communication in Development and Regeneration.** Edited by H. H. Althaus and W. Seifert. 865 pages. 1987.

Vol. 3: **Nicotinic Acetylcholine Receptor: Structure and Function.** Edited by A. Maelicke. 489 pages. 1986.

Vol. 4: **Recognition in Microbe-Plant Symbiotic and Pathogenic Interactions.** Edited by B. Lugtenberg. 449 pages. 1986.

Vol. 5: **Mesenchymal-Epithelial Interactions in Neural Development.** Edited by J. R. Wolff, J. Sievers, and M. Berry. 428 pages. 1987.

Vol. 6: **Molecular Mechanisms of Desensitization to Signal Molecules.** Edited by T. M. Konijn, P. J. M. Van Haastert, H. Van der Starre, H. Van der Wel, and M. D. Houslay. 336 pages. 1987.

Vol. 7: **Gangliosides and Modulation of Neuronal Functions.** Edited by H. Rahmann. 647 pages. 1987.

Vol. 8: **Molecular and Cellular Aspects of Erythropoietin and Erythropoiesis.** Edited by I. N. Rich. 460 pages. 1987.

Vol. 9: **Modification of Cell to Cell Signals During Normal and Pathological Aging.** Edited by S. Govoni and F. Battaini. 297 pages. 1987.

Vol. 10: **Plant Hormone Receptors.** Edited by D. Klämbt. 319 pages. 1987.

Vol. 11: **Host-Parasite Cellular and Molecular Interactions in Protozoal Infections.** Edited by K.-P. Chang and D. Snary. 425 pages. 1987.

Vol. 12: **The Cell Surface in Signal Transduction.** Edited by E. Wagner, H. Greppin, and B. Millet. 243 pages. 1987.

Vol. 13: **Toxicology of Pesticides: Experimental, Clinical and Regulatory Perspectives.** Edited by L. G. Costa, C. L. Galli, and S. D. Murphy. 320 pages. 1987.

Vol. 14: **Genetics of Translation. New Approaches.** Edited by M. F. Tuite, M. Picard, and M. Bolotin-Fukuhara. 524 pages. 1988.

Vol. 15: **Photosensitisation. Molecular, Cellular and Medical Aspects.** Edited by G. Moreno, R. H. Pottier, and T. G. Truscott. 521 pages. 1988.

Vol. 16: **Membrane Biogenesis.** Edited by J. A. F. Op den Kamp. 477 pages. 1988.

Vol. 17: **Cell to Cell Signals in Plant, Animal and Microbial Symbiosis.** Edited by S. Scannerini, D. Smith, P. Bonfante-Fasolo, and V. Gianinazzi-Pearson. 414 pages. 1988.

Vol. 18: **Plant Cell Biotechnology.** Edited by M. S. S. Pais, F. Mavituna, and J. M. Novais. 500 pages. 1988.

Vol. 19: **Modulation of Synaptic Transmission and Plasticity in Nervous Systems.** Edited by G. Hertting and H.-C. Spatz. 457 pages. 1988.

Vol. 20: **Amino Acid Availability and Brain Function in Health and Disease.** Edited by G. Huether. 487 pages. 1988.

NATO ASI Series H